# Multiscale Modeling for Process Safety Applications

# Multiscale Modeling for Process Safety Applications

Arnab Chakrabarty

Sam Mannan

Tahir Cagin

ELSEVIER

AMSTERDAM • BOSTON • HEIDELBERG • LONDON
NEW YORK • OXFORD • PARIS • SAN DIEGO
SAN FRANCISCO • SINGAPORE • SYDNEY • TOKYO

Butterworth-Heinemann is an imprint of Elsevier

B H

Butterworth-Heinemann is an imprint of Elsevier
The Boulevard, Langford Lane, Kidlington, Oxford OX5 1GB, UK
225 Wyman Street, Waltham, MA 02451, USA



ISBN: 978-0-12-396975-0

**British Library Cataloguing-in-Publication Data**
A catalogue record for this book is available from the British Library

**Library of Congress Cataloging-in-Publication Data**
A catalog record for this book is available from the Library of Congress

For information on all Butterworth-Heinemann publications visit our website at http://store.elsevier.com/



*Publisher:* Joe Hayton
*Acquisitions Editor:* Fiona Geraghty
*Editorial Project Manager:* Lindsay Lawrence
*Production Project Manager:* Lisa Jones
*Designer:* Matthew Limbert

Typeset by TNQ Books and Journals
www.tnq.co.in

# Contents

# Preface

Computational modeling has emerged as a powerful partner to experimental and theoretical studies. It is a familiar subject in several fields but has so far been considered of secondary importance in process safety applications, which mostly relies on plant personnel experience and accumulated learning of the community from historical incidences. The current standard practices in assessment of hazardous scenarios or relevant parameters can be significantly aided by the wider implementation of detailed computational methods, at various time and length scales. Execution of these approaches, as explored in this book, has become more practical owing to the significant increase in computational power and ease of access to appropriate computational tools. Researchers across the globe have worked in exploiting these methods for process safety applications, but until now, the ensemble of these efforts has remained mostly as isolated studies. The main objective of this book is to group these existing efforts under a common platform by providing the current status of this novel area and discussing the potential implementation of multiscale modeling approaches for process safety applications.

Professionals, in both industry and academia, can use this book, as can graduate researchers working in any domain where process safety challenges are implicitly or explicitly embedded. To address such challenges, the book has attempted to cover applications of quantum mechanics, molecular dynamics, quantitative structure—property relationship, quantitative structure—activity relationship, computational fluid dynamics, finite element analysis, chaos theory, statistical analysis, and dynamic simulation as applicable. The problems dealt with are consequence modeling, risk assessment, and related parameter estimation for potential hazards resulting from fire, explosion, and dispersion of toxic gasses.

In this book, the bridge between various modeling methodologies is not discussed. In that sense, the individual chapters are essentially independent. Nor is the mathematical background of these multiscale modeling approaches discussed in detail. Fortunately, for each of these topics, several good books are available on the market. These sources can be used to supplement one's training in mathematical fundamentals.

Typically, problems at the level of quantum mechanics and molecular dynamics have dealt with safety-related concerns for a specific molecular system. At the molecular modeling level, it is more about material characterization in a given environment. Limited by time and length scale, these methods are not suitable for answering safety concerns for the whole process domain. However, knowledge gained at these scales, such as estimation of parameters at the molecular scale, can be passed on to a modeling environment with a lesser resolution for an atomistically informed holistic model. Another example of using two different scales in modeling approaches for addressing the same concern could be the use of a conservative low-resolution model for screening analysis and then the performance of detailed modeling of selected scenarios, similar to the Pareto principle.

Emerging tools and increasing computational power modeling in process safety have improved much from the case of a spherical cow, a metaphor used for highly simplified scientific models of complex physical phenomena. Accordingly, the use of detailed computational modeling approaches in combination with accumulated knowledge of safety experts can only contribute in a positive manner toward improving the safety performance of chemical facilities. This book is written with the hope that it will contribute toward promoting the use of multiscale modeling methodologies in addressing process safety problems.

# Acknowledgments

It is our pleasure to acknowledge here some debts of gratitude. Without the tremendous help of the following, given below in no particular order, completion of this book would have faced much hindrance:

Monir Ahammad, Xiaodan Gao, Ning Gan, Zhe Han, Brian Harding, Logan Hatanaka, Sunder Janardanan, Xinrui Li, Yan-Ru Lin, Ruochen Liu, Yi Liu, Ha Nguyen, William Pittman, Samina Rahmani, Olga Reyes, Josh Richardson, Camilo Rosas, Nitin Roy, Sonny Sachdeva, Bin Zhang, and Jiaqi Zhang, all at Mary Kay O'Connor Process Safety Center, Texas A&M University, for their help with preparation of the chapters.

Sébastien Canneaux, Frédéric Bohr, LISM, and Eric Hénon, ICMR, at the University of Reims Champagne-Ardenne, for preparation of the KiSThelP tutorial problem.

Fiona Geraghty, Cari Owen, Natasha Welford, Lindsay Lawrence, Lisa Jones, Matthew Limbert and the Elsevier production team, for their patience and enormous help throughout the life of the project.

Our debts toward our families are too personal to acknowledge here entirely. As we struggled to complete our manuscript, they have suffered some neglect. It is to them, we dedicate this book with love and affection.

CHAPTER

# INTRODUCTION

# 1

*Concern for man himself and his safety must always form the chief interest of all technical endeavors. Never forget this in the midst of your diagrams and equations.*
**Albert Einstein**

Numerous incidents such as the Flixborough explosion resulting from the release of flammable hydrocarbons in June 1974, the methyl isocyanate release in Bhopal in December 1984, and the Macondo disaster in 2010 continue to remind us of the important role of process safety in the design and operations of process facilities. Over the years, process safety has emerged as a discipline in itself and has continued to play a dominant role in any process technological development. While a focus on process safety model development, experiments at various scales have gained momentum over the years and had a positive impact on the industry, the safety incident databases are anything but stagnant (MARSH, 2014). Even recently, in August 2013, in response to recent catastrophic chemical facility incidents in the United States, President Obama issued an executive order to enhance the safety and security of chemical facilities, and consequently reduce the risks associated with hazardous chemicals, for owners, operators, workers, and communities.

Given the complexity of existing and emerging technologies, interactions of process parameters, and intangible and hidden correlations, quantitatively defining the underlying root of a process safety phenomenon, or predicting the same scenario, requires a great deal of experience, extensive knowledge, and a structured approach. Additionally, while experimentally quantifying process safety parameters is often insightful, it is resource intensive and often not comprehensive owing to the number of experiments that would be required to achieve the same. Experimentally validated multiscale modeling methodologies are powerful tools that have the potential to offer answers which are otherwise often expensive or unreachable through experiments. For example, think about large-scale jet fires or deflagration-to-detonation events. Typical current practice in modeling process safety scenarios is to substitute the lack of understanding by introducing conservativeness to a safety model. Unfortunately, conservative safety parameters not only increase plant costs, but may not ensure the safety of a facility. The insufficient details or coarser resolution of the solution has the potential to offer more risk than is mitigated by adding conservativeness to it. As we have demonstrated in a later chapter, mixtures exhibiting minimum flash point behaviors pose such a risk, and therefore exercising conservativeness alone may not be sufficient.

With significant rise of computational power in recent times along with better understanding of the underlying physics, implementation of higher-resolution models in order to gain insights and refine existing models to assess hazardous scenarios has become more practical and achievable. Owing to the multiscale nature of processes, multiscale modeling has emerged as a new set of tools that provides insight into important features at multiple times and lengths of a physical phenomenon. Thus, in several applications, it has become crucial to incorporate information from a range of length and time scales into a model (Cameron and Gani, 2011). As an outcome of modeling product and process issues, a growing number of modeling efforts are resulting in multiscale modeling approaches. The strategies discussed throughout the book, albeit in the context of process safety, attempt to capture inherently important properties at various scales of a system and correlate them accurately to the system's macroscale properties. A reasonable amount of work has been done in that respect in other areas, as demonstrated in several publications (Lépinoux, 2000; Kwon et al., 2007; Derosa and Cagin, 2010; Maekawa et al., 2008). Multiscale modeling techniques have been successfully employed in material design to rationally develop and accurately predict the performance of systems with the building blocks of their macroscopic-level performance residing at much smaller scales. Similar to other fields, it is only appropriate to extend the advancement in materials theory at different time and length scales to address less understood safety concerns by incorporating an adequate level of detail. In the context of process safety, while some sporadic work exists to assess problems at different time and length scales, to date those have been mostly isolated efforts. The primary objective in this book is to group these existing efforts on a common platform. This book aims to provide a review of the current status in this area, discuss potential implementation of multiscale modeling, and help refine existing computational approaches used for safety analysis. It attempts to provide an overall picture of how safety issues are addressed at all scales of modeling, and discusses the latest methods in the field.

Chapter 2 serves as the introduction to process safety. It touches upon the status of current industrially accepted state-of-the-art modeling approaches in process safety analysis. As appropriate, the chapter introduces or reintroduces the reader to process safety fundamentals such as the physics, consequences, and risks of fire, explosion, and toxic hazards in process industries. Concepts of flammability, ignition phenomena, fire, dispersion of flammable and toxic gases, deflagration and detonation, risk assessment of fire, toxic, and explosion hazards, are also covered in this chapter.

Chapters 3–6 deal with the use of modeling methodologies as applicable to process safety applications within various time and length scales. Chapter 3 demonstrates the applicability, relevance, and benefits of molecular modeling methods such as quantum mechanics, molecular dynamics, quantitative structure–property relationship (QSPR), and quantitative structure–activity relationship (QSAR) in process safety applications at various capacities. Chapter 4 moves into greater time and length scales and examines the effectiveness and benefits of implementing computational fluid dynamics (CFD) in the development of consequence models of process facilities. It looks into use of CFD in assessing the consequence of various types of fires such as jet, pool, and flash fires. It also looks at the modeling of explosion and blast waves using CFD. In a similar fashion, Chapter 5 illustrates the practicality of using finite element methods in process safety applications. Finite element methods have been implemented in understanding flare systems, storage and transportation of flammable materials, and other concerns in process hazard analysis. Accessing larger time and length scales phenomena are demonstrated through implementation of dynamic process simulations in Chapter 6. The transient nature of a process is typically crucial in modeling plant start-up and shutdown phenomena. In the same chapter, chaos theory and statistical analysis are introduced within the context of addressing process safety concerns. Chaos theory can be applied to investigate runaway reactions and has the potential to provide early warning detection of same. Statistical analysis has been utilized to monitor real-time plant data. Multivariate statistical analysis can be applied to plant data that can in turn help in incident investigation.

The chapters referred to above insights into implementing modeling approaches for the accurate estimation of consequences from fire, explosion, or toxic emission at various scales. Risk assessment, the succeeding part of a consequence modeling study that estimates the likelihood of a consequence, is covered in Chapter 7. The chapter illustrates approaches to gain insight into risk-related parameters through multiscale modeling. Approaches such as Bayesian logic, Bayesian-LOPA methodology for risk assessment are analyzed in this chapter. Process and material characteristics that directly affect equipment failure rate, such as fracture, corrosion and similar phenomena are examined in this chapter in the context of quantitative risk assessment. For example, beyond purely monetary value, corrosion profoundly affects the safety of advanced products and processes. Chapter 8 deals with inherently safer design, an emerging and increasingly important process safety topic. Addressing safety concerns at the design level has gained momentum in recent years and is a promising path forward for the process safety community. The subject of industrial hygiene is covered in Chapter 9. While the chapter typically addresses molecular modeling approaches in assessing the toxicity of plant chemicals, the perspective is different than that of Chapter 3, part of which also addresses assessing toxicity through implementation of molecular modeling.

The book concludes with Chapter 10, followed by exercise problems at various scales related to the topics discussed in the book. The aim is to represent multiscale modeling methodology as a set of crucial tools for answering questions and gaining insights in process safety applications.

## REFERENCES

Cameron, Ian T., Gani, Rafiqul, 2011. Product and Process Modelling. Elsevier.

Derosa, Pedro, Cagin, Tahir, 2010. Multiscale Modeling. CRC Press.

Kwon, Young, Allen, David H., Talreja, Ramesh R., 2007. Multiscale Modeling and Simulation of Composite Materials and Structures. Springer Science & Business Media.

Lépinoux, Joël, 2000. Multiscale Phenomena in Plasticity: From Experiments to Phenomenology, Modelling and Materials Engineering. Springer Science & Business Media.

Maekawa, Koichi, Ishida, Tetsuya, Kishi, Toshiharu, 2008. Multi-scale Modeling of Structural Concrete. CRC Press.

MARSH, 2014. The 100 Largest Losses (In the Hydrocarbon Industry) 1974-2013. Uk.Marsh.com. http://uk.marsh.com/NewsInsights/Articles/ID/37406/The-100-Largest-Losses-in-the-Hydrocarbon-Industry-1974-2013.aspx (Accessed March 26.).

CHAPTER

# PROCESS SAFETY

2

## 2.1 FIRE

The concept of fire is one which the common person is usually familiar. One might describe a fire as "hot" or "capable of destroying buildings," but what really is a fire? What needs to be known about fire hazards to prevent incidents? These complex questions highlight a key concept about fire safety: fire is a complicated phenomenon, but to prevent harm to workers and facilities, a fundamental understanding of fire hazards at all scales is required.

Fire, explosions, and toxic releases are considered to be the three most commonly encountered types of hazards that cause severe incidents and loss in process industries. Compared with toxic releases and explosions, fires generally cause less damage but occur more frequently (Coco and Marsh Risk, 2001). However, because a fire often precedes an explosion or toxic release, the damages and loss from fires can have dire consequences. It is, therefore, important to understand the phenomena of fire to properly address the hazards presented by it.

Fire is a specific case of oxidation reaction that involves the rapid reaction of a fuel with an oxidizer. This chemical reaction results in a net release of heat, also known as an exothermic reaction. In a fire, an ignition source supplies energy to these aforementioned combustion reactions initially, and if the excess heat from the oxidation reactions is sufficient to drive the combustion of more material, the flame becomes self-sustaining (Crowl and Louvar, 2011).

Generally speaking, the oxidizing reactions that drive the fire occur in the vapor phase. Therefore, flammable solids and liquid must first heat up and vaporize before fully combusting. For a liquid, this is as simple as heating and evaporating or, in more extreme cases, boiling. For solid flammable materials, it is more complicated and usually first involves thermal degradation, known as pyrolysis, then vaporization and combustion (Drysdale, 2011). Although general information about fires is insightful, further discussion is required to fully understand the behavior and hazards of fire at all scales.

### 2.1.1 THE FIRE TRIANGLE

For a fire to occur, three components must be present: fuel, an oxidizer (often oxygen), and a source of ignition, as depicted in Figure 2.1. If any side of the fire triangle is removed, a fire will not form (Crowl and Louvar, 2011). If a flammable material is stored in the presence of oxygen but no ignition source is present, a fire cannot occur. Similarly, if a fuel is heated under an inert gas, a fire will not have oxygen and will not burn. Last, it is simple to see that a fire will not occur if there is no fuel. Regardless of the size of a system, without the three components of the fire triangle, a fire will not take place.

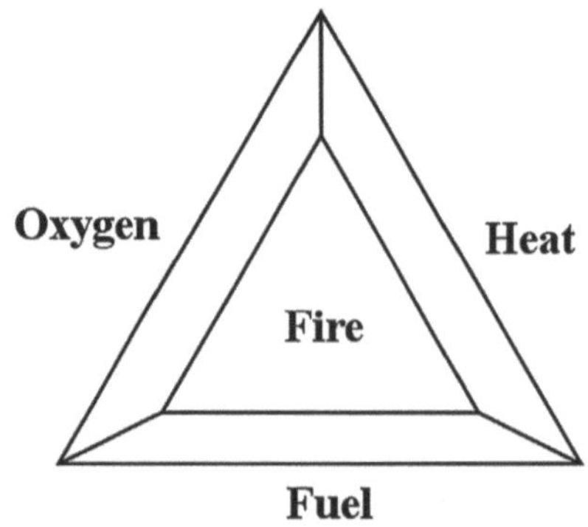


**FIGURE 2.1**

The fire triangle (American Institute of Chemical Engineers, Center for Chemical Process Safety, 2003).

While the absence of any side of the fire triangle prevents ignition, the presence of all three components of the fire triangle does not necessarily ensure that a fire will occur. Many factors relating to the three fire triangle component dictate whether a fire will occur, including the amount of each component present in the fire triangle (Crowl and Louvar, 2011). For example, a reduction in the oxygen concentration past a certain point will extinguish liquid flammable fires. The concentration of oxygen in air is an important factor in the existence of a fire.

Similarly, if the heat source used to initially begin combustion is not strong enough, a fire may fail to start. The minimum ignition energy (MIE) for a material is heavily dependent on material type, physical distribution, and the physical conditions, but most hydrocarbons have a MIE in the range of 0.01–2 mJ (Glassman and Yetter, 2008). However, increases in pressure can decrease the MIE, so it is important to account for a reduced ignition energy in pressurized systems (Glassman and Yetter, 2008). These factors play a role in estimation of ignition probability (CCPS, 2014; Moosemiller, 2011), which in turn predicts the chances of a fire as oppose to explosion or vice versa in the case of a flammable release.

The physical phase of a fuel also affects whether it can be ignited. As previously discussed, a flammable vapor mixture does not need to vaporize before catching fire, but solid and liquid must first be vaporized to burn, requiring a higher ignition energy (Crowl and Louvar, 2011). In addition, solids generally go through a thermal decomposition known as pyrolysis to produce a volatile flammable vapor (Drysdale, 2011). Consequently, the energy required to sustain a solid, liquid, or vapor fire varies.

Last, the physical geometry of the flammable fuel affects the chances that a fire will manifest. This is easily visualized when comparing the flame from a match burning top-to-bottom to one that burns from the bottom-to-top. It is clear that the fire triangle is greatly influenced by a large number of complicated factors.

### 2.1.2 IGNITION PHENOMENA

Fires are initiated by an external heat source but the source can sometimes be silent and discrete. Although some fires can be started by a simple ignition source such as a spark or hot surface, protecting against ignition sources is not the only precaution that must be taken to prevent ignition.

Ignition source awareness ensures that the proper precautions can be taken to control ignition sources and prevent an incident in the future.

The autoignition of a flammable mixture occurs when the temperature of the environment is high enough to provide the heat required for combustion. In studying these phenomena, it is useful to define the autoignition temperature for a vapor mixture, which is the minimal temperature at which a flammable mixture can undergo a rapid combustion process without any other ignition source (Crowl and Louvar, 2011). By maintaining a flammable mixture's temperature below its autoignition temperature, the hazard of spontaneous ignition is reduced greatly. Once above this temperature, the flammable mixture can have enough thermal energy to ignite, even without a source of heat.

Another source of ignition that is often overlooked is the auto-oxidization of a flammable liquid. This occurs when a flammable liquid has a high boiling point and is stored without temperature control. The slow oxidation of the material over time causes the temperature of the liquid to increase slowly. In volatile liquids, this heat is released by small amount of vaporization, but in low-volatility liquids, this heat can build up until it is sufficient to ignite vapor above the flammable liquid (Crowl and Louvar, 2011). Temperature controls or other precautions must be taken with similar low-volatility flammable organic liquids to avoid auto-oxidation.

### 2.1.3 FLAMMABILITY LIMITS OF GASES AND VAPORS

Flammable vapors and gases are a major fire hazard and, unlike most solids and liquids, can ignite with very little ignition energy. The ease with which flammable gas and vapor mixtures ignite warrants hazard awareness when handling these materials. However, real-world mixtures of gases and vapors are handled at a range of temperatures, pressures, and compositions, and therefore, understanding how these factors affect flammable gases and vapors is important.

The concentration of flammable vapor plays a key role in whether a fire will occur. If the concentration of a flammable vapor in air is too high, it is said to be too rich to burn. Similarly, if the vapor concentration is too low, it is said to be too lean to burn. These upper and lower concentration bounds of flammability in air are defined as the upper flammability limit (UFL) and lower flammability limit (LFL), respectively (Crowl and Louvar, 2011).

When a mixture of flammable gases is handled, the LFL and UFL of the mixture are different than the LFL and UFL of the separate components. One simple way to calculate LFL and UFL of flammable gas mixtures is to use Le Chatelier's equation, shown in Eqns (2.1) and (2.2) (Chatelier, 1891).

$$\mathrm{LFL_{mix}} = \frac{1}{\sum_{i=1}^{n} \frac{y_i}{\mathrm{LFL}_i}} \tag{2.1}$$

where $\mathrm{LFL}_i$ is the LFL of chemical species $i$ if it was pure, and $\mathrm{LFL_{mix}}$ is the LFL of the mixture consisting of chemical species $i$ through $n$.

$$\mathrm{UFL_{mix}} = \frac{1}{\sum_{i=1}^{n} \frac{y_i}{\mathrm{UFL}_i}} \tag{2.2}$$

where $\mathrm{UFL}_i$ is the UFL of chemical species $i$ if it was pure, and $\mathrm{UFL_{mix}}$ is the UFL of the mixture consisting of chemical species $i$ through $n$.

It should be noted the Le Chatelier's equation comes with assumptions that are reasonably valid for the LFL and less valid for the UFL. It is assumed that each chemical species has the same heat capacity as if they were pure, that there is a constant number of moles throughout combustion, that kinetics for each chemical species is independent from each other, and the each chemical species has an identical temperature rise under adiabatic conditions (Mashuga and Crowl, 2000).

The LFL and UFL are also dependent on temperature and pressure. Generally speaking, for most hydrocarbons the LFL tends to decrease with increasing temperature, while the UFL tends to increase (Zabetakis et al., 1959). One approximation for the LFL dependency is shown in Eqn (2.3), and another for the UFL in Eqn (2.4).

$$\mathrm{LFL}(T) = \mathrm{LFL}(25\ ^{\circ}\mathrm{C}) - \frac{100\ C_p}{\Delta H_c}(T - 25) \tag{2.3}$$

$$\mathrm{UFL}(T) = \mathrm{UFL}(25\ ^{\circ}\mathrm{C}) + \frac{100\ C_p}{\Delta H_c}(T - 25) \tag{2.4}$$

where $C_p$ is the specific heat capacity at constant pressure for the flammable material in kcal/mol °C (100 $C_p$ is often approximated at 0.75 kcal/mol °C), $T$ is temperature in degree Celsius, and $\Delta H_c$ is the heat of combustion in kcal/mol (Zabetakis et al., 1959).

Pressure effects on the LFL and UFL should also be considered to gain a full understanding of flammability limits. For most materials, the LFL is affected little by pressure changes. The UFL, however, increases with increasing pressure. One approximation of this dependency is shown in Eqn (2.5) (Zabetakis, 1965).

$$\mathrm{UFL}(P) = \mathrm{UFL}(P = 1\ \mathrm{atm}) + 20.6(\log(P) + 1) \tag{2.5}$$

where $P$ is the absolute pressure measured in MPa.

Another commonly used flammability limit is similar to the LFL and UFL, except instead of measuring the flammability limits in air, the lower oxygen limit (LOL) and upper oxygen limit (UOL) measure the flammability limits of a fuel in oxygen. Similar to the LFL and the UFL, the LOL and UOL are dependent on temperature and pressure. The LOL is usually similar to the LFL of the mixture, and the UOL can be found with relatively good agreement from Eqn (2.6) (Hansen and Crowl, 2010).

$$\mathrm{UOL} = \frac{\mathrm{UFL}[100 - C_{\mathrm{UOL}}(100 - \mathrm{UFL_O})]}{\mathrm{UFL_O} + \mathrm{UFL}(1 - C_{\mathrm{UOL}})} \tag{2.6}$$

where $\mathrm{UFL_O}$ is the oxygen concentration at the UFL and $C_{\mathrm{UOL}}$ is a fitting parameter (set to −1.87 by Hansen and Crowl).

One last useful flammability measurement that will be discussed is the limiting oxygen concentration (LOC). An easy way to visualize the LOC is to imagine a flammable gas mixed with air. As oxygen is removed from the flammable mixture, the fire is eventually extinguished at an oxygen concentration defined as the LOC. Similar to the other limits discussed, there are approximations to account for derivations in temperature and pressure of the system. Equation (2.7) summarizes the effects of temperature and pressure on the LOC of a flammable gas mixture (Hansen and Crowl, 2010).

$$\mathrm{LOC} = \left(\frac{\mathrm{LFL} - C_{\mathrm{LOC}}\mathrm{UFL}}{1 - C_{\mathrm{LOC}}}\right)\left(\frac{\mathrm{UFL_O}}{\mathrm{UFL}}\right) \tag{2.7}$$

where $C_{LOC}$ is a fitting constant (experimental results shows that a value of $-1.11$ is appropriate for most hydrocarbons).

As the length scales of an enclosure decrease, the flammability limits of a flammable mixture are likely to change. As the volume of a flammable mixture decreases, the surface-to-volume ratio increases. As a result, heat loss in micro-scale flames is dominated by convection and conduction rather than by radiant heat (Ju and Maruta, 2011). A commonly reported flammability limit for micro-scale combustion is the quenching diameter, which measures the inner diameter of the enclosure that makes flame propagation within the container impossible. At this diameter, heat transfer to the walls by conductive and convective heat losses is so great that the combustion reactions driving the fire cannot be sustained. One simulation tested an array of flames propagating through channels and showed that a conservative estimate for the quenching diameter (overly small) in micro-channels is about 6 times the expected flame length (Daou and Matalon, 2001; Ju and Xu, 2006). In contrast, the walls of a micro-scale enclosure can also reinforce the flame rather than quench it. Since the enclosure and structures near the flame are smaller at the micro-scale, their thermal inertia is drastically reduced, allowing heat from the flame to transfer to structures as it burns, which in turn allows surrounding structures to reheat the unburned gases (Ju and Maruta, 2011). This enhances the flame and aids in combustion.

### 2.1.4 TYPES OF FIRES

Depending on the conditions present and the materials involved, a fire can manifest itself in many different forms, all with different behavior and specific hazards. The goal of this section is to understand common types of fire aids in identifying potential hazards and ways to mitigate them.

Fire behavior is dependent on a myriad of conditions, including operating temperature, operating pressure, and weather conditions. However, when calculating important fire hazard parameters, two main quantities prove to be the most important. These are the amount of flammable material released and the rate at which it is released. This information can be used to determine other valuable information, such as heat release rates and heat fluxes, which can directly be used when determining expected damages and loss from fires. The mechanism for the release of flammable material is highly dependent on specific process conditions, but common types of fires and the conditions in which they appear are covered in this section.

#### *2.1.4.1 Diffusion fires*

Diffusion fires are a subset of fires in which the fuel and oxidizer are physically separated with a region of rapid combustion between them. In this sense, many solid material fires are diffusional in nature with pyrolysis vapor diffusing away from a material surface where it then reacts with oxygen (Drysdale, 2011). More traditional examples of diffusion fires include jet fires, which are a release of flammable material from an orifice into ambient conditions, where it ignites and forms a fire jet.

##### Jet fires

Jet fires occur when a pressurized flammable material is released from an opening to a lower pressure exterior. Jet fires can be further categorized as either laminar or turbulent. The transition between these two types of fire is categorized by a dimensionless quantity called the Reynolds number, shown in Eqn (2.8).

$$Re = \frac{\rho v D}{\mu} = \frac{\text{inertial effects}}{\text{viscous effects}} \tag{2.8}$$

where $Re$ is the Reynolds number, $\rho$ is the density of the fluid exiting the orifice, $v$ is the velocity of fluid flow out of the orifice, $D$ is the characteristic length of the orifice, and $\mu$ is the viscosity of the fluid exiting the orifice.

The dimensionless Reynolds number makes it simple and unambiguous to classify jet flames, regardless of scale. Laminar jet fires have Reynolds numbers less than 2000 at the orifice opening, while turbulent jet fires occur at substantially higher Reynolds numbers (Hottel and Hawthorne, 1949). These flammable jets have been shown experimentally to have a flame height dependency on the square root of the volumetric flow rate. However, because these fires usually occur at low Froude numbers, buoyant factors play a large role in flame behavior (Jost, 1939). Therefore, different materials have the potential to have noticeably different laminar jet flames.

At the micro-scale, most jet flames are classified as laminar because of their low Reynolds numbers. However, as the scale of jet flames decrease to the micro-scale, new complications arise. Mixing becomes increasingly poor and dominated by diffusion, which can lead to flame instabilities and complicated transient behavior (Miesse et al., 2005). Additionally, the smaller size of a jet flame means a greater surface area to volume ratio and an increased amount of heat loss, making micro jet flame behavior deviate from conventional laminar jet flames even more.

The second set of jet fire, turbulent jet fires, occurs at Reynolds numbers much greater than 2000 (Hottel and Hawthorne, 1949). As the velocity of the fuel increases, the flame begins to break up at the end (opposite of the fuel release). As the velocity increases further, this point of flame breakup gets closer and closer to the nozzle but never reaches it (Hottel and Hawthorne, 1949).

Interestingly, the efficiency of fuel burning is also increased in turbulent jet fires. High turbulence correlates to less soot formation and less heat loss in the fire overall, creating a flame that burns more completely and at higher temperatures (Delichatsios and Orloff, 1988). The flame height dependencies differ between turbulent and laminar jet fires. Instead of being related to volumetric flow rate, the height of turbulent jet fires is linearly dependent on the release orifice (Kanury, 1975).

Another important hazard parameter to calculate is the mass flow rate of flammable material during a jet fire. Since this value will affect burning rates and incident severities, determining the amount of flammable material release during a jet flame is important. Analytical and semianalytical models have been derived in the past for the release of pressurized fluids through orifices by analyzing the transport phenomena taking place. One such model is derived from a momentum balance around the orifice and relates the mass loss rate to general process conditions and material properties (American Institute of Chemical Engineers. Center for Chemical Process Safety. 2003). This model, shown in Eqn (2.9), assumes adiabatic expansion, which seems reasonable because the flammable gas is released and expands very quickly.

$$M = C_d \rho_a A_h \sqrt{\frac{2P_p}{\rho_p}\left(\frac{k}{k-1}\right)\left[1-\left(\frac{P_a}{P_p}\right)^{k-1/k}\right]} \tag{2.9}$$

where $M$ is the mass flow rate, $C_d$ is the coefficient of discharge (0.85 typically for gas release), $A_h$ is the area of the discharge opening, $\rho_a$ is the density of the ambient air, $\rho_p$ is the density of the process flammable fluid under pressure (the fluid that will be released), $P_a$ is the ambient pressure, $P_p$ is the pressure of the flammable process fluid under pressure, and $k$ is the isentropic expansion factor defined in Eqn (2.10) as

$$Pv^k = c \tag{2.10}$$

where $P$ is the pressure of the flammable fluid, $v$ is the specific volume of the flammable fluid, $k$ is the isentropic expansion factor, and $c$ is a constant.

However, there exists a specific situation where a maximum mass release is achieved and increasing the process pressure of the fluid or reducing the ambient pressure does not increase the mass release rate further. When the maximum release rate is achieved, it is called choked flow. Finding the maximum mass flow rate with respect to downstream pressure yields the criteria for choked flow shown in Eqn (2.11).

$$\left(\frac{P_a}{P_p}\right)_{\text{choked}} = \left(\frac{2}{k+1}\right)^{k/(k-1)} \tag{2.11}$$

When choked flow condition is achieved, Eqn (2.9) is simplified and expressed as Eqn (2.12) and is no longer dependent on ambient conditions.

$$M_{\text{max}} = C_d A_h \sqrt{P_p \rho_p k \left(\frac{2}{k+1}\right)^{(k+1)/(k-1)}} \tag{2.12}$$

where $M_{\text{max}}$ is the choked release rate.

After the release rate of the jet fire is established, the heat release rate is calculated simply by multiplying the mass flow rate by the heat of combustion for the flammable fluid. This heat release rate assumes complete combustion making it a conservative estimate. If the ambient air is relatively stationary and fairly laminar, the flame length can be calculated by Eqn (2.13) (Beyler, 2002).

$$L = 0.2\dot{Q}^{\frac{2}{5}} \tag{2.13}$$

where $\dot{Q}$ is the heat release rate, and $L$ is the flame length.

As ambient turbulence increases, the flame length become affected more and this model fails to provide useful results. It should also be noted that if the path of the jet flame is obstructed, it can greatly change the size and shape of the flame.

## Natural fires

Natural fires are a type of diffusional flame in which natural organics make up a fuel bed and are heated to the point of pyrolysis by an external heat source. Similar to other solid flammables, these pyrolysis compounds are vaporized and transported toward the combustion region of a fire where they are burned (Drysdale, 2011). In this sense, the solid fuel is separated from the air until it burns, classifying it as a diffusional flame.

A typical natural fire has a flame height related to its diameter. Natural fires with small diameter are tall, while very large-diameter fires have significantly shorter flames and regions of material that is not on fire at all (Heskestad, 1991). A tall bonfire compared with a low, sprawling forest fire shows how this dependency works in the real world. This is explained experimentally by a reduction in the dimensionless heat release rate for increased flame diameters. For very large-diameter fires, the dimensionless heat released (introduced later in the section “Pool fires”) is significantly less than with smaller-diameter fires.

Many natural fires have three distinct regions as described originally by McCaffery (1979). The region just above the burning material where combustion occurs is termed the persistent flame. Just above the material, a rich flammable pyrolysis vapor moves into the combustion region where the fire burns with a relatively constant intensity. As the reacting combustion products move up, they enter a new region of the fire called the intermittent flame. Here, flame flickering occurs and the presence of a flame can change from moment to moment. This flickering is caused by the formation of eddies at the fire–ambient interface. The contents of the fire move up, while the surrounding air remains stationary, causing disturbances in the surface of the flame that manifest as eddies and flame flickering. As the fire eddies move up, they move through the flame and eventually past the top of the visible flame front in a process known as eddy shedding. Above the visible fire is a region termed the buoyant plume. The buoyant plume is a mixture of entrained ambient air and heated combustion products that moves up. The heated combustion products are less dense than the surrounding air so buoyant forces cause them to rise. The buoyant plume also widens out and cools as it rises and interacts with the cooler ambient conditions (Drysdale, 2011).

### Pool fires

Another class of diffusional flame is the pool fire. As the name implies, a pool fire occurs when a pool of material forms and produces enough vapor to create an ignitable fuel source above the pool. Pool fires are generally a liquid, although gas and solid pool fires (such as a dense pool of gas or a solid polymer burning) also exist. Like some other types of diffusional flames, pool fires have three distinct regions: the fuel-rich core, the intermediate zone, and the downstream plume (Bouhafid et al., 1898). These three regions are similar to the regions discussed in other natural diffusion fires.

The fuel-rich core lies directly above the burning material surface and contains vaporized flammable materials. The boundaries of this region are formed by the air-entrained eddies above in the intermediate zone. This region contains a relatively small amount of combustion because the vapor is so rich (Smith and Cox, 1992). As the flammable vapor moves away from the material surface, it enters the intermediate zone where combustion is rapid. The movement of the flame up entrains cool, oxygen-rich air that fuels the fire as it moves upwards. Soot and products from incomplete combustion can be formed here as well. It is in this intermittent zone of a pool fire where the majority of heat is generated (Smith and Cox, 1992).

As the heated combustion products and vortices continue to rise, they exit the visible region of the flame and continue through to the downstream plume, which contains heated combustion products mixing with ambient air. As expected, this downstream plume broadens and cools as it rises from the fire. As the temperature drops, the kinetics of the combustion reactions slow exponentially and the formation of combustion products stops (Smith and Cox, 1992).

If the burning rate of the pool fire is assumed to be relatively constant (steady state), a simple approximation can be used to determine the steady-state diameter of the pool fire resulting from a given leak rate given in Eqn (2.14) (Spouge, 1999).

$$D_{SS} = \left(\frac{4\dot{V}_{\text{Leak}}\rho}{\pi \dot{m}''}\right)^{0.5} \tag{2.14}$$

where $D_{ss}$ is the steady-state diameter of the pool fire, $\dot{V}_{\text{Leak}}$ is the volumetric flow rate, $\rho$ is the density of the fuel, and $\dot{m}''$ is the constant mass burning rate (Spouge, 1999).

Heskestad proposed a simple model for the height for a pool fire dependent on the heat released by the fire and the diameter of the pool fire (Heskestad, 1981, 1983). However, it should be noted that this model represented in Eqn (2.15) is a simplification of pool fire phenomena and does not directly take into account many possible factors such as the vapor flow rate away from the burning material to the combustion region, effects of gravity, and time-dependent transient effects of the pool fire to name a few.

$$H = 0.23\dot{Q}^{2/5} - 1.02D \tag{2.15}$$

where $D$ is the pool fire diameter, and the heat release rate, $\dot{Q}$, can be written for a simplified pool fire as Eqn (2.16).

$$\dot{Q} = \dot{m}\Delta H_c = \dot{m}''A\Delta H_c \tag{2.16}$$

where $\dot{m}$ is the mass loss rate for the pool fire, $\dot{m}''$ is the mass loss flux for the pool fire, and $A$ is the area (American Institute of Chemical Engineers, Center for Chemical Process Safety, 2003). The fundamental heat, mass, and momentum transport make actual pool fire behavior very complicated, so resorting to dimensionless analysis is useful for pool fire analysis. Furthermore, the use of dimensionless variables allows fires at any scale to be comparable using the same measureable quantities. An important dimensionless parameter in the study of pool fire behavior is the dimensionless Froude number shown in Eqn (2.17).

$$Fr = \frac{v^2}{gD} = \frac{\text{inertial effects}}{\text{bouyant effects}} \tag{2.17}$$

where $Fr$ is the Froude number, $v$ is the fluid velocity, $g$ is gravity, and $D$ is the characteristic length of the pool fire (the diameter).

The Froude number is useful in establishing whether buoyant effects from the heated combustion products or the velocity of the rising gases dominate the fluid behavior of the flame.

The models derived by Orloff and Ris show that the pool fire can move laterally back and forth over the boundaries of the burning fuel, but the amount of lateral movement at the base of the fire is dependent on the Froude number. At low values of Fr, the boundary of the fire is able to move farther laterally than at higher Fr values (Orloff and Ris, 1982, 1983). With high Fr numbers, it seems the relatively high gas velocity up prevents the flame from deviating laterally compared with lower relative gas velocities associated with more rapid burning. Numerous studies have been conducted to determine the source of the pulsing behavior of diffusional flames with some progress; however, the true source of the instability of pool fires is still debated (Hertzberg, Cashdollar et al., 1978; Buckmaster and Peters, 1986; Bejan, 1991). Another dimensionless variable that appears to be useful in pulsating pool fires is the Strouhal number, shown in Eqn (2.18).

$$St = \frac{fD}{v} \tag{2.18}$$

where $St$ is the Strouhal number, $f$ is the frequency of eddie shedding, $D$ is the characteristic length, and $v$ is the fluid velocity.

The Strouhal number contains information about the frequency of oscillations within the fluid flow. Substantial research has been done to correlate the Strouhal number and Froude number with relatively good results. Hamins showed that the relation between the Froude and Strouhal numbers as shown in Eqn (2.19) was valid for various flames (Hamins et al., 1992).

$$St \propto Fr^{-0.38} \tag{2.19}$$

In addition to how a pool fire behaves, understanding the pool fires dimensions is important to fully understand the hazards they present. To first define the size of a fire, it is necessary to define the fire boundary. It is common to represent a fire as the region that emits visible light, but for turbulent flames with pulsating regions, this representation can become difficult. One way to combat this is to use a length measurement in which 50% of the flame would occur above that level and 50% of the flame would occur below that level. Using a dimensionless version of this flame length and a dimensionless heat release quantity, defined in Eqns (2.20) and (2.21), Heskestad developed a correlation for flame height as shown in Eqn (2.22) that fits experimental data closely (Heskestad, 1983; Zukoski et al., 1984).

$$N = \left(\frac{rC_pT_0}{H_c}\right)^3 \dot{Q}_D^2 \tag{2.20}$$

where $N$ is a dimensionless parameter, $r$ is the mass based stoichiometric air-to-fuel ratio, $C_p$ is the heat capacity of ambient air, $T_0$ is the ambient air temperature, $H_c$ is the fuel heat of combustion, and $\dot{Q}_D$ is dimensionless heat released as defined in Eqn (2.21) (Heskestad, 1983).

$$\dot{Q}_D = \frac{\dot{Q}}{\rho_0 C_p T_0 (gD^5)^{0.5}} \tag{2.21}$$

where $\dot{Q}$ is the heat release rate of the fire, $\rho_0$ is the density of the ambient air, $g$ is the acceleration due to gravity, and $D$ is the characteristic length of the fire (Heskestad, 1983).

$$\frac{Z_f}{D} = -1.02 + 15.6N^{1/5} \tag{2.22}$$

where $Z_f$ is the flame height, and $D$ is pool diameter.

Although Heskestad's correlation seems to fit the majority of data very well, acetylene ($C_2H_2$) data do not match Eqn (2.22) nearly as well as other mixtures (Hamins et al., 1992). Because acetylene burning forms a significant amount of soot, length scales could be redefined within the fire to more closely fit Eqn (2.22) (Delichatsios et al., 1992). Therefore, caution should be used when applying Heskestad's correlation to sooty flammable mixtures.

The key hazard from pool fires is the heat produced. Heat in a pool fire can have three ultimate destinations. First, heat can be transferred from the buoyant flame and plume to the adjacent ambient air. Second, heat can be transferred from the flame to the fuel source. Third, heat can be transferred across open spaces via radiation (Hamins et al., 1995).

The heat transferred from the buoyant plume to the surrounding ambient air is calculated through a simple energy balance and shown in Eqn (2.23).

$$Q_{\text{ambient}} = v_p A_p \rho_p C_p \Delta T \tag{2.23}$$

where $Q_{\text{ambient}}$ is the heat transferred to the ambient air, $v_p$ is the velocity of the bouyant plume, $A_p$ is the average area of the plume, $\rho_p$ is the average density of the plume, $C_p$ is the specific heat of the plume, and $\Delta T$ is the change in temperature of the plume from the hottest point of the plume to its coldest.

This equation is derived from a simple energy balance in which it is assumed that all the heat required to change in temperature of the buoyant plume is transferred to the ambient air with no energy losses or entropic considerations, so it is an approximate calculation (Hamins et al., 1995).

Aside from heating the ambient air, a fire can also radiate heat over open space to heat other objects. Radiant heating is especially hazardous because it acts in all directions, unlike conductive and convective heating that are limited in this sense. A series of studies with heptane measured the radiant heat released by using a series of radiometers to obtain vertical and radial heat flux profiles. These profiles were then used to estimate radiant heat from a fire. The results show that the fraction of heat released as radiant energy in a heptane pool fire is significant and varies depending on the diameter of the pool fire. Under 2 m, the fraction of radiant heat is relatively constant, but as the diameter grows above 2 m, the fraction of radiant heat drops. The cause for this drop at 2 m is attributed to the soot formation within the fire. Cleaner burning fuels do not exhibit a drop in radiant heat as they grow because soot particles in the flame do not block radiant heat. However, as the size of a heptane pool fire increases, a lack of available oxygen causes incomplete combustion and the formation of soot that blocks radiant heat (Hamins et al., 1995). This is similar to how standing in the shade on a hot day blocks heat from the sun. These data show that fires that form soot and carbonaceous particles are more likely to be colder than nonsooty counterparts.

A more direct way of determining the radiant heat released from a fire to an object is to use Eqn (2.24) (Shokri and Beyler, 1989). Equation (2.24) requires only two parameters to calculate the heat flux and provides a relatively straightforward method for determining radiant energies.

$$q'' = EF_{12} \tag{2.24}$$

where $q''$ is the radiant heat flux, $F_{12}$ is the dimensionless view factor for the radiant energy, and $E$ is the total emissive energy flux from the fire given in Eqn (2.25) as

$$E = 58\left(10^{-0.00823D}\right) \tag{2.25}$$

where $D$ is the diameter of the fire.

The view factor is a complicated parameter that essentially determines the fraction of radiant energy from a fire that reaches a target. Fortunately, previous algebraic analysis of these systems has yielded plots to obtain the view factor easily. Primarily, the view factor is dependent on the ratio of flame height to radius, distance from the center of the fire to the object, and the height of the target object (to be heated by a fire). A series of plots for various combinations of these parameters is found in *Guidelines for Fire Protection in Chemical, Petrochemical, and Hydrocarbon Processing Facilities* (American Institute of Chemical Engineers, Center for Chemical Process Safety, 2003). Once the maximum emissive heat released from the fire is calculated assuming complete combustion of all flammable material, this can be multiplied by the view factor for the specific scenario to find the heat flux on an object.

Aside from heat transfer from the plume to the ambient air and radiant heating from the fire to surrounding objects, the third destination for heat from a fire is to the fuel. When a fire burns, heat is transferred back toward the flame in a process called feedback. Feedback causes increased fuel vaporization, which in turn increases the burning rate and, thus, increases fuel vaporization further. A general expression for the heat transferred to the fuel can be written as Eqn (2.26), although it is of limited use because the heat lost ($\dot{Q}_{\text{lost}}$) and heat associated with setting up a thermal boundary layer ($\dot{Q}_{\text{liquid}}$) are complex (Hamins et al., 1995).

$$\dot{Q}_{\text{fuel}} = \dot{m}H_g + \dot{Q}_{\text{lost}} + \dot{Q}_{\text{liquid}} \tag{2.26}$$

where $\dot{Q}_{\text{fuel}}$ is the heat transferred to the fuel, $\dot{m}$ is the mass loss rate, $H_g$ is the heat of gasification, $\dot{Q}_{\text{lost}}$ is the heat lost to the surroundings, and $\dot{Q}_{\text{liquid}}$ is the heat associated with heating the liquid in the pool fire and establishing a boundary layer.

Without further detailed knowledge of the heat transfer in the fuel, it is difficult to calculate the rate of vaporization and thus the rate of burning in a pool fire. Empirical relations from literature suggest that mass flux is directly related to the $B$ number of the material as shown in Eqn (2.27).

$$B = \frac{\Delta H_c}{\Delta H_g} \tag{2.27}$$

where $B$ is the $B$ number, $\Delta H_c$ is the heat of combustion, and $\Delta H_g$ is the heat required to gasify a material.

The wind effects of pool fires have been empirically observed and used to develop a model for the angle of flame and the dimensionless wind velocity. When the wind impinges on a fire, the vertical axis of the flame will tilt and rotate an angle θ around the base of the fire in the direction of the wind velocity. This correlation is outlined in Eqns (2.28) through (2.30) (Society of Fire Protection Engineers, 2002):

$$cos\theta = 1 \quad \text{for} \quad u^* \leq 1 \tag{2.28}$$

$$cos\theta = \frac{1}{\sqrt{u^*}} \quad \text{for} \quad u^* > 1 \tag{2.29}$$

where θ is the angle of the flame due to the wind, and $u*$ is the dimensionless wind velocity given in Eqn (2.30) as:

$$u^* = \frac{u_w}{(g\dot{m}''D/\rho_v)^{1/3}} \tag{2.30}$$

where $D$ if effective fire diameter, $u_w$ is the wind speed measured at a height of 1.6 m, $g$ is the acceleration due to gravity, $\dot{m}''$ is the mass burning rate, $D$ is the pool fire diameter, and $\rho_v$ is the vapor density of the flammable vapor at the boiling point (Society of Fire Protection Engineers, 2002).

### Fireballs

Another form of fire that is generally classified as a diffusion flame is that of a fireball. A fireball occurs when there is a sudden loss of containment, resulting in the release of a vapor of liquid that immediately ignites (American Institute of Chemical Engineers, Center for Chemical Process Safety, 2003). It is essentially a region of highly flammable material that almost instantaneously comes into contact with the surrounding air and then catches fire. A boiling liquid expanding vapor explosion (BLEVE) is a type of fireball in which a contained liquid boils, building pressure and eventually rupturing the vessel, immediately releasing all of its contents and igniting.

Studies have been done to determine the behavior of fireballs including research done by Spouge to determine correlations for fireball diameter. Equation (2.31) shows how the diameter of a fireball varies with the amount of flammable material instantaneously released (Spouge, 1999).

$$D = 5.8M^{0.333} \tag{2.31}$$

where $D$ is the fireball dimension in meters, and $M$ is the mass of flammable material in kilograms.

Furthermore, the duration of burning has also been studied. The results from this are presented as Eqns (2.32) and (2.33), which form a piecewise expression for the time of burning based on the mass of flammable material released (Spouge, 1999).

$$t = 0.45M^{0.333} \quad \text{for} \quad M < 37,000 \text{ kg} \tag{2.32}$$

$$t = 2.6M^{0.167} \quad \text{for} \quad M > 37,000 \text{ kg} \tag{2.33}$$

where $t$ is amount of time a fireball will burn for in seconds, and $M$ is the mass of released flammable material in kilogram.

A study on the fraction of radiated heat was conducted, and Eqn (2.34) shows the empirical relation that is dependent on the vapor pressure of the fuel (Roberts, 1982).

$$\chi_r = 0.27P^{0.32} \tag{2.34}$$

where $\chi_r$ is the fraction of heat released in the form of radiant energy, and $P$ is the vapor pressure of the fluid when failure occurs in MPa absolute. Equation (2.34) was derived from empirical data below vapor pressure. In this range, the data fit the correlation well, but the validity of the results above this pressure is questionable.

#### 2.1.4.2 Premixed fires

In addition to diffusional fires, another class of fires exists, called premixed fires. As the name suggests, a premixed fire is a fire in which a mixture of flammable vapor and air is mixed prior to being ignited. This differs from diffusion fires, which require the fuel and oxidizer to be physically separate until combined in the combustion region. Further, premixed flames can be classified as either a deflagration or a detonation, the latter being a premixed flame with a burning velocity equal to or greater than the speed of sound (Drysdale, 2011).

A common example of a premixed flame is that of a Bunsen burner. Air and flammable gas is mixed and directed toward the nozzle of the burner, where it is combusted. Before the gas burns, it is preheated in a region just before entering the combustion region where the majority of burning takes place. Last, the fuel moves past the combustion region to the post flame region where hot gases cool as they move farther from the flame front. The temperature and concentration profile of the three regions associated with the flame front are shown in Figure 2.2 and are present in all premixed flames (Lewis and von Elbe, 1987).

In the case of the Bunsen burner, the burning velocity of the fire is fixed by the flow rate at which the flammable gas mixture is fed into the combustion region. However, freely propagating premixed flames can also occur. If the effects of heat loss are ignored for the burning of a premixed flame,

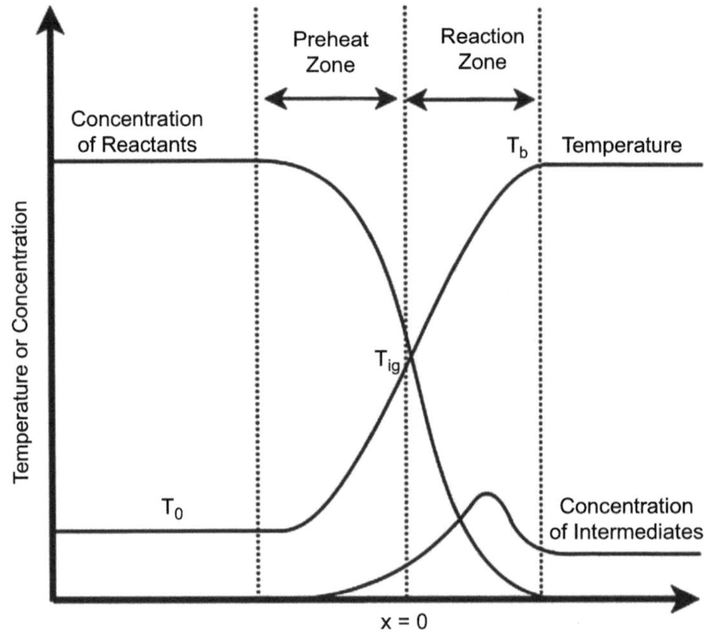


**FIGURE 2.2**

The preheat region, combustion region, and post flame region in a premixed flame (Drysdale, 2011).

transport equations can easily be solved for this system to give Eqn (2.35), the burning velocity for a freely propagating premixed flame (Kanury, 1975). Equation (2.35) is an approximation and does not reflect the flammability limits experimentally seen with premixed flames due to the assumption that the energy loss from combustion is negligible.

$$S = \left( \frac{2k}{\rho_0^2 c_p^2 (T_f - T_0)} \dot{Q}'''_{\text{ave}} \right)^{0.5} \tag{2.35}$$

where $S$ is the burning velocity, $k$ is the ambient thermal conductivity, $\rho_0$ is the ambient density, $c_p$ is the specific heat of the ambient, $T_f$ is the final temperature of the combustion products, $T_0$ is the initial temperature of flammable gas, and $\dot{Q}'''_{\text{ave}}$ is the average heat release rate in the reaction zone.

To make matters more complicated, the burning velocity is dependent on composition. Generally, the highest burning velocity with relation to composition occurs when the fuel concentration of a flammable mixture is marginally above the stoichiometric ratio. As the concentration deviates in either direction, the burning velocity will decrease (Lewis and von Elbe, 1987).

The burning velocity is also dependent on pressure effects and turbulent effects; however, these phenomena are not well understood and require significantly more research to understand. The scale of premixed flames can also dictate their behavior. At the micro-scale, thermal feedback is significantly more pronounced. To elaborate, because the enclosure and structures near the flame are smaller at the micro-scale, their thermal inertia is drastically reduced, allowing heat from the flame to transfer to structures as it burns, which in turn allows surrounding structures to reheat the unburned gases (Ju and Maruta, 2011). This enhances the flame and aids in combustion. In contrast, micro-scale combustion also causes a higher amount of radicals to be lost to the surrounding because a small volume of material is burning, but a relatively large surface area is available for radical loss. Similarly, this large surface area-to-volume ratio makes heat loses to conduction and convection dominant over radiant heat loses (Ju and Maruta, 2011).

#### Flash fires

A flash fire is a release of flammable vapor (or liquid that vaporizes) that premixes with air and expands, eventually igniting. Once ignition occurs, the burning velocity travels from the point of ignition toward the release point, potentially igniting the source as well. Since this ignites the entire volume of the vapor mixture, the volume of the flame is large and the majority of damage is due to flame impingement (American Institute of Chemical Engineers, Center for Chemical Process Safety. 2003). Large-scale flash fires are particularly dangerous as they can cover a large area and, once ignited, occur quickly and have the potentially to ignite the flammable release source. When premixed flammable mixtures are confined to smaller scales, flash fires may become more violent, and even escalate to an explosion.

### 2.1.5 FIRE RISK ANALYSIS

A fire risk analysis (FRA) is a tool that helps determine the overall risk for damages from fire hazards in a process. An FRA is not necessary for all processes, but for cases where there are numerous fire hazards or hazards that may be missed, an FRA can help greatly (Society of Fire Protection Engineers, 2002).

The general methodology of an FRA contains numerous steps. To truly determine the risk of fire at a plant, information about the process must first be collected. Although process information can be difficult to obtain for new or developing plants, obtaining a rough picture of process operations is needed to conduct the FRA (Society of Fire Protection Engineers, 2002). Once information on the process has been collected, hazard identification can begin. This process should be systematic and draw from the previously collected process information. Other methods for general hazard identification have already been developed and can be used for FRA, such as HAZID (hazard identification), HAZOP (hazard and operability study), checklists, and what-if scenario checklists (Society of Fire Protection Engineers, 2002).

After a list of potential fire hazards has been developed, they need to be further analyzed using another tool called a fire hazard analysis (FHA). An FHA takes previously discovered hazards and determines the severity and outcome of an incident. If the hazard was a jet flame, the FHA would calculate the outcome (property, production, injury, or loss) of the expected jet flame. This analysis does not necessarily need to be conducted for every hazard identified, and it should in fact focus more on hazards that seem to be more severe or problematic (Society of Fire Protection Engineers, 2002). After the possible outcomes of the identified fire hazards are determined, the likelihood of these incidents is found. If the outcomes from the FHA are shown to be severe enough, a likelihood analysis might not be necessary. For instance, if an expected outcome involves multiple deaths or injuries, it might be decided that no matter how low the likelihood of the event, the outcome is not acceptable. In that case, a consequence-based decision-making process would be useful (Society of Fire Protection Engineers, 2002).

If it is decided that a likelihood analysis is necessary, the frequency of each outcome from the FHA must be determined. This is done using fault trees, logic gates, and other probability-based calculations. These types of calculations require intimate knowledge about how equipment works together within the process and the modes of failure for these devices (Society of Fire Protection Engineers, 2002). This makes fault trees and logic trees very useful for the likelihood analysis of a process.

After the frequency of the expected outcomes is determined, the overall risk for the process can be found. This calculation is straightforward and shown in Eqn (2.36).

$$\text{Risk} = \sum_{i=1}^{n} f_i E_i \tag{2.36}$$

where $f_i$ is the frequency of a specific incident, and $E_i$ is the expected loss due to that specific incident.

Risk is defined as the overall expected outcome of incidents, taking into account all of possible incidents and their likelihoods. For each outcome found in the FHA, the expected loss (in dollars, injuries, or deaths) is multiplied by the frequency. This is done for every outcome from the FHA and then the values are summed to give the expected outcome for the process. This is the expected loss due to the process based on likelihood analysis and the outcome analysis (Society of Fire Protection Engineers, 2002).

By this point, all the information from the FRA has been condensed into a number, the total expected outcome for the process, known as the risk, but this is useless without some type of reference for risk. In some cases, this threshold risk is determined by the government and in other cases it simply depends on risk preference of the company or decision-making body involved. If the calculated risk is higher than the acceptable risk, changes must be made to reduce the severity or likelihood of incidents from occurring (Society of Fire Protection Engineers, 2002).

One common method for setting a tolerable risk level is simply called "as low as reasonably practicable" (ALARP). In this method, it must be shown that all reasonable measures to reduce risk have been taken, and doing any more is not feasible (either financially or technically). A second method is a simple cost–benefit analysis. If mitigation of hazards does not substantially help lower the risk or if reducing the risk is too expensive, it would not be necessary according to this method (Society of Fire Protection Engineers, 2002).

## 2.2 EXPLOSION

The term "explosion" is used to describe a rapid release of energy. The energy released in an explosion is normal from chemical energy, fluid expansion energy, or vessel strain energy. Damage caused by explosions varies according to different energy dissipation mechanisms, such as a pressure wave, projectiles, thermal radiation, and acoustic energy. Taking overpressure as an example, 5–7 psig (pounds per square inch gauge) will nearly cause complete destruction of houses (Clancey, 1972).

This section first briefly introduces basic concept about explosion, for example, classification and characterization of explosion, deflagration, and detonation. Then it demonstrates some models for explosions caused by different energy types, explosion types, and explosion damage. Finally, it presents several ways to prevent explosion.

**Table 2.1 Parameters Significantly Affecting the Behavior of Explosions (Crowl and Louvar, 2011)**

| Parameter | Description |
|---|---|
| 1 | Ambient conditions (pressure and temperature) |
| 2 | Composition of explosive material |
| 3 | Physical properties of explosive material |
| 4 | Nature of ignition source: type, energy, and duration |
| 5 | Geometry of surroundings: confined or unconfined |
| 6 | Amount of combustible material |
| 7 | Turbulence of combustible material |
| 8 | Time before ignition |
| 9 | Rate at which combustible material is released |

Although extensive efforts, including theoretical, semiempirical, and empirical studies have been undertaken toward explosion (Stull, 1977; Baker et al., 1983; Crowl, 2010), it is difficult to characterize explosion behaviors. Based on previous research, Crowl and Louvar summarized some parameters significantly affecting the behavior of explosions as stated in Table 2.1.

### 2.2.1 DEFLAGRATION AND DETONATION

There are two kinds of explosion from combustion of flammable gas: deflagration and detonation. The criterion to distinguish them is to compare the reaction front propagation speed with the speed of sound in the unreacted gases. Assuming the unreacted gases as ideal gases at 20 °C, it is a detonation if the reaction front propagation speed is faster than 344 m/s; otherwise, it is a deflagration. Figure 2.3 demonstrates the physical differences between a detonation and a deflagration (Crowl and Louvar, 2011). In a detonation, the reaction front moves faster than sound, driving the shock front immediately preceding it. In a deflagration, the reaction front moves slower than sound, while the pressure front moves away from the reaction at the speed of sound. Accompanied by a strong, sharp overpressure, detonation causes much more catastrophic consequences than deflagration.

A deflagration can evolve into a detonation by continuously offering energy to pressure waves. This is called a deflagration to detonation transition (DDT). DDT is now one of major unsolved problem in theoretical combustion theory. It is a complicated process, involving highly nonlinear interactions among flames, shocks, boundary layers, and turbulence (Oran and Gamezo, 2007). Several theoretical and numerical efforts were made to understand the DDT. For example, computational analysis of shock–flame interactions was performed in acetylene–air, ethylene–air, ethylene–air, and hydrogen–air with one, two, or three dimensions. Oran and Gamezo state that "hot spots" created by a series of shock–flame interactions will produce a detonation through the Zeldovich gradient mechanism. Figure 2.4 demonstrates time sequence of density fields showing the overall evolution of a Mach 1.5 incident shock interacting with a low-pressure acetylene–air flame.

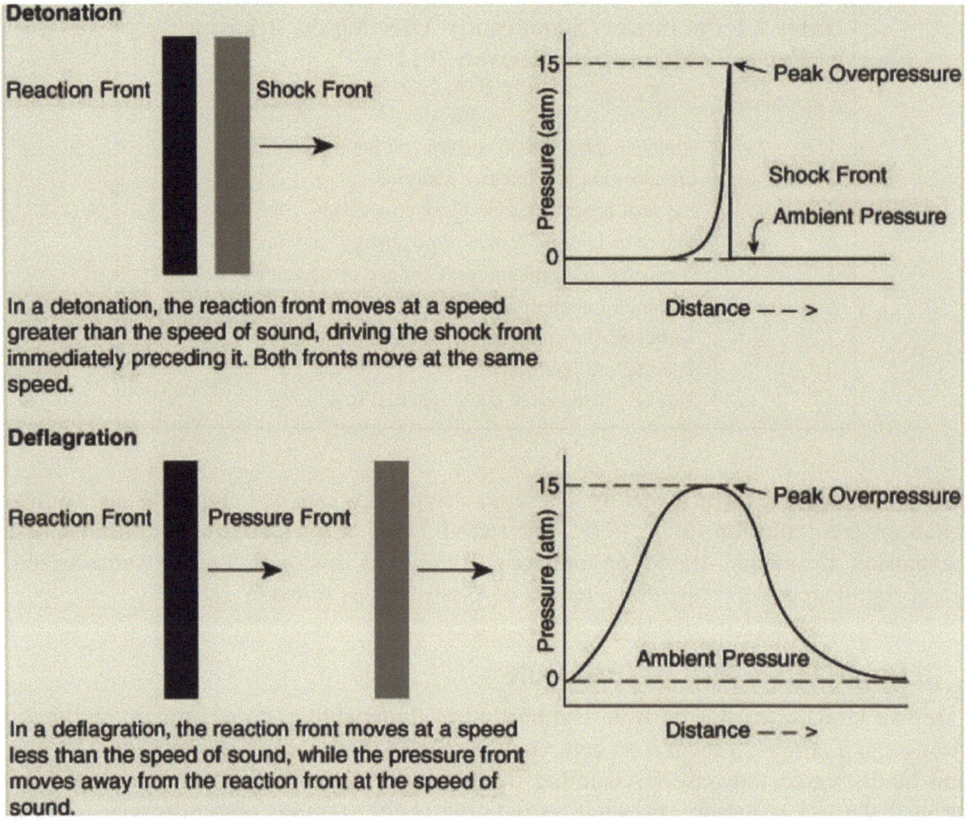


**FIGURE 2.3**

Comparison of detonation and deflagration gas dynamics (Crowl and Louvar, 2011).

## 2.2.2 EXPLOSION ENERGY

The energy released from an explosion normally comes from chemical energy, mechanical energy, or nuclear energy. The former two are much more common and will be discussed here in more detail.

### *2.2.2.1 Energy of chemical explosions*

Kinney and Graham (1985) discussed that the energy quantities characterizing a high explosive are the energy of explosion, the heat of explosion, and the heat of combustion. The energy of explosion is necessary to estimate the effects of an explosion:

$$W = -\int_1^2 PdV \tag{2.37}$$

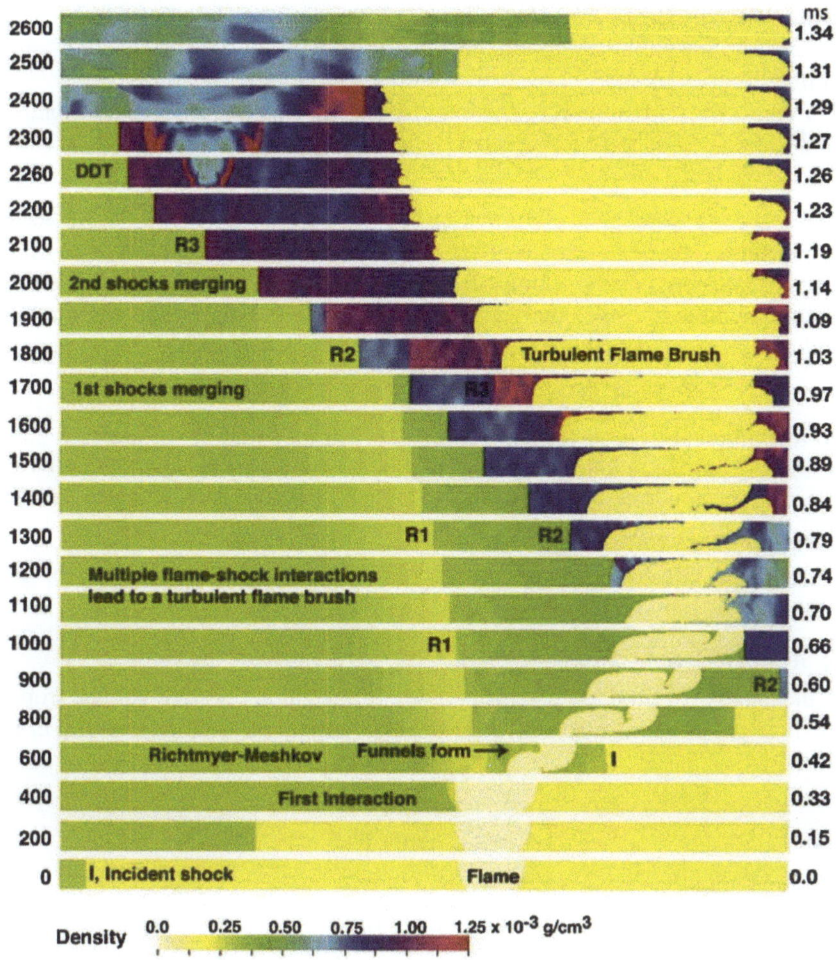


**FIGURE 2.4**

Time sequence of density fields showing the overall evolution of a Mach 1.5 incident shock interacting with a low-pressure acetylene–air flame (Oran and Gamezo, 2007).

where $P$ is the absolute pressure, $V$ is the volume, and $W$ is the work of expansion. However, it is difficult to evaluate the integral. This equation is simplified by using the following relationship (Moore, 1962):

$$-\int_1^2 PdV \approx \Delta A \tag{2.38}$$

where $\Delta A$ is the Helmholtz free energy change.

In many situations, the Gibbs free energy is used instead of the Helmholtz free energy due to the limited availability of data on the Helmholtz free energy and because of the small difference between Gibbs and Helmholtz free energy.

The heat of explosion corresponds to the internal energy change for the explosion:

$$\Delta E = \left(\Delta E_f^0\right)_p - \left(\Delta E_f^0\right)_r \tag{2.39}$$

where $(\Delta E_f^0)_p$ is the standard heat of formation for products, and $(\Delta E_f^0)_r$ is the standard heat of formation for reactants.

For chemical explosions, energy comes from thermal heating of the reaction products and the change in the total number of moles by reaction. Usually, the change in the total number of moles is very small and can be ignored. Taking the combustion of propane in air as an example:

$$C_3H_8 + 5O_2 + 18.8N_2 \rightarrow 3CO_2 + 4H_2O + 18.8N_2$$

The initial total number of moles is 24.8, while the final total number of moles is 25.8. Therefore, the energy released from the reaction can be easily calculated by using heat of combustion.

#### 2.2.2.2 Energy of mechanical explosions

Energy of mechanical explosions usually comes from a pressurized gas. For different situations, four methods can be used to estimate the energy released.

##### Brode's equation (Brode, 1959)

It determines the energy required to raise the pressure of the gas at constant volume from atmospheric pressure to the final gas pressure, also known as the difference in the internal energies of the gas. Since the internal energy is:

$$E = \frac{PV}{\gamma - 1} \tag{2.40}$$

The Brode equation is:

$$E = \frac{(P_2 - P_1)V}{\gamma - 1} \tag{2.41}$$

where $E$ is the energy of explosion, $P_1$ is the ambient pressure, $P_2$ is the final pressure, $V$ is the volume, and $\gamma$ is the heat capacity ratio for the gas:

$$\gamma = \frac{C_p}{C_v} \tag{2.42}$$

where $C_p$ is the heat capacity at constant pressure, and $C_v$ is the heat capacity at constant volume.

##### Isentropic expansion

This determines the energy released by the gas expanding isentropically. Baker's equation (Baker, 1973) is based on the work done on the surroundings in an isentropic expansion:

$$E = \frac{P_1V_1 - P_2V_2}{\gamma - 1} \tag{2.43}$$

where $E$ is the energy of explosion, and 1 and 2 denote the initial and final conditions. During an isentropic expansion:

$$P_1V_1^\gamma = P_2V_2^\gamma = \text{constant} \tag{2.44}$$

Therefore, the energy released during an isentropic expansion can be expressed as:

$$E = \left(\frac{P_2V_1}{\gamma - 1}\right)\left[1 - \left(\frac{P_1}{P_2}\right)^{(\gamma-1)/\gamma}\right] \tag{2.45}$$

### Isothermal expansion

This expression determines the energy released by the gas expanding isothermally. Kinney's equation (Kinney, 1962) is based on work done on isothermal expansion:

$$E = nR_gT_1 \ln\left(\frac{P_2}{P_1}\right) = P_2V \ln\left(\frac{P_2}{P_1}\right) \tag{2.46}$$

where $R_g$ is the ideal gas constant, $n$ is the number of moles, and subscripts 1 and 2 denote the initial and final conditions.

### Thermodynamic availability

The thermodynamic availability represents the maximum mechanical energy available. After analyzing batch thermodynamic availability, Crowl (1992) obtained the following equation to predict the maximum amount of energy that can be released from a gas contained within a vessel:

$$E = P_2V\left[\ln\left(\frac{P_2}{P_1}\right) - \left(1 - \frac{P_1}{P_2}\right)\right] \tag{2.47}$$

## 2.2.3 EXPLOSION TYPES

Lees classified explosions in the process industries into the following types (Mannan, 2005):

1. Physical explosion: The catastrophic rupture of a pressurized gas/vapor–filled vessel by means other than reaction, or the sudden phase-change from liquid to vapor of a superheated liquid (CCPS, 1994).
2. Condensed phase explosion: An explosion that occurs when the material is present in the form of a liquid or solid (CCPS, 1994).
3. Vapor cloud explosion (VCE): The explosion resulting from the ignition of a cloud of flammable vapor, gas, or mist in which flame speeds accelerate to sufficiently high velocities to produce significant overpressure (CCPS, 1994).
4. BLEVE: A type of rapid phase transition in which a liquid contained above its atmospheric boiling point is rapidly depressurized, causing a nearly instantaneous transition from liquid to vapor with a corresponding energy release. A BLEVE is often accompanied by a large fireball if a flammable liquid is involved, because an external fire impinging on the vapor space of a pressure vessel is a common BLEVE scenario. However, it is not necessary for the liquid to be flammable for a BLEVE to occur (CCPS, 1994).

5. Confined explosion with reaction: An explosion of fuel–oxidant mixture inside a closed system (e.g., vessel or building).
6. Vapor escapes into and explodes in buildings (VEEBs).
7. Dust explosion.

New types of explosion such as aerosol explosion should be taken into consideration as well. The following sections discuss three types of explosions: VCEs, BLEVEs, and dust explosions.

#### 2.2.3.1 Vapor cloud explosion

A VCE occurs when the combustion of a cloud of flammable vapor gives rise to an overpressure. Many of the lethal incidents in the chemical industry have resulted from VCEs, for example, the explosion at the Flixborough plant in 1974, the Piper Alpha explosion in 1988, and the Texas City explosion in 2005.

Work on fundamental models has been described by a number of authors (Taylor, 1946; Kuhl et al., 1973; Strehlow, 1975, 1979; Gugan, 1979; Wiekema, 1980; Baker et al., 1983; Van den Berg, 1985; Williams, 1985). Other work has been directed to the correlation of experimental results and to the development of semiempirical models (Van Wingerden et al., 1989, 1990, 1994). As described by Van den Berg (1985), the TNT (trinitrotoluene) equivalent model has been widely applied to VCEs. In this empirical model, the explosion is taken to be equivalent to that of a TNT explosion with the same energy of explosion. The basic TNT equivalent model was proposed by Brasie and Simpson (Brasie and Simpson, 1968), and the expression was written as:

$$W_{\mathrm{TNT}} = \frac{\alpha W \Delta H_c}{E_{\mathrm{TNT}}} \tag{2.48}$$

where $E_{\mathrm{TNT}}$ is the energy of explosion of TNT (kJ/kg), $\Delta H_c$ is the heat of combustion of the hydrocarbon (kJ/kg), $W$ is the mass of hydrocarbon (kg), $W_{\mathrm{TNT}}$ is the equivalent mass of TNT (kg), and $\alpha$ is the yield factor.

Brasie and Simpson (1968) obtained values of the yield factor $\alpha$ of 0.03–0.04 by analyzing three VCE incidents and proposed the value to be 0.02–0.05, which is widely used now. Harris and Wickens (1989) also present a TNT equivalent model. In this model, the mass of gas participating in the VCE is restricted to the confined and congested part of the plant. This model is expressed as

$$M_{\mathrm{TNT}} = \frac{\alpha M_{\mathrm{gas}} E_{\mathrm{gas}}}{E_{\mathrm{TNT}}} \tag{2.49}$$

where $E_{\mathrm{gas}}$ is the heat of the hydrogen. $E_{\mathrm{TNT}}$ is the energy of explosion of TNT, $M_{\mathrm{gas}}$ is the mass of gas (in the confined/congested region), $M_{\mathrm{TNT}}$ is the mass of TNT, and $\alpha$ is the yield factor.

By taking $\alpha = 0.2, E_{\mathrm{gas}} = 46\ \mathrm{MJ/kg}, E_{\mathrm{TNT}} = 4.6\ \mathrm{MJ/kg}$

$$M_{\mathrm{TNT}} = 2M_{\mathrm{gas}} \tag{2.50}$$

Assuming the density of the hydrocarbon–air mixture is 0.08 kg/m$^3$, the equation becomes

$$M_{\mathrm{TNT}} = 16 \times 10^{-5} V \tag{2.51}$$

where $V$ is the volume of gas (in the confined/congested region) (m$^3$).

Based on explosion experiments, Baker (1973) found that the overpressure can be estimated using an equivalent mass of TNT:

$$z_e = \frac{R}{(M_{\mathrm{TNT}})^{1/3}} \tag{2.52}$$

where $z_e$ is the scaled distance (m), and $R$ is the actual distance (m).

From Figure 2.5, for every scaled distance, a corresponding "side-on" overpressure can be found. The correlation can also be expressed as

$$\frac{P_0}{P_a} = \frac{1616\left[1 + \left(\frac{z_e}{4.5}\right)^2\right]}{\sqrt{1 + \left(\frac{z_e}{0.048}\right)^2}\sqrt{1 + \left(\frac{z_e}{0.32}\right)^2}\sqrt{1 + \left(\frac{z_e}{1.35}\right)^2}} \tag{2.53}$$

where $P_0$ is the peak side-on overpressure, and $P_a$ is the ambient pressure.

Harris and Wickens (1989) found that the peak overpressure at the edge of the vapor cloud is 4 bar, and the scaled distance of an overpressure of 4 bar on Figure 2.5 is 16 (m/te$^{1/3}$), the authors proposed the use of an effective scaled distance of

$$z_e = 16 + \frac{R}{(M_{\mathrm{TNT}})^{1/3}} \tag{2.54}$$

where $z_e$ is the scaled distance (m), and $R$ is the actual distance (m).

#### 2.2.3.2 Boiling liquid expanding vapor explosion

A BLEVE is defined by the CCPS as an explosion resulting from the failure of a vessel containing a liquid at a temperature significantly above its boiling point at normal atmospheric pressure (CCPS, 1994).

Abbasi and Abbasi published a literature survey on BLEVE (Abbasi and Abbasi, 2007) and detailed the typical BLEVE sequence of events in a few steps, as shown conceptually in Figure 2.6 (Laboureur et al., 2012). A BLEVE happens when a reservoir filled with liquid is submitted to an aggression that results in the reservoir rupture. Based on a compilation of 88 BLEVE events that occurred in the world since 1926, Abbasi and Abbasi (2007) identified the two major causes of BLEVE rupture as fire or mechanical damage. When the reservoir releases due to rupture, the vapor expands and the liquid becomes superheated, triggering the liquid boiling. This phenomenon has three major consequences: formation and evolution of a fireball, generation and propagation of a blast wave, and fragment projection. In this section, the propagation of blast wave is discussed.

A blast wave is produced when a pressure vessel ruptures due to the rapid expansion of the contents of the vessel. As opposed to a gas reservoir rupture, in a BLEVE, flash vaporization occurs in addition to vapor expansion. Blast wave signals, recorded at different distances from a BLEVE experiment (Van der Voort, M. M., et al, 2012) with 40-l liquid $CO_2$ bottle can be observed in Figure 2.7; however, the physical justification of the different peaks (i.e., whether the peaks are due to the vapor expansion, the flash vaporization or both) is still not understood.

In the literature, most blast wave models focus on the prediction of the first peak amplitude, and generalized thermodynamic models are used to predict overpressure. The procedure first consists of

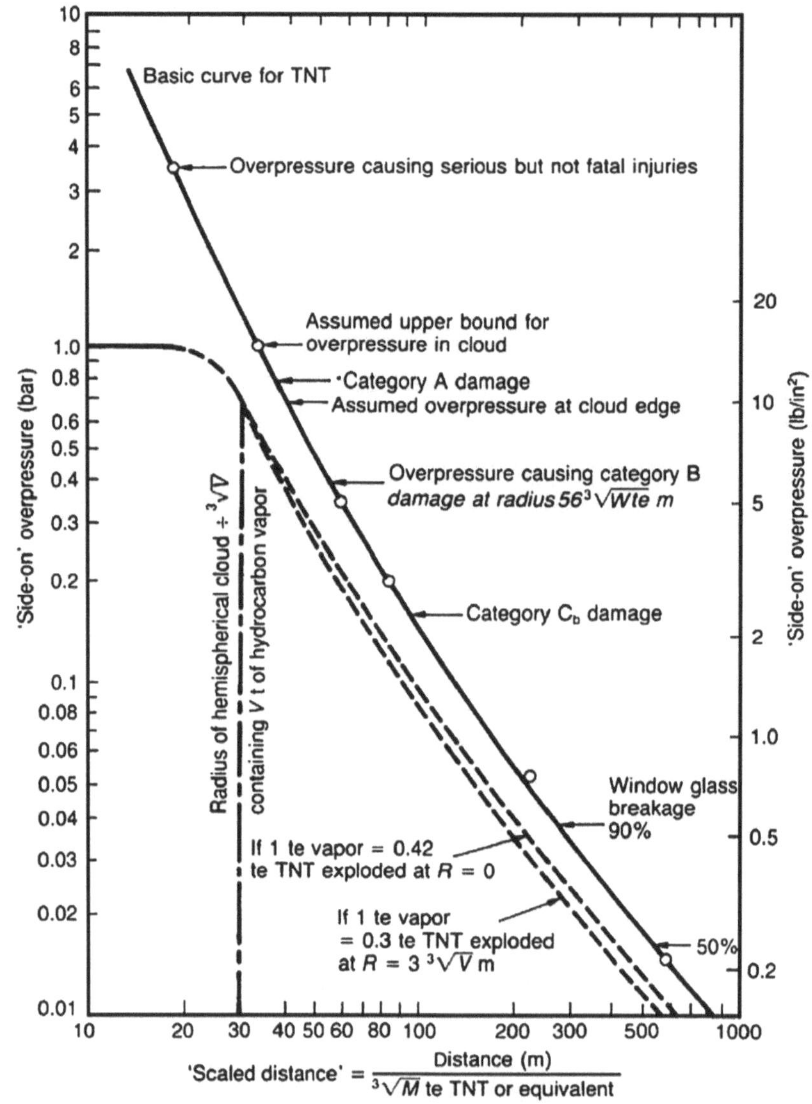


**FIGURE 2.5**

Correlation between scaled distance and explosion peak side-on overpressure (Harvey, 1979).

calculation of expansion energy based on the change in thermodynamic state of the substance stored in the vessel from the initial state to the final state. The initial state is the moment before the explosion and final state is where the fluid is at boiling temperature and atmospheric pressure. The expansion energy ($E$) can be expressed differently depending on the model, but all the models use it in the calculation of a scaled distance. This scaled distance can be expressed as Sach's scaled distance, associated with the distance from the source and the atmospheric pressure, as defined in Eqn (2.55).

FIGURE 2.6

BLEVE principle (Laboureur et al., 2012).

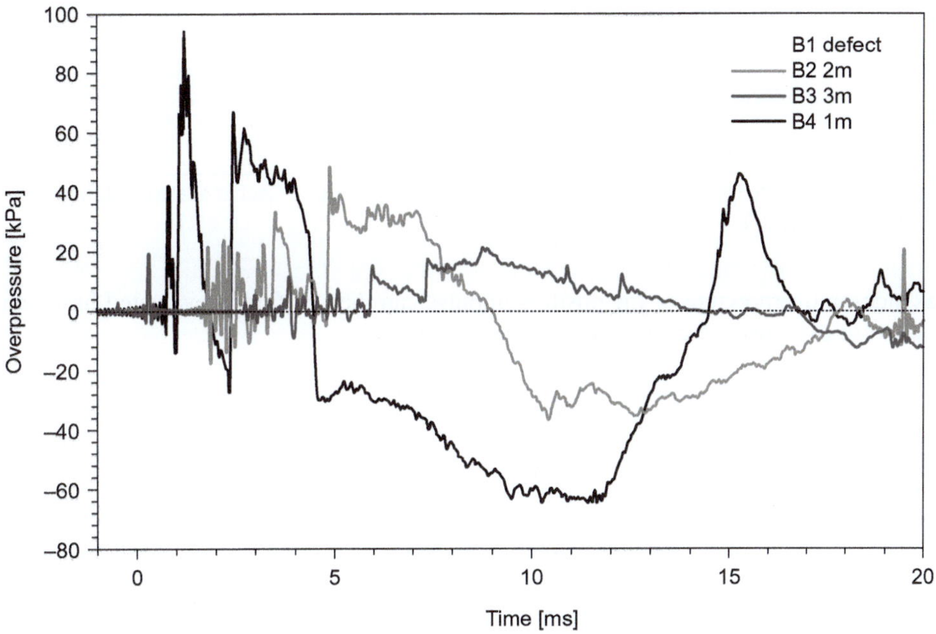


FIGURE 2.7

Overpressure–time records observed at 1, 2 and 3m distance from a BLEVE experiment (Van der Voort, M. M., et al, 2012) with 40-l liquid $CO_2$ bottle.

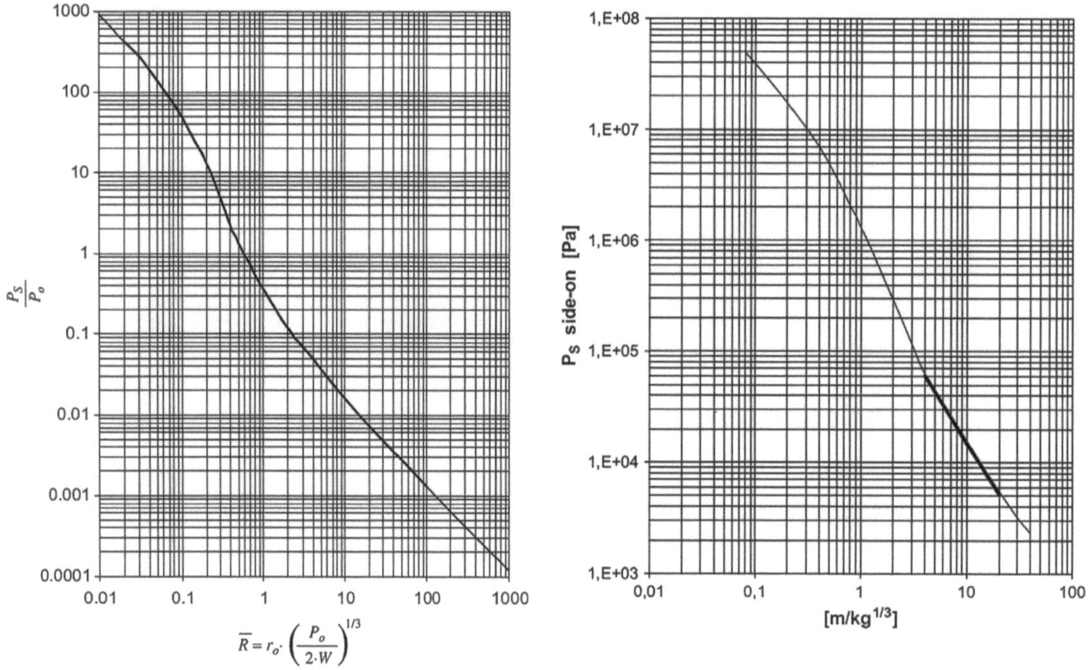


**FIGURE 2.8**

Overpressure graphs. Left: TNT equivalent (Silvestrini et al., 2009). Right: Sach's scaled distance (Van Den Bosh and Weterings, 1997).

The peak side-on overpressure or the positive impulse can then be finally evaluated based on scaled peak or impulse overpressure curves (see Figure 2.8, right).

$$\overline{R}_{\text{Sach}} = r\left(\frac{P_{\text{atm}}}{2E}\right)^{1/3} \tag{2.55}$$

$$\overline{R}_{\text{TNT}} = \frac{r}{m_{\text{TNT}}^{1/3}} \tag{2.56}$$

In a literature review of the expansion energy models (Abbasi and Abbasi, 2007; Laboureur et al., 2013), Laboureur et al. listed three models based on Sach's scaled distance: the Yellow Book of TNO (Van Den Bosh and Weterings, 1997), Roberts (2000), Genova et al. (2008), and four other models based on the TNT equivalent mass scaling: Prugh (1991), Planas-Cuchi et al. (2004), Casal and Salla (2006), and Birk et al. (2007).

The TNO method (Van Den Bosh and Weterings, 1997) models differentiate the expansion energy of the liquid and vapor phases, as expressed in Eqn (2.57). For each phase, the energy is defined as the product of the fluid phase mass with the difference in internal energies, assuming an isentropic expansion. The total expansion energy is then multiplied by a factor related to the geometry of the reservoir (generally a factor of 2 for ground effect). A special procedure for the calculation of the overpressure in near field has been developed by the TNO and is detailed elsewhere (Birk et al., 2007).

$$E = m_l(u_{l1} - u_{l2}) + m_v(u_{v1} - u_{v2}) \tag{2.57}$$

The TNO method was improved (Roberts, 2000) by considering that at the final state, a part of the liquid has flashed to vapor; both phase fractions were calculated from ratios of entropy differences. Finally, Genova et al. (2008) assumed that the expansion energy is mainly due to the liquid flash as a thermal phenomenon, related to excess of heat stored in the liquid. The expansion energy was modeled as Eqn (2.58), where $\beta$ is an empirical coefficient, set to 7% following a comparison with experimental results.

$$E = \beta m_l C_p (T_{\text{rupt}} - T_b) \tag{2.58}$$

Based on the literature review of the models based on a TNT equivalent mass, Prugh (1991) first defined the expansion energy by assuming an isentropic expansion and an ideal gas behavior expressed in Eqn (2.59). The volume $V^*$ used in Eqn (2.59) is the sum of the vapor volume and the volume occupied by the flashed fraction of liquid (see Eqn (2.60)).

$$E_{\text{TNT}} = \left(\frac{P_{\text{rupt}} V^*}{\gamma - 1}\right)\left(1 - \frac{P_{\text{atm}}}{P_{\text{rupt}}}\right)^{(\gamma-1)/\gamma} \tag{2.59}$$

$$V^* = V_v + V_l f\left(\frac{\rho_l}{\rho_v}\right) \tag{2.60}$$

Planas-Cuchi et al. (2004) stated that modeling the expansion energy as an isentropic expansion leads to an overestimation. They proposed that the real expansion energy ranges between the energy based on an isentropic expansion (*U*) and the energy based on an adiabatic and irreversible expansion where the only work performed is the one associated with the variation of volume (*V*). Iteration on the final vapor fraction that satisfies Eqn (2.61) is therefore needed. The expansion energy from both the Prugh and Planas-Cuchi models is multiplied by a factor of 0.4 for ductile failure or 0.8 for brittle failure; taking into account that a part of this energy is transformed into kinetic energy that propels the reservoir fragments.

$$-P_{\text{atm}}\Delta V = \Delta U \tag{2.61}$$

Casal and Salla (2006) defined the liquid superheating energy (SE) as the difference between the enthalpy of the liquid prior to rupture and the enthalpy of the liquid at the saturation temperature at atmospheric pressure. The expansion energy is then modeled as a fraction of the SE: 5% if an irreversible expansion is considered or 14% for an isentropic expansion. Finally, Birk et al. (2007) concluded, based on analysis of medium-scale experiments, that the liquid part does not contribute to the blast wave generation. Their new model is based on the TNO model but uses only the vapor part to calculate the expansion energy.

Laboureur et al. (2013) compared the overpressure models reviewed here with different scales of experiments published in the literature, involving a few tens of grams to a ton. From the different scale comparisons, the models of Genova and Casal with an irreversible expansion were found to fit the best the measured overpressures, whatever the reservoir scale, as observed in Figure 2.9. But these two models underestimate the predicted overpressure for some of the experiments, and if a conservative

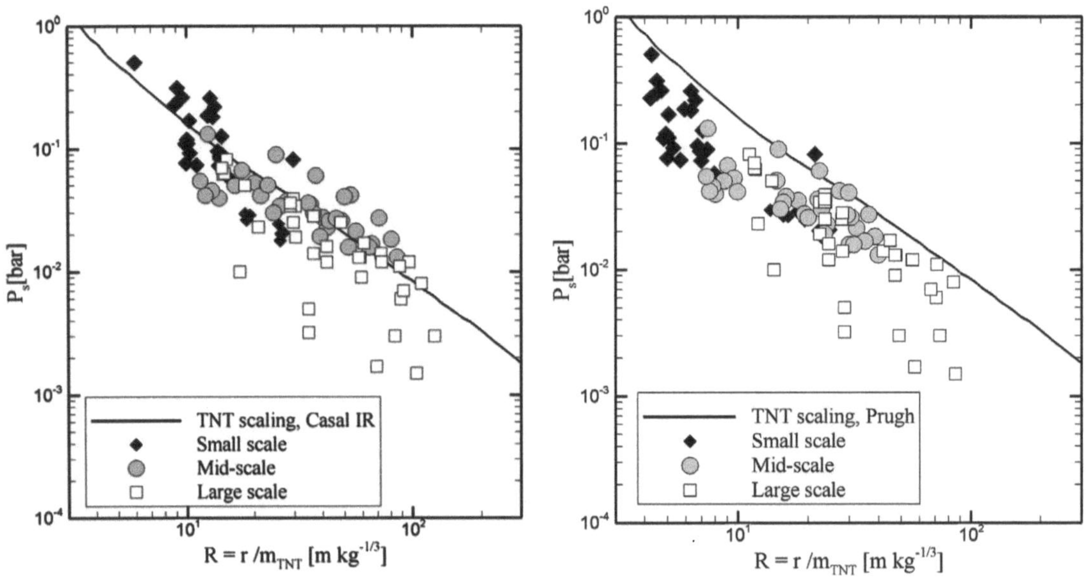


**FIGURE 2.9**

Comparison of all the experiments with the models of casal with an irreversible expansion (left) and Prugh (right).

approach is preferred, the model of Prugh should be chosen. These models estimate the side peak overpressure. The final overpressure was measured as lower, especially for reservoirs with a large length-to-diameter ratio.

### 2.2.3.3 Dust explosion

In addition to gases, combustible dusts can cause disasters. Combustible dusts are defined as "a combustible particulate solid that presents a fire or deflagration hazard when suspended in air or some other oxidizing medium over a range of concentrations, regardless of particle size or shape" (NFPA, 2002). Five elements are required to produce conditions conducive for a dust explosion: combustible dusts, confined area, oxygen, ignition source, and suspended cloud. Nagy and Verakis (1983) present plots showing that explosion characteristics are influenced by chemical composition, particle size, moisture content, oxygen concentration, inert gas, and admixed dust concentration. Also, other features in the dust explosion system, such as particle shape, dispersion method, agglomeration, ignition source, confinement, and so on, may also be critically important for dust explosion mechanism as well as in determining the overall explosion process (Cloney et al., 2014). Correspondingly, an organizational framework for studying dust explosion phenomena is developed by combining two factors: the pseudo-geometric scales and the evolution of the physical process, as shown in Figure 2.10.

In Figure 2.10, evolution of the physical process is identified as dust/oxidant, dispersion, ignition, flame propagation, and system interaction based on dust explosion development, while the pseudo-geometric scale is labeled from 1 (smallest scale) to 5 (largest scale) based on geometric scales:

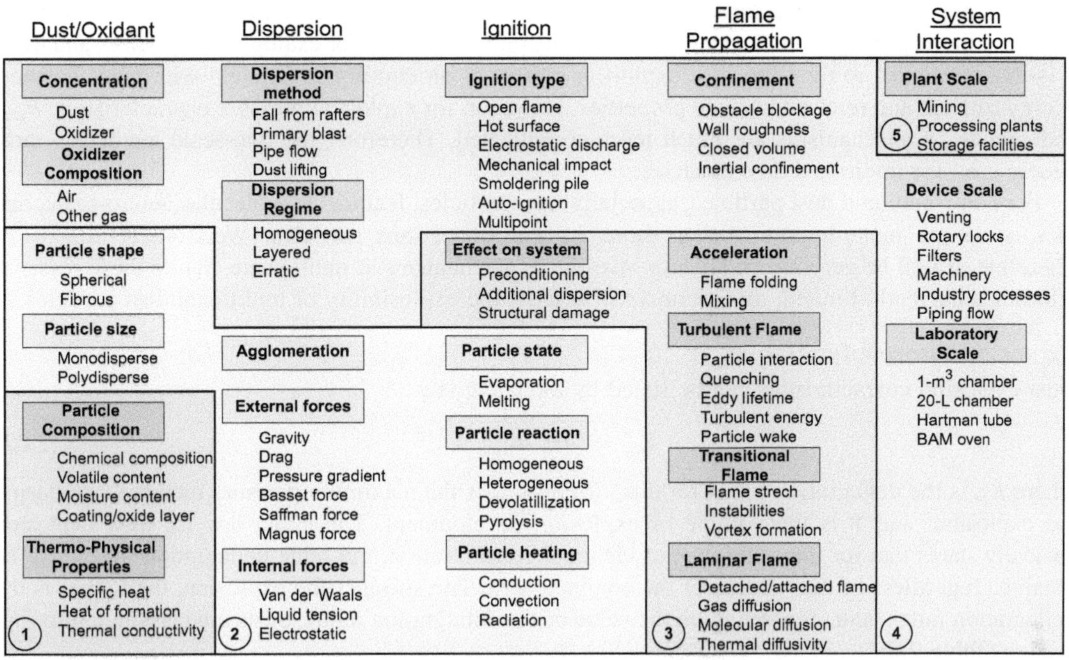


**FIGURE 2.10**

Organizational framework map for dust explosion model development based on evolution of the physical process (columns, left to right) and on pseudo-geometric scale (labeled: smallest scales (1) to largest scales (5)) (Cloney et al., 2014).

molecule (<1 nm); subparticle (1 nm to 1 μm); particle (1 μm to 1 mm); laboratory testing (1 mm to 1 m); device scale (1 m to 10 m); and plant scale (10 m to 1 km). For example, the parameters of the smallest scale, leveled as 1, varies from molecular level (thermal conductivity) to subparticle level parameters (chemical composition). Currently, researchers tend to simplify dust explosion as two-scale phenomena: subsystem scales and system scales or particle scale and cloud scale (Bind et al., 2012).

To estimate explosibility of various dust explosion scenarios, different methodologies are suggested to apply to different dust explosion systems and explosibility characteristics. Important explosibility characteristics of dust suspensions are (Mannan, 2005):

1. Explosibility classification
2. Minimum ignition temperature (MIT)
3. Minimum explosive concentration (MEC)
4. Minimum ignition energy (MIE)
5. Maximum permissible oxygen concentration to prevent ignition
6. Explosion pressure characteristics
   a. Maximum explosion pressure ($P_{max}$)
   b. Maximum rate of pressure rise [$(dp/dt)_{max}$]
   c. Average rate of pressure rise

Some explosibility characteristics only describe the very initial steps in dust explosion without involving flame propagation. For these explosibility characteristics, for example, MIT, MEC, and MIE, it is more practical to treat the dust explosion system as an equilibrium system with mass balance, energy balance and relevant particle properties. However, for explosion pressure characteristics, $P_{max}$ and $(dp/dt)_{max}$, mechanisms are much more complicated. Therefore, the two-scale method is most widely used for traditional dust particles.

For non-traditional dust particles, especially nanoparticles, features at molecular, micro-scale, and meso-scale are more important than those for traditional dust particles (Worsfold et al., 2012). Therefore, it will be very informative to investigate mechanisms at multi-scale. While these research efforts are currently limited, this section will discuss the explosibility of traditional dust particles.

### Explosibility classification

Dust explosion characteristics can be stated by the cubic law.

$$K_{St} = (dP/dt)_{max} V^{1/3} \tag{2.62}$$

where $K_{St}$ is the deflagration index for dust, $(dP/dt)_{max}$ is the maximum pressure increase rate during the explosion, and $V$ is the volume of explosion development. The cubic law for dust explosion basically states that for the same dust of identical concentration, the same deflagration index will be obtained regardless of the volume of the confined area. The stronger the explosion, the higher is the deflagration index, and, depending on the value of the deflagration index, dusts are classified into four classes (Table 2.2).

The main variable involved is the maximum pressure increase rate, which is discussed separately later.

### Minimum ignition temperature

MIT is the lowest temperature needed to ignite the dust layer or dust cloud. Ignition occurs when the heat generation by the reaction exceeds the heat losses. It can be determined by following ASTM E1491, Standard Test Method for Minimum Autoignition Temperature of Dust Explosion (ASTM, 2012).

Using an energy balance on a particle

$$m_s C_{ps} \frac{dT}{dt} = Q_{generated} - Q_{lost} \tag{2.63}$$

where $m_s$ is the particle weight, and $C_{ps}$ is the heat capacity.

**Table 2.2 Classes for Dusts Based on $K_{St}$ Value**

| Deflagration Index, $K_{St}$ (bar m/s) | Explosion Features | Class |
|---|---|---|
| 0 | No explosion | St-0 |
| 1–200 | Weak | St-1 |
| 200–300 | Strong | St-2 |
| >300 | Very strong | St-3 |

At the ignition temperature, using the "Van't Hoff and Taffanel Le Floch" conditions

$$\begin{cases} Q_{\text{lost}} = Q_{\text{generated}} \\ \dfrac{dQ_{\text{lost}}}{dT} = \dfrac{dQ_{\text{generated}}}{dT} \end{cases} \tag{2.64}$$

$$Q_{\text{lost}} = h\pi d_p^s (T_s - T_g) + \varepsilon\sigma\pi d_p^2 \left(T_s^4 - T_{\text{wall}}^4\right) \tag{2.65}$$

$$Q_{\text{generated}} = \dot{m}_c \Delta H_{\text{comb}} \tag{2.66}$$

where $\Delta H_{\text{comb}}$ is the combustion enthalpy, $h$ is the coefficient of heat convection, $d_p$ is the particle diameter, $T_s$, $T_g$, and $T_{\text{wall}}$ are the temperature of solid, gas, and wall, respectively, $\varepsilon$ is the particle emissivity; and $\sigma$ is the Stefan–Boltzmann constant. By neglecting the radiation term, $\varepsilon\sigma\pi d_p^2(T_s^4 - T_{\text{wall}}^4)$, the ignition temperature could be expressed as (Bouillard et al., 2010)

$$T_{\text{ign}} = \sqrt{\frac{\rho_s d_p^2 E_a \Delta H_{\text{comb}}}{6\tau_b Nu \lambda_g R}} \tag{2.67}$$

where $T_{\text{ign}}$ is the ignition temperature, $\rho_s$ is the density of particle, $d_p$ is the diameter of particle, $E_a$ is activation energy, $\Delta H_{\text{comb}}$ is the combustion enthalpy, $\tau_b$ is combustion, $Nu$ is the Nusselt number, $\lambda_g$ is the gas conductivity, and $R$ is the gas law constant. Therefore, the ignition temperature is lower when the density of the particle is lower, the combustion enthalpy is smaller, or the activation energy is lower.

### Minimum explosive concentration

MEC is the minimum concentration of dust needed to cause an explosion. It is determined according to ASTM E 1515, Standard Test Method for Minimum Explosible Concentration of Combustible Dusts (ASTM, 2007a). Usually, when the particle size decreases, the volatile matter content, oxygen concentration, moisture content, and admixed inert dust concentration increase, and the MEC decreases.

The influence of the particle size on the MEC has been studied (Eckhoff, 2003). Based on Jaeckel's theory (Jaeckel, 1924) and "Van't Hoff and Taffanel Le Floch" conditions, the MEC is defined as:

$$MEC \cdot X \cdot \Delta H_{\text{comb}} > \left(C_{pg}\rho_g + C_{ps} MEC\right)(T_{\text{ign}} - T_u) + Q_{\text{lost}} \tag{2.68}$$

where $C_{pg}$ is the heat capacity of the gas, $\rho_g$ is the density of the gas, $T_u$ is the temperature of the unburnt gases, $C_{ps}$ is the heat capacity of solid, $T_{\text{ign}}$ is the ignition temperature, $Q_{\text{lost}}$ is the radiation and convection heat loss, $X$ is the reaction conversion, and $\Delta H_{\text{comb}}$ is the combustion enthalpy.

When the radiation and convection heat losses are neglected, the MEC can be expressed as

$$MEC = \frac{C_{pg}\rho_g (T_{\text{inf}} - T_u)}{X \Delta H_{\text{comb}} - C_{ps}(T_{\text{ign}} - T_u)} \tag{2.69}$$

Combined with the previous equation concerning $T_{\text{ign}}$,

$$T_{\text{ign}} = \sqrt{\frac{\rho_s d_p^2 E_a \Delta H_{\text{comb}}}{6\tau_b Nu \lambda_g R}} \tag{2.70}$$

the trend of the MEC with the particle size can be estimated.

## Minimum ignition energy

MIE is the energy required to sustain ignition and inflammation at an initial temperature $T_u$ while the burning medium is at a final temperature $T_{ad}$, the flame adiabatic temperature (Ballal and Lefebvre, 1979; Williams, 1985; Stephen, 2000). It is determined according to ASTM E 2019, Standard Test Method for Minimum Ignition Energy of a Dust Cloud in Air (ASTM, 2007b). Usually, when the particle size decreases and the volatile matter content, oxygen concentration, moisture content, and admixed inert dust concentration increase, the MIE decreases.

According to the definition of MIE, for a sphere with an initial diameter $D_0$ and a final diameter $D_c$, MIE can be written as

$$MIE = \overline{\rho}_g \frac{\pi}{6} D_c^3 \overline{C}_p T_{ad} - \overline{\rho}_u \frac{\pi}{6} D_0^3 \overline{C}_p T_u \tag{2.71}$$

where $\overline{\rho}_g, \overline{\rho}_u$ are the average density for burnt and unburnt gas, and $\overline{C}_p$ is the average heat capacity.

By considering the mass conservation, it can also be written as

$$MIE = \overline{\rho}_g \frac{\pi}{6} D_c^3 \overline{C}_p T_{ad} - \overline{\rho}_u \frac{\pi}{6} D_0^3 \overline{C}_p T_u \tag{2.72}$$

By using the ideal gas law, the relationship between density at a given temperature of burnt gas and unburnt gas can be expressed as

$$\frac{\overline{\rho}_g}{\overline{\rho}_u} = \frac{PM_g}{RT_{ad}} \frac{RT_u}{PM_u} \approx \frac{T_u}{T_{ad}} \tag{2.73}$$

where $M_u$ and $M_g$ are molar weights for the unburnt and burnt gas.

Hence, the MIE can be expressed as:

$$MIE \approx \frac{\pi}{6} \overline{\rho}_u \overline{C}_p \frac{T_u}{T_{ad}} (T_{ad} - T_u) D_c^3 \tag{2.74}$$

## Explosion pressure characteristics

Explosion pressure characteristics $P_{max}$ and $(dp/dt)_{max}$ define the severity of dust explosions. Therefore, properly modeling dust explosions is essential to determine explosion behaviors and help safety system design. A two-scale reaction engineering model was presented and validated for aluminum and starch dust explosions (Bind et al., 2012). In this model, the combustion behavior of the cloud is first defined by the interparticle distance. If the interparticle distance is at least twice the flame radius of the particle, the individual combustion will take place; otherwise, it will be the group combustion or cloud combustion (Annamalai et al., 1994).

Therefore, for aluminum particles, there is little probability that group combustion will occur in case of aluminum particle combustion and the rate of consumption of particle phase could be (Bind et al., 2012):

$$\mathrm{R}''' = \alpha \times \frac{6}{d_p} \times R'(d_p, T, P, C_{O_2}) \tag{2.75}$$

where $\alpha$ is volume fraction of fuel, $R''(d_p, T, P, C_{O_2})$ is the surface reaction rate, $R'''$ is the volumetric reaction rate, and $d_p$ is the particle diameter. $R'''$ can be estimated by incorporating the particle scale dynamics though an appropriate correlation that models the particle scale combustion.

For organic particles, because of the volatile components, the interparticle distance is very small, and the combustion rate is determined by cloud combustion. Then the density-based solver of commercial CFD software Fluent can be used to solve the following equations (Bind et al., 2012):

Continuity:

$$\frac{\partial \rho}{\partial t} + \nabla \cdot (\rho \vec{v}\,) = 0 \tag{2.76}$$

Momentum:

$$\frac{\partial}{\partial t}(\rho \vec{v}\,) + \nabla \cdot (\rho \vec{v}\ \vec{v}\,) = -\nabla p + \nabla \cdot (\bar{\bar{\tau}}_{eff}) + \rho \vec{g} \tag{2.77}$$

Energy:

$$\frac{\partial}{\partial t}(\rho E) + \nabla \cdot (\rho \vec{v}\ E + p \vec{v}\,) = +\nabla \cdot \left( \kappa_{\text{eff}} \nabla T - \sum_j h_j \vec{J}_j + \bar{\bar{\tau}}_{\text{eff}} \vec{v} \right) + S_h \tag{2.78}$$

Species:

$$\frac{\partial}{\partial t}(\rho Y_i) + \nabla \cdot (\rho \vec{v}\ Y_i) = -\nabla \cdot ((\rho D_{i,\text{eff}} + \mu_t / Sc_t) \nabla Y_i) + R_i''' \tag{2.79}$$

where $\rho$ is the cloud density, $\vec{v}$ is the cloud viscosity, $\kappa_{\text{eff}}$ is the efficient cloud thermal conductivity, $p$ is pressure, $Y_i$ is the mass fraction of vapor phase of species '$i$', E is energy, $D_{i,\text{eff}}$ is the efficient diffusivity of species '$i$', T is temperature, $h$ is the coefficient of heat convection, $S_h$ is the entropy, $Sc$ is the Schmidt number, and $\bar{\bar{\tau}}_{eff}$ is the efficient shear stress.

This modeling work is validated by experimental data in a 20 L spherical apparatus. However, for a larger explosion tunnel, the geometric layout of explosion confinement is another important factor effecting dust explosion behaviors. Then, computational fluid dynamics (CFD) is also needed to build a three-dimensional (3D) hexahedral mesh of the tunnel (Humphreys et al., 2010).

### 2.2.4 EXPLOSION PREVENTION

To prevent or limit the potential damage from explosions, a threefold strategy is used: prevent formation of flammable mixtures, prevent the initiation of explosion, and minimize the damage after an explosion has occurred. This section will mainly discuss some widely used methods to prevent explosions: inerting, controlling static electricity, and ventilation.

#### 2.2.4.1 *Inerting*

Inerting is a process of reducing the concentration of oxygen below the LOC. LOC is about 10% for gases and 8% for dusts. Usually, an inert gas, such as nitrogen, carbon dioxide, or steam, is added into a combustible mixture. Vacuum purging, pressure purging, combined pressure–vacuum purging, and sweep-through purging can be used in the inerting process.

### Vacuum purging

Vacuum status is always reached in a vacuum purging procedure. From the perspective of safety, vacuum purging is only applied to vessels that are designed to withstand vacuum. A typical vacuum purging procedure consists of following steps:

1. Use a vacuum to pump out the gas inside vessels until the pressure reached the lower desired value $P_L$
2. Pump inert gas into the vessel until the higher desired pressure is reached. Usually, the higher pressure is the atmospheric pressure
3. Repeat step 1 and step 2 as many times as needed until the oxygen concentration is below the required value.

The process for vacuum purging is shown in Figure 2.11:

The concentration of oxygen after $j$ purge cycles, vacuum, and relief, $y_j$ is calculated as the following equation to estimate the number of repeating cycles needed.

$$y_j = y_0 \left( \frac{n_L}{n_H} \right)^j = y_0 \left( \frac{P_L}{P_H} \right)^j \tag{2.80}$$

where $P_L$ and $P_H$ are the vacuum-purged pressure and the initial pressure, $n_L$ and $n_H$ are the vacuum-purged and the initial moles of gas, and $y_0$ is the initial oxygen concentration.

The total moles of inert gas added for $j$ cycles, $\Delta n_{\text{inertgas}}$ is

$$\Delta n_{\text{inertgas}} = j(P_H - P_L)\frac{V}{R_g T} \tag{2.81}$$

where $V$ is the volume of the vessel, $R_g$ is the gas constant, and $T$ is the temperature.

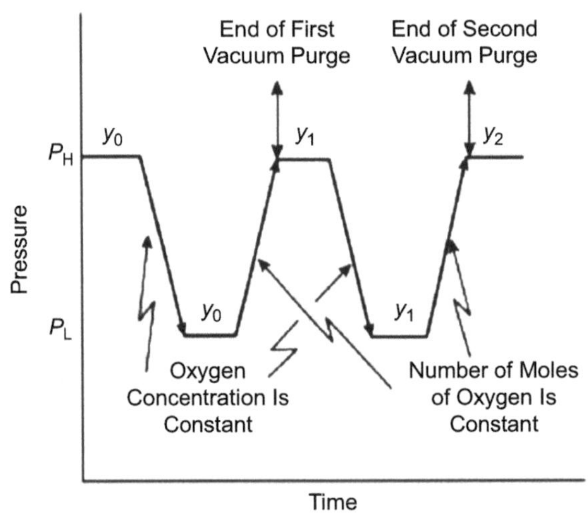


**FIGURE 2.11**

Vacuum purge cycles (Crowl and Louvar, 2011).

## Pressure purging

The pressure purging procedure is very similar to the vacuum purging procedure, except that a pressurized status will be reached during the procedure rather than a vacuum status. Consequently, this method is applied to vessels that can withstand high pressures. Here are the typical steps for pressure purging:

1. Pump inert gas into the vessel until the desired high pressure $P_H$ is reached.
2. Then draw a vacuum on the vessel until the lower pressure $P_L$ is obtained. Usually, the value of the lower pressure is equal to the atmospheric pressure.
3. Repeat step 1 and step 2 as many times as needed until the oxygen concentration is below the required value.

The process for pressure purging is shown in Figure 2.12.

The same equations as vacuum purging can be used when $P_L$ and $P_H$ are the initial and pressurized pressures.

## Combined pressure–vacuum purging

While the vacuum purging needs less inert gas to reach the desired oxygen concentration than the pressure purging, the pressure purging helps save time for the operating cycles because the pressurized procedure is much faster than the vacuum procedure. In some cases, pressure purging and vacuum purging can be combined to reach a balance between inert gas consumption and time consumption. A combined pressure–vacuum purging consists of the following steps:

1. Pressurize the vessels with inert gas until a high pressure value $P_H$, which is higher than the atmospheric pressure.

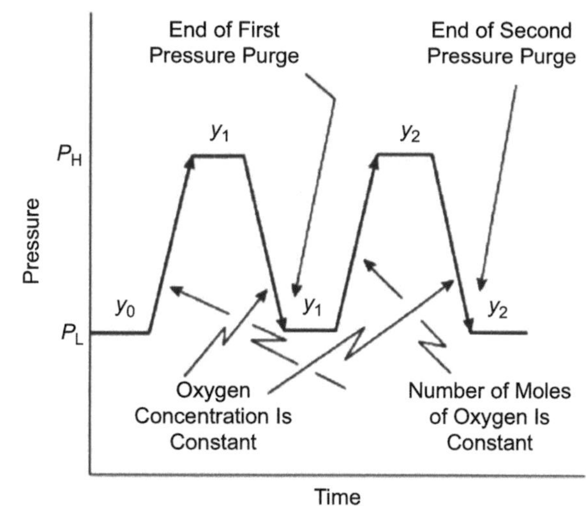


**FIGURE 2.12**

Pressure purge cycles (Crowl and Louvar, 2011).

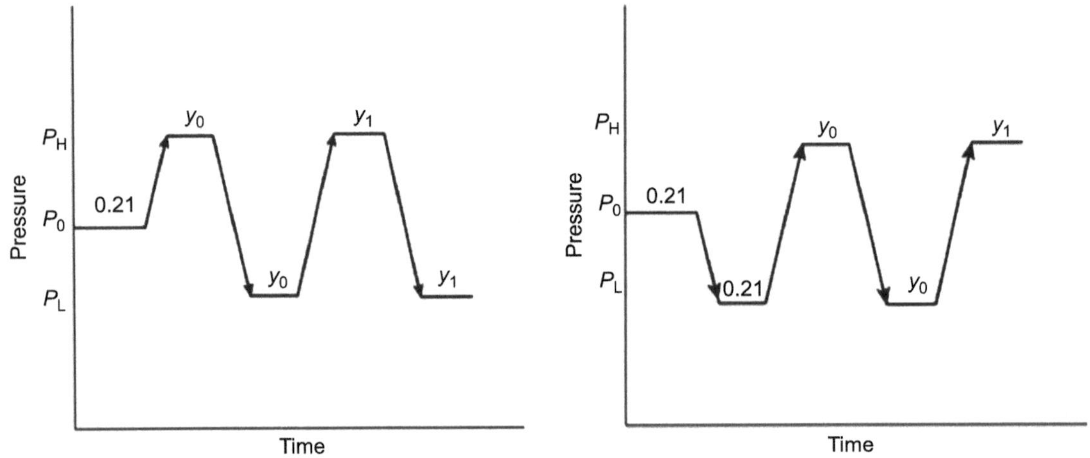


**FIGURE 2.13**

Vacuum–pressure purging with initial pressurization (left) and vacuum–pressure purging with initial evacuation (right) (Crowl and Louvar, 2011).

2. Vacuum the pressurized system until a low pressure value $P_L$, which should be lower than the atmospheric pressure.
3. Repeat step 1 and step 2 to reach the desired oxygen concentration. For steps 1 and 2, whichever comes first is fine.

The process for combined pressure–vacuum purging is shown in Figure 2.13.

## Sweep-through purging

The sweep-through purging is a process of adding inert gas at one opening and withdrawing the mixed gas from another opening. When assuming perfect mixing, constant temperature, constant pressure, and same rate of adding and withdrawing, the material balance around the vessel is

$$V\frac{dC}{dt} = C_0 Q_V - CQ_V \tag{2.82}$$

where $V$ is the vessel volume, $C$ is the concentration of oxidant within the vessel (mass or volumetric units), $C_0$ is the inlet oxidant concentration (mass or volumetric units), $Q_V$ is the volumetric flow rate, and $t$ is time.

The inert gas needed to reduce the oxidant concentration from $C_1$ to $C_2$, $Q_Vt$, is determined as

$$Q_V t = V \ln\left(\frac{c_1 - c_0}{c_2 - c_0}\right) \tag{2.83}$$

where $c_0$ is the oxidant concentration of the inert gas; it is usually 0.

#### 2.2.4.2 Controlling static electricity

Controlling static electricity is another widely used way to prevent explosion because the static electricity is treated as an ignition source. Many methods are applied in industries: relaxation; bonding and grounding; dip pipes; increasing conductivity with additives; and handling solids without flammable vapors (Crowl and Louvar, 2011).

#### 2.2.4.3 Ventilation

Proper ventilation can efficiently prevent fires and explosions. The primary concept behind ventilation is to lower the concentration of flammable gas or dust cloud until it is lower than the LFL or MEC. The relationship between concentration $C_{\text{ppm}}$, the ventilation rate $Q_v$, and the mass flow rate $Q_m$ from the leak can be expressed as (Crowl and Louvar, 2011):

$$kC_{\text{ppm}} = \frac{Q_m R_g T}{Q_v PM} \times 10^6 \tag{2.84}$$

where $T$ is the temperature, $P$ is the pressure, $M$ is the molecular weight of the material, $R_g$ is the gas constant, and $k$ is the nonideal mixing factor; it is usually between 0.1 and 0.5.

Also, the mass flow rate $Q_m$ can be measured according to $\langle C \rangle$

$$\langle C \rangle = \frac{Q_m}{\pi \sigma_y \sigma_z u} \tag{2.85}$$

where $\langle C \rangle$ is the concentration of gas measured at a point (y, z), $u$ is the wind speed, and $\sigma_y$, $\sigma_z$ are dispersion coefficients on $y$ and $z$ directions; they can be determined by the geometry and weather condition of a place. When using the value of LFL as $C_{\text{ppm}}$, the minimum ventilation rate $Q_V$ can be obtained.

## 2.3 TOXIC EFFECTS

One of the major hazards that could be present in a facility is the release of a toxic chemical. In general, this type of hazard depends on various different elements such as the conditions of exposure and on the properties of the substance that is being released. To determine the effects, one should focus on both: brief exposures at high concentrations and prolonged exposures at low concentrations. For instance, release of a very toxic chemical such as chlorine should be treated with caution as it could disperse through large areas and have much larger effect than those of fires or explosions leading to catastrophic incidents. Although the toxic releases occur less frequently than a fire or explosion still, the scenario should always be considered of high importance.

The First Report of the Advisory Committee on Major Hazards (ACMH) states (Harvey, 1976):

> "With toxic materials, the sudden release of very large quantities could conceivably cause even larger numbers of casualties than a flammable escape. In theory such a release could, in certain weather conditions, produce lethal concentrations in places 20 miles from the point of release but the actual number of casualties (if any) would depend on the population density in the path of the cloud and the effectiveness of the emergency arrangements that might include evacuation."

As stated, dispersion of toxic substances can travel long distances and, consequently, the long-term toxic effects due to the exposure at low concentrations of certain chemicals can affect a large part of the population.

This section covers the different routes of how toxic substances can enter an organism. A clear distinction between these mechanisms is established. Then a brief overview of toxic substances and how these can be identified in Material Safety Data Sheets (MSDSs) are shown. Subsequently, a description of a toxic assessment is given. In this part, it is briefly described how Environmental Protection Agency (EPA) establishes a distinction between cancer and noncancer effects. Additionally, in this section, the effects of toxic substances, limit values set for these in the United States of America are shown and the different approaches employed by federal agencies such as the Occupational Safety and Health Administration (OSHA) and the EPA to regulate these chemicals is presented along with the assessment of the hazards of toxic releases. It should be kept in mind that this chapter presents an engineering approach on how to avoid toxic releases and illustrates how the federal agencies work together to establish and implement regulations and how should these regulations should be interpreted for assessments of short- and long-term effects.

## 2.3.1 HOW TOXIC SUBSTANCES ENTER THE ORGANISM

In general, toxic substances can enter biological organisms via any of the following routes:

- Dermal absorption
- Injection
- Inhalation
- Ingestion

For every possible route, there is a method of control. For instance, ingestion can be controlled by reinforcement of rules on eating, drinking, and smoking; inhalation can be controlled by ventilation, respirators, and personal protection equipment; and injection and dermal absorption by proper protective clothing. Moreover, the smaller the particle size, the more probable it is for the particle to translocate from any of the entry routes into the circulatory and lymphatic systems following to the organs and tissues. Some particles, depending on their size and composition, may produce irreversible damages to cells or organelle injury. Figure 2.14 shows an example of a cell compared with different sizes of particles.

## 2.3.2 PARTICLE CLASSIFICATION

Particles can be classified based on dimensionality, morphology, composition, uniformity, and agglomeration. In general, motion of particles is not limited, and they can interact with humans with no difficulty; however, the smaller the particle is, the greater the health hazard is. At this point, it is important to clarify that not all particles are toxic, as toxicity is a function of chemical composition and shape. Additionally, classification of toxicity is limited and further research should be addressed for each material and morphology. In the following sections, a typical classification of the particles is given.

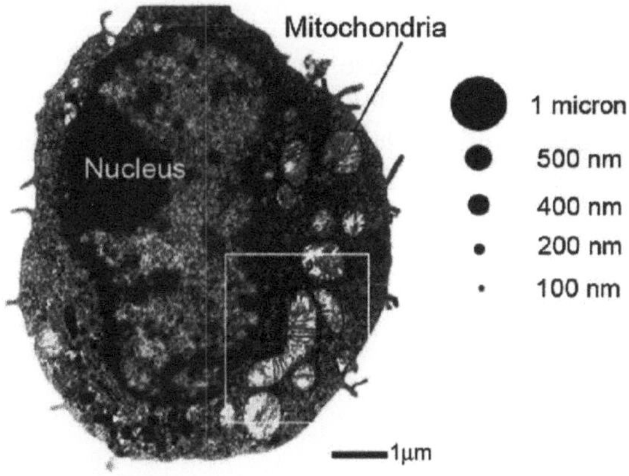


**FIGURE 2.14**

Size comparison between cells and other particles (Buzea et al., 2007).

#### 2.3.2.1 Dimensionality

A number of dimensions are used to classify particles. In other words, this is a generalization of the "aspect ratio" concept (Buzea et al., 2007). There are three different categories; one-dimensional (1D), two-dimensional (2D), and 3D materials.

- 1D materials: In the nanometer scale, these are thin films or surface coatings. An example of 1D materials is carbon nanotubes.
- 2D materials: These types of materials have two dimensions in the nanometer scale. An example of 2D materials is asbestos fibers.
- 3D materials: These types of materials have three dimensions in the nanometer scale. Examples of 3D materials are colloids and nanoparticles.

#### 2.3.2.2 Particle morphology

The morphology plays an important role when classifying a particle. For instance, flatness, sphericity, and aspect ratio are the more important characteristics to be taken into account.

#### 2.3.2.3 Particle composition

In general, particles can be composed of a single material or a composite of several materials (Buzea et al., 2007).

#### 2.3.2.4 Particle uniformity and agglomeration

Particles can exist as dispersed aerosols, suspensions/colloids, or in an agglomerated state. In the agglomerate state, particles may behave as larger particles, and their behavior will depend on the size of the agglomerate.

### 2.3.3 TOXIC SUBSTANCES

In essence, all chemical substances can cause adverse health effects if the chemical is induced in high doses; however, when the dose is low, typically no health effect is observed. There are some factors that should be kept in mind when dealing with toxic substances. Most of them can be found in the MSDS of the chemical substance and the following is a list of some of the factors (Mannan, 2012):

1. Generation of the substance
2. Toxic concentrations
3. Effects of exposure
4. Detectability by odor
5. Precautions in handling
6. Leak detection
7. First aid

Chemicals evolve along with technology and the use or production of toxic chemicals. Several of the toxic substances that present hazards for people and the environment are produced, and some of them are generated as byproducts of the process. For instance, carbon monoxide, produced by an incomplete combustion process, is an asphyxiant gas. In general, to achieve complete combustion is very difficult as there are many reactions that can be involved during the progress of the reaction. Another example is nitrogen oxides, which can be produced during welding processes. Nitrogen oxides present an irritant effect but on inhalation could be fatal.

As mentioned, chemicals evolve along with technology, and every year the number of chemicals produced increases. Studies performed by Langley (1978) identified that in the decade prior to 1978, a total of 4 million new chemicals were identified. From such a large quantity, it was estimated that the United Kingdom had 300–400 new chemicals per year, and the United States tends to duplicate such values.

Some of the problems related to exposure of toxic chemicals vary based on the time of exposure. For instance, noxious effects can result from continuous exposure to a substance that presents a level of toxicity that has not been appreciated. However, some toxic chemicals may cause cancer and malfunction of different human organs. In the following paragraphs, a more detailed explanation about different hazardous particles is given.

The human body needs several metals for proper functioning. However, it has been proved (Shah, 1998) that exposure to high levels of these metals can have an adverse health effect. In general, the smaller the particle, the higher are the chances for it to get into the organism. Therefore, activities such as dealing with or manufacturing nano- and micro-sized particles should be considered as a serious occupational hazard. In general, the inhalation of dusts is well known to have an adverse health effect. Moreover, the health effect they can cause is a function of the material itself, the exposure duration, and the dose. For instance, exposure to metal dust, such as platinum, nickel, or cobalt, can cause asthma, while the inhalation of other metallic dusts can lead to pulmonary fibrosis and lung cancer. In the following paragraphs, some examples of dusts and the effect of being exposed to them are discussed.

*Lead*: When lead is inhaled or ingested, it circulates in the blood and gets deposited in bone and other tissue. Chronic exposure to lead can cause impairment of mental functions, visuomotor performance, attention span, anemia, and kidney disease, among others (Shah, 1998).

*Cadmium*: High-dose exposure to cadmium leads to lung irritation, nausea, and vomiting. Long-term and low-dosage exposure in humans can cause lung emphysema and central and nervous system and liver damage. Lung cancer has also been related to exposure to cadmium. It is important to mention that cadmium is present in industrial and consumer waste, which leads to accumulation of cadmium in soil. Consequently, plants and crops are growing in a contaminated soil, leading to a contamination of vegetables and animals (Shah, 1998).

*Iron*: Even though small doses of iron are essential to ensure survival, it has been proven experimentally that high-dose exposure to iron in animals can lead to an increased risk of developing cancer. In humans, high-dose exposure to iron has been linked to cause sarcomas at the sites of deposition, liver cancer, and neurological diseases, such as Parkinson‘s and Alzheimer‘s diseases.

*Beryllium*: Inhalation of beryllium can lead to lung damage, hypersensitivity, and beryllium disease. Long-term exposure can increase risk of lung cancer in humans (Paris et al., 2004).

*Aluminum*: Long-term exposure to aluminum leads to anemia, bone disease, dementia, and breast cancer diagnosis (McGrath, 2003). Further, studies have proven that the brain plaque of people with Alzheimer‘s disease has high levels of aluminum (Shah, 1998).

## 2.3.4 TOXICITY ASSESSMENT

Toxicity assessment is used to identify the potential adverse health effects that a chemical can cause and determine how the appearance of these effects depends on the exposure level (dose). As mentioned, the toxic effects of a chemical depends on the route, dosage, and duration of exposure. Therefore, toxic chemicals must have a list of potential adverse effects and the frequency of such effects is noted based on the dose, route, and duration of exposure. According to the EPA (1991), the toxicity assessment process is divided into two parts. The first part characterizes and quantifies the noncancer effects of the chemical. The second part addresses the cancer effects of the chemical.

As mentioned earlier in this section, all chemicals can cause adverse health effects. However, for this to occur, exposure to the chemicals has to occur at a sufficiently high dose. If the dose is not sufficiently high, depending of the chemical, normally there is no adverse health effects. Therefore, for toxic substances, it is important to establish a threshold dose at which adverse effects become evident. As mentioned, not all particles produce adverse health effects; the toxicity depends on various factors, such as size, composition, crystallinity, and aggregation, among others. Moreover, the toxicity of particles to organisms depends on the organism genetic complement, which provides either ways to adapt or fight toxic substances. It has been found that the inhalation of certain nanoparticles is associated with certain diseases such as asthma, lung cancer, and Parkinson‘s and Alzheimer‘s diseases. Further, research has proved the presence of nanoparticles in the circulatory system in several cases of heart disease, arrhythmia, or arteriosclerosis.

### *2.3.4.1 Noncancer effect*

In general, the threshold dose is established from toxicological data. These data come from studies on humans and/or animals and consists of determining the highest dose that does not cause any observable adverse effect, called no observed adverse effect level (NOAEL), and the lowest dose that produces an effect, called the lowest observed adverse effect level (LOAEL). As there is an interval between the LOAEL and the NOAEL, a threshold should lie in between these values (EPA, 1991).

To determine a threshold value, a reference dose (Rd.) is estimated from the NOAEL. If there is no NOAEL reported, then the Rd. value is derived from the LOAEL. In any case, the NOAEL or LOAEL is divided by an uncertainty factor. In cases when the data are sufficiently reliable the uncertainty factor can be as small as 1. However, when there are limited data, the uncertainty factor can be as high as 10. In general, the Rd. is an estimate of a daily exposure of the human population that is likely to be affected by the hazardous effects of a chemical substance during a lifetime (Allport et al., 2003).

#### *2.3.4.2 Cancer effect*

In this case, toxicity assessment process has two components. The first is a qualitative evaluation of the weight of evidence that the chemical can or cannot cause cancer in humans. The second part of this assessment is only for chemicals that are believed to be capable of causing cancer in humans; in other words, the main goal is to describe the carcinogenic potency of a chemical. This is done by quantifying how the number of cancers on humans and/or animals increases as the dose increases (EPA, 1991).

### 2.3.5 RISK ASSESSMENT

In general, the information that can be obtained from toxicity assessments could be used to assess the toxic risk of a facility. However, as shown before, there are two different approaches to assess the risk due to toxic chemicals. The first one is proposed by federal agencies in the form of regulations. These regulations help to limit all the different companies that are covered by them. The other approach is the one carried out by the manufacturer for facility design and operation purposes.

There are several other methodologies for risk assessment in the federal government. Some of them are managing the process (NAS/NRC, 1983), assessment and management of chemical risks (Rodricks and Tardiff, 1984), toxicological risk assessment (Clayson et al., 1985), and the risk assessment of environmental hazards (Paustenbach, 1989; Bridges, 1985; Nabholz, 1991; Becking, 1995; Frederick, 1995; Mannan, 2012).

Furthermore, according to the National Academy of Sciences (NAS) report, risk assessment consists of four different stages (National Research Council, 1983; Cohrssen and Covello, 1999):

- Hazard identification: At this stage, a list of chemicals that can have an adverse effect is made. Additionally, a list of possible health problems that can arise due to exposure is considered. Here, several factors are accounted, such as the type and severity of the effects, route of exposure, and frequency of exposure.
- Dose–response assessment: Determine how a specific chemical can affect adversely human being's health at different doses.
- Exposure assessment: Determine the working capabilities of a facility. Questions addressed are, for example, How many hours is a worker exposed to a toxic substance? How many people are exposed during a certain period of time?
- Risk characterization: In this stage, the extra risk of health problems in the population that has been exposed to the chemical is analyzed.

At this point, various issues regarding toxic chemicals have been addressed. For instance, dose–response and TRA have been studied briefly. It is clear that there is a need for establishing thresholds to avoid adverse health effects. Still, the level of exposure to be expected has to be defined. This is performed by using the suitable risk criteria.

Toxic risk assessment has been used to define hygiene standards, especially in the United States. For example, for inorganic arsenic, OSHA used a linear dose–response model. In such a case, OSHA has stated that for every 1000 workers over a working lifetime, there could be a maximum of eight deaths at a level of 10 mg/m$^3$, 40 at 50 mg/m$^3$, and 400 at 500 mg/m$^3$. On the basis of this standard, the dose was reduced from 500 to 10 mg/m$^3$. To justify this decision, OSHA stated (Mannan, 2012):

> "The level of risk from working a lifetime of exposure at 10 mg/m$^3$ is estimated at approximately 8 excess lung cancer deaths per 1000 employees. OSHA believes that this level of risk does not appear to be insignificant. It is below risk levels in high-risk occupations but it is above risk levels in occupations with average levels of risk."

## 2.3.6 HYGIENE STANDARDS

As discussed before in the beginning of this chapter, there are two different types of toxic exposures: brief exposures at high concentrations and prolonged exposures at low concentrations. In general, the American Conference of Governmental Industrial Hygienists (ACGIH) establishes thresholds for different toxic substances. For instance, these thresholds are determined to avoid intoxication and, therefore, permit the body to detoxify and eliminate the toxic substance without showing any adverse health effect.

Threshold limit values (TLVs) refer to airborne concentrations that correspond to conditions under which no adverse health effect is expected during a worker's lifetime. It is assumed that the exposure is going to occur 8 h per day and 5 days per week (Crowl and Louvar, 2011).

There are three different types of TLVs. They are described in Table 2.3:

**Table 2.3 Definitions for Threshold Limit Values (TLVs) (Crowl and Louvar, 2011)**

| TLV Type | Definition |
|---|---|
| TLV-TWA | TLV time-weighted average<br>The concentration for a conventional 8-h workday and a 40-h work week, to which it is believed that nearly all workers may be repeatedly exposed, day after day, for a working lifetime without adverse effect. |
| TLV-STEL | TLV short-term exposure limit<br>A 15-min TWA exposure that should not be exceeded at any time during a workday, even if the 8-h TWA is within the TLV-TWA. The TLV-STEL is the concentration to which it is believed that workers can be exposed continuously for a short period of time without suffering (1) irritation, (2) chronic or irreversible tissue damage, (3) dose rate–dependent toxic effects, or (4) narcosis of sufficient degree to increase the likelihood of accidental injury, impaired self-rescue, or materially reduced work efficiency. Exposures above the TLV-TWA up to the TLV-STEL should be less than 15 min and should occur no more than four times per day, and there should be at least 60 min between successive exposures in this range. |
| TLV-C | TLV ceiling<br>The concentration should not be exceeded during any part of the working exposure. |

Table 2.3 describes exposure under working conditions, but exposures under emergency conditions are determined using probit equations. However, this approach is very limited, because data are available and published for a limited amount of chemicals. A secondary approach is to specify a toxic concentration criterion above which the interaction with the toxic substance can have adverse health effects. Some of these methods and criteria are described next (Crowl and Louvar, 2011):

- Emergency response planning guidelines (ERPGs) for air contaminants by AIHA
- Immediately dangerous to life or health (IDLH) levels by NIOSH
- Emergency exposure guidance levels (EEGLs) and short-term public emergency guideline levels (SPEGLs) by NAS/NRC
- Permissible exposure limits (PELs) by OSHA
- Toxicity dispersion (TXDS) by New Jersey Department of Environmental Protection
- Toxic end points by EPA

#### *2.3.6.1 ERPG*

The concentration ranges given by the ERPGs are provided as a consequence of exposure to a specific substance. There are three different levels for ERPGs.

ERPG-1 is the maximum airborne concentration below which a person could be exposed for 1 h and there is a transient adverse health effect or perceiving an objectionable odor.
ERPG-2 is the maximum airborne concentration below which a person could be exposed for up to 1 h without having an irreversible or a serious adverse health effect that could avoid the person to take any protective action.
ERPG-3 is the maximum airborne concentration below which all the population could be exposed for up to 1 h without developing life-threatening health effects.

#### *2.3.6.2 IDLH*

This criterion allows the identification of acute toxicity doses for certain industrial gases. NIOSH defines IDLH as "that poses a threat of exposure to airborne contaminants when that exposure is likely to cause death or immediate or delayed permanent adverse health effects or prevent escape from such an environment" (NIOSH, 2005). It is important to mention that the IDLH level is considered a maximum concentration above which only a breathing apparatus, such as a self-contained breathing apparatus (SCBA), is permitted. If these values are exceeded, all workers who do not have protection must leave the area immediately.

#### *2.3.6.3 EEGL*

EEGL is the concentration of a gas, vapor, or aerosol that is acceptable and allows exposure to perform specific tasks during emergency conditions that last between 1 and 24 h. In general, when a person is exposed to an EEGL concentration, possible effects such as irritation of the central nervous system that do not produce long-term adverse health effects or that would impair performance of a task.

SPEGL is the acceptable concentration for exposures of the general public. In general, SPEGL values are set between 10% and 50% of the EEGL and takes into account different types of population (i.e., the elderly, children, infants).

#### 2.3.6.4 PEL

PEL values are comparable to the TLV values as they take into account a time-weighted average exposure of 8 h. In fact, the main difference between the PELs and the TLVs is the agency from which they come. PELs are developed and published by OSHA.

#### 2.3.6.5 TXDS

New Jersey uses a consequence analysis methodology to estimate the catastrophic quantities of toxic substances. This is required by the New Jersey Toxic Catastrophe Prevention Act (TCPA). In general, acute toxic concentrations are defined as the exposed concentration at which a toxic substance can cause acute health effects in the affected population and one fatality of 20 or fewer during a 1-h exposure.

#### 2.3.6.6 RMP

In general, EPA's RMP requires that if a facility is handling a toxic substance that is listed as an extremely hazardous substance, dispersion modeling should be performed. This list of substances can be found in the document titled *Consolidated List of Chemicals Subject to the Emergency Planning and Community Right-To-Know Act (EPCRA), Comprehensive Environmental Response, Compensation and Liability Act (CERCLA) and Section 112(r) of the Clean Air Act*. In this study, a toxic end point must be used as a reference. EPA has an order of preference for these simulation scenarios. The first one is the ERPG-2 and the second one is the level of concern, which can be found in the EPCRA. According to the EPA, the level of concern is "the maximum concentration of an extremely hazardous substance in air that will not cause serious irreversible health effects in the general population when exposed to the substance for relative short duration" (EPA, 1996).

### 2.3.7 HAZARD ASSESSMENT METHODOLOGY

Before heading deeper into hazard analysis, it is important to clarify what is a hazard. According to OSHA, "a hazard is the potential for harm. In practical terms, a hazard often is associated with a condition or activity that, if left uncontrolled, can result in an injury or illness" (OSHA, 2002).

To perform a hazard analysis is usually very challenging. The level of complexity for this type of analysis is subject to the level of detail the hazard analysis team wants to go into. For instance, if there are potential hazards that could affect a nearby population, such detail is essential. On the other hand, it is possible to disregard a scenario that will not cause any harm to any person. For example, the team is evaluating many processes and operations, but they realize that from 10 operations, only three have the potential to harm people. It may be sufficient to analyze those three scenarios, because the other seven scenarios do not have a chance to affect the nearby population. There are several types of hazard analysis. All of them are different and present a different procedure of how to assemble a team and how to analyze such hazards.

According to OSHA, one of the easiest and possibly most effective methods is the job hazard analysis (JHA), also known as job safety analysis (JSA). In general, a JHA focuses on job tasks as a way to identify hazards before they occur. At the same time, it finds a relationship between the worker, the tasks, the tools, and the work environment. The main purposes of the JSA are to identify hazards that are not controlled and to take appropriate measures to either prevent the hazard or reduce it to an acceptable risk level. Table 2.4, identifies the major type of hazards that can be identified on a JHA (OSHA, 2002).

**Table 2.4 Hazards Identified by a JHA (OSHA, 2002)**

| Hazards | Hazard Descriptions |
|---|---|
| Chemical (toxic) | A chemical that exposes a person by absorption through the skin, inhalation, or through the bloodstream that causes illness, disease, or death. The amount of chemical exposure is critical in determining hazardous effects. Check material safety data sheets (MSDSs) and/or OSHA 1910.1000 for chemical hazard information. |
| Chemical (flammable) | A chemical that, when exposed to a heat ignition source, results in combustion. Typically, the lower a chemical's flash point and boiling point, the more flammable is the chemical. Check MSDS for flammability information. |
| Chemical (corrosive) | A chemical that, when it comes into contact with (corrosive) skin, metal, or other materials, damages the materials. Acids and bases are examples of corrosives. |
| Explosion (chemical reaction) | Self-explanatory. |
| Explosion (over pressurization) | Sudden and violent release of a large amount of gas/energy due to a significant pressure difference such as rupture in a boiler or compressed gas cylinder. |
| Electrical (shock/short circuit) | Contact with exposed conductors or a device that is incorrectly or inadvertently grounded, such as when a metal ladder comes into contact with power lines. A 60-Hz alternating current (common house current) is very dangerous because it can stop the heart. |
| Electrical (fire) | Use of electrical power that results in electrical overheating or arcing to the point of combustion or ignition of flammables, or electrical component damage. |
| Electrical (static/ESD) | The moving or rubbing of wool, nylon, other synthetic fibers, and even flowing liquids can generate static electricity. This creates an excess or deficiency of electrons on the surface of material that discharges (spark) to the ground resulting in the ignition of flammables or damage to electronics or the body's nervous system. |
| Electrical (loss of power) | Safety-critical equipment failure as a result of loss of power. |
| Ergonomics (strain) | Damage of tissue due to overexertion (strains and sprains) or repetitive motion. |
| Ergonomics (human error) | A system design, procedure, or equipment that is error provocative. (A switch goes up to turn something off.) |
| Excavation (collapse) | Soil collapse in a trench or excavation as a result of improper or inadequate shoring. Soil type is critical in determining the hazard likelihood. |
| Fall (impacts) | Conditions that result in falls from (slip, trip) height or traditional walking surfaces (such as slippery floors, poor housekeeping, uneven walking surfaces, exposed ledges). |
| Fire/heat | Temperatures that can cause burns to the skin or damage to other organs. Fires require a heat source, fuel, and oxygen. |
| Mechanical/vibration (chaffing/fatigue) | Vibration that can cause damage to nerve endings, or material fatigue that results in a safety-critical failure. (Examples are abraded slings and ropes, weakened hoses and belts.) |
| Mechanical failure | Self-explanatory; typically occurs when devices exceed designed capacity or are inadequately maintained. |
| Mechanical | Skin, muscle, or body part exposed to crushing, caught-between, cutting, tearing, shearing items or equipment. |
| Noise | Noise levels (>85 dBA 8-h TWA) that result in hearing damage or inability to communicate safety-critical information. |

**Table 2.4 Hazards Identified by a JHA (OSHA, 2002)—cont'd**

| Hazards | Hazard Descriptions |
|---|---|
| Radiation (ionizing) | Alpha, beta, gamma, neutral particles, and x-rays that cause injury (tissue damage) by ionization of cellular components. |
| Radiation (nonionizing) | Ultraviolet, visible light, infrared, and microwaves that cause injury to tissue by thermal or photochemical means. |
| Struck by (mass acceleration) | Accelerated mass that strikes the body causing injury or death. (Examples are falling objects and projectiles.) |
| Struck | Against injury to a body part as a result of coming into contact of a surface in which action was initiated by the person. (An example is when a screwdriver slips.) |
| Temperature extreme (heat/cold) | Temperatures that result in heat stress, exhaustion, or metabolic slow down such as hypothermia. |
| Visibility | Lack of lighting or obstructed vision that results in an error or other hazard. |
| Weather phenomena (snow/rain/wind/ice) | Self-explanatory. |

### 2.3.8 SOURCE TERM

When dispersion of a substance occurs, two stages can be identified. The first is the source term, and the second one is atmospheric dispersion. In general, the source term occurs right after the release, when the behavior of the fluid is dominated by the storage conditions and the conditions of release. After the influence of the source starts decaying, the atmospheric dispersion starts ruling over the source term. Figure 2.15 illustrates the possible emissions situations.

There are two different types of releases. The first one is the continuous release, and the second one is the instantaneous release. In general, there is a criterion to determine what type of release is occurring. A continuous release is defined by the following relation:

$$\frac{u_x T_0}{x} \geq 0.15 \tag{2.86}$$

where $u_x$ is the friction velocity. According to Mannan (2012), Eqn (2.86) implies that at a certain distance $x$ where the release can be regarded as continuous recedes closer to the source as the duration of the release decreases to $T_0$. If distances are much greater than the ones given by Eqn (2.86), there will be an overestimate in the concentration. Now, to determine if the release is instantaneous, the following relation must be applied:

$$\frac{u_x T_0}{x} \leq 0.04 \tag{2.87}$$

It can be noticed that there is a gap between continuous and instantaneous release. In this gap, the release behaves neither as continuous nor as instantaneous; it is a transient release. The release thus can be described as quasi-continuous or quasi-instantaneous. In general, quasi-continuous release can be treated as follows: the total amount of substance released $Q_0$ over the duration of the release $T_0$ gives an effective release rate $q_0$. With this release rate, it is possible to obtain the concentration at a certain distance. If the release is treated as an instantaneous release, the only parameter that has to be used is the total amount of substance released $Q_0$ (Mannan, 2012).

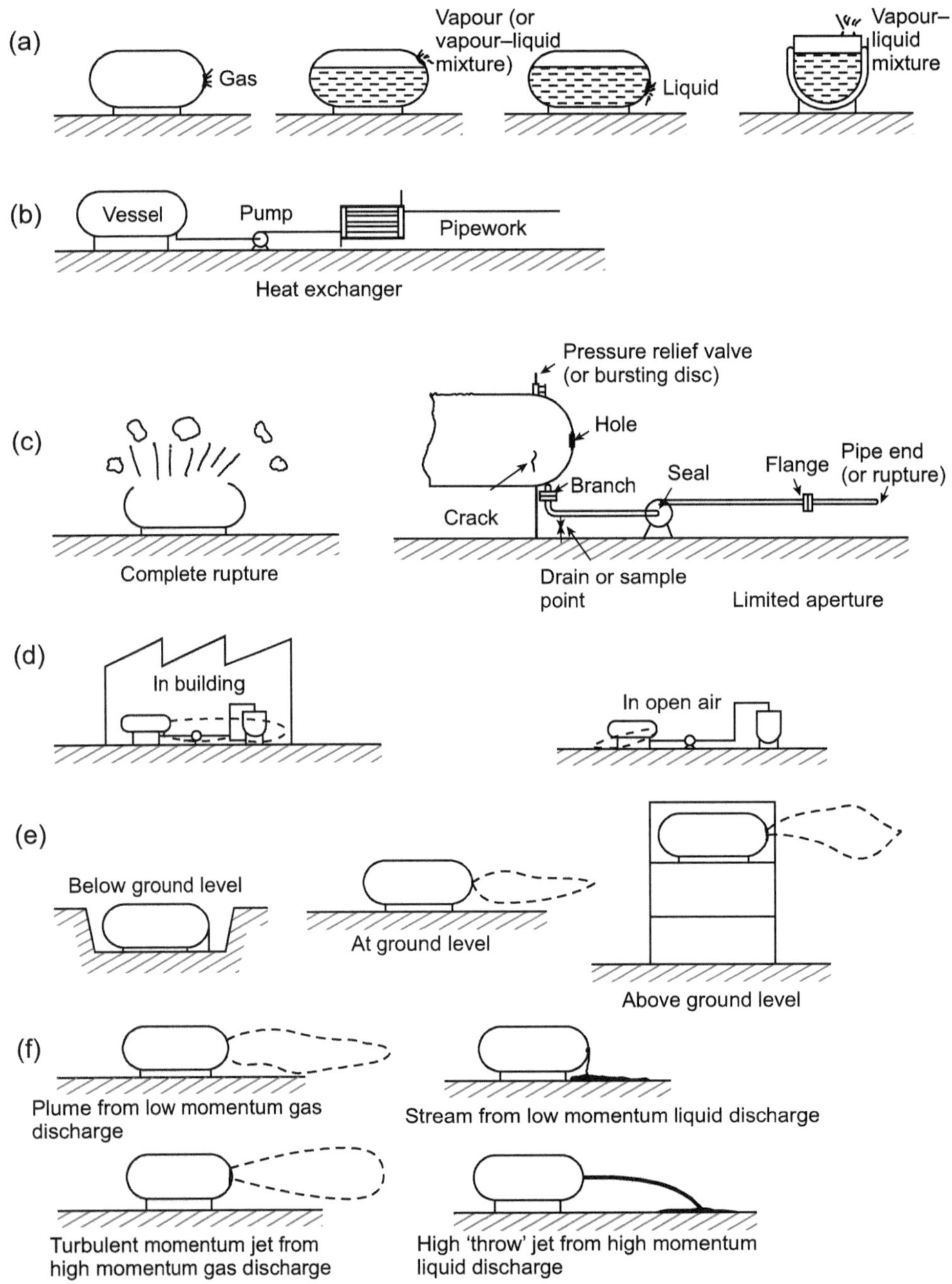


**FIGURE 2.15**

Some emissions situations (Mannan, 2012).

### 2.3.9 GAS DISPERSION

In most cases, the vapors of the principal toxic liquids exhibit heavy gas behavior and great strides have been made in heavy gas dispersion modeling. It is important to mention that there are four categories in which most of the models can be found. These classes are:

- Workbooks/correlations
- Integral models
- Shallow layer models
- Computational fluid dynamics (CFD) models

Most of the commercial or publicly available models belong to one of these categories and can be found in Table 2.9 (Luketa-Hanlin, 2006).

#### *2.3.9.1 Workbooks/correlations*

These types of models are recognized, especially due to the simplicity of their calculations (compared with different groups). There are many models of this type that can predict the concentration profile of gases that are being dispersed downstream. These types of models have been validated against experimental data, such as Britter and McQuaid's and the Gaussian Plume Model (GPM) (Luketa-Hanlin, 2006).

#### *2.3.9.2 Integral models*

These types of models are much more complex than workbook/correlation models. They include the solutions of basic differential equations. However, it is important to note that most of these types of models tend to idealize atmospheric conditions and some terrain descriptions (they assume that the terrain is plain and the dispersion is not going to be affected by any obstruction), and some of these models do not allow for the inclusion of geometry, making it difficult to simulate different obstacles and allowing for uncertainty in the calculations. When a complex scenario (with obstacles) is going to be simulated, most of the results obtained in this type of simulation will not be entirely trustworthy (Luketa-Hanlin, 2006). Many models of this kind have been successfully validated against experimental data.

#### *2.3.9.3 Shallow layer models*

This group has characteristics of the CFD models and the integral models. For example, some models will allow the inclusion of geometry in the simulation. The properties such as temperature, concentration and cloud height are modeled on horizontal coordinates in a depth integral sense. Basically, most of these types of models are used as research tools (Luketa-Hanlin, 2006).

#### *2.3.9.4 Computational fluid dynamics*

This group uses the Navier–Stokes equations to predict how the dispersion of a gas is going to behave and presents the results in three dimensions. These models allow for the inclusion of geometry and correctly specify the scenario. However, these models are highly sensitive to the scenario setup. The user has to be extremely aware and have a wide understanding of this peculiarity. The major problems with CFD models include their large use of computer resources and time and their need for a person who has expertise in the CFD simulation field (Webber et al., 2010; Britter and McQuaid, 1988; Cormier, 2008; Hansen et al., 2010).

Appropriate use of these models allows the replication of experiments that are performed in wind tunnels. Furthermore, wind tunnel experiments allow to include obstruction in a terrain and to study the effect of slopes, buildings, barriers, and water spray curtains. In general, the concentrations obtained by these models (mostly used for heavy gas dispersion) are very different to the results obtained in passive gas dispersion models (such as the Gaussian Model).

### 2.3.10 CONCENTRATION FLUCTUATIONS

In general, most models predict a concentration average. Such average is the most expected concentration for the prescribed input conditions at certain point. For example, if various experiments are performed identically (with the same atmospheric and release conditions), this average would represent the average of the amount of experiments performed. In reality, this is extremely difficult to perform as there is always turbulence in the environment. Therefore, if the same experiment is performed twice, it is very likely to obtain two different concentration averages in the same place. This is what dispersion models do. They do not show the concentration fluctuation around a point; they average the concentration fluctuation in that point and the response that is given is the simple average of these fluctuations (CCPS, 1996).

From this, it has been shown that at a fixed point the concentration fluctuates considerably in a vapor cloud. Griffiths and Harper (1985) mentioned that these concentration fluctuations have influence on the effective toxic load (Wilson and Simms, 1985; Wilson, 1991; Blewitt et al., 1987; Davies, 1987, 1989) and propose some methods to treat this influence.

### 2.3.11 MITIGATION: TERRAIN, BARRIERS, SPRAYS, SHELTER, AND EVACUATION

When a dispersion event occurs in a facility, there are several parameters that can affect the behavior of the dispersion. For instance, if a heavy gas is being dispersed; such gas will tend to be near the ground. Particular attention should be given to these types of scenarios, as these types of releases tend to affect large areas (Busini et al., 2012). In this specific case, the terrain will play an important role. For example, if there is a sharp slope upwards it will be more difficult for the gas to disperse. On the contrary, if there is a sharp slope downward, the gas will disperse with ease. An important parameter to take into account when studying this type of scenarios is the surface roughness. The surface roughness will create certain friction between the gas cloud and the floor. This force will slow down the velocity of dispersion and, consequently, the size of the flammable cloud. This situation can be observed when there is a high surface roughness (TNO, 2005).

There are different methods to mitigate dispersion and reduce the size of the toxic cloud. For instance, the inclusion of barriers in the system may diminish the size of the cloud. In general, barriers are installed near the location of a possible leak. The purpose of the barriers is to contain as much gas as possible and make it more difficult for the gas to go around it and keep dispersing. Several types of barriers are used; some substances are regulated to be stored in vessels and those vessels, at the same time, must be in dikes.

From the discussion regarding the different types of models, it is clear that the only model capable of predicting how a gas is going to be dispersed in presence of obstacles, like in a facility, is the CFD model (Mannan, 2012). Therefore, when simulations are going to be performed for a certain facility, it is preferable to use CFD models. However, as CFD models demand a great amount of time, a balance between precision and estimate should be made. In other words, using CFD models for simple geometries can be a waste of time, and this type of scenarios could be studied using integral models. Figure 2.16 illustrates how different outcomes are obtained when there are obstacles included in the scenario and when the scenario does not have any obstruction.

As can be observed in Figure 2.16, the obstacles present in the scenario significantly reduce the downwind distance the cloud could reach. This occurs mainly because there is an increasing turbulent phenomenon around the obstacles, which changes the flow field around them. It can also be observed from this figure that the cloud finds its way out of the obstructed area; however, it takes more time for this to occur. Additionally, the cloud size becomes smaller.

As it has been stated by Busini et al. (2012), there is a relationship between the height and position of the wall with the cloud dimensions in the open field. However, one cannot construct an infinite height/length barrier to reduce the risk. Busini addressed this issue and determined that for an obstacle to influence the cloud dispersion the wall has to have an adequate width and the height of the wall should be the same as that of the cloud to induce variations in the maximum affected distance. For instance, if the cloud is higher than the barrier, the barrier will not present much of an obstruction to the cloud.

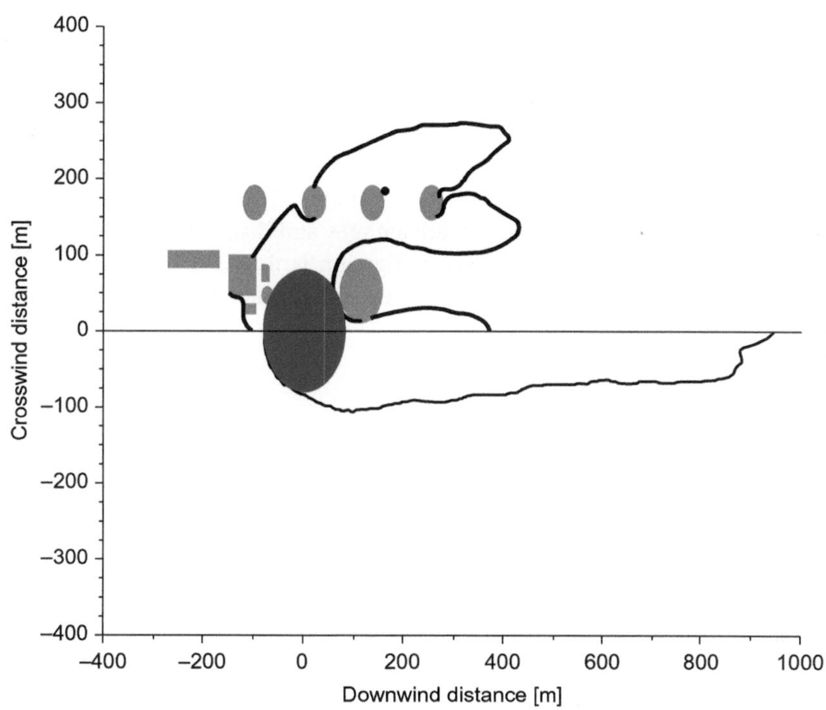


**FIGURE 2.16**

Comparison between obstructed and unobstructed dispersions of a heavy gas cloud. The black area represents the pool, and the lighter areas represent obstacles (Busini et al., 2012).

Therefore, before installing a barrier, design studies should be performed to determine the appropriate height and width of the barrier for it to fulfill its purpose (Busini et al., to be submitted for publication).

In the past decades, water sprays have gained a lot of attention from both academia and industry. It has been proved to be an effective engineering measure to mitigate the hazards of a dispersing cloud. Some of the reasons for the success of water sprays are their availability, the simplicity of use, and their efficiency (Uznaski and Buchlin, 1998). According to Hald et al. (2005), water curtains can be classified as full cone, full square cone, hollow cone, and flat-fan. All of them present a different flow pattern. Now, it is important to study the key parameters that can make water curtain successful when mitigating gas dispersion. Initially, one must focus on the atmospheric conditions. For instance, the ambient temperature and pressure should allow the water to be in the liquid phase and avoid getting into the freezing zone. Another important factor is the wind direction.

Several studies have been conducted to understand how dispersion is affected by water curtain. However, most of the studies have been performed with flammable gases such as LNG. Kim et al. (2013) discussed that there is less risk for the population and a couple of them have been performed with highly toxic substances such as hydrogen fluoride (HF) (Blewitt et al., 1987). In the first set of studies just mentioned, they concluded that water curtain has been proved to be effective in mitigating flammable vapor concentration below the flammability limit. In other words, the use of water curtain can reduce the dimensions of a flammable cloud. Now, if the same principle is applied to toxic substances, it is possible to reduce the size of the vapor cloud as well. Blewitt et al. (1987) reached the same conclusions in their experiments as well. They measured the effectiveness of the water curtain by its ability to make the HF dispersion end up near the source. For this system to operate efficiently, huge amounts of water should be released to the environment so as to capture more acid. Cornwell et al. (1998) performed a comparative quantitative analysis in evaluating different mitigation systems for HF. In this study, advantages and disadvantages of water spray were described. Table 2.5 presents such advantages and disadvantages.

Another important feature of this system of mitigation is that if a substance reacts with water, different chemicals could be used. Further, the effectiveness of water curtain depends on the following parameters:

- Water/chemical droplet size
- Nozzle type used
- Direction of the water/chemical flow

**Table 2.5 Advantages and Disadvantages of a Water Curtain (Cornwell et al., 1998)**

| Advantages | Disadvantages |
|---|---|
| • Reduces the amount of acid from the dispersed cloud.<br>• Dilutes the acid that falls on the ground, avoiding the possibility of evolving clouds.<br>• Adds turbulence in the environment, which favors the process of dilution. | • It is an active mitigation system (requires human intervention for it to start operating).<br>• Human error might be present in the sequence of activation of the system.<br>• A release must be identified. |

**Table 2.6 HF Removal Efficiencies at Different Release Conditions (Cornwell et al., 1998)**

| Hole Size | Total Airborne Mass Release Rate (Hydrocarbon and HF) (kg/s) | Airborne HF Mass Release Rate (kg/s) | Water Curtain Activation Time (min) | Water Curtain Mass Rate (kg/s) | Water/HF Mass Ratio | Modified Airborne HF Mass Release Rate (kg/s) | Percent Reduction in Airborne HF Mass Release Rate (%) |
|---|---|---|---|---|---|---|---|
| 6 in | 330 | 260 | 1 | 100 | 0.38 | 260 | 0 |
| 1 in | 10 | 8 | 1 | 100 | 12.5 | 5.6 | 30 |
| ¼ in | 0.7 | 0.5 | 3 | 100 | 200 | 0.08 | 84 |

- Water/chemical pressure
- Vapor cloud
- Properties of the toxic substance

In Table 2.6, a study performed by Cornwell et al. shows different scenarios where a water curtain was used to prevent the dispersion of HF to reach high distances.

Currently, there are several efforts to model how the dispersion phenomenon behaves with spray curtains. Again, in this case, the most desirable model to use is CFD as most of them have the ability to add water curtains in the system. An example of CFD models that could be used are FLACS and FLUENT. As previously discussed, the heavy gas dispersion models available can now treat situations where migratory features exist. Since the typical release occurs in a work environment, the gas dispersion will be strongly affected by the presence of buildings, which tend to enhance the dispersion. An illustration of a hazard assessment of a toxic release in the presence of buildings is given elsewhere (Deaves, 1987).

Another way to mitigate a toxic release is to find shelter. However, this mitigation measure should be used with caution. As has been described by Mannan (2012), once a human being starts performing physical activity, larger volumes of air are inhaled. Consequently, if there is a toxic substance mixed within the air, greater amounts of such substance are going to be inhaled and injuries could be reached much faster. Table 2.7 shows how the inhaled air increases with physical activity.

Moreover, as can be observed from Table 2.7, as the activity requires more effort, more oxygen is needed by the body. Therefore, before designating buildings for shelter, careful considerations should be made in designing the routes that could lead people to shelter and installing adequate ventilation systems. Finally, evacuation should be considered when very toxic substances have the potential to affect large populated areas with relatively small doses. Some substances that require evacuation if a release occurs are (Mannan, 2012):

- Chlorine: Lethal dosage is 40–60 ppm for 30–60 min of exposure
- Acrylonitrile: Lethal dosage is 200–400 ppm for 30 min of exposure
- Hydrogen cyanide: Lethal dosage is 100 ppm for 1 min of exposure
- Parathion: 30 mg (dermal)

**Table 2.7 Inhalation Rate at Different Type of Activities of a Man (Henderson and Haggard, 1943; Mannan, 2012)**

| Activity | Inhalation Rate[a] (l/min) | Oxygen Consumption Rate[b] (l/min) |
|---|---|---|
| Rest in bed, fasting | 6 | 0.240 |
| Sitting | 7 | 0.300 |
| Standing | 8[c] | 0.360 |
| Walking, 2 mile/h | 14 | 0.650 |
| Walking, 4 mile/h | 26 | 1.200 |
| Slow run | 43 | 2.000 |
| Maximum exertion | 65–100 | 3.000–4.000 |

[a]*Measured at 0 °C and 760 mm Hg.*
[b]*Measured at 20 °C.*
[c]*This value is quoted by Meyer (1938) and Prentiss (1937).*

## 2.3.12 PLANT LAYOUT

Prior to the use of toxic substances, a facility siting study (plant layout) must be performed. Such a study is designed to minimize the consequences in case a release occurs and helps in the planning of evacuation routes and emergency response procedures. Generally, large emissions and fugitive emissions must be taken into account. Additionally, two different sets per scenario must be performed to comply with EPA's Risk Management Plan (RMP) (Crowl and Louvar, 2011).

The first type of scenario is the worst-case scenario. Here, all the contents of the vessel must be released to the atmosphere in 10 min (almost catastrophic failure). Moreover, the atmospheric conditions must be set to a relatively low wind speed (1.5 m/s) and a very stable Pasquill stability (F). Passive mitigation systems allow the dispersed cloud to concentrate near the facility and extend the hazardous distances (EPA, 1996).

The second type of scenario is a more realistic scenario, also known as the alternative release scenario. Credit can be taken for both active and passive mitigation systems when developing the alternative release scenario. One must address the key parameters that may affect the dispersion of the toxic substances. For instance, consider the situation where a gas is being discharged from an orifice in the vessel. In this case, if we have a bigger hole, the discharge rate is going to be relatively bigger than if we have a smaller hole. Therefore, it is important to be aware of the selection of the orifice size prior to setting up any type of calculations. The following are the most important parameters the reader should be aware of (Crowl and Louvar, 2011):

1. Orifice area
2. Operating pressure and atmospheric pressure
3. Properties of the substance
4. Temperature of the gas

**Table 2.8 Guidelines for Selection of Process Incidents (Crowl and Louvar, 2011)**

| Incident Characteristics | Guideline |
|---|---|
| **Realistic release scenario** | |
| Process pipes | Rupture of the largest-diameter process pipe as follows:<br>• Diameters smaller than 2 in, assume a full bore rupture.<br>• Diameters between 2 and 4 in, assume rupture equal to that of a 2-in-diameter pipe.<br>• Diameters greater than 4 in, assume rupture area equal to 20% of the pipe cross-sectional area. |
| Hoses | Assume full-bore rupture |
| Pressure relief devices relieving directly to the atmosphere | Use calculated total release rate at set pressure. Refer to pressure relief calculation. All material released is assumed to be airborne |
| Vessels | Assume a rupture based on the largest diameter process pipe attached to the vessel. Use the pipe criteria |
| Other | Incidents can be established based on the plant's experience, or the incidents can be developed from the outcome of a review or derived from hazard analysis studies |
| **Worst case incidents** | |
| Quantity | Assume release of the largest quantity of substance handled on-site in a single process vessel at any time. To estimate the release rate, assume the entire quantity is released within 10 min |
| Wind speed/stability | Assume F stability, 1.5 m/s wind speed, unless meteorological data indicate otherwise |
| Ambient temperature/humidity | Assume the highest daily maximum temperature and average humidity |
| Height of release | Assume that the release occurs at ground level |
| Topography | Assume urban or rural topography, as appropriate |
| Temperature of released substance | Consider liquids to be released at the highest daily maximum temperature, based on data for the precious three years, or at process temperature, whichever is higher. Assume the gases liquefied by refrigeration at atmospheric pressure are released at their boiling points |

In general, when performing facility siting studies, the area of the largest pipe connected to the vessel is considered. Still, this will give overestimates of the consequences; however, it will be a conservative estimate. Table 2.8 illustrates typical guidelines for the selection of process incidents.

## 2.3.13 COMPUTER AIDS

The computations either of the consequences of a toxic release or of the risk to the population around an industrial site or along a transport route is a natural application for computer codes, and a considerable number of codes are available. Table 2.9 summarizes some of the dispersion models used nowadays.

**Table 2.9 List of Models**

| Model Name | Model Type | Supported by |
|---|---|---|
| ALOHA | Integral | Publicly available (CAMEO, EPA) |
| CANARY | Integral | Quest Consultants Inc. |
| CFX | 3D-CFD | ANSYS |
| DEGADIS | Integral | Publicly available (e.g., Trinity Consultants, Lakes Environmental) |
| DRIFT | Integral | ESR Technology, UK |
| FDS | 3D-CFD | Publicly available, NIST |
| FEM3A | 3D-CFD | University of Arkansas |
| FLACS | 3D-CFD | Gexcon AS, Norway |
| FLUENT | 3D-CFD | ANSYS |
| GASTAR | Integral | CERC, UK |
| HGSystem (HEGADAS) | Integral | Shell, UK |
| SLAB | Integral | Publicly available (e.g., EPA, Trinity Consultants, Lakes Environmental) |
| SLAM | Shallow layer | Risø, Denmark |
| SCIPUFF | Lagrangian | L3, Communications Titan Group, Trinity Consultants |
| STAR-CD | 3D-CFD | CD-AdapcoIoMosaic |
| SUPERCHEMS EXPERT | Integral | IoMosaic |
| TSCREEN (Britter-McQuaid model) | Box | Publicly available (e.g., EPA, Lakes Environmental) |
| PHAST | Integral | Det Norske Veritas (DNV), Norway |

## 2.4 PRESENT APPROACH TO PROCESS SAFETY

### 2.4.1 RISK AND HAZARD

"Hazard" and "risk" are often used interchangeably and treated as equivalent terms in informal communication. However, each term has its own distinct and precise usage in formal risk communication. A "hazard" is something with the potential to cause an undesired consequence by virtue of one or more of its inherent properties. Something with this potential is said to be "hazardous." The severity of the consequences and the likelihood of those consequences are not addressed by "hazard." The term "risk" is used to describe a measure of the probability of a consequence occurring and the severity of that consequence. Therefore, a "hazard" can be thought of as a potential source of harm and "risk" is a measure of the size of that potential harm (Modarres, 2006).

Hazards arise in a process and in life from a variety of sources including but not limited to (Crowl and Louvar, 2011):

- Objects or persons at elevation (potential for falls from a height)
- Toxic substances
- Flammable or explosive substances
- Substances stored under pressure
- Substances stored at extremes of hot or cold
- Conduction or build-up of electric charge

Two facilities can have identical hazards, possibly as a result of operating nearly identical processes, but have very different associated risk levels. Risk is traditionally defined as the product of the expected frequency and severity of a consequence (Crowl and Louvar, 2011):

$$R = f * S \tag{2.88}$$

where $f$ is the frequency of the event, usually expressed annually or monthly, and $S$ is the severity of the consequence, usually expressed in monetary terms.

If barriers exist to prevent the consequence, the probability of these barriers succeeding or failing can be incorporated into the calculation of the event frequency.

$$f = (\text{IEF}) * \prod_{i=1}^{n} (\text{PFD}_i) \tag{2.89}$$

where IEF is the initiating event frequency, and $\text{PFD}_i$ is the probability of failure on demand of the $i$th barrier of $n$ total barriers. If multiple hazards or consequences are possible, the total risk can be seen as the sum of the risks from the individual hazards and consequences (Modarres, 2006).

$$R = \sum_{i=1}^{n} f_i S_i \tag{2.90}$$

Incidents and losses may be initiated by events or factors outside of system that may be beyond the control of the operators of the system. Similarly, losses may occur to the system or they may be caused by the system and inflicted on one or more recipients.

Risks are broadly grouped into five main categories: health, safety, security, financial, and environmental. Health risks principally involve disease and illness. Safety risks involve disasters, incidents, products, and technologies. Security risks pertain to war, terrorism, social unrest, crime, and misappropriation or misuse of information. Financial risks involve currency fluctuations, market fluctuations, interest rates, market share, bankruptcy, supply interruptions, and business interruptions. Finally, environmental risks involve contamination or damaging of the ecosystem (air, land, water, and space) with trash, toxins, noise, and odors (Modarres, 2006).

Risk analysis involves identifying, characterizing, managing, and disseminating information with regard to potential losses. Risk analysis, therefore, typically involves three elements or stages: risk assessment, risk management, and risk communication.

Risk assessment is the component traditionally thought of when risk analysis is referenced. It involves the estimation of event probabilities, frequencies of occurrence, and consequence severities. This assessment is normally handled by first identifying hazards, generating scenarios, identifying barriers where they exist, assessing the likelihood of the scenarios, and then estimating the consequences. A scenario deemed too unlikely to be credible may not warrant consequence analysis.

Risk management is principally concerned with managing uncertainties but is a highly complicated decision-making process. Resources are always limited, and it is infeasible for every hazard to be mitigated through the use of extremely expensive or complex systems. Additionally, it is impossible to reduce risk to zero. There will always be some residual risk that cannot be eliminated. As such, companies and management are required to define levels of *acceptable risk*. Once a risk is reduced to below this threshold value, further mitigation is unlikely as any additional resources are likely to be directed toward addressing other risks. If a risk has not been reduced to the level deemed acceptable but a further reduction of risk is cost prohibitive the risk can be considered conditionally acceptable (Modarres, 2006).

A risk can be referred to as being ALARP when it has been reduced to the point where further reductions in risk will cost more than they are expected to save through risk reduction.

Assessing risk also requires the expenditure of time and resources. A full, detailed, quantitative risk assessment will require more time and resources than a simple qualitative assessment. Management must decide how much effort to expend on assessing the risk from various hazards and scenarios. A hazard that is expected to produce low consequences is unlikely to be the subject of detailed analysis. A high consequence scenario, on the other hand, may warrant a thorough analysis.

Risk management is the most important part of risk analysis as only a proper allocation of resources will result in optimized risk levels and overall process safety.

Risk communication involves the dissemination of information with regard to the existence of risks, their likelihood and consequences, how the risks were identified and assessed, the assumptions and uncertainties involved in the analysis, and what steps are being taken or can be taken to manage the risks.

## 2.4.2 METHODOLOGY IN RISK ASSESSMENT

Some major steps are common to all risk assessments. These include:

- Hazards identification
- Barrier identification
- Barrier performance assessment
- Exposure assessment
- Risk characterization

The methods for carrying out the assessment will vary depending on the type(s) of risk being assessed.

*Hazards identification*: The first step in the risk assessment process, as the name suggests, involves the identification of hazards, which result from all potential sources in the system. Some hazards are natural and some are artificially made. Some hazards are inherent to a material, or process. Some hazards are emergent, resulting or arising from a unique combination of materials, conditions, and circumstances. Once identified the hazards must be characterized in terms of their amount, intensity, and severity.

*Barrier identification*: Barriers are systems or structures that function as "obstacles [for] containing, removing, neutralizing, preventing, mitigating, controlling, or warning" of hazards (Modarres, 2006). Barriers can be active or passive. Passive barriers provide protection merely through their presence and do not require activation or the operation of some function to prevent or mitigate the hazard. Passive barriers include dikes, casings, natural circulation, and ventilation, among others. Active barriers require that they be activated or triggered to perform their function. Activation can be manual, through the action of an operator or other personnel, or automatic, usually triggered by some kind of sensor. Active barriers include cooling systems, alarms, automatic shutdown systems, drainage pumps, fire water systems, and foam fire suppression systems among others.

*Barrier performance analysis*: No barrier or protective system can be perfect. There will always be a chance that a particular barrier will fail when it is needed. An event that requires the action or the function of the system to avert a hazard is called a "demand" as it places a demand or a need to function on a system that normally sits and waits. The chance or the probability that the system will fail to function or function inadequately is known as the "probability of failure on demand,"

or PFD. The PFD for a system can change over time because of age, degradation, damage, or overuse. Proper maintenance will slow down, but usually not prevent, this degradation.
The PFD for a system must be estimated through one of a variety of means, some of which are discussed in Yañes et al. (2002). There will always be uncertainties in the assessment of barrier performance and reliability. These uncertainties should be included in the final assessment and considered in the analysis.
*Exposure assessment*: This step attempts to predict the likely consequences of a barrier failure. If some or all of the barriers to a hazard fail, the effects of the hazard may be felt, either in full or to a lesser, mitigated degree.
*Risk characterization*: In the final step of the risk assessment, the magnitude of the consequences from various hazards and scenarios are estimated based on available information. These consequence estimates are combined with probability and frequency calculations to estimate the expected losses from each individual hazard and all identified hazards collectively.

The estimation of probabilities, frequencies, and consequences can be either quantitative or qualitative.

#### 2.4.2.1 Nodes in risk assessment

Many of the risk assessment methodologies are involved and can be complicated. It is difficult, if not impossible, to apply many of them to an entire process from start to finish at the same time. It is therefore common, even necessary, to break up a process into sections, or "nodes." A node generally consists of one major process vessel, often a reactor or a column, along with the associated process lines, instrumentation, and supporting equipment such as heat exchangers and settling tanks. Process can be divided into as many or as few nodes as necessary to facilitate the analysis. The analysis should, however, be able to consider how problems in one node might impact another.

#### 2.4.2.2 Teams and information required for a risk assessment

All forms of risk assessment require extensive information about the process to be analyzed to be successful. The first step in almost any risk assessment is the collection of data and documentation on the process including detailed flow diagrams, instrumentation diagrams, chemical inventories, maintenance logs, and more. A thorough risk assessment will also require a diverse team that is knowledgeable with regard to the process and the likely outcomes or consequences of various abnormal events. Risk assessment teams should include leaders or facilitators, engineers, maintenance personnel, and operators.

### 2.4.3 QUANTITATIVE RISK ASSESSMENT

Quantitative risk analysis (QRA) attempts to estimate the probability or frequency of an event and its consequences. QRA seeks to arrive at a numerical value for the level of risk presented by a hazard. There will be uncertainty in these estimates that must also be characterized and quantified through established and accepted statistical means. An accurate QRA requires a large amount of data and information. In cases where data are sparse, quantitative analysis is difficult and uncertainties can be large. QRA also tends to be much more resource and time intensive than qualitative and semiquantitative approaches. Because of this, QRA is preferred in situations where the data, resources, and time are available in addition to justification with regard to the high risk and complexity of operations.

## 2.4.4 SCALABILITY IN RISK ASSESSMENT METHODOLOGIES

As a general rule of thumb, qualitative and semiquantitative approaches to risk assessment will scale better than quantitative methods. A true quantitative analysis is rigorous and computationally heavy. The complexity of the analysis will increase rapidly as the size and scope of the project increases. As a result, quantitative analysis is usually highly targeted and used to answer important questions with regard to scenarios of potentially high consequences.

Qualitative approaches tend to scale better between smaller and larger projects as the simplifying assumptions used in them prevents the calculations and data requirements from becoming overly burdensome.

## 2.4.5 PROBABILITY BASED APPROACHES

Probabilistic risk assessment (PRA) is a procedure for examining the operation and interaction between parts of complex systems. A thorough PRA accounts not only for equipment but also for human elements, software, and control systems, all of which can fail. The PRA has the ability to identify the components that make the largest contributions to risk in the system. There are several steps to a PRA (Modarres, 2006).

1. Objectives and methodology definition
2. Information assembly and system familiarization
3. Identification of initiating events
4. Scenario development
5. Logic modeling
6. Failure data collection and performance assessment
7. Integration and quantification of risks
8. Uncertainty analysis
9. Sensitivity analysis
10. Risk ranking and importance analysis
11. Interpretation of results

*Objectives and methodology definition*: The scope and goal of the analysis as well as the expected final deliverables should be selected at the beginning of the analysis. Once the analysis is scoped and objectives set, a technique should be chosen from the options available, some of which are discussed here.

*Information assembly and system familiarization*: All the information needed for the analysis must be collected and then reviewed. Before beginning the analysis, all of the components of a system must be identified and understood.

*Identification of initiating events*: Generate a list of events that could set a hazardous event or consequence in motion. Functional block diagrams of the system may be helpful for this purpose. Initiating events should then also be grouped based on the required responses to the initiating events.

*Scenario development*: In principle, scenario development seeks to identify all possible pathways for the propagation of a hazardous event. This is not always possible or practical; however, there can be an infinite or nearly infinite number of scenarios depending on the system and hazard being analyzed. It is therefore sometimes necessary to select a representative set of scenarios that are thought to encompass most major or significant outcomes and consequences.

*Logic modeling*: Certain types of analyses, including fault trees and event trees, require the development of a logic model that models the dependencies of various systems on their base components.

*Failure data collection and performance assessment*: Estimation of failure rates and event frequencies will require data on the reliability and performance of all of the base components that make up the system. These data will need to be collected as an input to the calculations.

*Integration and quantification of risks*: The analysis is conducted and the final numerical outputs of the analysis are generated. Boolean reductions may be conducted at this time. Minimum cut sets should be identified, and the results of the analysis should be used to determine which components make the biggest contributions to risk in the system.

*Uncertainty and sensitivity analysis*: Uncertainty analysis will assess the expected margin for error in the final results based on the quality and quantity of the data and the assumptions used in the analysis. Sensitivity analysis examines the model and parameters used in the analysis as well as the assumptions used to conduct the analysis to determine the impact that each has on the final outcome. The analysis or model is said to be "sensitive to those parameters or assumptions that have the greatest impact on the results." Sensitivity analysis and uncertainty analysis help focus attention and resources on those aspects of a PRA that might warrant further attention or consideration to improve the quality of the PRA.

*Risk ranking and importance analysis*: Components and systems should be ranked according to their importance to the system and its reliability. This can be done through a variety of means. Some importance measures are based purely on the mathematical contribution of that component to the overall risk of the system, while other methods rank components based on their contribution to the uncertainty of the analysis, called "uncertainty ranking." Some methods rank the importance of a component by looking only at that component, while others rank the importance of a component based on its importance compared with other components and subsystems.

*Interpretation of results*: The interpretation of results is actually a continuous process in the PRA in which the outcomes of each stage of the analysis are examined and opportunities to improve the accuracy of the analysis are looked for. Interpretation of the results should continue until the analysis arrives at what is felt to be a reasonably stable and accurate input with acceptable uncertainties.

#### 2.4.5.1 Fault tree analysis

Fault tree analysis seeks to determine how systems or hazard barriers can fail by starting with a highly specific top event or failure and work back to determine how that top level event could occur. As such, fault tree analysis is a deductive process in which the risk analysis team attempts to determine how an event might occur. Fault trees can be used to evaluate simple top level events for small systems or larger plants; however, fault trees for larger systems can easily grow to include thousands of events, spread out over hundreds of pages. Fault trees for simple events, like a tank overfilling, for example, can be constructed and evaluated in a few minutes if the required information is available. Large fault trees are frequently drawn and evaluated using computer programs.

The number of events that must be dealt with in a fault tree is controllable by limiting the resolution of the analysis. Nonbasic events can be left undeveloped in the analysis, especially if a detailed analysis of the event is not considered likely to be beneficial, in the interest of saving time or resources.

Fault trees are constructed from a variety of logic gates and event types. Some of most commonly used symbols in fault tree construction are shown in Figure 2.17.

There are several steps to developing a fault tree that can be expressed as:

1. Precisely define a top event.
2. Determine the conditions required for the top event.
3. Define the physical limits of the system and the scope of the analysis.
4. Determine which events will be considered nonallowed or too rare to be considered in the analysis.
5. Define the equipment configuration and the normal condition of the process.
6. Define the level of resolution, or how far the analysis will proceed.
7. Draw the tree, starting at the top and continuing until intermediate events have been explored and only basic events, external events, and undeveloped events remain.
8. Evaluate the tree, starting at the bottom with basic events, external events, and undeveloped events.

Drawing the tree can be conducted using the following sequence:

1. Select an event (top or intermediate).
2. Determine the conditions (events) required for this event and select appropriate gates (AND, OR, NAND, NOR).
3. Determine whether each condition is an intermediate event, basic even, external event, or undeveloped event.
4. Repeat steps one to three until all intermediate events are explored and developed. (Note: an event can be left undeveloped if it is not considered worth further exploration.)

The tree can be evaluated in the following ways:

1. Determine the failure probabilities of all basic, external, and undeveloped events.
2. Determine all minimal cut sets and/or success paths for the fault tree.
3. Calculate the probability of each cut set or success path to determine the probability of the top level event occurring.

or

2. Perform the necessary calculations across the logic gates associated with these events to determine the probability of the next level of intermediate events.
3. Repeat step 2 until the top level event is reached and its probability is calculated.

#### *2.4.5.2 Event tree analysis*

Event trees are most appropriate when the successful operation of the plant or a safety outcome to an emergency requires the chronological and independent operation of barriers and systems (Modarres, 2006). The sequential nature of the events becomes important in modeling the propagation of an incident and the role of various systems and barriers in determining the ultimate outcome (Crowl and Louvar, 2011). The event tree starts with what is appropriately named an initiating event and follows

| | | |
|---|---|---|
|  | AND Gate : | The resulting output event requires the simultaneous occurrence of all input events. |
|  | OR Gate : | The resulting output event requires the occurrence of any individual input event. |
| 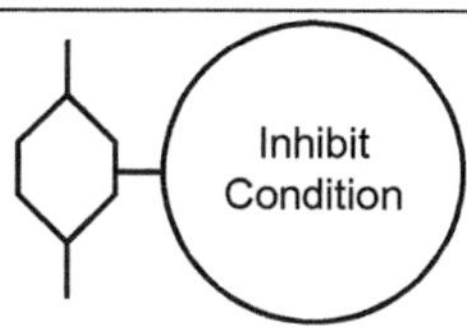  | INHIBIT Event : | The output event will occur if the input occurs and the inhibit event occurs. |
| 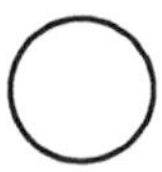 | BASIC Event : | A fault event that needs no further definition. |
|  | INTERMEDIATE Event : | An event that results from the interaction of a number of other events. |
| 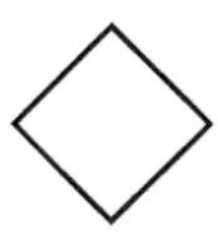 | UNDEVELOPED Event : | An event that cannot be developed further due to lack of suitable information. |
|  | EXTERNAL Event : | An event that is a boundary condition to the fault tree. |
| 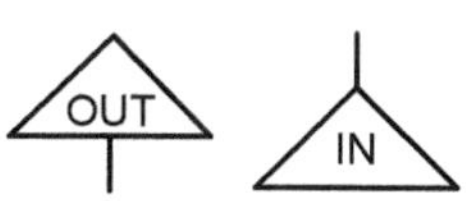  | TRANSFER Symbols : | Used to transfer the fault tree into and out of a sheet of paper. |

**FIGURE 2.17**

Common fault tree symbols.

*Reproduced from Crowl and Louvar (2011).*

an inductive process to predict probable consequences (Modarres, 2006; Crowl and Louvar, 2011). The steps in an event tree analysis are typically thought of as (Crowl and Louvar, 2011):

1. Identify or select an initiating event.
2. Identify the systems that deal with or mitigate the initiating event.
3. Determine the order of the events.
4. Construct the event tree.
5. Determine the resulting event sequences and outcomes.

The event sequence and the sequence outcome are typically listed on the right side of the event tree. Trees are typically constructed with a successful (or positive) outcome taking the upper branch and a failure (or a negative outcome) taking the lower branch. Many event trees are constructed with two final outcomes: success and failure, safe and unsafe. However, because some barriers can act to mitigate or limit, rather than prevent, an incident, an event tree that includes such systems and barriers can be constructed with one or more intermediate outcomes. Branching of the tree does not always occur at every event for every pathway (Bahr, 1997). For example, two events in a sequence might be the operator manually shutting down the system and an automatic system shutdown in response to an alarm or trigger. The system can only be shut down once in the sequence. Therefore, if the operator is intended to take action first and successfully shuts the system down, then automatic shutdown will not function. The system is already in a shutdown state. The event tree for these two barriers would look like Figure 2.18.

Evaluation of event trees is handled using Boolean expressions to determine the probability of each outcome of the event tree.

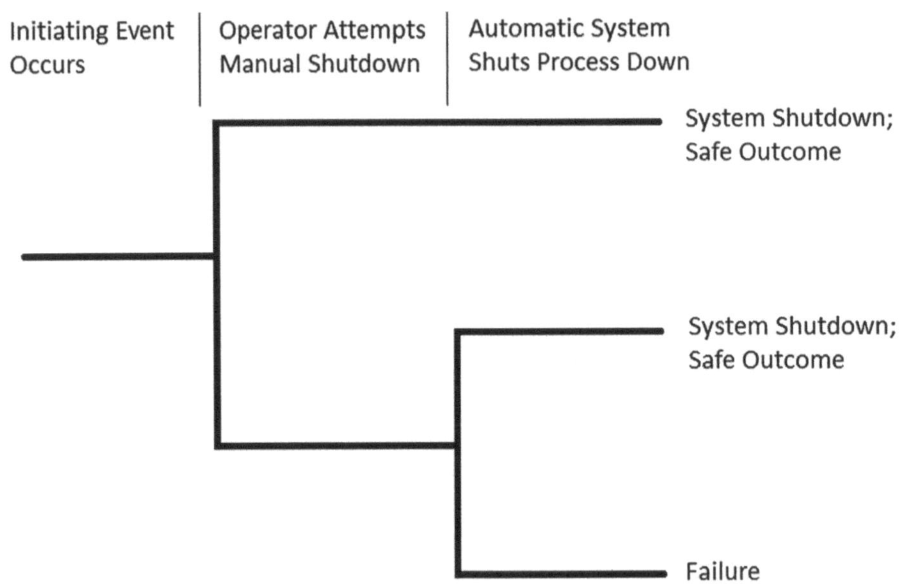


**FIGURE 2.18**

Example event tree for nonbranching events.

There is no limit on the number of events/barriers that can be used or evaluated in an event tree analysis. The event tree can have as many branching events as are deemed necessary for the analysis. The technique is therefore infinitely scalable with regard to the complexity of the analysis.

The technique is also flexible. It can be applied to a wide range of systems and initiating events as it only requires the identification of a hazardous event and barriers designed to counter that event. This allows the method to be applied to tanks overfilling, reactors overheating, attacks on the facility, loss of outside power to part of the facility or the entire facility, and software glitches.

### 2.4.5.3 Bow-tie plots

A bow-tie plot can be looked at as a combination of a fault tree and an event tree. If the top level event of a fault tree is the same as the initiating event of an event tree, the plots can be arranged end-to-end and combined to form a bow-tie diagram. The bow-tie diagram is named because of the characteristic shape formed by it as the triangular shapes of the event tree and fault tree narrow to the process mishap in the middle as illustrated by Figure 2.19.

More generally, the bow-tie plot combines the deductive and inductive processes of the fault tree and event trees to allow a single diagram and analysis to trace a failure event from its root causes to its ultimate consequences. The bow-tie plot traces the path of a failure through the process control systems to the top level event and then through the protective layers of the mitigation system. The bow-tie plot traces from the basic failure events to the final outcome and consequences (Bahr, 1997). Because the bow-tie plot is a combination of a fault tree and an event tree, it enjoys the flexibility and scalability of the other two approaches.

### 2.4.5.4 Failure modes and effects analysis

Failure modes and effects analysis (FMEA) lists the ways (modes) in which components and elements in the system can fail. The likelihood and effects of each failure mode on the system are determined. Mitigating steps that can or should be taken are recommended based on the resulting scenario probabilities and consequences. FMEA is best applied to simple systems or systems with a limited number of parts or failure modes. It works well for simple systems and single pieces of equipment but it is not readily applicable to larger systems like whole processes and facilities.

The FMEA technique consists of a methodical study of component failures. Typically, this review starts with a diagram of the process that includes the most probable components that could fail and consequently create a safety hazard for the process. According to OSHA, the following list represents a typical FMEA and the process that should be used to perform this analysis.

- Potential mode of failure
- Consequence of the failure
- Effect on other components

**FIGURE 2.19**

Structure of a bow-tie plot.

- Effect on whole system
- Hazards class (qualitative)
- Probability of failure
- Detection methods
- Compensating provision/remarks.
- Recommendations

### 2.4.5.5 Bayesian networks

Thomas Bayes (1702–1761) developed the Bayes theorem, which facilitates the calculation of the probability of causes based on observation of effects. During the 1980s, the model was extended to relate the probabilistic relationships among statistically or causally related variables. Bayesian networks (BNs) were developed from probabilistic graphical structures designed to represent these relationships (Jordaan, 2005).

A BN is a graphical description of the direct dependencies among a set of variables. The directed graph is constructed with variable nodes and arcs connecting them. Each variable node has a node probability table or a conditional probability table of probabilities for each state of the node variable. There are "parent nodes" and "child nodes" encoded with the assumption that the value or state of the "parent node" variable influences the value or state of the "child node" variable. As a result, there must be no cycles back to a prior node in an arc on a graph to avoid circular reasoning. A "root" node has no parent nodes. It is the start of a chain and no other variables in the network influence it. A "leaf" node has no child nodes and therefore influences no other variables. In a directed chain of nodes, a node is an "ancestor" of another if it appears earlier in the chain and is a "descendant" of a node if it appears later in the chain. The set of parents of a node "X" is designated by Parents(X) or pas(X). All nodes that are not root or leaf nodes are "intermediate" nodes. The root nodes therefore represent original or initiating events and the leaf nodes are final effects or consequences (Jordaan, 2005; Fenton and Neil, 2012).

In Figure 2.20, nodes or variables labeled A, B, and D are root nodes. G and H are leaf nodes. A and B are parents of C, C and D are parents of E, which is a parent of F and G. F is a parent of H.

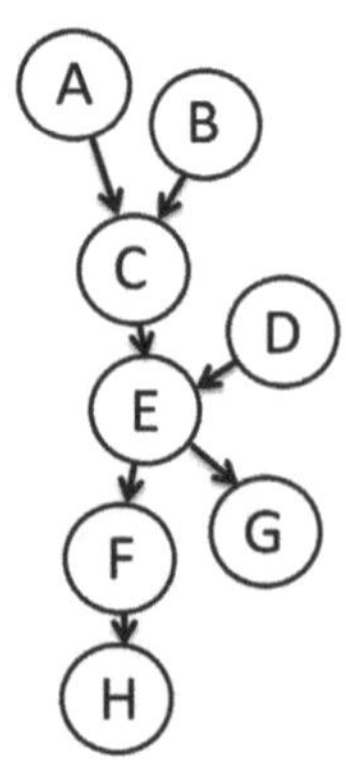


**FIGURE 2.20**

Example of a Bayesian network with a directed chain.

The BN allows for us to take knowledge of the state of H to improve our knowledge of the state of F and to use knowledge with regard to the state of F and G to improve our knowledge E, working back through the chain. In the figure, C, E, and F are intermediate nodes. Nodes A, B, C, and D are ancestors of E, while F, G, and H are descendants of E.

BNs assume the **Markov property** is in effect. That is, the only dependencies that are in effect are the ones shown by the directed arcs and that unconnected nodes are conditionally independent.

When the state of a BN variable is observed or known, the "credibility" of the network changes and it must be conditioned by the new information. This conditioning process can be referred to as probability propagation, inference, or belief updating. Information flows in both directions, not just in the directions of the arcs. This results in different reasoning types in BN calculations: diagnostic, predictive, and intercausal. Diagnostic reasoning moves in the opposite direction of the arcs, it starts with effects and updates causes. Predictive reasoning moves in the direction of the arc and uses information on causes to update probabilities of effects. Intercausal reasoning involves effects with more than one cause. Observing the state of an effect with two or more causes increases the likelihood of all the causes; however, if the state of one cause is also known, the state of that cause can make the other cause(s) less likely. The other causes are therefore "explained away" (Korb and Nicholson, 2003; Jordaan, 2005).

Any nodes in a network can be "evidence" nodes, the state of which is known, or "query" nodes, the state of which is influenced by the evidence.

BN calculations are now normally handled by computer programs. But some of the calculations involved can be provided here. The probability of a set of values or states for all of the variables shown in Figure 2.20, *P(A, B, C, D, E, F, G, H)* can be given by (Jensen and Nielsen, 2009):

$$P(A,\ B,\ C,\ D,\ E,\ F,\ G,\ H) = P(A)P(B)P(C|A,B)P(D)P(E|C,D)P(F|E)P(G|E)P(H|F) \quad (2.91)$$

However, because the Markov property is in effect, the probability of an event depends only on the state of events above it in a chain. Take the probability of *C*, *P(C)*, as an example.

$$P(C|A,B,D,E,F,G,H) = P(C|A,B) \quad (2.92)$$

As a further example, if we assume *F* and *H* are both variables with only two possible states, true or false, we can use knowledge of *H* to learn more about *F*. Say event *H* is the failure of a system to operate and event *F* is the failure of a component in that system, which can, but does not necessarily, cause the system to fail. Table 2.10 is the probability table for *F*, a component failure. Table 2.11 is the probability table for *H*, a system failure, given *F* (Fenton and Neil, 2012).

If we observe a system failure, we can update our estimate of the probability of the component failure, *F*. $P(\varepsilon)$ is the marginal probability of the observed parameter:

$$\begin{aligned} &P(\varepsilon) = P(H = \text{True}) = \\ &P(H = \text{True}|F = \text{True})P(F = \text{True}) + P(H = \text{True}|F = \text{False})P(T = \text{False}) \quad (2.93) \\ &P(\varepsilon) = 0.8 * 0.1 + 0.1 * 0.9 = 0.17 = \text{Probability of System Failure} \end{aligned}$$

**Table 2.10 Probability Table for Component Failure**

| | |
|---|---|
| False | 0.9 |
| True | 0.1 |

**Table 2.11 Probability Table for System Failure Given a Component Failure**

| Component failure | | False | True |
|---|---|---|---|
| System failure | False | 0.9 | 0.2 |
| | True | 0.1 | 0.8 |

Given that the system failed, we can update the probability that the component has failed based on:

$$P(F = \text{True}|H = \text{True}) = \frac{P(H = \text{True}|F = \text{True})P(F = \text{True})}{P(\varepsilon)}$$
$$P(F = \text{True}|H = \text{True}) = \frac{0.8 * 0.1}{0.17} = 0.47 \tag{2.94}$$

Therefore, the observation that the system has failed increases the probability that the component has failed from 0.1 to 0.47.

As the prior example illustrates, BNs can be constructed to analyze systems of as little as two events, but they can also be used to analyze systems far more complicated than the system in Figure 2.20, with hundreds or thousands of events. Similarly to fault trees, once BNs exceed a certain size it becomes prohibitively difficult to construct the network and evaluate the system without the use of a computer; however, various computer programs have been developed for using BNs. Even with a computer, a large, complex network will be more time-consuming to construct. However, once the network is built, the availability of cheap computing power makes the method easy to use, even for relatively large systems.

## 2.4.6 CONSEQUENCE-BASED APPROACHES

There is an approach in risk analysis known as "consequence-only decision making." In some cases the consequences of a scenario or situation are deemed to be so severe that a company or facility may decide to take additional steps to prevent or mitigate an event without determining the likelihood or expected frequency of the event. This is usually rare, but consequence estimation remains an important component in risk analysis.

### *2.4.6.1 Fire consequence modeling*

A wide variety of models and equations have been developed over time to model various types of fires and predict the consequences thereof. An introduction to some of these models is presented in Section 2.1.4. Explosion consequence modeling is discussed in detail in Section 2.2.3. This section discusses some of the impacts and consequences of fires.

#### Impact on personnel

The primary danger to a person within the line-of-sight to a flame is thermal radiation, which will most often result in burns to exposed skin as well as the ignition or melting of clothing. Burns are classified in degrees, of which there are three, based on the depth of the burn, the extent of blistering, and the type of damage sustained. A first-degree burn is the least severe with only superficial burns, similar to most sunburns. A third-degree burn is the most severe with burns more than 2 mm deep, which affect nerve endings and cause a loss of sensation (Crowl and Louvar, 2011).

Skin will begin to burn at 45 °C (113 °F) and becomes nearly instantaneous at 72 °C (162 °F). Time to injury can be predicted for bare, unprotected skin using tables available in the literature if the expected heat flux is known (CCPS, 2010).

Smoke is usually a secondary concern with fires that occur in a large, open, outdoor space. Smoke can be the primary concern in the case of indoor fires or in cases where individuals are within a large smoke plume resulting from an outdoor fire but largely unaffected by the thermal radiation from the fire (CCPS, 2010).

Smoke plumes can limit visibility to less than 1 m when present in high concentrations; however, it is the toxicity of the smoke that will likely present the greatest danger. The smoke is composed of carbon particles, unburnt fuel, and combustion products (CCPS, 2010), including CO, $CO_2$, $NO_X$ compounds, and $SO_X$ compounds. Most of these compounds are toxic and all are capable of displacing oxygen.

### Impact on structures

Building frames and support structures make heavy use of steel, aluminum, concrete, fiberglass reinforced plastic (FRP), and other materials, all of which can be damaged or affected by fires. The primary concern with metals and concrete is the loss of yield strength above certain temperatures. All structural elements in a building will be under some kind of load, whether it be compressive, tensile, or shear. Buildings are designed with safety factors and the load of a structural element should never exceed 60% of the expected yield strength. The temperature at which a structural material experiences a 50% reduction in its yield strength therefore tends to be especially significant. This temperature is estimated to be approximately 600 °C (1112 °F) for steel and 150 °C (302 °F) for aluminum, but multiple alloys of steel and other metals exist that will exhibit different behavior at elevated temperatures. The properties of the material actually used in the construction of a structure should be considered when estimating the consequences of a fire.

Based on this weakening with increasing temperature and a large amount of experimental data, the time to failure for various structural elements can be estimated if the heat flux or the type of fire is known. The exact time to failure will, however, depend on the degree of heating (heat flux), the extent of exposure or flame impingement, the geometry of the material (e.g., the thickness) and any mitigation efforts undertaken using cooling water or passive insulation.

Fluid-filled pipes and vessels will be less likely to fail due to the ability to use the liquid like a heat sink. Pipes and vessels with liquid flowing through them will be less likely to fail due to the ability of the liquid to draw off heat and take it to other areas.

The exact nature of the failure will depend on the nature of the heating and other factors. Tanks and pipes can burst or rupture due to increasing pressure from the boiling liquid within (causing a BLEVE) or through a combination a weakening of the vessel and an increase in the pressure of a confined gas (without the presence of liquid). Pipes and other structures can collapse or buckle under their own weight with support structures weakened by flame. An infinite number of scenarios for a fire exist (CCPS, 2010).

### Impact on Electrical Equipment

Electrical equipment tends to be much more susceptible to damage than process equipment and structures. Scheffey et al. (1990) indicate that operating electronic equipment can be expected to fail at temperatures as low as 50 °C. They predict permanent damage to nonoperating equipment at 150 °C and failure of PVC cable at 250 °C. Some combustion products from the fire, including HCl and other acids can damage electrical contacts, damage silicone rubber and other insulators, and in other ways damage electronic equipment (Society of Fire Protection Engineers, 2002).

#### Impact on the environment

The environment can be damaged by the fire itself through the destruction of the ecosystem and the release of toxic or corrosive combustion products to the air. However, environmental damage can also be sustained by failing to contain the run-off of effluent from the fire, which may be contaminated, and by using environmentally harmful fire suppression agents, including substances like halon (CCPS, 2010).

### 2.4.6.2 Probit analysis: dose–response modeling

Dose–response curves can and have been constructed for a variety of exposure-dependent hazards including exposure to heat, pressure, radiation, impact, sound, and toxic substances. However, these curves do not lend themselves to computational analysis and so various methods have been developed to convert dose–response curves into more usable analytical equations.

One of the most common of these, which lends itself to single-exposure response assessment by providing a straight-line equivalent to the dose–response curve, is the probit method. A "probit" is a probability unit. The probit method introduces the probit variable, *Y*, which is related to the probability, *P*, by Eqn (2.106) below:

$$P = \frac{1}{(2\pi)^{1/2}} \int_{-\infty}^{Y-5} \exp\left(-\frac{u^2}{2}\right) du \tag{2.95}$$

While this equation can be used to convert probit values into probabilities, it is common for conversion tables to be constructed (Crowl and Louvar, 2011).

The probit variable is calculated using the following equation:

$$Y = k_1 + k_2 \ln(V) \tag{2.96}$$

where $V$ is a causative variable and $k_1$ and $k_2$ are fitting parameters. The probit variable can be a single variable or a combination of variables, including time, depending on the hazard being assessed. Table 2.12 shows several common probit correlations (Crowl and Louvar, 2011).

## 2.4.7 QUALITATIVE AND SEMI-QUANTITATIVE APPROACHES

Qualitative analysis tends to be much faster but less rigorous than quantitative analysis. It does not attempt to arrive at a numerical value as with quantitative analysis and it does not expend as much effort into assessing uncertainties and errors. As a result it is far less expensive to conduct.

Semiquantitative approaches attempt to numerically estimate probability, consequence, and risk, but generally do so by making simplifying assumptions. As such, a semiquantitative approach does not have the rigor of a true quantitative analysis and often provides only order-of-magnitude estimates of risk. These methods provide a kind of half-step between fully qualitative and quantitative techniques.

**Table 2.12 Probit Correlations for Selected Exposures**

| Type of Injury or Damage | Causative Variable | Probit Parmeters $k_1$ | $k_2$ |
|---|---|---|---|
| **Fires** | | | |
| Burn deaths from flash fire | $t_e I_e^{(4/3)}/10^4$ | −14.9 | 2.56 |
| Burn deaths from pool burning | $tI^{(4/3)}/10^4$ | −14.9 | 2.56 |
| **Explosions** | | | |
| Deaths from lung hemorrhage | $p^o$ | −77.1 | 6.91 |
| Eardrum ruptures | $p^o$ | −15.6 | 1.93 |
| Deaths from impact | $J$ | −46.1 | 4.82 |
| Injuries from impact | $J$ | −39.1 | 4.45 |
| Injuries from flying fragments | $J$ | −27.1 | 4.26 |
| Structural damage | $p^o$ | −23.8 | 2.92 |
| Glass breakage | $p^o$ | −18.1 | 2.79 |
| **Toxic release** | | | |
| Ammonia deaths | $\Sigma C^{2.0}T$ | −35.90 | 1.85 |
| Carbon monoxide deaths | $\Sigma C^{1.0}T$ | −37.98 | 3.70 |
| Chlorine deaths | $\Sigma C^{2.0}T$ | −8.29 | 0.92 |
| Ethylene oxide deaths | $\Sigma C^{1.0}T$ | −6.19 | 1.00 |
| Hydrogen chloride deaths | $\Sigma C^{1.0}T$ | −16.85 | 2.00 |
| Nitrogen dioxide deaths | $\Sigma C^{2.0}T$ | −13.79 | 1.40 |
| Phosgene deaths | $\Sigma C^{1.0}T$ | −19.27 | 3.69 |
| Propylene oxide deaths | $\Sigma C^{2.0}T$ | −7.42 | 0.51 |
| Sulfur dioxide deaths | $\Sigma C^{1.0}T$ | −15.67 | 1.00 |
| Toluene deaths | $\Sigma C^{2.5}T$ | −6.79 | 0.41 |

$t_e$ = *effective time duration (s).*
$I_e$ = *effective radiation intensity (W/m²).*
t = *time duration of pool burning.*
I = *radiation intensity from pool burning (W/m²).*
$p^o$ = *peak overpressure (N/m²).*
J = *impulse (Ns/m²).*
C = *concentration (ppm).*
T = *time interval (min).*
*Reproduced from Crowl and Louvar (2011)*

In many cases, a qualitative or semiquantitative analysis may be undertaken as a first screening for hazards and loss scenarios and a full quantitative analysis may then be undertaken to further analyze specific high-risk or high-consequence events. To implement any one of these techniques, a team of experts should be formed. Such a team has to represent various disciplines, such as production, mechanical, safety, and technical. After selecting the members of the team, they are provided with basic information on hazards of materials, process technologies, procedures, previous hazards reviews and other useful information about the procedures in the facility.

### 2.4.7.1 Layer of protection analysis

Layer of protection analysis (LOPA) is a semiquantitative method for assessing risk using simplified calculations to estimate the frequencies and consequences of failures. The method starts with an initiating event or failure and then considers the effects of independent protection layers (IPLs) on the likelihood or consequence of the initiating event. IPLs can come in a variety of forms including basic process control systems, safety instrumented systems, alarms, passive mitigation, plant emergency response, and community emergency response. In an LOPA, IPLs are generally arranged in an order that progresses from inherent and process control layers to mitigation, containment, and emergency response layers as this is the order in which the layers would be expected to function in the event of an incident. Figure 2.21 below shows a likely progression of IPLs in an LOPA.

Similar to an event tree, there is no limit on the number of IPLs that can be used in a LOPA and LOPA can be applied to large and small systems, ranging from individual pieces of equipment to entire processes and plants. The IPLs chosen/identified, however, will likely be different based on the scale of the system. An LOPA on a tank will likely involve operator actions, alarms, and valves. An LOPA on a plant-wide or process-wide event is more likely to involve process controls systems, complex safety instrumented systems, and on-site and off-site emergency response.

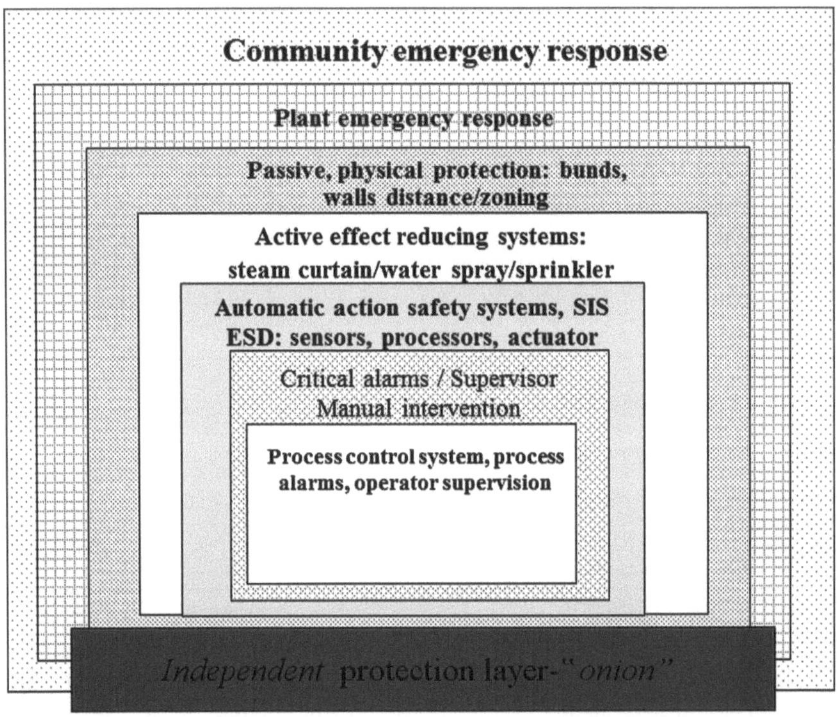


**FIGURE 2.21**

Graphic representation of layers of protection.

IPLs can reduce the likelihood of an event, its consequences, or both, depending on the layer. No IPL can be perfect and will always have some probability of failure. The probability that an IPL fails can be estimated from historical data or expert opinion. Risk analysis teams must also be careful to ensure that the IPLs considered are truly independent and not subject to common cause failures.

#### *2.4.7.2 Risk matrix*

As discussed previously, risk is the product of the consequence of an adverse event and its frequency. It is possible to create a graph of consequence vs frequency and plot lines on that graph that all correspond to the same level of risk. These lines or curves are called "iso-risk curves" because every point on the line has the same level of risk. The product of the consequence value and the frequency value is the same for every point on that line. An example of this is shown in Figure 2.22.

When plotted using rectilinear axes, the iso-risk curve is a curved line that cannot touch the vertical asymptote at zero frequency or the horizontal asymptote at zero consequence. When plotted using logarithmic axes, however, the iso-risk curves become straight lines.

This logarithmic risk plot can then be broken up into a matrix with various categories, severity levels. Frequency categories can include "frequent," "possible," "rare," "remote," "improbable," and "impossible." Severity categories can include "incidental," "minor," "serious," "major," and "catastrophic." The plot can then be divided into zones of "acceptable," "conditionally acceptable," and "unacceptable risk."

The result is called a "risk matrix." Risk will be "low" in the bottom right corner of the risk matrix, where probability and consequence are low, and "very high" in the upper left corner of the matrix, where probability and consequence are high. Figure 2.23 shows a typical 4 × 4 risk matrix. A risk matrix does not need to be 4 × 4. They can be of different frequency–severity combinations; 5 × 5,

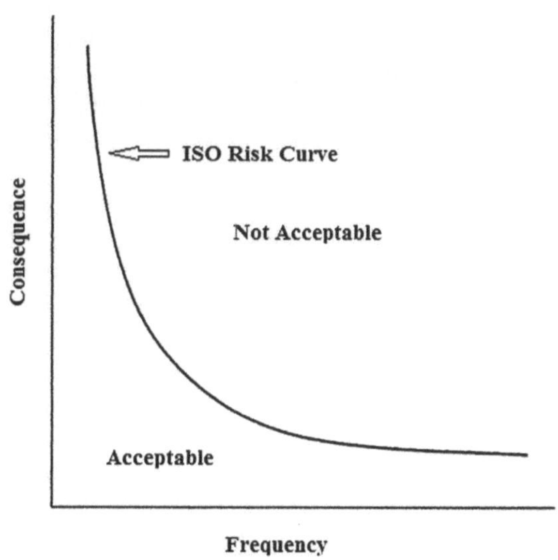


**FIGURE 2.22**

Iso-risk curves on rectilinear axes.

| | | Likelihood | | | |
|---|---|---|---|---|---|
| | | Frequent | Possible | Rare | Remote |
| Severity | Major | Very High | Very High | High | Moderate |
| | Serious | Very High | High | Moderate | Low |
| | Minor | High | Moderate | Low | Low |
| | Incidental | Moderate | Low | Low | Low |

**FIGURE 2.23**

A typical completed 4 × 4 risk matrix.

4 × 6, and 6 × 7 combinations as well as others are in use and can be effective risk management tools. The overall form of the risk matrix, the number of categories and what risk levels are deemed acceptable and unacceptable is determined by the management of the company or organization that uses the matrix. Because of the flexibility in the construction of the risk matrix, the method scales easily to the depth of analysis, the level of resolution desired and the range desired for consequences and event frequencies.

The risk matrix can be used qualitatively by having the risk assessment teams label events as "frequent" or "rare" with "major" or "incidental" consequences. It can also be used more quantitatively by precisely plotting specific frequency or consequence values on the matrix based on the underlying values and iso-risk curves used to construct the matrix.

A risk matrix can be applied to a problem of any scale as long as the matrix used for the analysis has been constructed appropriately. When using general levels like "frequent" or "rare" with "major" or "incidental" for a qualitative analysis, little to no scaling is required. When using the risk matrix in a more quantitative way, using actual frequencies and damages to place events on the matrix, the risk matrix can be applied to incidents of various scales by changing the definition of acceptable and unacceptable risk. When applying a risk matrix to a small, simple system, and acceptable risk level might be defined as $100/year or $200/year. For facility-wide scenarios, an acceptable risk level might be $1 million/year.

To use the risk matrix, a scenario or event is placed on the matrix according to its frequency of occurrence and severity of consequence. If the scenario falls into the "acceptable" zone, no action is required. If the scenario falls into the "conditionally acceptable" zone, action may or may not be taken depending of the nature of the hazard and the costs of risk reduction. If a scenario falls into the "unacceptable" zone, corrective action must be taken to reduce the risk of this event or scenario. Risk reduction measures can reduce the consequences of the event, the frequency of it, or both, as shown in Figure 2.24. Risk reduction measures are generally assumed to provide 1 or 2 orders of magnitude of risk reduction. Risk reduction measures and barriers should be added until the scenario or event falls in or near the acceptable zone.

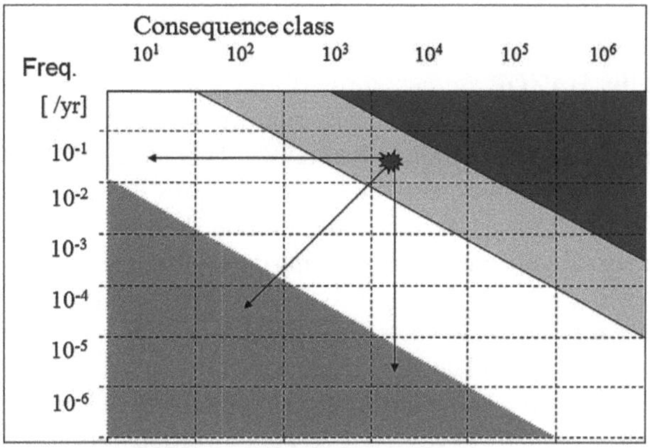


**FIGURE 2.24**

Effects of risk reduction measures on a scenario placement in the risk matrix.

### *2.4.7.3 HAZOP*

The hazards and operability (HAZOP) study is a highly structured, inductive, procedure for identifying hazards in a facility. The HAZOP method is regarded as effective and is therefore widely accepted and implemented.

The HAZOP study generates process deviations by combining guide words with process parameters. Some guideword and parameter combinations are impossible or invalid and can therefore be ignored. Valid deviations are further analyzed and considered.

A complete HAZOP study for a process can take a very long time. to keep team members engaged and provide for effective brainstorming, it is best to break the study into sessions that last only a few hours. Meetings should only be held two or three times a week. As a result however, a full HAZOP study may require several weeks or months to complete. The HAZOP will follow a set of steps:

1. Collect process information including flow sheets.
2. Divide the process into nodes.
3. Select study team.
4. Determine the design intent of a node.
5. Select a process parameter.
6. Apply a guide word to the parameter to generate a deviation.
7. If the deviation is valid, determine possible causes and barriers.
8. Evaluate consequences and likelihood of the deviation.
9. Recommend actions be taken as appropriate if the expected consequences are too severe or if the current barriers are thought to be inadequate.
10. Record the results.
11. Repeat steps 6 through 10 for all guidewords with the selected parameter.
12. Repeat steps 5 through 11 with all relevant parameters in the node.
13. Repeat steps 4 through 12 for all nodes in the process.

This technique analyzes each element of a system for all the ways in which key operation parameters can deviate from the intended design operation conditions that could create hazards and safety problems. Typically, HAZOP studies are performed by analyzing the piping and instrument diagram (P&ID) of the facility. Usual parameters that are selected to perform this study are flow, temperature and pressure. Then a list of deviation keywords is used, for example, more of, less than, part of. Each of these parameters will create different situations that could lead to diverse outcomes and scenarios that could affect the integrity of the facility. Furthermore, this technique allows the identification of the safeguards and protection systems currently used and the ones that are needed to lower the risk to a tolerable level.

Documentation of the results is the most important part of the HAZOP study. Detailed documentation of the study will be needed later for review by auditors and to ensure that action items from the study are taken up and resolved. There are multiple methods for documenting the study results and the methods used vary from company to company and organization to organization. A common means of documentation is a form that lists the study node, parameter, guideword, causes, consequences, and recommended actions. An example of such a form is presented in Figure 2.25.

Similar to other methods discussed so far, there is no limit to the number of nodes used in a HAZOP or the size of the facility or system to which it can be applied. However, the difficulty of the analysis and the time required for it will scale roughly linearly with the number of nodes in the system. The more nodes, the longer the analysis will take, and a HAZOP on an entire plant can require a team to meet weekly or twice weekly for months to complete.

#### *2.4.7.4 What-if analysis*

This is a relatively informal method of analysis where-in the words "What-If" are placed in front of a number of possible events, failures or process upsets to form questions. The risk analysis team then determines the consequences and potential impacts on the process that the failure or upset might produce. Corrective and preventative actions are taken where necessary (Crowl and Louvar, 2011).

Some examples of questions include:

- What-if the flow of coolant to a heat exchanger is stopped?
- What-if the tubes in the heat exchanger leak?
- What-if a fire starts near the tank?
- What-if the concentration of the feed to the reactor is wrong?
- What-if the power fails?
- What-if water is accidentally introduced to the system?

Depending on the process, there are a potentially infinite number of questions that can be asked. However, the informal and unstructured nature of the what-if process requires that the analysis team know and ask the right questions. If a question is not asked, a hazard can be missed.

The informal nature of this method and its lack of structure make it likely ill-suited for analyzing large or complex systems, and this method is much more likely to be used either in the early stages of a design or for smaller systems.

The main purpose of this technique is to elaborate a list of what-if questions regarding the hazards and safety of the operations. As each question is answered, hazard scenarios are identified, and based on the team brainstorming, recommendations are developed specifying actions to be taken and additional studies to be performed to address risks.

| Hazards and Operability Study | | | | | | | |
|---|---|---|---|---|---|---|---|
| Project Name: | | | Date: | | Page: | | Completed: |
| Process: | | | | | | | No Action: |
| Section: | | | | | Reference Drawing: | | Reply Date: |
| Item | Study Node | Process Parameter | Guide word | Possible causes | Possible Consequences | Action Required | Assigned To: |
| | | | | | | | |

**FIGURE 2.25**

Example form for recording study results and recommendations.

#### 2.4.7.5 Checklist

The checklist or process hazard checklist is a method generally used in the early stages of design and planning for the purposes of hazard identification and screening. Checklists can be useful for reminding operators or an analysis team of common or significant hazards in the process. The checklist design will vary depending on the purpose of the list and the scope of the analysis, which could be an entire process unit or a specific piece of equipment. Checklists suffer some limitations; however, as they can encourage a "check the box" mentality and will likely miss any hazards not on the list (Crowl and Louvar, 2011).

#### 2.4.7.6 What-if/checklist

As the name suggests, this method combines elements of the what-if method and the checklist method to overcome some of the limitations of each. The checklist component seeks to ensure that some known significant hazards are thought of and addressed. The what-if component of the analysis encourages the analysis team to go beyond the checklist and look for additional hazards to the process.

#### 2.4.7.7 Dow fire and explosion index

The Dow Fire and Explosion Index (F&EI) attempts to rate the hazards associated with the storage and handling of flammable and explosive substances in a purely systematic way that is as free as possible from independent judgment. The F&EI produces a score by which the relative size of the risk that a facility poses due to flammable hazards can be assessed. This is a relative measure that allows risks and processes to be compared. It is not an absolute or quantitative measure of risk.

The F&EI uses a material factor that is specific to the flammable material or mixture that is present in the unit or process. The material factor is then modified with multipliers, called penalty factors and credit factors, based on process characteristics, operating conditions, and safety precautions that have been put into place in the process. More detailed guidance on the F&EI can be obtained from Dow's Fire and Explosion Index Hazard Classification Guide, 7th edition, or from Crowl and Louvar (2011).

As might be inferred from the structure of the Dow F&EI analysis, it is designed to be applied to processes, not to individual pieces of equipment and not to large integrated facilities with multiple process trains. Examples of documenting the results of DFE&I are shown in Figures 2.26 and 2.27.

## 2.5 PROCESS SAFETY CHALLENGES AND LOOKING AT THE FUTURE

### 2.5.1 INTRODUCTION

Over the last few decades, process safety has become paramount in the chemical industries. To increase the safety of the existing operations, plant personnel have identified and reduced potential hazards by installing adequate measures. A majority of companies have used tools like HAZOP, LOPA, and FTA to mitigate the dangers associated with process operations. Current incidents seem almost always the result of a combination of organizational issues, lack of competency, and technical failures of equipment (Knegtering and Pasman, 2009). Increase in complexity and sophistication of the process installations and control systems have compromised the safety of the operations. The implementation of an effective safety program with the potential to overcome the current challenges is desired.

| FIRE & EXPLOSION INDEX SUMMARY SHEET | | | |
|---|---|---|---|
| Location | Date | Process Unit | Bldg |
| Prepared by | Approved by | Reviewed by | Reviewed by |
| Materials in Unit | | | |
| State of Operation<br>_ Design _ Start Up _ Normal Operation _ Shutdown | | Basic Materials for Material Factor | |
| Material Factor (See Table 1 or Appendices A or B) Note Requirements when unit temperature over 140 F (60 C) | | | |
| **1. General Process Hazards** | | **Penalty Factor Range** | **Penalty Factor Used** (for no penalty factor use 0.00) |
| Base Factor | | 1.00 | 1.00 |
| A. Exothermic Chemical Reactions | | 0.30 to 1.25 | |
| B. Endothermic Chemical Reactions | | 0.20 to 0.40 | |
| C. Material Handling and Transfer | | 0.25 to 1.05 | |
| D. Enclosed or Indoor Process Units | | 0.25 to 0.90 | |
| E. Access | | 0.20 to 0.35 | |
| F. Drainage and Spill Control | | 0.25 to 0.50 | |
| **General Process Hazards Factor (F1 = Sum of 1A through 1F + Base Factor)** | | | |
| **2. Special Process Hazards** | | **Penalty Factor Range** | **Penalty Factor Used** (for no penalty factor use 0.00) |
| Base Factor | | 1.00 | 1.00 |
| A. Toxic Materials | | 0.20 to 0.80 | |
| B. Sub-Atmospheric Pressure (<500 mm Hg) | | 0.50 | |
| C. Operation In or Near Flammable Range (_ Inerted) (_ Not Inerted) | | | |
| 1. Tank Farms Storage Flammable Liquids | | 0.50 | |
| 2. Process Upset or Purge Failure | | 0.30 | |
| 3. Always in Flammable Range | | 0.80 | |
| D. Dust Explosion (See Table 3) | | 0.25 to 2.00 | |
| E. Pressure (See Figure 2) | Operating Pressure:<br>Relief Setting: | | |
| F. Low Temperature | | 0.20 to 0.30 | |
| G. Quantity of Flammable / Unstable Material | Quantity:<br>Heat of Combustion: | | |
| 1. Liquids or Gases in Process (See Figure 3) | | | |
| 2. Liquids or Gases in Storage (See Figure 4) | | | |
| 3. Combustible Solids in Storage, Dust in Process (See Figure 5) | | | |
| H. Corrosion and Erosion | | 0.10 to 0.75 | |
| I. Leakage - Joints and Packing | | 0.10 to 1.50 | |
| J. Use of Fired Equipment (See Figure 6) | | | |
| K. Hot Oil Heat Exchange System (See Table 5) | | 0.15 to 1.15 | |
| L. Rotating Equipment | | 0.50 | |
| **Special Process Hazards Factor (F2 = Sum of 2A through 2L + Base Factor)** | | | |
| **Process Unit Hazards Factor (F1 x F2) = F3** | | | |
| **Fire and Explosion Index (F3 x Material Factor = Fire & Explosion Index)** | | | |

**FIGURE 2.26**

Example form for documenting the results of the dow fire and explosion index, page 1.

| LOSS CONTROL CREDIT FACTORS | | |
|---|---|---|
| **C1. Process Control Credit Feature** | Credit Factor Range | **Credit Factor Used** (for no credit factor use 1.00) |
| A. Emergency Power | 0.98 | |
| B. Cooling | 0.97 to 0.99 | |
| C. Explosion Control | 0.84 to 0.98 | |
| D. Emergency Shutdown | 0.96 to 0.99 | |
| E. Computer Control | 0.93 to 0.99 | |
| F. Inert Gas | 0.94 to 0.96 | |
| G. Operating Instructions / Procedures | 0.91 to 0.99 | |
| H. Reactive Chemical Review | 0.91 to 0.98 | |
| I. Other Process Hazard Analysis | 0.91 to 0.98 | |
| **Process Controls Credit Factor (C1 = Product of A through I)** | | |
| **C2. Material Isolation Credit Feature** | Credit Factor Range | **Credit Factor Used** (for no credit factor use 1.00) |
| A. Remote Control Valves | 0.96 to 0.98 | |
| B. Dump / Blowdown | 0.96 to 0.98 | |
| C. Drainage | 0.91 to 0.97 | |
| D. Interlock | 0.98 | |
| **Material Isolation Credit Factor (C2 = Product of A through D)** | | |
| **C3. Fire Protection Credit Feature** | Credit Factor Range | **Credit Factor Used** (for no credit factor use 1.00) |
| A. Leak Detection (Alarm and Shutdown) | 0.94 to 0.98 | |
| B. Fireproofing for Structural Steel | 0.95 to 0.98 | |
| C. Fire Water Supply | 0.94 to 0.97 | |
| D. Special Systems | 0.91 | |
| E. Sprinkler Systems | 0.74 to 0.97 | |
| F. Water Curtains | 0.97 to 0.98 | |
| G. Foam | 0.92 to 0.97 | |
| H. Portable Fire Extinguishers / Fire Monitors | 0.93 to 0.98 | |
| I. Cable Fire Protection (Instrumentation and Electrical Cables) | 0.94 to 0.98 | |
| **Fire Protection Credit Factor (C3 = Product of A through I)** | | |
| **Loss Control Credit Factor (C1 x C2 x C3)** | | |

| PROCESS UNIT RISK ANALYSIS SUMMARY | | |
|---|---|---|
| **Fire & Explosion Index** | | |
| **Radius of Exposure (ft)** | | |
| **Area of Exposure (sq. ft.)** | | |
| **Value of Area of Exposure** | $ value per sq. ft. | $ value total<br>$ - |
| **Damage Factor** | | |
| **Base Maximum Probable Property Damage** | | $ - |
| **Loss Control Credit Factor** | | |
| **Actual Maximum Probable Property Damage** | | - |
| **Maximum Probable Days Outage (Figure 9)** | | |
| **Business Interruption** | $ value per day | $ value total<br>$ - |

**FIGURE 2.27**

Example form for documenting the results of the dow fire and explosion index, page 2.

## 2.5.2 CURRENT REGULATIONS

Before 1936, occupational health was not given adequate importance by the government agencies. Many regulations were created in the next 30 years to enhance occupational safety. In 1970, there was a major breakthrough in process and occupational safety. The U.S. Congress enacted OSHA, which brought about revolutionary changes in industrial practices. To fulfill the government responsibilities mentioned in OSHA, the Occupational Safety and Health Administration was established. In the same year, the U.S. Environmental Protection Agency (EPA) was created to enforce regulations concerning human health and environment.

In 1985, the American Institute of Chemical Engineers established the Center for Chemical Process Safety (CCPS) to integrate process safety with chemical processing (Berger, 2009). Following this, the administrator of the EPA promulgated the EPCRA in 1987. The act instructed the officials to provide emergency response personnel with MSDS information. The EPA also conducted studies to determine the toxic substances in the communities under the Toxic Release Inventory Survey (Atlas, 2007). In 1990, the Clean Air Act Amendments were promulgated that addressed wide ranging environmental and safety issues. As mandated by the Clean Air Act Amendments, OSHA promulgated the "Process Safety Management" (PSM) standard in 1992. The 14 elements of the PSM standard are the basis of process safety programs in the industry for the management of hazardous chemicals. Similar to OSHA's PSM standard, EPA promulgated the "Risk Management Program" (RM Program) regulations in 1996. The RM Program was designed to safeguard the environment and the public from exposure to hazardous substances. Also, the Clean Air Act Amendments required the establishment of the U.S. Chemical Safety and Hazard Investigation Board (CSB) for investigation of chemical incidents (Willey et al., 2005).

Similar to the United States, a number of regulations were passed in the United Kingdom. The U.K. government took stringent measures after the Flixborough incident of 1974. Following the incident, the Health and Safety at Work Act, was passed to monitor the health and safety at workplace. As mandated by the Act, the Health and Safety Executive (HSE) was established in 1974. In 1978, the HSE published the First Canvey report, which highlighted the likely hazards of the ammonium nitrate installations in Canvey Island and suggested means to reduce risks (Griffiths, 1984). Results from hazards models, which included consequence analysis and risk assessments, were included in the report. The hazard models were improved and newer injury relationships were introduced in the Second Canvey report (HSE, 1981). The Control of Major Accident Hazards (COMAH) was developed by HSE to assist the industrial operators to reduce risks. to deal with the onshore hazards, Notification of Installations Handling Hazardous Substances (NIHHS) regulations were developed (The National Archives) in 1982. The regulation prescribed threshold values for the quantities of hazardous substances that can be stored at an onshore site. Any industrial site exceeding this limit had to notify the HSE. For onshore non-nuclear major hazards the Control of Industrial Major Accident Hazards (CIMAH) regulations were established in 1984. Sites were expected to prepare comprehensive safety reports, which included the probability of anticipated major incidents along with the expected effects on the local people. to increase the awareness of the external stakeholders and to inform them about the HSE decision-making, "Reducing Risks, Protecting People" was published (HSE, 2001).

### 2.5.3 RECENT MAJOR INCIDENTS IN INDUSTRY

Despite the stringency in process safety, the industry has been a victim of many devastating disasters. The following section provides a brief overview of the major incidents in process industry in the past 30 years. This is followed by a concise illustration of the common characteristics of most incidents.

#### 2.5.3.1 Flixborough disaster (Venart, 2004; Crowl and Louvar, 2011)

Nypro Ltd, Flixborough, UK, was exposed to fire and explosion on June 1, 1974. The ignition of a vapor cloud containing flammable cyclohexane was responsible for the incident. The incident caused 28 fatalities and injured 36 people. The structures in the vicinity of the incident were severely affected by the explosion and flying debris. About 1821 houses and 167 shops and factories were affected by the incident.

The Flixborough works of Nypro Limited was constructed to produce caprolactam. About 70,000 tons of caprolactam was produced on a yearly basis. The caprolactam was a basic raw material used in the synthesis of nylon. The production of caprolactam involved the use of cyclohexane at high pressures (7.9 atm) and temperatures (155 °C). Oxidation of cyclohexane to cyclohexanone and cyclohexanol was carried out in the presence of air and catalyst in a chain of six reactors.

A vertical crack in the stainless steel structure of reactor 5 had resulted in the disengagement of that reactor. A decision was made to continue the production process by connecting reactor 4 to reactor 6 using a dog-leg–shaped pipe section of 20-inch diameter. Flexible bellows were used, as the diameter of the feed line was 28 inches. The dog-leg pipe section connecting the reactors was structurally weak and was not adept to handle high pressures. On June 1, 1974, the bypass pipe section ruptured, releasing 30 tons of cyclohexane. The released liquid vaporized forming a flammable vapor cloud, which ignited after 45 s. The resulting explosion is considered to be the largest peacetime explosion in the United Kingdom.

#### 2.5.3.2 Seveso disaster (Crowl and Louvar, 2011)

Higher temperatures in a trichlorophenol reactor due to loss of control lead to large concentrations of 2,3,7,8-tetrachlorodibenzoparadioxin (TCDD), a potent toxin. The chemical was vaporized and released through the relief system into the atmosphere affecting the people in the town of Seveso, Italy, on July 10, 1976. Due to improper communication between the Icmesa Chemical Company (the plant owners) and the local authorities, the local people were not evacuated until several days later.

Icmesa Chemical Company produced hexachlorophene, a chemical used as a bactericide. The plant was situated in Seveso, which is 15 miles from Milan, Italy. TCDD was produced as a byproduct during the synthesis of one of the intermediate, trichlorophenol. TCDD is a dangerous toxin, which is very difficult to separate owing to its insolubility in water.

On July 10, 1976, one of the batches producing trichlorophenol experienced a runaway leading to high reactor temperatures. The temperature went above 200 °C, leading to a large production of TCDD. The internal pressure of the system increased to a large extent, leading to the failure of the safety devices. A mixture containing TCDD and other reactive chemicals was released into the atmosphere and formed a white cloud. The chemical mixed with the rainwater and reached the soil, contaminating it. The chemical in the air affected people, leading to 250 cases of chloracne. Apart from this, more than 2000 people had to have their blood tested.

#### 2.5.3.3 Bhopal disaster (Willey et al., 2005; Crowl and Louvar, 2011)

On December 3, 1984, accidental release of deadly methyl-isocyanate (MIC) into the atmosphere caused the death of 2000 civilians in the city of Bhopal, India. MIC was stored in a plant owned by Union Carbide. Exposure of MIC to water triggered an uncontrolled exothermic reaction, which led to pressure build up in the tank. The safety devices failed releasing the poisonous gas into the atmosphere.

Union Carbide's plant in Bhopal (Madhya Pradesh, India) produced insectides. MIC was used as an intermediate in the process. The company was aware of the reactive, toxic and flammable nature of the gas. Also, MIC reacts violently with water producing large amounts of heat. The absence of adequate cooling can lead to boiling of the liquid. Hence, proper refrigeration of the MIC storage tanks is a must.

The plant had many layers of safety systems. For example, the pressure relief system of the storage tanks was connected to the scrubber and flare systems. However, these systems were not operational during the time of the incident. On the night of December 2, 1984, one of the MIC storage tank was contaminated with water. The resulting exothermic reaction vaporized the MIC, which passed through the relief system. About 25 tons of the MIC vapor escaped into the atmosphere and spread to the adjacent town. Heavy vapors settled to the ground, affecting the people severely. Though none of the plant workers were affected, more than 20,000 civilians were injured.

### 2.5.3.4 Naples disaster

The petroleum depot near Naples, Italy, exploded in December 1985. The explosion produced a blast wave, which killed four people and injured 170. The severity of the explosions was such that it destroyed 12 large buildings, 448 small industrial units, and 220 small houses affecting property within a 5-km radius. Accidental spill of gasoline and the resulting vapor cloud was the main cause of the incident.

On December 21, 1985, there was accidental gasoline spill from one of the tanks in a petroleum depot near Naples. Gasoline overflowed through the roof of tank 17 for 1.5 h, releasing 700 tons of the liquid. The spilled liquid produced a vapor cloud of about 4500 $m^3$, which was ignited by an unknown source. Instantly, 20 tanks were set ablaze destroying terminal buildings and nearby industrial and residential structures. About 2000 residents had to be evacuated from the nearby site. The fire lasted for a week, demolishing all the equipment in the vicinity. Fixed roofs, cooling devices, and other fire protection devices were blown away by the blast (Mannan, 2011).

### 2.5.3.5 Chernobyl disaster (Saenko et al., 2011)

The Chernobyl Nuclear Power Plant was hit by one of the biggest nuclear disasters on April 26, 1986. The breakdown of the cooling system led to the overheating of the nuclear core followed by a major explosion. Exposure to intense radiations killed five firemen and two staff members and affected many others in the aftermath.

Located in the north of Ukraine, the Chernobyl Nuclear Power plant was one of the largest nuclear plants. The main reactor was heated to high temperatures by the nuclear fission reaction involving uranium. Hence, water was passed through the reactor core to cool the system to lower temperatures. The water used for heat removal was boiled to produce steam, which was passed through turbines. Graphite rods controlled the fission reaction and the power output by absorbing the fission products (neutrons).

On April 26, 1986, there was sudden unexpected power surge due to an operator error. Due to this, there was excessive heat transfer to the coolant water producing a shock wave. This shock wave was responsible for disruption of the cooling water pipes and instant vaporization of water to steam. Due to loss of cooling, the nuclear fuel in the core vaporized, leading to a massive explosion. The reactor and the surrounding building were completely destroyed. Also, reactor debris and radioactive materials were thrown in the vicinity. Twenty-eight the 237 firemen died in the next 4 months due to radiation exposure. The firemen who were not killed by the exposure suffered from acute radiation sickness and were hospitalized. The effect of the disaster was so devastating that 19 more deaths were seen until the year 2004.

#### 2.5.3.6 Piper Alpha disaster (Shallcross, 2013)

The Piper Alpha platform in the North Sea was set on fire due to the accidental passage of gas through a condensate injection pump, which was isolated from its pressure safety valve. The explosion, which occurred on July 6, 1988, resulted in the death of 165 people on the platform. This is one of the worst offshore oil platform disasters to date.

The 31-km$^2$ Piper oil field in the North Sea was discovered in 1973. By 1976, the Piper Alpha platform was built and produced 40,000 m$^3$ of oil per day. The Tartan and Claymore oil platforms were built in the vicinity to exploit the surrounding regions.

The process involved extraction of oil accompanied by water and gas. The oil was separated from the mixture using separators and transported to the export line. The water was treated to remove the oil traces and discarded into the sea. The gas is passed through a series of condensation and compression stages to extract the condensate, which was added to the oil in the export line. Piper Alpha had two condensate injection pumps (A and B). During the time of the incident, the condensate injection pump was under preventive maintenance and the pressure safety valve attached to the pump needed inspection. A work permit was issued and the valve was removed for recertification. As it was past the normal working hours, the supervisor failed to replace the disengaged valve and returned the work permit to the control room. However, the absence of the valve on the pump was not properly communicated to the control room officials. Unfortunately, the second condensate pump B failed the same evening. According to the officials, the preventive maintenance work on Pump A had not started and hence they decided to connect the pump to the system. A large amount of gas leaked and ignited, leading to fire and explosion. The walls used for construction did not have explosion proofing and collapsed, damaging the platform severely. Two operators attending to the gas leak were killed instantly. Only 61 of the 226 men on board survived.

#### 2.5.3.7 St. Herblain disaster

On the morning of October 7, 1991, there was an unconfined vapor cloud explosion in Nantes, France, near a petroleum depot. The faulty nature of a filling valve resulted in release of lead-free gasoline, which initiated the event. The blast claimed the life of one employee while severely injuring five others. The impact destroyed the foundations of the structures in a 100-m radius.

The tank farm, located to the west of greater Nantes area had a series of 11 storage tanks with a total capacity of 80,000 m$^3$. The tank farm was located along the Loire River and was supplied by a quay located 125 m away. The facility opened at 4 a.m. in the morning and had two employees to handle the loading operations.

On the morning of the 7th, normal loading operations resumed. The remote controlled opening of a valve located at the base of one of the tanks was improper and released lead-free gasoline. This produced about 23,000 m$^3$ of a large vapor cloud, which started spreading to the surrounding area. In 15 min, the cloud had increased its depth to 1.5 m and extended 50 m to reach the road. The humidity and the nature of the wind assisted the stabilization of the vapor cloud, leading to an ignition and explosion in the next 5 min. The explosion was equivalent to about 2.7 tons of TNT and destroyed two tanks, 15 tanker trucks, a wash station, and four cars (Mannan, 2011; French Ministry of Environment, 2006).

#### 2.5.3.8 Warffum disaster

A small-scale explosion rocked the Dutch petroleum company located in Warffum, Netherlands, in May 2005. The welding activities ignited a small residue of natural gas condensate present on a tank. Three people were affected by the blasts (Knegtering and Pasman, 2009).

Since February 2005, maintenance activities had been carried out at the Dutch petroleum company (Shell). The vapor recovery pipe of a tank was being renovated. The entire project was outsourced to external contractors who took care of the planning and the hardware installations. The tank was connected to a valve to limit the exposure of its contents to air.

On May 31, 2005, three GTI employees were working on the pipe when the tank exploded leading to a fatal fire. The presence of natural gas condensate on the tank's welding triggered the explosion. Two GTI contractors were killed while the third one was seriously injured. The fire fighter received first-aid for minor burns developed during the subsequent rescue operation.

#### 2.5.3.9 Texas city disaster

On March 23, 2005, a refinery in Texas City, USA, owned by BP and believed to be the third largest refinery in the United States, sustained one of the worst disasters in history (CSB, 2007). The blast was caused by discharge of hydrocarbon liquid from a refinery unit, which lead to the formation of flammable vapor cloud. Fifteen people were killed and 180 were injured in this tragic incident. The incident damaged property in a 1.2-km radius. The estimated financial loss was about $1.5 billion.

The refinery's isomerization (ISOM) unit was on a maintenance outage. It was being restarted on March 23, 2005. Liquid hydrocarbon was being charged to the tower. The charging time on that day was about 3 h, which was contradictory to the general startup procedures. The liquid overflowed the 52-m-tall tower and entered the overhead pipe. The alarms and control system were faulty and did not warn the operator. The liquid in the overhead pipe ran down the side of the tower toward the pressure relief valve located 45 m below. The three pressure relief valves released the flammable hydrocarbon liquid into the blow-down drum. The entire blow-down system was improperly designed and was not connected to any flare systems. The vent stack, 34 m in height, was open and released the flammable liquids. On reaching the ground, the liquid evaporated and formed a flammable vapor cloud. It is believed that the backfire of a diesel pickup truck located 7.6 m from the drum ignited the vapor cloud. Fifteen contract employees who worked in the vicinity of the drum bore the brunt of the explosion and were killed.

#### 2.5.3.10 Buncefield disaster

The Buncefield oil storage depot in Hemel Hempstead, Hertfordshire, UK, saw an explosion in December 2005. Like many other previous incidents, the disaster was caused by the unexpected discharge of liquid gasoline. The explosion was so huge that it caused massive fires, which injured more than 40 people. Apart from plant personnel, the explosion damaged property in a 10-km radius. This was one of the largest peace time explosions in Europe for 50–60 years.

Buncefield oil storage facility is a major transfer point for petroleum derivatives in the UK. On December 10, 2005, about 8400 $m^3$ of unleaded gasoline was brought to the site. It was decided that a part of the oil would be transferred to one of the tank (912). According to the attached pressure gauge, the tank was two-thirds full and had about 4971 $m^3$ of empty volume (Mannan, 2011).

On the morning of the 11th, tank 912 was filled with gasoline. From 3 a.m. in the morning, the tank gauge remained static showing that the tank was two-thirds full. As the gasoline filled the tank, its temperature increased. At about 5:35 a.m., tank 912 began overflowing. The installed CCTV cameras showed images of mist engulfing the facility site. After the incident, when the people in the vicinity were interrogated, they confirmed this fact. It was further revealed that a distinctive odor of gasoline swamped the surroundings in the early hours of December 10. Until 6 a.m. in the morning, about 300 metric tons of gasoline was released from the tank. The flammable vapor cloud ignited at 6 a.m. and there was a major explosion. This explosion was succeeded by other explosions, which led to fires causing damage to people and property. Figure 2.28(a) shows a picture of the first day of the fire.

### 2.5.3.11 Fukushima disaster (Steinhauser et al., 2014)

The tsunami resulting from a major earthquake was responsible for flooding and disrupting the cooling system of Fukushima Daiichi Nuclear Power Plant on March 11, 2011. The consequent meltdown of the fuel rods was responsible for a major explosion.

Tokyo Electric Power Company constructed Fukushima Daiichi Nuclear Power Plant between the years 1971–1979 in Japan. The plant consisted of six boiling water reactors and was protected by a 10-m sea wall. Reactors 4, 5, and 6 were nonoperational during the incident. An earthquake of magnitude 9 on the Richter scale hit the eastern coast of Japan on March 11, 2011. The seismic signals generated by the earthquake triggered an automatic shutdown of the first three reactors in the plant. The earthquake created a tsunami which breeched the sea walls of the nuclear power plant. The water intake buildings of the NPP were completely flooded and blocked by the floodwaters. Owing to this, the entire cooling system of the plant was disrupted. The battery-driven cooling pump meant to cool the reactor core in the event of a cooling failure was started. However, the failure of heat removal systems from the building was responsible for shortage in the supply of steam required to drive the pump. The pump stopped functioning leading to the meltdown of the reactor core. Reaction between zirconium and water at higher temperatures led to large amounts of hydrogen gas and hydrogen explosions. Though there were no reported casualties, the Fukushima disaster is considered to be one of the biggest environmental disasters in modern times.

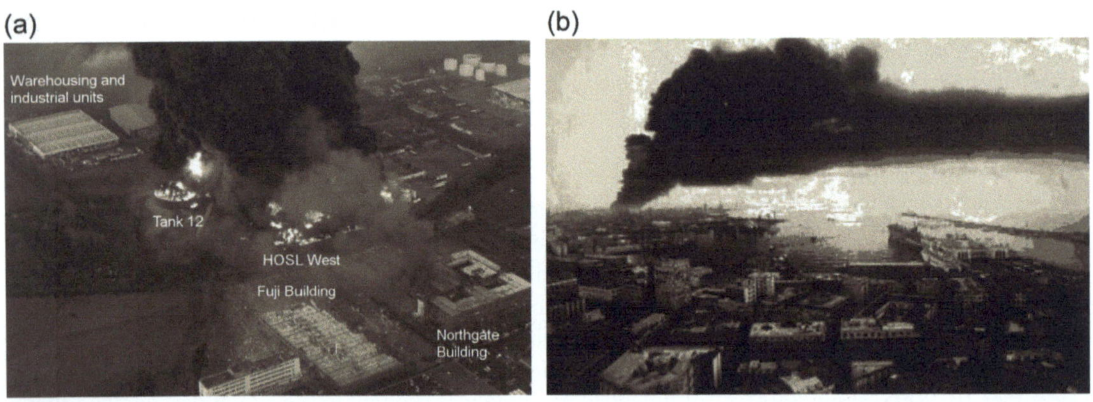


**FIGURE 2.28**

Disasters in industries. (a) Buncefield facility on the first day of the fire. (b) Smoke from Naples (Hailwood et al., 2009).

### 2.5.3.12 Common characteristics of major industrial incidents

Analysis of most major industrial incidents indicates that none of the disasters were initiated by unanticipated hazards or unpredictable situations. The concerned authorities were aware of the properties of the involved chemicals and the hazards associated with it. In fact, it could be postulated that many of these incidents could be averted if the safety principles were properly implemented.

Another factor that is common in many major industrial incidents is that incidents do not happen due to a single failure or defect. A number of problems culminate into a big disaster. Hence, it is very difficult to isolate a single flaw, which could have triggered the entire event. Reason (1990) used a Swiss cheese model to explain this phenomenon. In the Swiss cheese model, the organizational barriers against failure are considered as slices of cheese. The weaknesses in different parts of the system are represented as holes in these slices. When the holes in each of the slices align momentarily, an incident occurs. Figure 2.29 describes the influence of failing system barriers, operator error, preconditions, latent failures, and fallible decisions of management on the incident (Knegtering and Pasman, 2009).

Faulty control systems, inadequate maintenance, lack of awareness, constant focus on production, improper communication, and poor management are responsible for many process safety incidents.

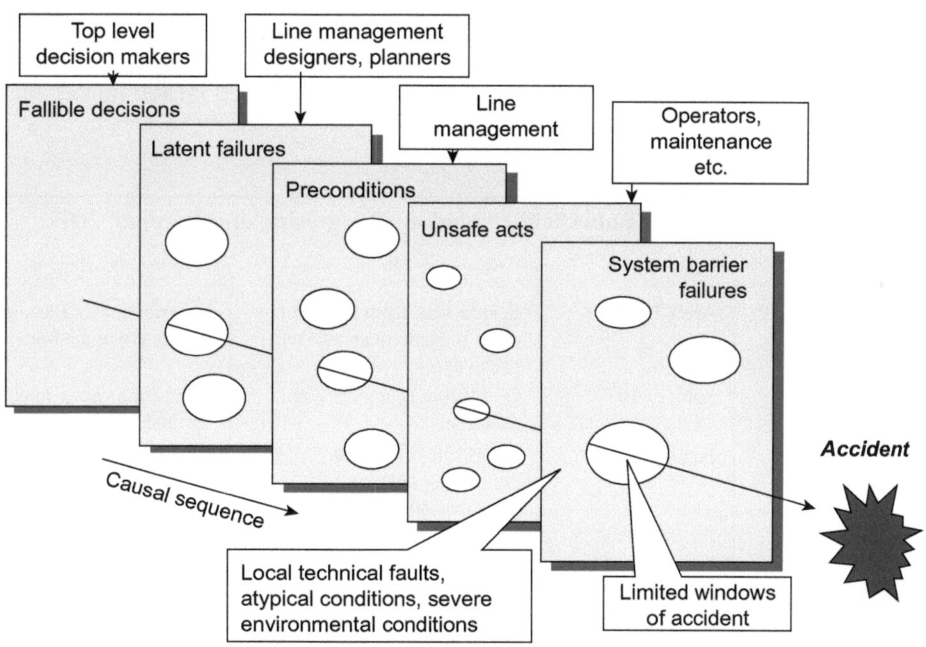


**FIGURE 2.29**

Swiss cheese model (Reason, 1990).

## 2.5.4 PROCESS SAFETY CHALLENGES AND OPPORTUNITIES

The main challenges and therefore opportunities to improving the safety performance in the process industries can be summarized as follows.

### *2.5.4.1 Organizations have no memory*

The integration of past knowledge and present situation helps avert dangers. It is generally seen that when individuals leave organizations, they leave with their knowledge. Their experiences and wisdom are lost forever and the organization as a whole is deprived of their intelligence. The past incidents and mishaps are overlooked and the same mistakes are repeated (Kletz, 1993). Most of the industries focus on developing consequence models or doing risk assessments to increase safety (Qi et al., 2012). The most common reason behind many incidents is that most organizations do not appreciate the fact that "Human memory is weak and corporate memory short" (Knegtering and Pasman, 2009). According to Kletz (1993), an organization loses vital information regarding past incidents every 10 years due to personnel changes. Many plant workers are unaware of the importance of certain safety practices or the need of some equipment, which was introduced after major incidents.

The lack of proper analysis of past incidents affecting other industries is another cause of failure. For example, before the Buncefield incident, there were many reported incidents, which involved overfilling of a gasoline tank. The Naples incident, mentioned in the previous section, is one such example. Another similar incident was the New Jersey Texaco incident, which occurred in 1983 (Hailwood et al., 2009). In this incident, about 80,265 tons of unleaded gasoline was released into the surroundings, which ignited under favorable conditions. Other such incidents are summarized in Table 2.13.

Similar to Buncefield, the Warffum incident was preceded by incidents involving explosion due to welding activities.

**Table 2.13 Incidents Similar to Buncefield Explosion (Knegtering and Pasman, 2009)**

| Location | Date | Comments: Background | Comments: Explosion |
|---|---|---|---|
| Houston, Texas, USA | April 1962 | Severe leak from a gasoline tank. Ignition near adjacent highway. | Described as a blast, but no details are available. |
| Baytown, Texas, USA | January 1977 | Overfilling of a ship with gasoline. | Congestion must have caused the leak. |
| Texaco, Newark, New Jersey, USA | January 1983 | Overfilling of a tank containing unleaded gasoline. 80,265 tons released. Ignition source 300 m away. | Relatively noncongested area. High overpressures reported, but not quantified. Three minor explosions preceded the main blast. |
| Jacksonville, Florida, USA | January 1993 | Overfilling of a tank containing unleaded gasoline. 132 tons released. | High overpressure produced. |
| Laem Chabang, Thailand | December 1999 | Overfilling of a gasoline tank. | High overpressure and congestion. |

As such, there is a dearth of an efficient system, which can document past knowledge and act as a retention tool. There is a grave need to examine past case histories and extract useful information from them. Currently, most industries have ample documents and data corresponding to safety principles and codes of conduct. However, the specific experience gained from designing or operating a piece of equipment is still not easily accessible to a beginner, who has just joined the plant. Owing to the inadequate interaction between the designer and worker in the field, experience from the field does not flow back sufficiently to new design (Knegtering and Pasman, 2009).

#### *2.5.4.2 Disadvantages of probability-based approaches*

As mentioned earlier, probability-based approaches are used widely in industrial risk assessment process. But a variety of limitations affect the effectiveness of these techniques. The following provides a discussion of some of these limitations and disadvantages.

##### Limitations of fault tree analysis

1. As the complexity of a process increases, the size of the fault tree increases. Current industrial processes demand fault trees with thousands of gates and numerous intermediate events. This increases the computation time and makes it onerous.
2. The fault tree analysis requires the knowledge of failure probabilities. These values are not known accurately and in some cases are not known at all.
3. Fault trees assume that the failure probability of one component is not affected by the failure of another component. Owing to this assumption, failure probabilities are underestimated in some cases, affecting the overall probability and risk calculations.
4. The development of a fault tree depends to a large extent on the engineer. Different people create different structures, which give different values of failure probabilities (Crowl and Louvar, 2011).
5. Fault tree analysis has been used extensively on a commercial scale. However, the use of fault trees for reduction of hazards in laboratory and pilot scale systems is relatively sparse owing to lack of initiative and shortage of experience.

##### Limitations of event tree analysis

1. Quantitative data are needed for each event to predict failure frequencies.
2. Detailed processes have huge event trees.
3. An event tree consists of an initiating event and terminates with a large number of consequences. The certainty of a concerned consequence arising from a selected failure is low (Crowl and Louvar, 2011).
4. Similar to fault tree methodology, the applicability of event tree analyses to laboratory and pilot scale plants is limited.

#### *2.5.4.3 Deficiencies in consequence-based approaches*

##### Fire consequence modeling

Major problems arise during the estimation of thermal heat radiation from a fire. The physical height of the flame depends to a large extent on the circumstances of release. The basic assumption that a fire resembles a uniform cylinder is not always true. Apart from this, smoke production affects the total fire area due to shielding effect. Second, the modeling of the extent of damage caused by a particular intensity of heat radiation to people and objects is complex. Most of the models are simple and neglect the effect of intervening objects (e.g., houses, trees) in their calculations of hazard distances. These calculations play a pivotal role in determining the hazard zones near an industry. The location of an industry (e.g., urban, rural, industrial) will influence the density of surrounding obstacles and thereby influence the hazard zone calculations. To make the models more comprehensive, the absorption of the thermal radiation by the atmosphere has to be included in the models. The radiative properties (e.g., absorptive, reflective) of the receiving surface are not considered while assessing the impact of a thermal radiation on the surroundings. The radiative properties of a surface determine the thermal inertia of a body, which governs the thermal response of a body. For example, the damage caused to a human depends to a large extent on (a) amount of skin exposed, (b) amount of clothing, (c) physiological response (sweating), (d) wind velocities, and (e) ability to take shelter (Raj, 2007).

### Explosion consequence modeling

The inability to predict confined vapor clouds is a major drawback of the TNT equivalency model. to overcome this limitation, a new model called TNO Multi-Energy Method was introduced. This model takes into account the degree of confinement of a volume and includes it in the calculations. The congested areas of a field are inspected to identify the potential blast sources. The blast strength is represented by a number (1–10). For example, a strong blast is characterized by strength of 10 and a blast arising from a quiescent part is considered to have blast strength of 1. For each of the blasts, the scaled distance is calculated using the following equation (Crowl and Louvar, 2011) and using Figure 2.30.

$$\overline{R} = \frac{R}{\left(E/P_o\right)^{1/3}} \tag{2.97}$$

where $\overline{R}$ is scaled distance, $R$ is distance from the charge (m), $E$ is charge combustion energy (J), and $P_o$ is ambient pressure (Pa).

Based on the value of the scaled pressure, the overpressure is calculated using the following formula,

$$p_o = \Delta\overline{P_s} \times P_a \tag{2.98}$$

where $p_o$ is scaled distance and $P_a$ is ambient pressure (Pa)

The selection of the severity factor for various confinements is a complicated process. Combining the results from each blast is another challenge. Current techniques involve adding the combustion energies from each source and assuming maximum initial blast strength of 10. However, this is valid only for blast sources located in close proximity. Also, the model cannot adequately depict partially congested geometries.

Currently, consequence models have widespread use in process industries. Though they are infinitely scalable, these models have not been adapted on a laboratory or pilot scale. to validate these models for a laboratory/pilot scale system, proper testing is required.

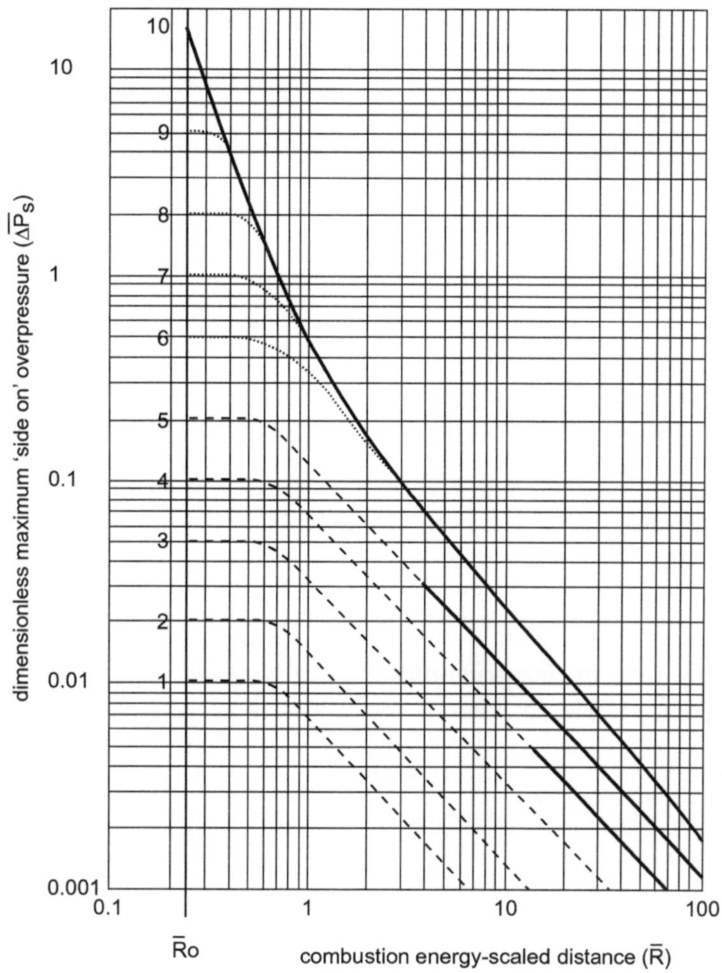


**FIGURE 2.30**

Scaled pressure versus scaled distance.

*CCPS (1994).*

### *2.5.4.4 Disadvantages and limitations of qualitative and semiquantitative techniques*

Layer of protection analysis

Major limitations exist in the process of inclusion of human actions and activities within LOPA. Most companies give independent protection layers (IPLs) more credit than can be substantiated. It is necessary that all companies provide thorough training to their operators so that they are capable of carrying out their actions.

Second, the use of dependent IPLs in the analysis process often leads to ineffectiveness. High priority should be given to include IPLs, which are independent of each other. For instance, common issues arise from the use of the same operator or operator group in multiple IPLs.

Also, the data for initiating events and IPLs are based on initial conditions. It is necessary that efforts are concentrated toward maintaining the IPLs at these conditions. It is often seen that the IPLs lose their reliability with time. The human IPL is one such example. The human IPL may change depending on factors such as work structure, recruitment process, and presence of temporary workers.

Another common problem experienced during the application of LOPA is the inappropriate use of data from handbooks and standards. People tend to use failure probabilities from literature without assessing the applicability of the data for a particular plant, site, or process. The validation of the selected data for a particular plant plays an important role in the successful implementation of LOPA (Myers, 2013).

#### HAZOP approach

One of the main disadvantages of HAZOP approach is that it is tedious and requires ample amount of time (Crowl and Louvar, 2011). Second, a HAZOP is not meant to and is incapable of detecting all the design flaws. For example, a weakness in the layout cannot be discovered by a HAZOP analysis. A HAZOP inherently assumes that the plant is supposed to function as designed. All the assumptions made during the design phase must be incorporated during construction and operation for ensuring the success of a HAZOP analysis. For instance, if a different material is used for construction problems may result. The HAZOP method focuses entirely on identification of hazards. Some of the identified hazards may have low probability or consequence. HAZOP approach in itself is not comprehensive and requires risk-based calculations for validation (Kletz, 2001).

### *2.5.4.5 Inherently safer design principles are not followed*

Inherently safer design (ISD) is a philosophy that focuses on safety issues in design and operation of equipment involved with hazardous chemicals (Hendershot, 2010; Kletz and Amyotte, 2010) as conceptually shown in Figure 2.31. ISD involves minimizing or substituting the use of hazardous

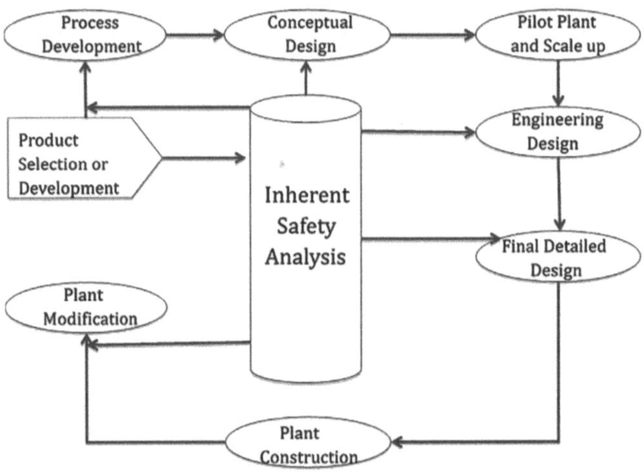


**FIGURE 2.31**

ISD in process design life cycle (Khan and Amyotte, 2002).

chemicals or operating at moderate conditions to reduce the overall risk. Reduction of complexity of operations is another salient feature of ISD. It is better to incorporate ISD in the initial design phase of process development. Generally, the engineering and financial decisions made in the later stages of development do not permit proper incorporation of ISD techniques in process development (Heikkilä, 1999). The industry has acknowledged the importance of ISD and has implemented various programs to implement the concepts. However, the implementation of ISD as a routine practice is hindered by many factors.

### Why ISD is not a routine

The first factor, which limits the application of ISD, is the lack of awareness of ISD concepts among designers and chemists (Khan and Amyotte, 2003). Since the application of ISD begins at the onset of a project, the design engineers and the chemists need to have a thorough understanding of ISD. Most of the designers are not adept in incorporating the ISD principles into the design. Apart from this, most industries opt for the high-performance materials to maximize profits. A case in point is the use of solvent-based paints over aqueous latex paints. The aqueous paints are safer but may show substandard performance under certain conditions. The innocuous nature of a chemical may not guarantee its amiability toward the surroundings. Many chemicals, which might be inherently safer, might be highly deleterious to the environment. For example, chlorofluorocarbons (CFCs) are used as a substitute for ammonia in refrigeration system. Though inherently safer, CFCs tend to deplete the ozone concentration in the environment. The process of developing ISD is technically challenging and demands research and development efforts. Especially, in the case of implementing it to an existing plant, the cost of the improvements may override the relatively small gains. Also, the additional time required to incorporate the changes might make it impracticable.

### Quantification of ISD

Measurement techniques and analysis tools have been constantly developed to measure ISD. Traditional methods were based on Boolean principles, which used tentative and subjective information (Gentile et al., 2003). These methods were replaced by sophisticated techniques like qualitative ranking schemes, quantitative risk analysis, and formal decision-making tools. These techniques vary from simple "what-if" analysis and critical examination to comprehensive decision analysis techniques (Khan and Amyotte, 2003).

Hazard index is another way of characterizing ISD. In this method, index values are attributed to chemicals and processes of a plant. The Mortality Index is one of the basic methods to measure ISD. In this method, the average number of human deaths caused by an explosion of one ton of a liquefied flammable gas is obtained from historical records. For example, the Mortality Index of chlorine is 0.3 while that of ammonia is 0.02 (Kletz and Amyotte, 2010). This means that 1 ton of ammonia is 15 times safer than 1 ton of chlorine.

$$\text{Mortality index} = \frac{\text{Total deaths caused by the explosion}}{\text{Total amount of liquefied flammable gas exploded}} \tag{2.99}$$

However, the index does not account for nature or frequency of leaks that can occur in a plant. For example, a certain chlorine plant might have lower number of pumps or drainage points (leak sources) compared with an ammonia plant. Hence, more sophisticated indices have been devised.

The Dow Fire and Explosion Index (F&EI) is another safety metric used in industrial design and optimization. This is used to analyze the fire, explosion, and chemical exposure hazards (Suardin et al., 2007). The details of this method were introduced elsewhere. The Mond Fire Explosion and Toxicity Index (FETI) is another type of index used commonly in the industry. It is similar to the F&EI with certain modifications. Additional hazard factors like special material hazard factor and layout hazard factors are introduced. to account for the material inventory a quantity factor is considered. Most importantly, a toxicity index is determined for every chemical based on the amount and the toxicological properties. The toxicity index is characterized on a scale of 1–7 (1 is low, 7 is severe) (Khan et al., 2001).

Another popular safety metric is the Safety Weighted Hazard Index (SWeHI). The SWeHI measures the radius of area under moderate hazard due to a certain plant. This can be given by the following expression (Khan et al., 2001):

$$SWeHI = \frac{B}{A} \tag{2.100}$$

where $B$ is quantitative damage caused by a plant and $A$ represents credits due to the existing safety and control measures.

The term $B$ comprises two parts: (a) damage due to fire and explosion ($B_1$), and (b) damage due to toxic release ($B_2$).

1. The estimation of *SWeHI* begins with the classification of industrial units as storage units, units involved in physical separations, units involving chemical operations, transportation units, other units (furnace, boilers).
2. This is followed by the calculation of energy factors and penalties for each case.

The following procedure describes the entire process (Khan et al., 2001). The calculations are performed for a storage unit. Similar calculations can be carried out for other units.

***Calculation of $B_1$ for storage units.*** Three energy factors are used to account for the chemical ($F_1$) and physical energies ($F_2$, $F_3$) of the storage units.

$$F_1 = \frac{0.1 \times M \times H_c}{K} \tag{2.101}$$

$$F_2 = 1.304 \times 10^{-3} \times PP \times V \tag{2.102}$$

$$F_3 = \frac{0.001 \times (PP - VP)^2 \times V}{(T + 273)} \tag{2.103}$$

where $M$ is mass release rate (kg/s), $H_c$ is heat of combustion (kJ/mol), $K$ is constant (3.148), $PP$ is processing pressure (kPa), $V$ is volume (cu·m), $VP$ is vapor pressure (kPa), and $T$ is temperature (°C).

The effect of the various parameters on damage potential is obtained using energy factors and penalties. For instance, the effect of temperature is modeled using penalty ($pn1$) where

$$pn1 = ft1(\text{flash point, fire point, autoignition temperature working temperature})$$

where *ft1* is a set of functions that determines the numerical value of the penalty.

if *(fire point > temperature > flash point)*
$pn1 = 1.34.$
*or*
*(Autoignition temperature > temperature > fire point)*
$pn1 = 1.55$
*or*
$pn1 = 1.1$
Similarly, two energy factors ($F_2$, $F_3$) and a single penalty (*pn2*) are used for modeling the pressure.
*If(VP > AP)*
*If(PP > VP)*
$F = F_2 + F_3$
$pn2 = f(\text{operating pressure})$
*or*
$F = F_2$
$pn2 = f(\text{operating pressure})$
*or*
$F = F_3$
$pn2 = f(\text{operating pressure})$
where *VP* is vapor pressure (kPa), *AP* is atmospheric pressure (kPa), *PP* is processing pressure (kPa), and *f* is function of operating pressure.

Other parameters like the presence of a hazardous unit (*pn*3), amount of chemical inventory (*pn*4), flammability/reactive characteristics of chemicals (*pn*5), congestion associated with the concerned unit (*pn*6) are also modeled using penalties. Apart from this, external factors like earthquakes (*pn*7) and incident vulnerability of an area (*pn*8) are also considered. For example, if the frequency of occurrence of an earthquake is 1/year, then a penalty of two is assumed. All the energy factors and penalties are used to obtain the hazard potential.

$$\begin{aligned}\text{Hazard potential} &= (F_1 \times pn1 + F \times pn2) \times pn3 \times pn4 \\ &\quad \times pn5 \times pn6 \times pn7 \times pn8\end{aligned} \tag{2.104}$$

$$B_1 = 4.76 \times (\text{Hazard potential})^{1/3} \tag{2.105}$$

***Calculation of* $B_2$ *for storage units.*** The parameter $B_2$ corresponds to the radius of the area affected by a toxic release. The load should have 50% probability of causing death. The air quality is assumed to be slightly stable (something between highly stable and highly unstable). The index $B_2$ is derived from factors like (a) amount of chemicals in the unit, (b) physical state of the chemicals involved, (c) toxic nature of the chemical, (d) operating conditions, and (e) site characteristics.

The calculation begins with estimation of the *G* factor. The expression used to derive *G* is

$$G = S \times m \tag{2.106}$$

where *m* is release rate (kg/s), and *S* depends on release conditions.

The value of *S* is obtained from the following table (Table 2.14).

The penalties for operating pressure and temperature are estimated in a manner similar to $B_1$. Apart from this, penalties for vapor density (*pnr*3), toxicity (*pnr*4), and site characteristics (*pnr*5) are evaluated as follows.

**Table 2.14** ***S*** **Values for Various Types of Chemical Release (Khan et al., 2001)**

| NFPA Rank | Liquid | Liquefied Gas | Gas | Solid |
|---|---|---|---|---|
| 4 | 4.00 | 8.00 | 13.40 | 0.1300 |
| 3 | 0.40 | 0.80 | 1.34 | 0.0130 |
| 2 | 0.20 | 0.40 | 0.67 | 0.0060 |
| 1 | 0.07 | 0.10 | 0.25 | 0.0025 |

$$pnr3 = 1.2 \times \frac{\text{Vapor density}}{\text{Air density}} \tag{2.107}$$

Dense vapors like ammonia tend to settle down and have a larger impact on the environment. They tend to disperse slowly leading to a buildup of toxic load.

$$pnr4 = \text{Maximum}\ (1,\ 0.6 \times \text{NH}) \tag{2.108}$$

where NH is NFPA ranking for health hazard.

$$pnr5 = fp(\text{population density}) \tag{2.109}$$

where *fp* is population density function.

Also, the calculated values of penalties for external factors (*pnr*6) and plant vulnerability (*pnr*7) are used.

Using the values of the *G* factor and the penalties, we get $B_2$

$$B_2 = 25.35 \times (G \times pnr1 \times pnr2 \times pnr3 \times pnr4 \times pnr5 \times pnr6 \times pnr7)^{0.425} \tag{2.110}$$

The larger of the two values ($B_1$, $B_2$) is considered as the hazard potential index (*B*).

***Quantification of A.*** The control measures in an industry play an important role in deciding the overall risk potential of the plant. These measures are directed toward reducing the damage potential of industrial units and decreasing the likelihood of disaster events. In the evaluation of *A*, credit values are given for emergency resource planning (*cr*1), disaster management plan (*cr*2), process control system (*cr*3), detection devices (*cr*4), emergency control measures (*cr*5), human error (*cr*6), equipment reliability (*cr*7), and damage control measures (*cr*8). Using these values, *A* is calculated using the expression,

$$A = 0.15 \times (1 + cr1) \times (1 + cr2) \times (1 + cr3) \times (1 + cr4) \times (1 + cr5) \times (1 + cr6) \times (1 + cr7) \times (1 + cr8) \tag{2.111}$$

*SWeHI* is then calculated using Eqn (2.100). Apart from these indices, the Chemical Exposure Index, Fuzzy Logic–based Inherent Safety Index (FLISI), and Hazardous Waste Index are commonly used in the industry. Although a large number of hazard indices are available, no index method can incorporate all the safety parameters (Suardin et al., 2007). Depending on the plant, an engineer has to select the appropriate methodology. Barring the Mortality Index method, all the techniques are complex and cannot be easily integrated with preliminary process design (Heikkilä, 1999). For example, characterization of the toxicity index in the Mond FETI method requires a lot of experience.

A majority of these methods require detailed process information, which is not available at the early design stage. For example, the identification of general process hazards involving enclosed process units in F&EI method. Methods like FLISI require sophisticated mathematical tools not easily available in the market. Hence, most of the companies are looking for simpler methods to measure ISD, which can be easily adaptable with computer optimization and simulation tools. If such a method is developed it can have a widespread use in industries.

#### Common misconceptions about ISD

Though the use of ISD is recommended, there are certain myths regarding ISD and its applications (Hendershot, 2010).

***ISD will remove all hazards.*** ISD addresses safety issues and reduces the overall risk of the plant. However, it does not mean that the plant is hazard free. The ISD may correspond to reduction in a particular hazard, but it does not mean that all other hazards are mitigated. Also, ISD should be implemented with extreme care. The installed improvements should not introduce newer forms of hazards or increase other existing hazards.

***Inherently safer technology is always the best way to manage a hazard.*** As mentioned in the earlier section, the introduction of safer strategies to reduce one hazard may initiate another hazard. In such circumstances, the implementation of ISD is not the best solution available. Additionally, in some cases a safer process is preferred over an inherently safer process owing to the societal benefits it engenders. For example, the use of nuclear energy to generate electricity is relatively cheaper than other sources of electricity generation. Also, the process emissions have limited impact on the environment. However, a nuclear power plant is not inherently safer and has the potential to cause a major disaster.

***ISD can be applied only at the early stages of design.*** The use of ISD in the early stages of research and development guarantees maximum benefits. However, this does not imply that its application at later stages is unrewarding. For example, following the Bhopal disaster, many companies have made a conscious effort to reduce the inventory of toxic and explosive chemicals, thereby reducing the hazards associated with it (Hendershot, 2010).

***The role of plant operators in ISD is minimal.*** Plant personnel have first-hand knowledge of the plant. They have good insight into the mechanical aspects of the plant. They might not be able to suggest a safer chemistry for the process, but they can identify the operational challenges imposed by the improved process. For example, a startup procedure, which requires an operator to manually operate a set of valves in a correct sequence, is less safe than a procedure operated by an automatic set of valves. Another alternative could be reducing the number of existing valves by process technology change. In both cases, the number of human errors is drastically reduced (Hendershot, 2010). However, plant operations personnel are rarely involved in the initial stages of the design and construction of the plant, and so their input to any ISD considerations or analysis is at a minimum. Efforts should be made to address this barrier as much as possible.

### 2.5.4.6 Not giving enough importance to leading indicators

It is said, "prevention is better than cure." So, we need to identify the signals before an incident occurs (Grabowski et al., 2007). Traditionally, the safety performance of many industries has been assessed by 'after the loss' measurements. Factors such as incident and injury rates, incidents, and dollar costs are used as parameters to analyze safety and prevent future incidents. Recently, safety and performance of plants have been measured using safety indicators.

Leading indicators are conditions, events, or measures preceding an adverse event, which have the potential to predict the arrival of this event. They are performance indicators or metrics, which have the capability of eliminating or minimizing risks by identifying and mitigating hazards. Generally, indicators measure the performance of six different functional areas (Øien et al., 2011).

1. Management, organization, and administration
2. Design of facility and processes
3. Training and qualification
4. Operation
5. Maintenance
6. Emergency preparedness planning

The indicators in a system and the onset of undesirable events are linked in a number of ways. Some indicators can be the harbinger of adverse incidents, while others can be a direct cause of them. In both cases, the links are easily perceivable and can be analyzed. The lack of efforts shown by industries in understanding process and chemicals hazards and the insufficiency in the consequence modeling tools have been responsible for making industries oblivious to the indicators. Many disasters could have been forestalled, if the leading indicators, missed signals, and dismissed alerts were properly managed. In the case of Buncefield, factors like vapor cloud formation due to liquid leak or the heat effects due to explosion were not investigated (Qi et al., 2012). Also, in recent years, terrorism has become an emerging threat to mankind. In the post-9/11 world, potential terrorist attacks on chemical plants have become a possibility to consider. While antiterrorism regulations have been promulgated by national governments, there is a great scope for research in this field.

#### 2.5.4.7 Lack of communication

In the past 30 years, the industry has increased its focus on energy saving, product quality, profit margins, as well as safety (Knegtering and Pasman, 2009). Owing to this, process installations are being used to their full capacities. Apart from this, the ever-growing need of increasing the flexibility of operations has led to complicated systems. The complex nature of the control systems is difficult to decipher. The transfer of technology from the Western world to other countries has further added to the challenges faced by local employees. In addition to this, outsourcing has led to dispersal of the core business (Qi et al., 2012). Enormous efforts need to be taken to maintain the standard levels of safety at all locations. to solve these problems, it is important that a company gives adequate importance to communication.

Communication plays a pivotal role in human interaction. The organization as a whole should be committed and willing to tackle the problem. The whole system should be divided into smaller groups with group leaders. The groups should be made to interact with each other, thereby improving collaboration and communication. Frequent group briefings before and after operations will prevent miscommunication among the workers. According to government regulations, workers should be made aware of the hazards and risks associated with the processes in the industry. Apart from this, the experienced personnel must properly train and coach new workers. The process-specific experience and technical know-how should be shared with the new employees.

## 2.5.5 CONCLUSION

### 2.5.5.1 *Process safety: a holistic approach*

A safe situation is sustained due the coherence of many related aspects. These aspects need to be investigated as a whole; tackling single aspects separately may not be sufficient to guarantee safety. There are numerous occasions in a plant, when single problems go undetected. However, these problems are easily handled by the sophisticated process systems. But, if the management constantly neglects such problems, the problems can pile up leading to a big failure. Even if the problem is noticed, it might not be deemed as urgent or safety critical because a single problem does not lead to an undesirable event in today's industry. Process safety incidents occur due to a number of reasons. To avert an adverse event, the industry must have an open eye for all possible incident causes and good overall control. All the aspects affecting process safety must be controlled. to achieve these objectives, human beings need to be more prudent and perceptive. Also, as humans have limited far-sightedness, reliable metrics and comprehensive computer simulations are required to predict the future.

### 2.5.5.2 *Need for further research*

Preventing process disasters requires constant vigilance. When a plant is not exposed to a major mishap for a long time, people tend to get complacent. They stop appreciating the importance of safety systems and control measures. This is a major reason for a disaster.

The advancement of industry, science, and technology has given rise to new problems. New incident scenarios have been identified. These scenarios are directly related to the organizational framework and the safety culture. They are responsible for the major incidents, which affect the people and surroundings. The constant change in the industry demands a continuous change in the process control techniques. Most importantly, a reliable and adequate measurement system is required. Similarly, a suitable information technology tool is required to incorporate previous information from incidents into the current design and operation of the plant. Research and identification of new ISD techniques and the use of safer chemicals may be able to solve a majority of problems faced by the new millennium. Maintaining sustained process safety performance without compromising on plant production is a formidable mission. In addition to diligent and consistent enforcement of current regulations on safety, there is great scope for fundamental and applied research in safety.

## REFERENCES


Abbasi, T., Abbasi, S., 2007. The boiling liquid expanding vapour explosion (BLEVE): mechanism, consequence assessment, management. Journal of Hazardous Materials 141 (3), 489–519.

Allport, D.C., et al., 2003. Supporting Sciences. MDI and TDI: Safety, Health and the Environment. John Wiley & Sons, Ltd, 277–430.

American Institute of Chemical Engineers, Center for Chemical Process Safety, 2003. Guidelines for Fire Protection in Chemical, Petrochemical, and Hydrocarbon Processing Facilities. Center for Chemical Process Safety of the American Institute of Chemical Engineers, New York, NY.

Annamalai, K., et al., 1994. Interactive processes in gasification and combustion—part III: coal/char particle arrays, streams and clouds. Progress in Energy and Combustion Science 20 (6), 487–618.

ASTM, 2007a. E1515: Standard Test Method for Minimum Explosible Concentration of Combustible Dusts.

ASTM, 2007b. E2019: Standard Test Method for Minimum Ignition Energy of a Dust Cloud in Air.

ASTM, 2012. E1491: Standard Test Method for Minimum Autoignition Temperature of Dust Explosion.

Atlas, M., 2007. TRI to communicate: public knowledge of the federal toxics release inventory*. Social Science Quarterly 88 (2), 555–572.

Bahr, N.J., 1997. System Safety Engineering and Risk Assessment: A Practical Approach. Taylor & Francis Group.

Baker, W.E., et al., 1983. Explosion Hazards and Evaluation. Access Online via Elsevier.

Baker, W., 1973. Explosion in Air. Texas.

Ballal, D., Lefebvre, A., 1979. Ignition and flame quenching of flowing heterogeneous fuel-air mixtures. Combustion and Flame 35, 155–168.

Becking, G.C., 1995. Use of mechanistic information in risk assessment for toxic chemicals. Toxicology Letters 77, 15–24.

Bejan, A., 1991. Predicting the pool fire vortex shedding frequency. Journal of Heat Trasnfer 113, 261–263.

Berger, S., 2009. History of AIChE's Center for chemical process safety. Process Safety Progress 28 (2), 124–127.

Beyler, C.L., 2002. Fire hazard calculations for large open hydrocarbon fuel fires. In: The SFPE Handbook of Fire Protection Engineering. National Fire Protection Association, Quincy, MA.

Bind, V.K., et al., 2012. A reaction engineering approach to modeling dust explosions. Chemical Engineering Journal 207–208, 625–634.

Birk, A., et al., 2007. Blast overpressures from medium scale BLEVE tests. Journal of Loss Prevention in the Process Industries 20 (3), 194–206.

Blewitt, D.N., et al., 1987. Effectiveness of water spray on motigating anhydrous hydrofluoric acid releases. In: International Conference on Vapor Cloud Modeling, Cambridge, MA.

Bouhafid, A., et al., 1898. Characterisation of thermal radiation from freely burning oil pool fires. Fire Safety Journal 15 (5), 367–390.

Bouillard, J., et al., 2010. Ignition and explosion risks of nanopowders. Journal of Hazardous Materials 181 (1), 873–880.

Brasie, W., Simpson, D., 1968. Guidelines for estimating damage explosion. Loss Prevention 2, 91–97.

Bridges, G.W., 1985. PSM cost/benefit survey results. Process Safety Progress 13 (23).

Britter, R.E., McQuaid, J., 1988. Workbook on Dispersion of Dense Gases. Health and Safety Executive, Sheffield, U.K.

Brode, H.L., 1959. Blast wave from a spherical charge. Physics of Fluids 2, 217.

Buckmaster, J., Peters, N., 1986. The infinite candle and its stability-a paradigma for flickering diffusion flames. In: Twenty-first Symp. (Int.) on Combust. Pittsburgh, PA, Combustion Institute: 1829.

Busini, V., et al., 2012. Influence of large obstacles and mitigation barriers on heavy gas cloud dispersion: a liquefied natural gas case-study. Industrial and Engineering Chemistry Research 51 (22), 7643–7650.

Busini, V., et al. Heavy gas dispersion in presence of large obstacles: selection of the modeling tools, (to be submitted for publication).

Buzea, C., et al., 2007. Nanomaterials and nanoparticles: sources and toxicity. Biointerphases 2 (4), MR17–MR71.

CSB, 2007. BP America Refinery Explosion, Final Investigation Report.

CCPS, 2010. Guidelines for Fire Protection in Chemical, Petrochemical, and Hydrocarbon Processing Facilities. Wiley.

Casal, J., Salla, J.M., 2006. Using liquid superheating energy for a quick estimation of overpressure in BLEVEs and similar explosions. Journal of Hazardous Materials 137 (3), 1321–1327.

CCPS, 1994. Guidelines for Evaluating the Characteristics of Vapor Cloud Explosions, Flash Fires, and BLEVEs, American Institute of Chemical Engineers. Center for Chemical Process Safety, New York.

CCPS, 1996. Guidelines for Use of Vapor Cloud Dispersion Models, second ed. Center for Chemical Process Safety/AIChE.

CCPS, 2014. Guidelines for Determining the Probability of Ignition of a Released Flammable Mass. John Wiley & Sons.

Chatelier, H.L., 1891. Estimation of firedamp by flammability limits. Annales Des Mines 8 (19), 388–395.

Clancey, V., 1972. Diagnostic features of explosion damage. In: Sixth International Meeting of Forensic Sciences, Edinburgh.

Clayson, D.B., et al., 1985. Toxicological Risk Assessment. CRC Press, Boca Raton, FL.

Cloney, C.T., et al., 2014. Development of an organizational framework for studying dust explosion phenomena. Journal of Loss Prevention in the Process Industries.

Coco, J.C., Marsh Risk, C., 2001. Large Property Damage Losses in the Hydrocarbon-chemical Industries: A Thirty-year Review. Marsh Risk Consulting.

Cohrssen, J.J., Covello, V.T., 1999. Risk Analysis: A Guide to Principles and Methods for Analyzing Health and Environmental Risks. Diane Publishing Company.

Cormier, B., 2008. Computational Fluid Dynamics for LNG Vapor Dispersion Modeling: A Key Parameters Study. Chemical Engineering. Texas A&M University, College Station, TX. Doctor of Philosophy.

Cornwell, J.B., et al., 1998. The Use of Comparative Quantitative Risk Analysis in Evaluating Proposed Hydrogen Fluoride Mitigation Systems. Plant Process Safety Symposium, Houston, TX.

Crowl, D.A., Louvar, J.F., 2011. Chemical Process Safety : Fundamentals with Applications. Prentice Hall, Upper Saddle River, NJ.

Crowl, D.A., 1992. Calculating the energy of explosion using thermodynamic availability. Journal of Loss Prevention in the Process Industries 5 (2), 109–118.

Crowl, D.A., 2010. Understanding Explosions. Wiley.com.

Daou, J., Matalon, M., 2001. Flame propagation in Poiseuille flow under adiabatic conditions. Combustion and Flame 124 (3), 337–349.

Davies, J.K.W., 1987. A comparison between the variability exhibited in small scale experiments and in the Thorney Island phase I trials. Journal of Hazardous Materials 16 (0), 339–356.

Davies, J.K.W., 1989. The application of box models in the analysis of toxic hazards by using the probit dose-response relationship. Journal of Hazardous Materials 22 (3), 319–329.

Deaves, D.M., 1987. In: Incorporation of the Effects of Buildings and Obstructions on Gas Cloud Consequence Analysis Proc. Int. Conf. On Vapour Cloud Modelling, Boston, MA.

Delichatsios, M.A., Orloff, L., 1988. Effects of turbulence on flame radiation from diffusion flames. Proceedings of the Combustion Institute 22, 1271–1279.

Delichatsios, M.A., et al., 1992. The effects of fuel sooting tendency and the flow on flame radiation in luminous turbulent jet flames. Combustion Science and Technology 84 (199).

Drysdale, D., 2011. An Introduction to Fire Dynamics. Wiley, Chichester, West Sussex.

Eckhoff, R., 2003. Dust Explosions in the Process Industries: Identification, Assessment and Control of Dust Hazards. Access Online via Elsevier.

EPA, 1991. Guidelines for Develomental Toxicity Risk Assessment. US Environmental Protection Agency, Washington, DC.

EPA, 1996. RMP Offsite Consequence Analysis Guidance. US Envrionmental Protection Agency, Washington, DC.

Fenton, N.E., Neil, M., 2012. Risk Assessment and Decision Analysis with Bayesian Networks. Taylor & Francis.

Frederick, C.B., 1995. Summary of panel discussion on the 'advantages/limitations/uncertainties in the use of physiologically based pharmacokinetic and pharmacodynamic models in hazard identification and risk assessment of toxic substances. Toxicology Letters 79 (1–3), 201–206.

French Ministry of Environment, 2006. Explosion followed by fire In a hydrocarbon tank farm, October 7, 1991. Saint Herblain, Loire Atlantique, France.

Genova, B., et al., 2008. Evaluation of the blast-wave overpressure and fragments initial velocity for a BLEVE event via empirical correlations derived by a simplified model of released energy. Journal of Loss Prevention in the Process Industries 21 (1), 110–117.

Gentile, M., et al., 2003. Development of an inherent safety index based on fuzzy logic. AIChE Journal 49 (4), 959–968.

Glassman, I., Yetter, R., 2008. Combustion. Elsevier Science.

Grabowski, M., et al., 2007. Leading indicators of safety in virtual organizations. Safety Science 45 (10), 1013–1043.
Griffiths, R.F., Harper, A.S., 1985. A speculation on the importance of concentration fluctuations in the estimation of toxic response to irritant gases. Journal of Hazardous Materials 11 (0), 369–372.
Griffiths, R.F., 1984. Chemical plant risk assessment: uncertainties and development needs. Environment International 10 (5–6), 523–530.
Gugan, K., 1979. Unconfined Vapour Cloud Explosions. Institution of Chemical Engineers.
Hailwood, M., et al., 2009. Conclusions drawn from the buncefield and Naples incidents regarding the utilization of consequence models. Chemical Engineering & Technology 32 (2), 207–231.
Hald, K., et al., 2005. Heavy gas dispersion by water spray curtains: a research methodology. Journal of Loss Prevention in the Process Industries 18 (4–6), 506–511.
Hamins, A., et al., 1992. An experimental investigation of the pulsation frequencty of flames. In: Twenty-Fourth Symp. (Int.) on Combust., The Combustion Institute: 1695.
Hamins, A., et al., 1995. Characteristics of Pool Fires. Fire Resistence of Industrial Fluids. American Society for Testing and Materials, Indianapolis, IN.
Hansen, T.J., Crowl, D.A., 2010. Estimation of the flammable zone boundaries for flammable gases. Process Safety Progress 29, 3.
Hansen, O.R., et al., 2010. Validation of FLACS against experimental data sets from the model evaluation database for LNG vapor dispersion. Journal of Loss Prevention in the Process Industries 23 (6), 857–877.
Harris, R., Wickens, M., 1989. Understanding Vapour Cloud Explosions-an Experimental Study. Institution of Gas Engineers.
Harvey, B.H., 1976. First Report of the Advisory Committee on Major Hazards. HM Stationery Office, London.
Harvey, B.H., 1979. Advisory Committee on Major Hazards, Second Report. HMSO, London.
Health and Safety Executive, 1981. Canvey: A Second Report. A Review of the Potential Hazards from Operations in the Canvey Island/Thurrock Area Three Years after Publication of the Canvey Report. HMSO, London.
Health and Safety Executive, 2001. Reducing Risks, Protecting People. HSE's Decision Making Process. HSE Books.
Heikkilä, Anna-Mari, 1999. Inherent Safety in Process Plant Design: An Index-based Approach. VTT Technical Research Centre of Finland.
Hendershot, D.C., 2010. A summary of inherently safer technology. Process Safety Progress 29 (4), 389–392.
Henderson, Y., Haggard, H.W., 1943. Noxious Gases. Reinhold, New York.
Hertzberg, M., et al., 1978. The Diffusion Flame in Free Convection. U.S. Bureau of Mines.
Heskestad, G., 1981. Peak gas velocities, and flame heights of Buoyancy-controlled turbulent diffusion flames. In: Eighteenth Symposium on Combustion. Pittsburgh, PA, the Combustion Institute.
Heskestad, G., 1983. Luminous height of turbulent diffusion flames. Fire Safety Journal 5 (2).
Heskestad, G., 1991. A reduced scale mass fire experiment. Combustion and Flame 83, 293–301.
Hottel, H.C., Hawthorne, W.R., 1949. Diffusion in laminar flame jets. In: 3rd Symposium (International) on Combustion. Baltimore, MD, pp. 255–266.
Humphreys, D., et al., 2010. CFD simulation of underground coal dust explosions and active explosion barriers. In: Coal Operators' Conference.
Jaeckel, G., 1924. Die Staubexplosionen. Zeitschrift für technische Physik 67–78.
Jensen, F.V., Nielsen, T.D., 2009. Bayesian Networks and Decision Graphs. Springer.
Johnson, D., Pritchard, M., 1991. Large Scale Experimental Study of Boiling Liquid Expanding Vapour Explosions (BLEVEs). British Gas plc, Research and Technology Division.
Jordaan, I., 2005. Decisions under Uncertainty: Probabilistic Analysis for Engineering Decisions. Cambridge University Press.
Jost, W., 1939. Explosions–und Verbrennungsvorga ägne in Gasen. Springer-Verlag, Berlin.
Ju, Y., Maruta, K., 2011. Microscale Combustion: Technology Development and Fundamental Research. Progress in Energy and Combustion Science.

Ju, Y., Xu, B., 2006. Effects of channel width and Lewis number on the multiple flame regimes and propagation limits in mesoscale. Combustion Science and Technology 178 (10–11), 1723–1753.

Kanury, A.M., 1975. Introduction to Combustion Phenomena. Gordon and Breach, London.

Khan, F.I., Amyotte, P.R., 2002. Inherent safety in offshore oil and gas activities: a review of the present status and future directions. Journal of Loss Prevention in the Process Industries 15 (4), 279–289.

Khan, F.I., Amyotte, P.R., 2003. How to make inherent safety practice a reality. The Canadian Journal of Chemical Engineering 81 (1), 2–16.

Khan, F.I., et al., 2001. Safety weighted hazard index (SWeHI): a new, user-friendly tool for swift yet comprehensive hazard identification and safety evaluation in chemical process industrie. Process Safety and Environmental Protection 79 (2), 65–80.

Kim, B.K., et al., 2013. Key parametric analysis on designing an effective forced mitigation system for LNG spill emergency. Journal of Loss Prevention in the Process industries 26 (6).

Kinney, G.F., Graham, K.J., 1985. Explosive Shocks in Air. Springer-Verlag, Berlin and New York, 282 p. 1.

Kinney, G.F., 1962. Explosive Shocks in Air. The Macmillan Company, New York.

Kletz, T.A., 1993. Organizations have no memory when it comes to safety. Hydrocarbon Processing 72 (6), 88.

Kletz, T.A., 2001. Some Problems and Opportunities that have been overlooked. In: Proc. of the Mary Kay O'Connor Process Saf. Cent. Annu. Symp. College Station, Texas.

Kletz, T.A., Amyotte, P., 2010. Process Plants: A Handbook for Inherently Safer Design, Second Edition.

Knegtering, B., Pasman, H.J., 2009. Safety of the process industries in the 21st century: a changing need of process safety management for a changing industry. Journal of Loss Prevention in the Process Industries 22 (2), 162–168.

Korb, K.B., Nicholson, A.E., 2003. Bayesian Artificial Intelligence. Taylor & Francis.

Kuhl, A., et al., 1973. Pressure waves generated by steady flames. In: Symposium (International) on Combustion, Elsevier.

Laboureur, D., et al., 2012. Experimental Characterization and Mondeling of Hazards: BLEVE and Boilover/ Etude expérimentale et modélisation de phénomènes aggravants: BLEVE et BOILOVER.

Laboureur, D., et al., 2013. BLEVE overpressure: multiscale comparison of blast wave modeling. Process Safety Progress.

Langley, E., 1978. Health and safety aspects of the proposed notification scheme for newchemicals. Chemistry and Industry 15, 504.

Lewis, B., von Elbe, G., 1987. Combustion, Flames and Explosions of Gases. Academic Press, Orlando, FL.

Luketa-Hanlin, A., 2006. A review of large-scale LNG spills: experiments and modeling. Journal of Hazardous Materials 132 (2–3), 119–140.

Mannan, S., 2005. Lee's Loss Prevention in the Process Industries: Hazard Identification, Assessment, and Control. Elsevier.

Mannan, M.S., 2011. The Buncefield explosion and fire–lessons learned. Process Safety Progress 30 (2), 138–142.

Mannan, S., 2012. Loss Prevention in the Process Industries, Hazard Identification, Assessment and Control. Oxford, U.K., Butterworthds/Heinemann.

Mashuga, C.V., Crowl, D.A., 2000. Derivation of le Catelier's mixing rule for flammability limits. Process Safety Progress 19 (2), 112–117.

McCaffery, B.J., 1979. Purely Buoyant Diffusion Flames: Some Experimental Results. National Bureau of Standards.

McGrath, K.G., 2003. An earlier age of breast cancer diagnosis related to more frequent use of antiperspirants/ deodorants and underarm shaving. European Journal of Cancer Prevention 12 (6), 479–485.

Meyer, A., 1938. Les Gaz de Combat. Charles- Lavauzelle, Paris.

Miesse, C., et al., 2005. Diffusion flame instabilities in a 0.75 mm non-premixed microburner. Proceedings of the Combustion Institute 30 (2), 2499–2507.

Modarres, M., 2006. Risk Analysis in Engineering: Techniques, Tools, and Trends. Taylor & Francis.

Moore, J.M., 1962. Plant Layout and Design. Prentice Hall.

Moosemiller, M., 2011. Development of Algorithms for Predicting Ignition Probabilities and Explosion Frequencies.
Myers, P.M., 2013. Layer of protection analysis–quantifying human performance in initiating events and independent protection layers. Journal of Loss Prevention in the Process Industries 26 (3), 534–546.
Nabholz, J.V., 1991. Environmental hazard and risk assessment under the United States toxic substances control act. Science of the Total Environment 109/110, 649–665.
Nagy, J., Verakis, H.C., 1983. Development and Control of Dust Explosions. CRC Press.
NAS/NRC, 1983. Risk Assessment in the Federal Government: Managing the Process. National Academy of Sciences, Washington, DC.
National Research Council, 1983. Risk Assessment in the Federal Government. Managing the Process. National Academy Press, Washington, DC.
NFPA, 2002. "NFPA 54." National Fuel Gas Code, eighth ed. MA (USA), Quincy.
NIOSH, 2005. NIOSH Pocket Guide to Chemical Hazards. US Department of Health and Human Services, Washington, DC.
Øien, K., et al., 2011. Building safety indicators: part 1–theoretical foundation. Safety Science 49 (2), 148–161.
Oran, E.S., Gamezo, V.N., 2007. Origins of the deflagration-to-detonation transition in gas-phase combustion. Combustion and Flame 148 (1), 4–47.
Orloff, L., Ris, J.d., 1982. Froude modelling of pool fires. In: Nineteenth Symp. (Int.) Combust. Pittsburgh, PA, the Combustion Institute: 885.
Orloff, L., Ris, J.D., 1983. Froude Modeling of Pool Fires. Factory Mutual Research, Norwood, MA.
OSHA, 2002. OSHA 3071, Job Hazard Analysis. US Department of Labor, Washington, DC.
Paris, C., et al., 2004. Lung cancer and occupational exposures (asbestos not considered). Revue du Praticien 54 (15), 1660–1664.
Paustenbach, D.J., 1989. The Risk Assessment of Environmental Hazards: A Textbook of Case Studies. Wiley, New York.
Planas-Cuchi, E., et al., 2004. Calculating overpressure from BLEVE explosions. Journal of Loss Prevention in the Process Industries 17 (6), 431–436.
Prentiss, A.M., 1937. Chemicals in War. McGraw-Hill, New York.
Prugh, R.W., 1991. Quantitative evaluation of "Bleve" hazards. Journal of Fire Protection Engineering 3 (1), 9–24.
Qi, R., et al., 2012. Challenges and needs for process safety in the new millennium. Process Safety and Environmental Protection 90 (2), 91–100.
Raj, P.K., 2007. LNG fires: a review of experimental results, models and hazard prediction challenges. Journal of Hazardous Materials 140 (3), 444–464.
Reason, J., 1990. The contribution of latent human failures to the breakdown of complex systems. Philosophical Transactions of the Royal Society of London. B, Biological Sciences 327 (1241), 475–484.
Roberts, A.F., 1982. Thermal radiation hazards from releases of LPG from pressurized storage. Fire Safety Journal 4, 197–212.
Roberts, M.W., 2000. Analysis of Boiling Liquid Expanding Vapor Explosion (BLEVE) Events at DOE Sites. Safety Analysis Workshop.
Rodricks, J.V., Tardiff, R.G., 1984. Assessment and Management of Chemical Risks. American Chemical Society, Washington, DC.
Saenko, V., et al., 2011. The chernobyl accident and its consequences. Clinical Oncology 23 (4), 234–243.
Scheffey, J.L., et al., 1990. Analysis of Quick Response Fire Fighting Equipment on Submarines-Phase II, Full Scale Doctrine and Tactics Tests. NRL Memorandum Report 6632. Washington, D.C.
Shah, C.P., 1998. Public Health and Preventive Medicine in Canada. University of Toronto Press, Toronto, Canada.
Shallcross, D.C., 2013. Using concept maps to assess learning of safety case studies–the Piper Alpha disaster. Education for Chemical Engineers 8 (1), e1–e11.

Shokri, M., Beyler, C., 1989. Radiation from large pool fires. Journal of Fire Protection Engineering 1 (4).
Silvestrini, M., et al., 2009. Energy concentration factor. A simple concept for the prediction of blast propagation in partially confined geometries. Journal of Loss Prevention in the Process Industries 22 (4), 449–454.
Smith, D.A., Cox, G., 1992. Major chemical species in buoyant turbulent diffusion flames. Combustion and Flame 91, 226–238.
Society of Fire Protection Engineers, 2002. SFPE Handbook of Fire Protection Engineering, third ed. National Fire Protection Association (NFPA).
Spouge, J., 1999. A Guide to Quantitative Risk Assessment for Offshore Installations. The Center for Marine and Petroleum Technology, Aberdeen, UK. Publication 99/100.
Steinhauser, G., et al., 2014. Comparison of the Chernobyl and Fukushima nuclear accidents: a review of the environmental impacts. Science of The Total Environment 470–471 (0), 800–817.
Stephen, R.T., 2000. An Introduction to Combustion: Concepts and Applications. McGrw-Hill Companies, Inc.
Strehlow, R.A., 1975. Blast waves generated by constant velocity flames: a simplified approach. Combustion and Flame 24, 257–261.
Strehlow, R.A., 1979. The Blast Wave from Deflagrative Explosions, an Acoustic Approach, DTIC Document.
Stull, D.R., 1977. Fundamentals of Fire and Explosion. American Institute of Chemical Engineers.
Suardin, J., Sam Mannan, M., El-Halwagi, M., 2007. The integration of Dow's fire and explosion index (F&EI) into process design and optimization to achieve inherently safer design. Journal of Loss Prevention in the Process Industries 20 (1), 79–90.
Taylor, G., 1946. The air wave surrounding an expanding sphere. Proceedings of the Royal Society of London. Series A, Mathematical and Physical Sciences 186 (1006), 273–292.
TNO, 2005. Methods for the Calculation of Physical Effects Due to Releases of Hazardous Materials (Liquids and Gases). Ministerie van Verkeer en Waterstaat, The Hague.
Uznaski, D.T., Buchlin, J.M., 1998. Mitigation of industrial hazards by water spray curtains. In: The 5th Conference of the International Emergency Management Society (TIEMS), Washington, DC.
Van den Berg, A., 1985. The multi-energy method: a framework for vapour cloud explosion blast prediction. Journal of Hazardous Materials 12 (1), 1–10.
Van Den Bosh, C., Weterings, R., 1997. Methods for the Calculation of Physical Effects (Yellow Book). The Hague (NL): Committee for the Prevention of Disasters.
Van der Voort, M.M., et al., 2012. "Blast from explosive evaporation of carbon dioxide: experiment, modeling and physics." Shock Waves 22 (2), 129–140.
Van Wingerden, C., et al., 1989. Vapor cloud explosion blast prediction. Plant/Operations Progress 8 (4), 234–238.
Van Wingerden, C., et al., 1990. Parameter Study into Blast Generated by Vapour Cloud Explosions. Safety and Loss Prevention in Chemical and Oil Processing Industries, 15.
Van wingerden, K., et al., 1994. Turbulent flame propagation in gas mixtures. In: Institution of Chemical Engineers Symposium Series.
Venart, J.E.S., 2004. Flixborough: the explosion and its aftermath. Process Safety and Environmental Protection 82 (2), 105–127.
Webber, D.M., et al., 2010. LNG Source Term Models for Hazard Analysis: A Review of the State-of-the-art and an Approach to Model Assessment. U.K. Health and Safety Executive, Buxton, U.K.
Wiekema, B., 1980. Vapour cloud explosion model. Journal of Hazardous Materials 3 (3), 221–232.
Willey, R.J., et al., 2005. The Bhopal tragedy: its influence on process and community safety as practiced in the United States. Journal of Loss Prevention in the Process Industries 18 (4–6), 365–374.
Williams, G., 1985. Combustion Theory.
Wilson, D.J., Simms, B.W., 1985. ExposureTime Effects on Concentration Fluctuations in Plumes. University of Alberta.

Wilson, D.J., 1991. Accounting for peak concentration in atmospheric dispersion for worst case hazard assessments. In: International Conference on Modelling and Mitigating the Consequences of Accidental Release, New Orleans, LA.

Worsfold, S.M., et al., 2012. Review of the explosibility of nontraditional dusts. Industrial & Engineering Chemistry Research 51 (22), 7651–7655.

Yañes, M., Jolgar, F., Modarres, M., 2002. Generalized renewal process for analysis of repairable systems with limited failure experience. Reliability Engineering and System Safety 77, 167–180.

Zabetakis, M.G., et al., 1959. Flame temperatures of limit mixtures. In: Seventh Symposium on Combusion. London, p. 484.

Zabetakis, M.G., 1965. Fire and explosion hazards at temperature and pressure extremes. In: AICHE Institute of Chemical Engineers Symposium Series 2, pp. 99–104.

Zukoski, E.E., et al., 1984. Visible structure of buoyant diffusion flames. In: Twentieth Symp. (Int.) Combut., the Combustion Institute, pp. 361–366.

CHAPTER

# MOLECULAR-LEVEL MODELING AND SIMULATION IN PROCESS SAFETY

3

## 3.1 INTRODUCTION

> Life ... is a relationship between molecules.
>
> **Linus Pauling (Hager, 1995)**

Molecular modeling is the science (or art) of representing molecular structures numerically and simulating their behavior with the equations of quantum and classical physics (Frenkel and Smit, 2002). Investigating chemical and material problems at an atomistic scale through capturing and understanding molecular interactions requires effective implementation of molecular modeling or computational chemistry techniques. The terms ("molecular modeling" and "computational chemistry") are collectively used to refer to the theoretical methods and computational techniques used to model or mimic the behavior of molecules. It is a branch of chemistry which implements computer simulation of atoms and electrons to assist solving chemical and material problems. Simply put, it is a merger of:

- Physics of a system at the atomistic level
- Mathematical representation of the same in a solvable form
- Set of computational algorithms for solving the developed set of differential algebraic equations.

The type of problems that can be addressed through implementation of computational chemistry tools and techniques can be broadly categorized into (Lewars, 2010):

- Optimization of molecular geometry
- Energetics of molecules and transition states
- Chemical reactivity
- Infrared (IR), ultraviolet (UV), and nuclear magnetic resonance (NMR) spectra
- Molecular interactions
- Physical properties of a molecule

Molecular modeling empowers us with the ability to dig deeper into the characteristics of hazardous chemicals and processes, which are otherwise difficult and expensive to obtain through experimental observations. Application of molecular modeling has significantly increased in recent years and, among others, has been used in the design and development of novel materials, drug discovery, protein engineering, microelectronics, and hydrogen storage. While it has the potential to replace unnecessary experiments, insight gained from it can guide a new set of experiments.

**Multiscale Modeling for Process Safety Applications.**

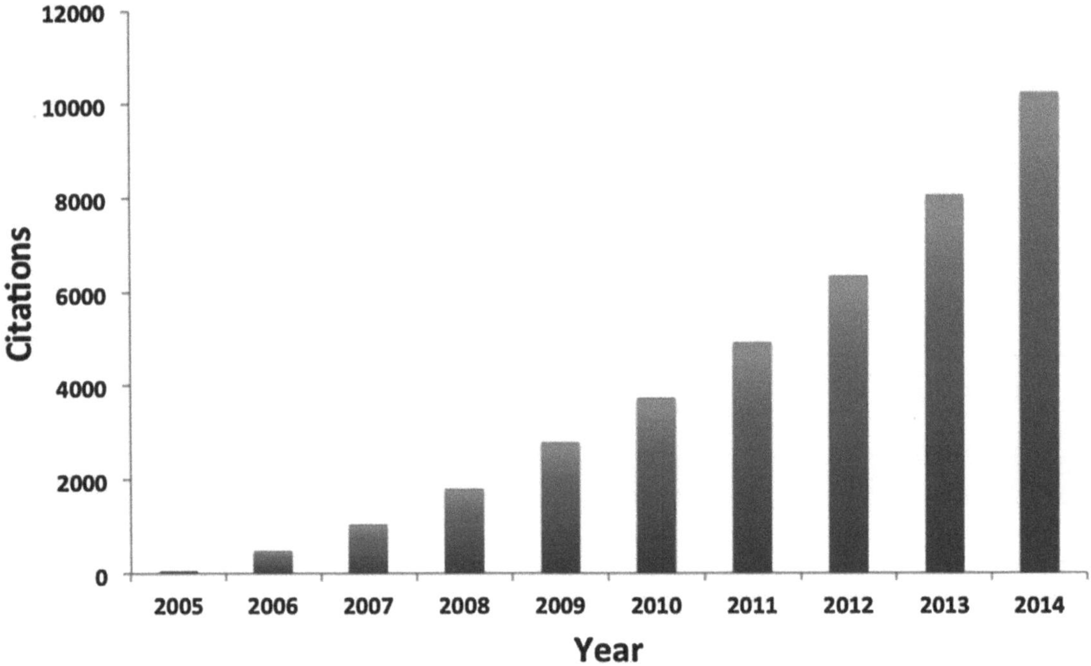


**FIGURE 3.1**

Citations in each year.

The realized potential of molecular modeling among researchers is illustrated through Figure 3.1. The figure is constructed based on a search in ISI Web of Science for citation record of highly cited English-language papers implementing molecular modeling studies within the engineering domain.[1] The trend of increasing citations, as illustrated in the figure, testifies to the increased utilization of molecular modeling tools and techniques for engineering studies.

For our purposes, within the realms of molecular modeling, the set of molecular modeling approaches can be broadly categorized in the following way, each of which can be implemented based on the problem objective:

1. Quantum mechanics/ab initio
2. Molecular mechanics
3. Molecular dynamics and Monte Carlo
4. Quantitative structure—property relationship (QSPR)/quantitative structure—activity relationship (QSAR)

[1]**TOPIC:** (Molecular Modeling) *OR* **TOPIC:** (Molecular Dynamics) *OR* **TOPIC:** (Quantum Mechanics) *OR* **TOPIC:** (Coarse grain) *OR* **TOPIC:** (Mesoscale) *OR* **TOPIC:** (QSPR) *OR* **TOPIC:** (QSAR) *OR* **TOPIC:** (COSMO-RS) **Refined by: ESI Top Papers:** (Highly Cited Papers) AND **DOCUMENT TYPES:** (ARTICLE) AND **LANGUAGES:** (ENGLISH) AND **RESEARCH DOMAINS:** (SCIENCE TECHNOLOGY) AND **RESEARCH AREAS:** (ENGINEERING) **Timespan:** 1864—2014.

The flow of this section starts with a brief introduction of the computational chemistry techniques, followed by application of the approaches for addressing problems related to industrial process safety. Figure 3.2 demonstrates some of the approaches used in modeling and simulation of materials at the molecular level.

Before digging deeper into the application of molecular modeling approaches, one needs to think about:

- Suitable methods for addressing the problem objective
- Suitable and accessible software for the purpose
- The availability of potential or force field parameters for molecular dynamics simulations
- Tradeoff between available computational resources and system size to be studied

Table 3.1 illustrates the computational cost associated with various levels of molecular modeling approaches. While solving Schrödinger's equation is never an option, choice of the remaining three depends on the answers to the concerns just specified.

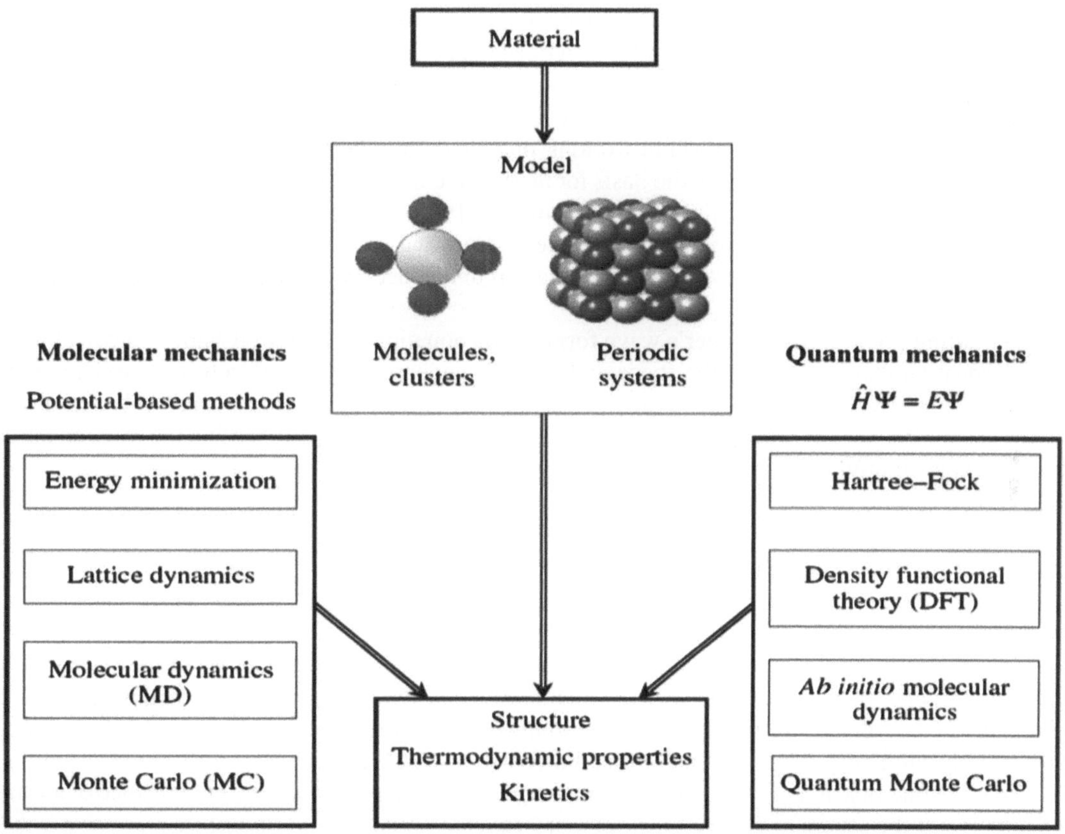


**FIGURE 3.2**

Various computational techniques used for modeling and simulation of materials at the atomistic scale (Stølen and Grande, 2004).

**Table 3.1 Various Computational Methods and Their Scaling With System Size (Lee, 2011)**

| Methods | Scaling | $N_{max}$ (atoms) |
|---|---|---|
| Schrödinger | $O(e^N)$ | 1 |
| Hartree–Fock | $O(N^4)$ | ~50 |
| DFT | $O(N) \sim O(N^3)$ | ~200 |
| Molecular dynamics | O(N) | $10^{7-11}$ |

### 3.1.1 QUANTUM MECHANICS/AB INITIO APPROACHES

The rules of application of quantum mechanics are unquestioned and the accuracy and precision of its predictions are unsurpassed in the entire history of science. Although heated debate continues about how quantum theory should be interpreted, there can be no debate about whether the theory is fundamentally correct (Baggott, 2011).

Quantum mechanical calculations are fundamental in nature and deals with physical phenomena at the atomistic scale where the units are in the order of Plank's constant ($6.62606957 \times 10^{-34}$ $m^2$ kg/s). Two equivalent formulations of quantum mechanics were derived by Schrödinger and Heisenberg. The Heisenberg approach (Heisenberg, 1925) came from formulation of the quantum theory through development of matrix mechanics while working on calculating the spectral lines of hydrogen. The other formation, which later became the basis for nearly all computational chemistry methods (Young, 2004), is Schrödinger's wave formulation (Schrodinger, 1926) of quantum mechanics.

His methods were rapidly adopted by other investigators and applied with such success that there is hardly a field of physics or chemistry that has remained untouched by Schrödinge's work (Pauling and Wilson, 1985).

The starting point of Schrödinger's wave formulation equation was the well-known classical wave motion equation that interrelates the space and time dependencies of any waveform through a wave function, generally represented as $\psi$(psi). It is also known as probability amplitude functions. Schrödinger introduced mathematical description of electron as a wave for enabling the wave equation to provide meaningful solutions. The equation is represented as

$$\hat{H}\,\psi = \mathrm{E}\psi$$

where $\hat{H}$ is the Hamiltonian operator, $\psi$ is the wave function, and E is the energy. It represents the quantum version of the traditional Hamiltonian equation and stands for the total energy of particles exhibiting wave properties. This means that the differential operator, for the total energy of a system, the Hamiltonian operator, operates on the wave function to yield the total energy of the system. The name of the equation has been given as wave equation as it is a second order differential equation with respect to the coordinates of the system, somewhat similar to the wave equation of classical theory. However, the similarity is not close (Pauling and Wilson, 1985).

The wave function $\psi$, is a function of the electron and nuclear positions. The significance of a wave function is that the square of a wave function represents the probability of finding an electron per unit volume. Hence, the wave function is a fundamental system property of a molecule and is the central entity in all quantum mechanical calculations. In order to obtain a physically relevant solution of the

Schrödinger's equation, the wave function must be continuous, single-valued, normalizable, and antisymmetric with respect to the interchange of electrons (Young, 2004). Theoretically, the mathematical foundation of quantum mechanics is precise enough to ensure exact estimation of molecular properties. However, in practice, the exact solution has been obtained for only one-electron systems. In lieu of that, a plethora of methods have been developed over the years for finding approximate solutions of Schrödinger's representation of multiple-electron systems.

The various methods in quantum chemical calculations refer to the different types of approximation of the solution of Schrödinger's wave formulation equation for it to be effective for multiple-electron system. Finding and describing approximate solutions of electronic Schrödinger's wave formulation equation have been major preoccupation of quantum chemists since the very beginning of quantum mechanics (Szabo and Ostlund, 2012).

Currently, there are two main branches of electronic structure approaches used by researchers with various degrees of assumptions. While statistical errors, as realized in various experimental measurements, are not present in quantum chemical calculations, inherent errors appear and are associated with the various degrees of assumptions required to facilitate computational calculations. Based on the problem complexity, the practical approach to electronic structure calculations involves approximation, and it is vital to understand these along with its effects on the solution. The wave function–based methods start with Hartree–Fock (HF) method, in which many body wave functions are approximated as combinations of single-particle wave functions. Density functional theory (DFT), on the contrary, uses charge density of the system as an alternative representation to the many-body wave function. Since most calculations work on differences in energy (between two states) of systems, the accuracy of calculations is increased by error cancellation, as absolute errors in individual calculations owing to approximations nullify each other while estimating differences.

The HF method captures most of the important interactions involved in a system but does not incorporate correlation energy accounting for electron–electron interaction. It is reasonably good for computing the structures and vibrational frequencies of stable molecules and some transition states, but it is insufficient for accurate modeling of the energetics of reactions and bond dissociation. Accordingly, the HF method is used as a starting point for first principle calculations. Perturbation theories such as Møller–Plesset (MP$n$ where $n$ is 2 for the simplest level of theory) have been developed to go beyond the HF level of accuracy by providing perturbative corrections to HF results. The computational cost of MP$n$ theories increase with values of $n$ and is typically not used beyond MP5 (Foresman and Frisch, 1996). However, one must note that MP$n$ theories cannot succeed where HF theory fails (Brazdova and Bowler, 2013). A comparison of different models in terms of accuracy and cost involved is presented in Figure 3.3.

In general, the relative accuracy of ab initio results can be presented as (Young, 2004):

$$\mathrm{HF} \ll \mathrm{MP2} < \mathrm{CISD} \cong \mathrm{MP4} \cong \mathrm{CCSD} < \mathrm{CCSD(T)} < \mathrm{CCSDT} < \mathrm{Full\ CI}$$

DFT (Kohn and Sham, 1965) handles the difficulty of solving equations involving electron–electron equation by replacing the term through effective potential through mean field approximation. The theory computes electron correlation via general functionals (functions of functions) of the electron density. The first theorem of density functional theorem proves that *the ground state energy from Schrödinger's equation is a unique functional of the electron density*. This one-to-one mapping of ground-state wave function of a molecule to its ground-state electron density leading to the fact that by solving ground-state electron density of a molecular system allows one to estimate all molecular

| Model | Name | Description | Error | Scaling | Cost |
|---|---|---|---|---|---|
| | | | equil geom | ~ basis | relative |
| | | | kcal/mol | functions | 1-10 |
| AM1 | Austin Model 1 | semi-empirical | 10–30 | $N^2$ | 1 |
| HF | Hartree–Fock | averaged electron–electron | 10–20 | $N^4$ | 2 |
| B3LYP | Becke, Lee, Yang, Parr | density functional | 5 | $N^4$ | 3 |
| MP2 | Moller–Plesset | perturbation, correlated, 2nd order | 10 | $N^5$ | 3 |
| MP3 | Moller–Plesset | perturbation, correlated, 3rd order | 5 | $N^6$ | 5 |
| CISD | CI singles, doubles | configuration interaction, correlated | 5.8 | $N^6$ | 5 |
| CCD | CC doubles | coupled cluster, correlated | 2.4 | $N^6$ | 5 |
| CCSD | CC singles, doubles | coupled cluster, correlated | 1.9 | $N^6$ | 6 |
| QCISD | QC singles, doubles | quadratic configuration interaction | 1.7 | $N^6$ | 6 |
| MP4SDQ | Moller–Plesset | perturb, corr, 4th order, ignore triples | 2.7 | $N^7$ | 6 |
| MP4 | Moller–Plesset | perturbation, correlated, 4th order | 1.3 | $N^7$ | 7 |
| G2 | G2, multilevel corrections | composite method | 1.2 | $N^7$ | 6 |
| G3* | G3, multilevel corrections | composite method, Truhlar, et al | 1.0 | $N^7$ | 7 |
| QCISD(T) | QC singles, doubles, (triple) | quadratic configuration interaction | 0.3 | $N^7$ | 7 |
| MP5 | Moller–Plesset | perturbation, correlated, 5th order | 0.8 | $N^8$ | 8 |
| MP6 | Moller–Plesset | perturbation, correlated, 6th order | 0.3 | $N^9$ | 9 |
| CCSD(T) | CC singles, doubles, (triple) | coupled cluster, singles, doubles, triple | 0.3 | $N^8$ | 8 |
| CCSDT | CC singles, doubles, triples | coupled cluster, singles–triples | 0.2 | $N^8$ | 9 |
| MP7 | Moller–Plesset | perturbation, correlated, 7th order | | $N^{10}$ | 10 |
| CCSDTQ | CC sings, doubs, trips, quads | coupled cluster, singles–quadruples | 0.01 | $N^{10}$ | 10 |

**FIGURE 3.3**

Comparison of various model performances in electronic structure calculations.

properties, including energetics and wave function. The second theorem on density functional theorem states that *the electron density that minimizes the energy of the overall functional is the true electron density corresponding to the full solution of the Schrödinger's equation.*

Despite the success of DFT, results from its calculation using LDA, in particular band gap calculation (difference in energy between highest occupied molecular orbital (HOMO) and lowest unoccupied molecular orbital (LUMO)) is inaccurate (Brazdova and Bowler, 2013). DFT approximates both exchange and correlation energy, while HF calculates exact exchange energy but ignores correlation energy. While hybrid functionals based on HF exchange energy rectify some of these deficiencies in DFT by adding a fraction of this exchange energy, the added component must be tested for the problem to be studied.

In quantum mechanics calculations, one needs to select a representation for wave function. This is done through specifying a basis set. This is analogous to specifying between cartesian or cylindrical coordinates for setting up a problem to represent position in space. Unfortunately, there is no perfect basis set, and a variety of basis sets are available to be chosen from. Ideally, in choosing a basis set, one must look for the following features (Brazdova and Bowler, 2013):

- Well adapted (i.e., few functions needed)
- Systematic convergence
- Good computational scaling with basis

**Table 3.2 Description of Some Commonly Used Basis Sets in Electronic Structure Calculations**

| Basis Set | Description |
|---|---|
| STO-3G | Minimal basis, qualitative results—large systems |
| 3-21G | More accurate results on large systems |
| 6-31G(d) | Moderate set, common use for medium systems |
| 6-31G(d,p) | Used where H is site is of interest, more accurate |
| 6-31+G(d) | Used with anions, excited states, lone pairs, etc. |
| 6-31+G(d,p) | Used with anions, excited states, lone pairs with H is site is of interest |
| 6-311++G(d,p) | Good for final accurate energies but expensive |

- Simple operations/calculations between basis functions
- Easy calculations of forces
- Bias free

Ideally, the basis set needs to be able to approximate the actual wave function sufficiently to give meaningful results. Table 3.2 provides description of some of the commonly used basis sets and their appropriate application domain.

Similar to any other computational approaches, electronic structure calculations require making some choices from a given set of models. Some generic recommendations are (Foresman and Frisch, 1996):

- Barring the computational cost, use B3LYP/6-31G(d) for geometries and zero point corrections and B3LYP with the largest practical basis set for energy calculations.
- For very large systems, performing multistage optimization such as performing HF optimization and frequency calculation followed by a B3LYP/6-31G(d) optimization may be more cost-effective.
- Somewhat in the similar lines of the above point, in cases where B3LYP/6-31G(d) is too expensive to perform, the best tradeoff is to use HF/3-21G(d) geometries and zero point corrections followed by the final B3LYP energy calculations.
- When AM1 geometries are practical for a molecular system, using B3LYP functional for the final single point calculation will still offer significantly more accurate results.

In general, to setup an ab initio or first principle–based calculations a researcher has to provide the following as an input:

- The molecular geometry (including spin state) and the system size
- Basis set and model selection for determination of the wave function
- Properties to be calculated
- Types of calculations and associated assumptions

These inputs can be provided in software GUI (Gaussian Inc.) or as explicit declarations inside input files (VASP).

### 3.1.2 QSPR/QSAR

In medicinal chemistry, regression modeling approaches at the molecular level, such as QSAR and QSPR, are exceedingly practiced. In varied studies, in a similar manner, attempts have been made to map the flammable activity of a molecule to its structural parameters (Gharagheizi, 2009). The QSPR takes the form of the following mathematical model:

$$\text{Activity} = f(\text{physiochemical properties and/or structural properties}) + \text{error}$$

The functional form is then found out through data mining of large data sets. Development of QSPR-based mathematical models to estimate flammability properties has gained popularity among researchers in recent times and is considered one of the modern flammability limits estimation methods. Structure–property (activity) relationships are qualitative or quantitative empirically defined relationships, between molecular structure and measurable properties or activities (Young, 2004). Expressed in a different way, QSPR and QSAR studies attempt to correlate several structural descriptors of a chemical compound with its physicochemical properties or biological activities. Ideally, to obtain a reliable and robust QSPR model, there should be an order of magnitude higher number of compounds than the number of parameters fitted (Young, 2004). The molecular descriptors are calculated to represent the molecular structure of the compound incorporating topological, charge, and geometric descriptor. Most of the molecular descriptors are easily determined, such as molecular weight, moment of inertia, and so on. The primary challenge in this approach lies in describing the molecular structure through appropriate molecular descriptor and selection of the appropriate modeling method (Yao et al., 2004). At present, molecular descriptors such as topological indices and quantum chemical parameters have been demonstrated to describe the structural property of a molecule (Karelson, 2000; Karelson and Lobanov, 1996; Todeschini and Consonni, 2008; Devillers and Balaban, 2000). Properties such as electronegativity, charge density, Fukui function, HOMO and LUMO energies, and electron affinity are examples of quantum chemical descriptors. A full set of list can be found in any book or book chapter on molecular descriptors (e.g., Todeschini and Consonni, 2008; Young, 2004). Since molecular descriptors are related to molecular structures, LFL values have also been estimated (Pan et al., 2009) using QSPR for organic compounds, where the estimation was carried out using the web-based DRAGON software program (Todeschini et al., 2006; Todeschini and Consonni, 2008, 2009; Tetko, 2003). The software, which has the capability of providing close to 5000 molecular descriptors, takes in molecular structure of a chemical as an input to calculate its molecular descriptors. As part of the process, which is briefed here next, the authors (Pan et al., 2009) optimized the geometry of all the molecular structures to obtain minimum energy configuration. The work used AM1 semiempirical quantum mechanics method and MM+ molecular mechanics force field in order to perform the geometry optimization through energy minimization of the molecular structure. But, based on accuracy requirement, minimum energy configuration can also be achieved by other quantum mechanical approaches such as the HF method or DFT.

Getting into the nitty gritty details of the QSAR or QSPR approaches is beyond the scope of this book. For the sake of clarity and convenience, we will briefly demonstrate the steps involved to develop QSPR or QSAR models for flammability properties. These are typically regression models that relate a set of predictor variables to a set of response variables. The underlying hypothesis behind a QSAR/QSPR model is that various physicochemical properties or biological activities of a compound are strongly coupled with its molecular structure. So once a reliable relationship has been established between a predictor and response variables, it can be used to predict the same properties for other compounds for which data do not exist. This is particularly helpful in designing novel materials but also of significant importance in regard to properties of existing chemicals. One may classify steps in developing QSPR-based approaches in several different ways. In the coarser side, only two broad segments (Pan et al., 2010) are defined:

1. Design and generate molecular descriptors
2. Constructing QSPR models with proper descriptors

A little more classification in the approach is presented (Yao et al., 2004):

1. Data collection
2. Molecular descriptor selection
3. Developing model correlation
4. Model evaluation and validation

An even better resolution set of key steps involved in the development of a generic QSPR/QSAR model can be shown as illustrated in Figure 3.4 (Knotts et al., 2001):

As one glances through the generic steps, one can see that accurate geometry optimization and deriving quantum chemical descriptors from that require detail-level modeling involving atoms and electrons. Since the values of many descriptors are related to the values of bond length and angles, having an optimized geometry of the chemical is required prior to performing any QSPR or QSAR calculations. This is where quantum mechanical or first principle–based calculations can be implemented to optimize its geometry by energy minimization techniques. If a suitable force field of the chemical of interest is available, molecular mechanics approach can also be implemented for geometry optimization of the molecule. The quantum mechanical approach is further used to generate the molecular descriptors from the optimized geometry and the obtained wave function. Once the geometry of the molecule is optimized, values of several descriptors are calculated using a suitable software. To select the best descriptors, first, descriptors with little or no variations across compounds

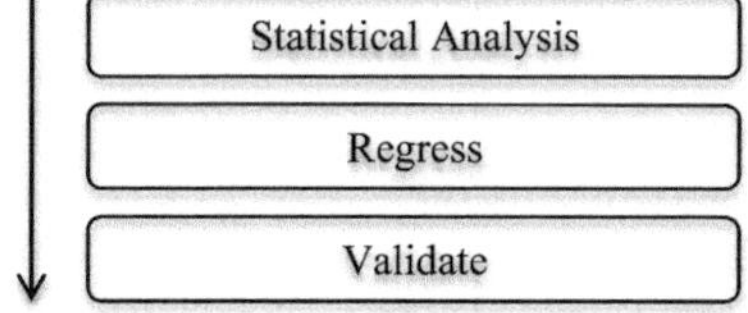


**FIGURE 3.4**

Typical steps in QSPR/QSAR in developing correlations.

*Reprinted with permission from Knotts et al. (2001).*

are removed. By finding pairwise correlations between the remaining descriptors, only one among every two descriptors that are strongly correlated is retained so as to remove duplicate contributors and develop a more efficient model. To choose correlated descriptors, one has to define a high correlation coefficient ($\rho$) as the cut off point; for example, in the case of Pan et al. (2010), it was defined to be $\rho > 0.96$. Examples of molecular descriptors can be atomic charges, dipole moment, HOMO and LUMO energies, etc. Considering the fact that the number of molecular descriptors typically can run into thousands, reducing the number of descriptors to the appropriate and most effective ones is an important step of the model development process. Through state-of-the-art hybrid approaches of coupling various regression analysis methods with genetic algorithm, such as that with multiple linear regression (Pan et al., 2010; Leardi et al., 1992) or partial least squares (PLS) regression (Leardi and González, 1998), optimal molecular descriptor set to represent mapping between the property set, such as flammability characteristics, and molecular descriptor set are identified. As a side note, in QSPR studies, chosing a appropriate regression technique for success of the model among choices such as multiple linear regression (MLR), PLS, and various types of artificial neural networks (ANN) is of prime importance (Pan et al., 2009). The final step in the development of the QSPR model is to validate the model through its predictive capability and robustness. Generally a good data set such as that from DIPPR 801, is divided into two sets for:

1. Training and QSPR model development
2. Model validation

The steps involved in developing the QSPR model are very similar in nature to the development of other QSPR models as developed for process safety applications as demonstrated in earlier sections.

### 3.1.2.1 Data set

To develop QSPR-based mathematical models on predicting severity of dust explosibility, the initial step involves accumulating explosibility characteristics data, such as deflagration index ($K_{st}$) and maximum pressure ($P_{max}$), of dusts. This is a standard step in any QSPR method, where the first step is to gather quality experimental data for the properties or parameters to be modeled. In addition to the data of $K_{st}$ and $P_{max}$, one also needs the data of the parameters whose function are $K_{st}$ and $P_{max}$. Parameters correlated to $K_{st}$ and $P_{max}$ in dust explosion scenarios typically are representative particle size, moisture content, humidity, oxygen availability, shape, concentration, ignition source, and others. In general, the severity of dust explosion increases with smaller particle size. While an experimental data set is required to train and validate QSPR models, once built, it can be used to estimate dust explosion–related parameters for molecules of similar structures, eliminating the need for further experiments of chemicals whose structure–property relationship can be addressed by the developed QSPR models. In the work by Reyes et al. (2011), the authors used experimental combustion and explosion data of 31 chemical dusts from online database GESTIS-DUST-EX (Ifa, Gestis-Dust-Ex, 2014). The authors in their work used the median values for choosing the dust particle size. For the purpose of this study and validate their approach, the authors in this case extracted more data from other median values and created two specific data sets. One data set is to train and develop the molecular information–based mathematical model through the QSPR approach, and the other data set was used as a test set to validate their approach. It must be borne in mind that, based on the particle size distribution, median value of a data set may or may not be the best representation of the particle size to be used as an input for QSPR modeling. As an example, if the particle size is not uniformly distributed across the frequency diagram leading to large standard deviation, the calculation method for determination of representative particle size needs to be modified accordingly.

#### 3.1.2.2 Geometry optimization

The QSPR methodology requires the incorporation of the structural information of the chemicals of interest. Since the aim is to build a mathematical model to map between the property of interest, explosivity characteristics in this case, and the chemical structure among others, for the purpose molecular modeling, optimization of the molecular structure yielding in its global minimum energy structure is of paramount importance. At the molecular level, the geometry can be optimized through implementing any of the following approaches, namely, first principle calculations, semiempirical quantum calculations, or implementation of classical force field methods using molecular dynamics as applicable to the chemical of interest. Owing to its fundamental nature, of the three methods, optimizing geometry of molecular structure using first principle calculations is expected to result in the most consistent and accurate results for any set of chemicals on interest, with a tradeoff with available computational resources as discussed in the introduction section of this chapter. In the particular study, by Reyes et al. (2011) the authors implemented DFT calculations at the B3LYP/6-31g level for optimizing the geometry of the molecules. The authors chose B3LYP function based on its effective performance for the systems involving organic compounds. A brief description along with pros and cons of various functionals in DFT methodology has been discussed at the beginning of this chapter.

#### 3.1.2.3 Molecular descriptors

The optimized molecular structure obtained in the above step is used as an input to determine molecular descriptor. To generate these descriptors, various software packages are available. Typically, these packages derive a three-dimensional optimized molecular structure of the chemical of interest and calculate the descriptors using DFT. The initial substep within this step is to select a large number of descriptors and then reducing it to only the relevant ones and having only significantly sensitive descriptors with respect to the properties of interest. In general, the molecular descriptors can be of various kinds (Baatia et al., 2013), such as:

1. Constitutional (e.g., number of atoms and bonds)
2. Geometric (e.g., volume and surface area)
3. Topological (e.g., function of atom connectivity)
4. Electrostatic (e.g., partial charges)
5. Quantum chemistry derived (e.g., molecular orbital energies)

From this list, one can appreciate the importance of optimizing the geometry of the molecule prior to evaluate the molecular descriptor values. In the work by Reyes et al. (2011) the authors used the descriptors models and VAMP electrostatics in the simulation model of the Material Studio software package (Accelrys Materials Studio for Materials Modeling & Simulation, 2014). Starting with a set of 200 molecular descriptors, and following a heuristic method, the authors could narrow down their QSPR model to a set of 34 descriptors. Reduction of descriptors is necessary to increase the efficiency of the QSPR model and often done in steps. Further reduction in the number of descriptors can be done through various methods such as ability to calculate the descriptor for all structures, relevancy of the descriptor based on the property of interest, descriptors insignificantly sensitive to molecular structures and the like. In the work by Fayet et al. (2010b), the authors calculated 300 molecular descriptors using the CodessaPro software starting with the optimized molecular structure as obtained through DFT and AM1 calculations. Detailed definition of various quantum mechanical calculation—based and other molecular descriptors descriptors can be found (Karelson and Lobanov, 1996; Karelson, 2000).

#### 3.1.2.4 *Correlation-based model development*

Depending on available data, its quality, and other factors, one can resort to several approaches in determining the correlations between the property of interest and the calculated molecular descriptors. For small data sets, a parametric space search technique such as GFA algorithm (genetic function approximation) can build several QSPR models from the available descriptors. In the study done by Reyes et al. (2011) application of the GFA technique resulted in two different QSPR models, each consisting of five parameters to estimate the $K_{st}$ and $P_{max}$ values.

### 3.1.3 MOLECULAR DYNAMICS

First introduced by Alder and Wainwright (1957) by representing molecules as hard spheres, molecular dynamics has long been the tool for researchers to predict bulk properties and gain insight of various molecular systems based on atomistic scale calculations. In 1964, Rahman carried out the first simulation with Lennard-Jones potential on liquid argon as opposed to the hard sphere concept, which was being used until then. The need for molecular dynamics arose as, unlike ideal gases and metallic crystals, the properties of liquids were not easy to derive owing to their nonordered structure and nonideal behavior. Early models of liquids involved hard spheres and disks to represent molecules of a liquid system. Due to the same reason while molecular dynamics emerged as one of the main tool in predicting bulk properties of molecular systems, another method based on probabilistic approach called Monte Carlo, was also developed. Monte Carlo deals systems independent of time and hence is not capable of predicting dynamic properties. Metropolis et al. (1953) laid the foundation for modern Monte Carlo.

The theoretical basis for molecular dynamics modeling is statistical thermodynamics. Knowing the structure and dynamics of a system at the atomistic level can predict the bulk property of a system at the macroscopic level through application of statistical mechanics on the information obtained from molecular dynamics. Molecular dynamics simulation generates information of a system at the atomistic level indicating dynamic changes in a structure through estimating positions and velocities of all atoms of a system. According to classical mechanics, given initial position and velocity of a particle and the forces acting on it at any moment, one can predict its new position and velocity from Newton's second law of motion. The force on each atom at every time step is calculated by estimating the first derivative of the potential energy of the particle with respect to its position at that very instant. When more than one particle is involved in such a system, which is usually the case, different forms of interaction between them contributes to the potential energy. Identification and evaluation of various types of interactions between different types of particles are done with a set of rules called force field. A specific force field is developed based on information either obtained through experiments or first principle/ab initio calculations as introduced in the previous section. The practice in force field development involves development of force field for specific types of systems, such as biological or polymeric systems. Most types of interactions are all covered by a force field. In classical mechanics based calculations (molecular mechanics and molecular dynamics), the potential energy of a system is represented by analytical functions, namely the interaction force fields. The parameters of these functions are optimized so as to reproduce the fundamental properties, density, lattice parameters, vibrational frequencies, and the like. The functional forms are based on quantum mechanical arguments (Exp-6 form for van der Waals interactions, Morse form for bond stretch and harmonic form for angle bending, periodic-truncated Fourier series forms for torsion). Hence, a force field is the mathematical expression that describes the dependence of potential energy of a molecular system to that of the atomic positions of its constituent atoms. A force field has the ability to estimate the change in potential of a system from the knowledge of its initial condition of individual atoms, that is, atom positions and velocities along with the information on the perturbation of the system. The energy of a system can be written as

$$E = E_{VAL} + E_{NB}$$

where

$$E_{VAL} = E_B + E_A + E_T + E_I$$

$$E_{NB} = E_{VDW} + E_C + E_{HB}$$

where $E_B$ is energy due to bond stretching (two body); $E_A$ is energy due to angle bending (three body); $E_T$ is energy due to torsion (four body); $E_I$ is energy due to out of plane configuration (four body); $E_{VDW}$ is energy due to van der Waals interaction; $E_C$ is energy due to columbic interaction; and $E_{HB}$ is energy due to hydrogen bonding.

While the first four terms are attributed to bonded interactions, the last two terms are due to nonbonded interactions. The calculation of the nonbonded interaction is computationally expensive and contributes significantly to the overall simulation time. The electrostatic interactions are also called long-range interactions as they decay inversely with the value of "r" and hence contribute even for a large value of "r."

Force fields are classified in different categories. There are second-generation force fields developed by high parameterization (e.g., CFF, PCFF COMPASS, etc.), rule-based force fields like Universal and Dreiding where parameters are decided by rules such as hybridization, classical or first-generation force fields like AMBER, CHARMM, and CVFFm which is based on parameterization but from experimental values as opposed to that of second generation, which is based on quantum input and special purpose force field.

Table 3.3 provides examples of functional forms as used in some of these force fields for estimating cost for departing from minimum energy configuration of a molecule.

In a molecular dynamics simulation, the system and its characteristics is defined by its structural behavior as captured by the force field within a given ensemble. The success of MD simulations is primarily due to the existence of statistical thermodynamics. It acts as the bridge between microscopic structure and the macroscopic bulk properties. One of the basic concepts in statistical mechanics is that if one waits long enough, the observer will be able to see almost all the microscopic states of the system for which the system will have the same set of macroscopic properties. This is termed as ergodicity. In other words, a set of macroscopic properties that completely defines a system has many microstates. These sets of macroscopic properties by which a system is defined completely are defined as different ensembles.

There are primarily four kinds of ensembles in statistical mechanics. Ensembles are nothing but a set of configurations (microstates) of the same set of molecules (making the system) while being consistent with the constraints with which the system is characterized macroscopically. The main four

**Table 3.3 Illustration of Energy Forms in Bonded Interactions and Their Corresponding Functional Forms**

| | Energy Forms | Example Functional Form(s) |
|---|---|---|
| Bond stretching | 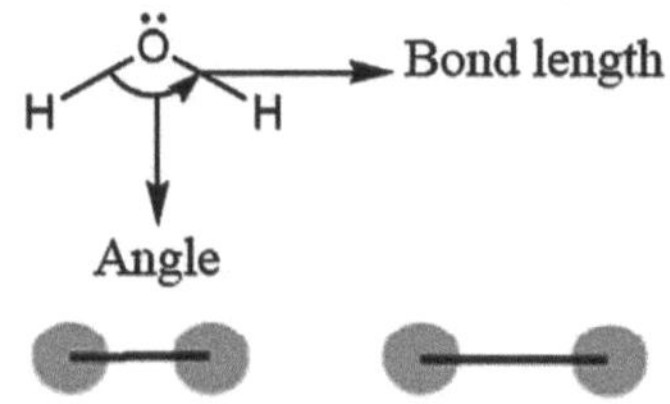  | $E = \frac{1}{2}k_e(R - R_e)^2$<br>$E = D_e(e^{-\alpha(R-R_e)} - 1)^2$ |
| Angle bending | 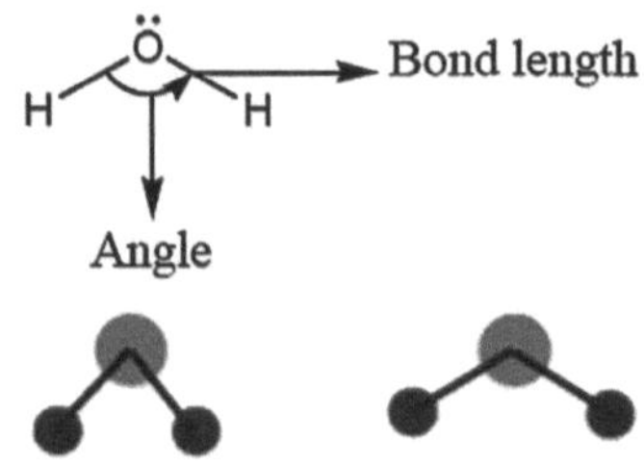  | $E_{IJK} = \frac{1}{2}C_{IJK}[\cos\theta_{IJK} - \cos\theta_J^0]^2$<br>$E_{IJK} = \frac{1}{2}K_{IJK}[\theta_{IJK} - \theta_J^0]^2$ |
| Torsion | 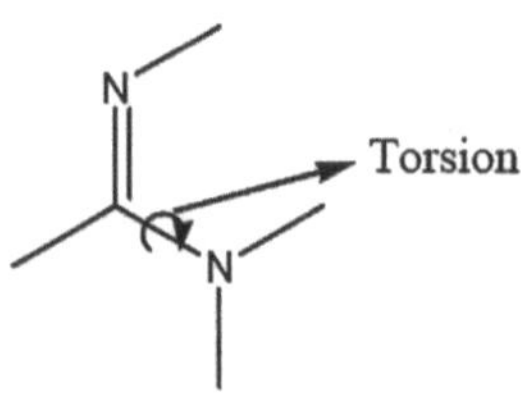  | $E_{IJK} = \frac{1}{2}V_{JK}\{1 - \cos[n_{JK}(\vartheta - \vartheta_{JK}^0]\}$ |

kinds of ensembles are namely microcanonical (N, V, E), canonical (N, V, T), grand canonical (T, V, $\mu$), and isothermal–isobaric (N, P, T) ensembles where N stands for number of particles, V stands for volume of the system, P is pressure, T is temperature, and $\mu$ is chemical potential. Different algorithms exist to ensure the constant property dynamics while running MD. All these ensembles are related to macroscopic property by the corresponding partition function.

- *NVE or microcanonical*: The value of macroscopic properties N, V, and E are constant in this ensemble. As these three macroscopic properties are kept constant, the values of pressure, temperature, and chemical potential are determined for the equilibrated system.
- *NVT or canonical ensemble*: The values of N, V, and T are constant in this ensemble.
- *NPT or isothermal–isobaric ensemble*: In this ensemble, as the name suggests, the different systems that are dealt with have the same values of N, P, and T.
- *TV $\mu$ or grandcanonical ensemble:* Here, the values of T, V, and $\mu$ are kept constant. Any standard book (Sandler, 2010; Hill, 2012; McQuarrie, 2000) can be referred to for a detailed description of these ensembles and their relation to the macroscopic property of the material.

## 3.2 FLAMMABILITY LIMITS

Flammability limits, namely LFL and UFL, of a material, are two of the most important variables used to assess hazards for handling, storing, and transportation of chemicals. Inaccurate estimates and understanding of mixture flammability limits have the potential of significant impact from maintaining hazardous condition ignorantly in a facility to spending significant capital expenditure in mitigating minimal consequence event. This calls for improved understanding and reliable estimates of LFL and UFL of individual chemical and their mixtures either through experiments or computational modeling. We have briefly touched on estimation of flammability limits through Le Chatelier's equation in Chapter 2 (equation 2.1) In addition to compositional changes, flammability limits also varies with temperature and pressure as demonstrated in Chapter 2.

One of the most extensive resources for finding flammability data is the work done by Zabetakis (Drysdale, 2011; Zabetakis, 1965; Mannan and Lees, 2005; Consultants, 2012), which is largely based on experimental results obtained from an apparatus developed at the U.S. Bureau of Mines (Mannan et al., 2006; Coward and Jones, 1952). The work by Coward et al. (Fayet et al., 2009a; Coward and Jones, 1952; Gustin, 1998; Etchells, 1997) and Kuchta (Kaszniak, 2010; Kuchta, 1985, 1986; Mannan and Lees, 2005) also contributed much needed original flammability information for chemicals of various classes. However, owing to its expensive, time-consuming, and hazardous nature, dependence on test methods, and apparatus used (Crowl, 2012; Britton, 2006), development of reliable theoretical prediction methods for estimating flammability characteristics is of significant importance (Drysdale, 2011; Pan et al., 2010; van den Bosch and Weterings, 2005; Vidal et al., 2004; Lees, 2005). For example, LFL values of several chemicals reported in earlier literature were in great disagreement with test results from an experimental program whose objective was to test the accuracy of such available values (Markowski et al., 2010; Liekhus et al., 2000). A theoretical model, if appropriately developed, can be used to:

- Validate experimental results where disagreement is present from various sources
- Estimate flammable properties of mixtures and newly designed chemicals
- Help decide in designing alternative safer chemicals

Lumping various modeling (Yeoh and Yuen, 2009; Pan et al., 2009) methods of QSAR and QSPR as one type of methodology, namely molecular modeling techniques, computational modeling approaches on estimation of flammability limits can be categorized primarily into two mainstream approaches:

- Macroscale modeling approaches
- Molecular modeling approaches

Since flammability limits are expressed as volume percentages, it has been observed that between members of a homologous series, it monotonically decreases with increasing molecular weight (Annual Book of Astm Standards, 2002; Drysdale, 2011). Rough estimations of LFL have been proposed using various empirical relations based on experimental data. Spakowski used heat of combustion (Crowl, 2012; Crowl and Louvar, 2011; Bodurtha, 1980), Jones used stoichiometric concentration of the flammable product (Crowl and Louvar, 2011; Bodurtha, 1980), and Britton (2002) used a combination of heat of oxidation, heat of combustion, and material-specific constant to calculate LFL. The respective empirical equations are as given in Table 3.4.

Flammability limits are affected by:

1. Incorporating effects of temperature changes
2. Incorporating effects of pressure changes
3. Incorporating effects of composition changes for mixtures

In higher-level modeling or macroscale modeling, attempts have been made to develop theories so as to incorporate these specific effects. One concern, in the context of changing composition, is capturing the effect of introducing inert and is of crucial importance. It can be extremely useful, specially in an industrial setting where, for example, it can help controlling safe environment in the vapor space of a reactor through controlled flow of inert into it with precise knowledge of its contribution toward attenuation of mixture flammability. Inerting is frequently used in industry to mitigate fire- and explosion-related hazards (Liaw et al., 2012). The study by Ma (2011) examined addressing these issues by developing a Le Chatelier—like method but with the ability to incorporate the effect due to the presence of inert(s). Some of the common problems in estimating flammability limits, as pointed out by Hansel et al. (1992) are:

- Single fuel and inert in air
- Multiple fuels and an inert
- Multiple inerts with fuel(s)

The work on compilation of flammability data covers flammability information on chemicals addressing the first concern to some extent (Ma, 2011). The second concern can be managed by taking a conservative approach and use regular Le Chatelier method of estimating flammability limits with calculating accurate LFL and approximate UFL. Having multiple inert in a chemical mixture makes things little more difficult to estimate accurate flammable properties. Attempts have been made through concept of pseudofuel, such as the work by Zabetakis (1965) and Haessler (1989), using

**Table 3.4 Estimation of Flammability Limits Through Empirical Relations**

| Method | Empirical Relation |
|---|---|
| Spakowski | $LFL = -4354/\Delta H_{combustion}$ |
| Jones | $LFL = 0.55C_{stoichiometry}$ |
| Britton | $LFL = c_0 \dfrac{S}{\Delta H^2_{combustion}}$ |
| Le Chatelier | $LFL = \dfrac{100}{\sum_i \frac{V_i}{LFL_i}}$ |

*$\Delta H_{combustion}$ is the heat of combustion of the fuel in kJ/mol; $C_{stoichiometry}$ is the stoichiometric concentration of the flammable product; S is the heat of oxidation that is defined as the amount of energy available to break the molecular oxygen bond as necessary to initiate the chain reaction of the combustion process; $c_0$ is a constant based on the type of material, that is, hydrocarbon or CHON and needs to be estimated separately; $V_i$ is the volume percentage of ith chemical component in the mixture composition; and $LFL_i$ is the LFL of the ith chemical component in the mixture composition.*

extended Le Chatelier rule as attempted by Kondo et al. (2006, 2007), and using critical adiabatic flame temperature (CAFT) (Hansel et al., 1992) and several other methods as reviewed by Ma (2011).

To incorporate the effect of diluent or inert, first we need to keep in mind that each species within the chemical mixture contributes to flammability of the mixture either by generating heat or by absorbing it. Generation of heat helps to maintain the flame temperature, whereas absorbing energy helps the reactants to overcome reaction barrier and oxidized, based on the observation that oxygen consumption by mass is relatively constant for most fuels as demonstrated (Huggett, 1980). Since inerts do not participate in combustion, their primary role is to absorb energy leading to a need for increased heat release rate to sustain the flame temperature. Ma's approach (Ma, 2011) uses two assumptions, namely constant CAFT and validity of oxygen calorimetry at both limits (i.e., the reaction proceeds to completion at the UFL and is similar to that in the LFL limit). By performing an energy balance of a stochiometrically balanced equation, at the both limits, keeping in mind that the heat generation at LFL is dictated by the fuel available as opposed to availability of oxygen in the UFL limit, the equation set for lower and upper flammable limit can be derived.

The reaction incompleteness at the UFL is not considered here and generally ignored. Similar to any other models, the model is limited by the inputs it needed. In another work by Liaw et al. (2012), a similar approach was taken to estimate the flammability limits. Based on energy balance, it also took into account the thermal radiation loss while estimating the limits. The work, however, assumed, like so many others, a constant value of CAFT.

Incorporating the effect of quenching potential of inerts is an important concept toward better development of better models for estimation of flammability.

One of the more fundamental methods involves knowledge of chemical equilibrium (Melhem, 1997; Palucis et al., 2007), using direct minimization of Gibbs free energy. In the absence of experimental data, extrapolation of existing equilibrium data of pure and chemical mixtures to novel materials is risky and can lead to underestimation or overestimation of hazardous scenarios. The approach implicitly accounts for the effect of change in temperature, pressure, and diluent and thus provides several advantages over empirical and semiempirical methods for assessing flammability limits. One of the key differences the methodology embraced is modification of cubic equation of state for both phases, so as to enable direct analytical minimization of the Gibbs free energy at high pressure and nonideal systems. Other approaches to obtain chemical equilibrium information for the same purpose are activity coefficient or equation of state-based model and can be useful in several occasions; they can also be challenging and/or imprecise for several reasons as clarified below:

- Most common models such as Peng–Robinson and Soave–Redlich–Kwong (SRK) are applicable only for nonassociating fluid mixtures. Self-associated systems such as water and ethanol cannot be modeled with these approaches (Elliott et al., 1990). Given the intensive resource requirement of implementing detailed molecular simulations at the time, researchers have proposed alternative simple methods (Elliott et al., 1990) while accepting the superiority of a detailed model.
- With the advancement of technology and introduction of new chemicals, binary interaction parameters as required for activity coefficient models may not be available for various chemicals.
- In an attempt to generalize models over a large range of fluids, thermodynamic modeling of real fluids and mixtures tends to introduce empirical formulation, which does not work for molecules with significant difference in size or functionality (Lee and Sandler, 1987).

- Group contribution models such as UNIFAC suffers from significant inaccuracies when several strong nonalkyl groups are considered (Panayiotou, 2003). These methods are also unable to differentiate between isomers for intramolecular interactions (Panayiotou, 2003). Even though these are parameterized against a large database, these methods cannot handle every type of molecule, especially when an unusual combination of functional group occurs.
- Theoretically founded data interpolation Gibbs excess energy models such as Wilson, nonrandom two liquid (NRTL), or universal quasi-chemical (UNIQUAC) methods need a sufficient amount of reliable experimental data and usually fail for novel chemical compounds (Klamt, 1995).

Flammability envelope estimation through Gibbs energy minimization procedure has also been used for carboxylic acids and acetate (Palucis et al., 2007). In addition to the robustness of the model, one also needs accurate input data for individual components for accurate estimation of properties based on that. The sensitivity of using inaccurate data is illustrated in the exercise chapter through an example (Mashuga and Crowl, 2000).

## 3.2.1 MOLECULAR MODELING OF FLAMMABILITY LIMITS

### *3.2.1.1 First principle calculations*

The role of first principle calculations in estimation of flammability limits is primarily demarcated within accurate estimation of properties or parameters associated with the flammability of a chemical or its mixture, which can in turn serve as an accurate input for facilitating molecular level approaches of QSAR- or QSPR-based calculations or alternatively serve as inputs to well-established theories.

Let's illustrate a little. For example, we know that LFL values can be estimated by the knowledge of heat of combustion as provided in Table 3.4. Similar to calculating accurate enthalpy of formation through implementation of first principle calculations (da Silva et al., 2006), heat of combustion calculations can also be performed. So, while the functional form for estimating flammability limits, such as those referred in Table 3.4, is not altered and is directly derived from a compound's molecular structure, through empirically fitting experimental results, the inputs required for it to function, such as heat of combustion of a chemical in this particular case, can be made more accurate and reliable by providing first principle calculations—based results. In addition, as we have mentioned before, first principle—based calculations provide an accurate estimation of optimized geometry for the chemical compound of interest and thus provide reliable estimation of quantum chemical descriptors as utilized in QSPR and QSAR calculations. Before we go into more details of how we can make that happen, for a moment let us stop and look into various factors that should be kept in mind before we move into performing ab initio—level calculations.

Within an electronic level simulation, considering even details of a system consisting of few 100 atoms is resource intensive. One might be able to afford to extend the size of the system studied through sacrificing some rigorousness in the same level of theory, say for example, a different choice of model such as choosing HF as opposed to Moller—Plesset fourth order (MP4) will typically reduce the required computational resource by approximately 70% (as scaling of basis functions is $N^4$ in HF as opposed to $N^7$ for MP4), whereas the error in the energy estimation of a minimized structure will likely to increase from about 1.3 kcal/mol to about 15 kcal/mol, but even then that is not significant.

The detail-level modeling at the electronic structure level benefits us by capturing electronic interaction effects accurately, leading to reliable estimates of parameters, which, if possible, can only be determined through experiments. In general, first principle calculations can serve as a reliable input provider as opposed to conduct experiments for calculations, which can then be used to calculate flammability estimates. The different types of calculations, which can be benefitted by first principle or quantum calculations in estimating flammability limits, can primarily be classified as:

1. Chemical equilibrium calculations
   a. Binary interaction parameter
2. Quantum mechanical descriptors for QSAR calculations
3. Property calculations for higher-level theory
   a. Heat of combustion

#### *3.2.1.2 Chemical equilibrium*

Precise knowledge of chemical equilibrium composition in vapor phase for various temperatures and pressures is of significant importance for the estimation of flammability limits. Even though some of the present approaches, as mentioned in Table 3.4, use the knowledge of heat of combustion and stoichiometric ratio for estimation of LFL values, we have previously mentioned how the knowledge of vapor mole fraction is used in Le Chatelier's (1891) approach of estimation of flammability limits. Estimation of flammability limit through chemical equilibrium information has the obvious advantage of implicitly accounting for:

- Pressure effects
- Temperature effects
- Diluent effects

An accurate estimation of chemical equilibrium through Gibbs free energy minimization under constrained optimization is expected to result in a more detailed and accurate approach in estimation of flammability limit, specially for mixtures where no prior experimental data are available (Melhem, 1997). While using chemical equilibrium information for estimation of flammability limits is not new (Melhem, 1997), first principle calculations can be used to provide information where:

- No experimental data are available
- Existing data are scattered

This not only helps in maintaining a nonhazardous mixture composition in the vapor space of a vessel under normal process conditions, but accurate chemical equilibrium model also help predict worst case hypothetical scenarios.

Quantum mechanics—based calculations such as conductor-like screening model, or COSMO, and its derivatives can help estimate thermodynamic properties of liquids (Klamt et al., 2010; Diedenhofen and Klamt, 2010). Approaches such as this introduce the solvent effect into quantum mechanical calculations. The approach uses quantum mechanics for treating solute and solvent, while statistical mechanics are used for treating molecular surface interactions. Unlike other quantum calculation approaches, which uses vacuum as the environment, the starting point of COSMO theory is in state of molecules embedded in a conductor for fluid phase simulations. To combine quantum calculations with this approach, at least DFT level methods or Moller—Plesset second order perturbation theory is recommended so to predict the electrostatics reasonably accurately. The calculations involve the following steps (Klamt, 1995):

1. Define starting geometry for the solute under consideration.
2. Around the solute, define a cavity as the boundary of the conductor and subdivide it into small segments.

3. Generate initial electron density in the present configuration using quantum mechanical calculations.
4. Calculate solute electrostatic potential on the grid of the cavity segments.
5. The conductor screening charge density is calculated from the conductor boundary condition such that the total electrostatic potential arising from the solute and the polarization charges are neutral.
6. The external charges are updated from the polarization charges, and step 3 is repeated.
7. Repeat steps 3 through 6 until self-consistency is achieved.
8. A correction is applied to the total energy to compensate for the outlying charge errors arising from a small portion of electron density outside the defined boundary.
9. Gradient of the total energy of the system is calculated analytically from the converged electron density and polarization charges and system coordinates are updated toward a lower energy structure.
10. Repeat steps 1 through 9 until self-consistency is achieved with respect to solute geometry.

However, for fairly polar solutes, a major part of the polarizability of the solvent behaves nonlinearly. COSMO-RS, an extension of COSMO, allows avoidance of this questionable dielectric approach that is unable to account for the nonlinear behavior of dipole reorientation at the molecular surface in the presence of a strong electric field. In the COSMO-RS approach, the deviation from the ideally screened molecules is described by misfit interactions of the ideal screening charges on neighboring molecules in the fluid. More details can be found (Klamt et al., 1998). In order to better estimate various properties of chemical including vapor pressure, through accurate estimation of chemical equilibrium of various mixtures, the dielectric continuum solvation model COSMO and its extension beyond dielectric approximation such as COSMO-RS have been carefully parameterized (Klamt et al., 1998). DFT was used in the study to generate reliable molecular potentials and was used to optimize some of the (18 in total) parameters of COSMO-RS. Overall COSMO-RS uses the intermediate results from quantum mechanical calculations on individual molecules to predict thermodynamic properties of their mixtures. The study parameterized elemental properties of H, C, O, and Cl, making it a powerful tool for estimating chemical potential of any solute or solvent without any requirement of experimental data. In addition to the advantage of a first principle model—based estimation of chemical potential of any chemical mixture, the effect of temperature is included in a natural way in this approach. General predictive method for vapor pressure and enthalpies using COSMO solvation model has also been developed (Lin et al., 2004), based on calculation of solvent free energy with contributions from electrostatic, dispersion, and cavity formation. The quantum mechanical calculations part played a role in estimating the electrostatic component of the solvent free energy. The model parameters, as given below were determined based on study of 371 pure substances of varying structure, functionality, and size:

1. van der Waals atomic radii to describe molecular shape
2. Universal interaction parameters, two in number, to describe electrostatic interactions
3. Atom specific dispersion coefficients (two in number)
4. One universal parameter to scale atomic exposed surface area
5. A parameter defining the ratio between the hard core to atomic radii

COSMO-SAC (Wang et al., 2007), another variant of COSMO, from the DFT/COSMO combination as has been illustrated above, projects the three-dimensional optimized charge density distribution onto a two-dimensional histogram and referenced as the sigma profile $p(\sigma) = \frac{n_i(\sigma)}{n_i} = \frac{A_i(\sigma)}{A_i}$, where the subscript $i$ refers to pure liquid component $i$, and $n_i(\sigma)$ refers to the number of segments with charge density $\sigma$ with total surface area $A_i$.

However, it has been pointed out that, while activity coefficient model can address nonideality in liquid phase, for vapor phase capturing nonideality arising from high-pressure scenarios, equation of state models needs to be implemented (Melhem, 1997). To overcome limitations of activity coefficient models, using COSMO-RS and COSMO-therm for high-pressure and temperature phase equilibrium systems, Panayiotou proposed (Panayiotou, 2003) a model based on surface contact thermodynamic a model such as COSMO-RS approach by incorporating free volume, a basic idea underlying the lattice fluid model (Panayiotou and Sanchez, 1991; Panayiotou and Vera, 1980), into the COSMO model. As a consequence, the approach is suited for developing equation of state model that can address both supercritical and subcritical stages. The work neglected the change in adjustment of wave functions that occurs to minimize interaction energy during a phase change.

### 3.2.2 PRACTICAL CONSIDERATIONS

We have discussed several approaches at various scales from using heat of combustion information, equilibrium approaches to quantum mechanical descriptor–based QSPR approaches for computationally estimating flammable limits. Depending on the methodology implemented, one may face various types of challenges and limitations that a user must bear in mind. While using empirical or semiempirical methods, for scenarios such as the existence of a cool flame (e.g., when ether is present in a mixture), process condition with elevated temperature, and pressure, sudden depressurization and the like can result in inaccurate estimation of flammability limits (Vidal et al., 2004; Ma, 2011; Melhem, 1997). Using ab initio–based results is reliable but needs large amounts of computation and such paths are typically recommended when no data are available or the results are in disagreement. However, together, this set of multiscale modeling approaches has the ability to act as an efficient alternative to performing extensive set of experimental tests to estimate flammability limits of a mixture with various composition. With emerging technologies and newly designed material, the need for estimation of flammability data can be greatly benefitted by using appropriate combination of multiscale modeling approaches.

Now given that, let us look into some practical considerations one should keep in mind while implementing multiscale modeling–based methodologies for flammability limits estimates. At the macroscale, generally speaking, the sources of error in flammability limit estimates by modeling can arise from:

1. Simplicity of model
   a. Inability to incorporate change in temperature effect
   b. Inability to incorporate change in pressure effect
   c. Inability to incorporate change in composition effect
2. Implementation of model outside the domain of applicable chemicals
3. Inaccurate experimental data used for model development

Let us start by examining the limitations of widely used empirical and semiempirical approaches, that is, on the inability of capturing or incorporating effect on flammability limits owing to changes in temperature, pressure, and composition of a mixture.

#### 3.2.2.1 Composition effect

A direct consequence of the inability to handle composition change in a model implies that it cannot be used in scenarios where introduction of inert in a controlled fashion is possible to effectively control the flammability of that mixture. Le Chatelier (1891) proposed a simple formula in volume percent for estimation of mixture flammability limits through the following set of equations for a system with $N$ components:

$$\mathrm{LFL_{mix}} = \frac{100}{\sum_{1}^{N} \frac{X_i}{\mathrm{LFL}_i}} \tag{3.1}$$

$$\mathrm{UFL_{mix}} = \frac{100}{\sum_{1}^{N} \frac{X_i}{\mathrm{UFL}_i}} \tag{3.2}$$

where $X_i$ is the vapor mole fraction.

While the set of formulas for mixing rules in the context of flammability is still widely used today, derivation of these incorporates a number of assumptions that are generally valid while estimating LFL values but less so for UFL estimation (Mashuga and Crowl, 2000). The formula is empirical in nature and since it contains several assumptions, as enlisted below, on making its way to final derivation (Mashuga and Crowl, 2000), applicability of those can be questioned for some scenarios. The assumptions required to derive Le Chatelier's formula are:

1. Constant values of
   a. Product heat capacities
   b. Number of moles of gas
2. Combustion kinetics of one species is not affected by the presence of other
3. Adiabatic temperature rise at the flammability limit is same for all species

Along with the assumptions as mentioned above, Le Chatelier used only three data points to prove Eqn (3.1), allowing potential for drawbacks in certain scenarios. Additionally, the simplicity of the formula may also prove inadequate while estimating flammability limits for mixtures containing fuels such as acetone or ether, leading to "cool flames" (Melhem, 1997; Ma, 2011).

Cool flames are associated with low temperature gas phase oxidation in a fuel-rich region above upper explosion limit (Mewis and Pasman, 1995). So, cool flame exists beyond the UFL limit.

Another concern that is not addressed by every empirical or semiempirical formula in the context of composition is the presence of inert. The presence of inert introduces absorption of reaction energy. This in turn leads to reduced chances of the reaction supplying enough energy to sustain self-propagation. Adding more inerts thus affects the flammability of a chemical compound. As a result, even though Le Chatelier's rule gives a good initial estimate, further validation is necessary (Crowl, 2012). A similar statement can be made for other empirical and semiempirical formula used for flammability estimates. However, equilibrium-based approach such as this (Melhem, 1997) can address composition changes and thus is better at estimating flammability limits in similar situation. ASPEN process simulator uses the Gibbs free energy minimization algorithm as described (Melhem, 1997) to estimate thermodynamic equilibrium.

#### 3.2.2.2 Temperature effects

An increase in temperature translates into requirement of less energy for a combustion reaction to be sustained. Higher initial temperatures results in enlarged flammability envelopes. Zabetakis (1965) provided an empirical relation to extrapolate LFL and UFL estimates at NTP to other temperatures as given:

$$\mathrm{LFL}(T) = \mathrm{LFL}(298\ \mathrm{K}) - \frac{0.75}{\Delta H_c}(T - 298)$$

$$\mathrm{UFL}(T) = \mathrm{LFL}(298\ \mathrm{K}) + \frac{0.75}{\Delta H_c}(T - 298)$$

LFL and UFL are shown to be equally affected in the opposite direction by change in temperature. What is important to consider, though, is that these relationships are only valid for straight chain hydrocarbons (Melhem, 1997). Impact of initial temperature is illustrated in Figure 3.5 (Melhem, 1997).

#### 3.2.2.3 Pressure effects

Changes in pressure affect flammability limits. The work by Zabetakis (1965) reported the following relationships with limits in volume percentage and $P$ in atmospheres:

$$\mathrm{LFL} = 4.9 - 0.71 \log P$$

$$\mathrm{UFL} = 14.1 + 20.4 \log P$$

Unlike the case of change in temperature, it is clear from the functional form that changes in pressure have a higher impact on UFL as opposed to LFL. Empirical models based solely on heat of combustion or stoichiometric ratio cannot capture these effects. However, Gibbs free energy minimization technique can appropriately capture that information as shown in Figure 3.6 (Melhem, 1997).

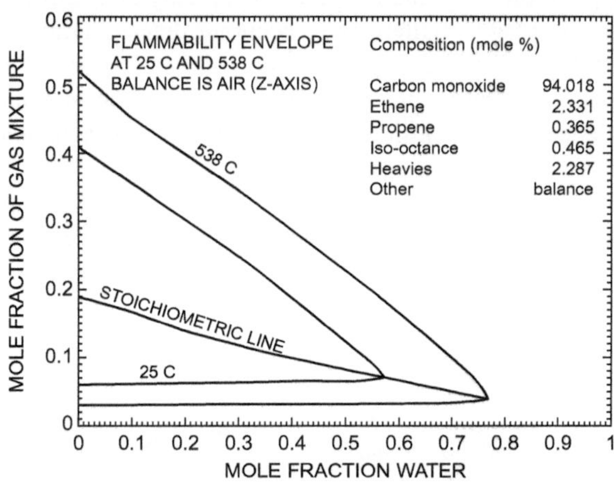


**FIGURE 3.5**

Impact of initial temperature and water content on flammability limits.

*Reprinted with permission from Melhem (2004).*

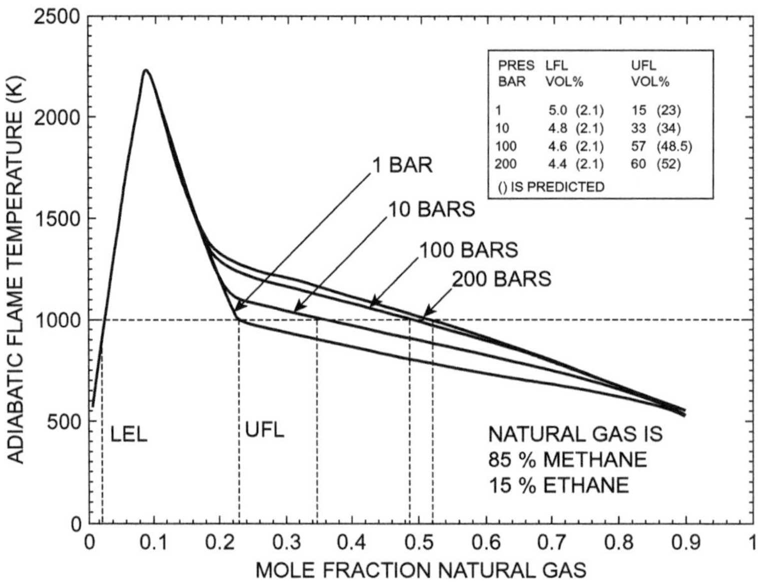


**FIGURE 3.6**

Effect of pressure changes on flammability limits.

Concerns are different when one looks into implementation of molecular level modeling in flammability limit estimation techniques. Among these, some relates to QSPR or QSAR techniques where as others relate to first principle– or ab initio–based calculations. For example, floppy molecules or nonrigid molecules poses difficulty to model through QSPR or QSAR techniques (Young, 2004). Examples of floppy molecules include ammonia, ethylene, ethane, etc. These molecules exhibit multi-minima, which creates challenges in accurate determination of structure–property or structure–activity relationships. Figure 3.7 (Jensen and Bunker, 2006) demonstrates two configurations of the ammonia molecule possible through tunneling of the nitrogen nucleus. Interested readers on floppy or nonrigid molecules can look into some great references available on the subject (Jensen and Bunker, 2006).

Owing to the nature of QSPR model development, it has a domain of applicability based on the choice of chemical data used as a training set. In other words, a model developed on data from organic compounds is only valid on similar property prediction of organic compounds. The implementation of

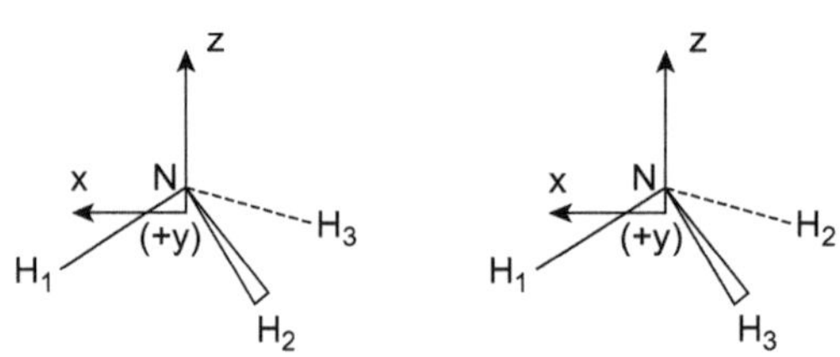

**FIGURE 3.7**

Two equivalent configurations of ammonia, possible through tunneling of the nitrogen nucleus.

various algorithms for development of QSPR model also affects the relative accuracy of a model. In terms of performance, while developing QSPR model for LFL prediction of organic compounds, Pan et al. (2009) demonstrated obtaining better results with implementation of support vector machine (SVM) than MLR, PLS, or ANN.

In ab initio— or first principle—based calculations, there are typically four sources of error (Young, 2004):

- Born—Oppenheimer approximation
- Incomplete basis set
- Incomplete correlation
- Omission of relativistic effects

Despite claiming an enormous amount of cpu hours, memory, and disk space, with advancement of newer and faster machines, owing to its fundamental nature, ab initio methods has the potential to deliver accurate results next to no other approach. While improvement in computer speed and efficiency has certainly helped ab initio— or first principle—based calculations to access more and more larger systems and faster solutions, because of its demand of resources, careful sacrifice of accuracy is often done in favor of increasing efficiency. For example, minimally correlated methods such as MP2 and GVB are often used when correlation is important (Young, 2004). Regardless of the fact that several approaches are considered as ab initio or first principles, the results are affected by functional forms chosen for the atomic and molecular wave functions and within ab initio approaches, variations of results are expected.

The estimates of flammability limits helps define volume of a flammable cloud over a facility based on dispersion analysis. It also helps in providing a picture of interaction of congested zones and flammable vapor cloud. LFL values define the outer boundary of a flammable cloud and hence the volume of it. Multiscale modeling of dispersion analysis of flammable clouds is discussed in Chapter 2 and Chapter 4. The estimates obtained here are used as part of the inputs.

## 3.3 FLASH POINT

Flash point is widely used to assess hazard associated with a flammable liquid to determine its fire and explosion risk. It is defined as the minimum temperature needed for a material to start emitting sufficient vapor and form a flammable mixture with air (Crowl and Louvar, 2011; Engineers. Center for Chemical Process Safety, 1993; Lees, 2005; Moghaddam et al., 2012) at atmospheric conditions. For a liquid mixture consisting of a single flammable liquid with miscible nonflammable liquids, the flash point of the mixture occurs at a temperature where the partial pressure of the flammable liquid in the mixture is equal to the vapor pressure of the pure liquid at its flash point temperature. It is used to determine a chemical's storage temperature and handling requirements for its safe transportation. Several series of explosions has, in an unfortunate way, demonstrated the importance of flash point values (Santon). On the other extreme, conservative models for flash point can potentially overestimate a flammable hazard and thereby lead to an expensive mitigation approach. Estimation of flash point with reasonable accuracy thus has significant importance in process hazard analysis studies. Computational modeling for flash point can often get tricky owing to exhibition of minimum flash point behavior (MFPB) by some chemicals. It is exhibited when the flash point of a mixture is below the flash point of the individual components. Several mixture rules, as illustrated in Section 3.1, owing to their definitions of estimating flash point of a mixture, limits the result within the range of minimum and maximum flash point of individual component of a mixture. A conservative approach in estimating flash point is typically to use the minimum flash point among the individual constituents as the flash point for the whole mixture. As we will observe and as the definition goes, chemicals exhibiting MFPB behavior, is an exception to this understanding and has the potential to pose a problem if unaddressed.

Experimentally, there are generally two ways to determine flash point of a chemical:

- Open cup
- Closed cup

Fundamentally, closed cup measurements are conservative as all the vapors are available and none escapes. Various standardized test methods have been developed basing these methods and review of these are presented in ASTM E 502 (Harrison). In addition, experimental measurements of flash point is expensive, resource intensive, and, especially for toxic, volatile, and explosive chemicals, it is even more difficult to measure (Bagheri et al., 2012; Annual Book of ASTM Standards, 2002; Katritzky et al., 2001; Mannan and Lees, 2005). Design of test apparatus, as in the case of estimating flammability limits, also plays a role in contributing toward uncertainties in flash point estimation (Rowley et al., 2010). Therefore, development of reliable computational models to estimate flash point can provide significant value including determination of flash point for new chemicals without the need for synthesizing it. In the next section we will look into detail on the current state of the art approaches and methodologies at various scales which helps estimates flash points with different extent of limitations. Several of the theoretical approaches on estimation of flash points have been reviewed (Vidal et al., 2004; Catoire and Naudet, 2004) to a reasonable extent. Mathematical models for estimating flash points can be divided into two broad categories:

1. Empirical and semiempirical correlations
2. Molecular modeling approaches

Empirical and semiempirical approaches generally refer to estimation of vapor pressure through such methods and that in turn is fed into an advanced and more computer-intensive approaches. In the section that follows, we will emphasize various methods with a different level of details incorporated in their models to estimate flash point. The origin of estimation of flash point through mathematical modeling is almost a century old; in 1923, Mack et al. reported calculating flash point of pure organic substances based on its LFL as estimated by Thornton (1917). This is done by using vapor pressure curve of a pure component. Thornton noticed that irrespective of bond types, the heat of combustion difference between successive members of paraffins, olefins, and other $C_xH_yO_z$ type compounds is reasonably constant for the radical $CH_2$. In other words, it meant that the heat of combustion of a molecule was directly proportional to the number of oxygen atoms it required for complete its combustion. Mack et al. used these formulations to estimate LFL of a flammable gas and calculated corresponding vapor pressure required for that component to meet its LFL. The flash point was then defined as the temperature in which the organic substance will have the required vapor pressure. However, in addition to being only validated with a limited number of single compounds, this approach also yielded results deviating as much as 14 K for isomeric compounds (Liu, 2010).

In general estimating flash points for a mixture with a single flammable liquid is done as follows:

1. Note mole fraction of the flammable liquid in liquid solution ($x_i$), flash point temperature of the pure flammable liquid ($T_f$), and total pressure
2. Calculate or find the saturation vapor pressure $P_{i,sat}$ of the flammable liquid at its flash point $T_f$.

3. Iteratively find the flash point of the mixture so that the saturation vapor pressure $P_{i,sat}$ of the flammable liquid at the mixture flash point is equal to the partial pressure ($p_i$) of the flammable vapor as calculated by vapor liquid equilibrium information of the mixture

Other computational approaches for flash point estimation includes correlating flash point with boiling point data (Hshieh, 1997; Patil, 1988; Satyanarayana and RAO, 1992; Prugh, 1973; Butler et al., 1956), specific gravity data (Satyanarayana and Katati, 1991; Metcalfe and Metcalfe, 1992), and a combination of boiling point and enthalpy of vaporization data (Catoire and Naudet, 2004). The basis for correlating flash point with boiling point of a substance lies in the fact that flash point is expected to be correlated to vapor pressure and hence its boiling point (Voutsas et al., 2002). Most of these models developed, dealt with pure organic compounds (Mack et al., 1923; Hshieh, 1997) and built empirical relations between the variables. The polynomial functions used to fit the data varied in form, from second order with respect to the boiling point or $T_b$ (Satyanarayana and Katati, 1991; Patil, 1988) to second order with respect to inverse of $T_b$ (Hshieh, 1997). See Table 3.5 for more details. While these methods can provide a rough estimate of flash point in an efficient manner, the downside of these approaches lies in the fact that these are heavily dependent on reliable experimental data and extrapolation of the data beyond the domain of materials utilized for model development. As a result, implementation must be done with their limitations in mind. As an example (Hshieh, 1997), for the compound phenyl-triethoxysilane, with a predicted flash point of 92 °C, using Patil's correlation (Patil, 1988), as provided below, is lower by 28 °C than its actual flash point. Such deviation observed in the particular example will result in overly conservative set of decisions. A worse scenario, predictions resulting in higher flash point values than actual and thereby creating a potential for devastating consequences.

**Table 3.5 Various Empirical Relations for Predicting Flash Points**

| References | Flash Point ($T_f$) Estimation from Boiling Point ($T_b$), Density ($d$), and Heat of Vaporization ($\Delta H_{vap}$) |
|---|---|
| Patil (1988) | $T_f = 4.656 + 0.844T_b - 0.000234T_b^2$<br>Units of $T_f$ and $T_b$ are in K |
| Butler et al. (1956) | $T_f = 0.683T_b - 119$<br>Units of $T_f$ and $T_b$ are in °F |
| Satyanarayana and Katati (1991) | $T_f = -83.3362 + 0.5811T_b + \left({}^{0.0001118}/_{T_b}\right) + 38.734d$<br>Units of $T_f$ and $T_b$ are in °C and units of $d$ are in g/cm$^3$ |
| Metcalfe and Metcalfe (1992) | $T_f = -84.794 + 0.6208T_b + 37.8127d$<br>Units of $T_f$ and $T_b$ are in °C and units of $d$ are in cm$^{-3}$ |
| Hshieh (1997) | $T_f = -54.5377 + 0.5883T_b + 0.00022T_b^2$<br>Units of $T_f$ and $T_b$ are in °C |
| Satyanarayana and Rao (1992) | $T_f = a + b[(c/T_b)e^{-(c/T_b)}]/(1 - e^{-(c/T_b)})^2$<br>Units of $T_f$ and $T_b$ are in K. $a$, $b$, and $c$ are constants |
| Catoire and Naudet (2004) | $T_f = (1.477T_b^{0.79686})\Delta H_{vap}^{0.16845}n^{-0.05948}$<br>Units of $T_f$ and $T_b$ are in K, units of $\Delta H_{vap}$ is in kJ/mol, and $n$ is the total number of carbon atoms |

Boiling point of a substance is primarily related to its nonbonded interactions such as van der Waals forces. A flash point of a substance, however, is not only a function of its ease of being vaporized at a certain temperature but also its ability to react with oxygen. Hence, any empirical formulation, which correlates only boiling point to its flash point, has potential to generate misleading information. The same argument holds for any correlation, such as the one using specific gravity or heat of vaporization alone or in combination with boiling point to predict flash point. For mixtures, a reliable way must be known to estimate the mixture boiling point, specific gravity, and heat of vaporization prior to estimation of flash points. Due to all these legitimate concerns of applicability, some developers of these models even recommended (Hshieh, 1997) the use of experimental data (if available) over predictions based on empirical equations. Theoretical models (Liu, 2010) such as those specific for hydrocarbon mixtures in air (Affens and McLaren, 1972; White et al., 1997; Affens, 1966), model using Raoult's law for alkanes in air (Affens, 1966) have also been used to estimate flash point of specific set of chemicals. Because of the possibility of MFPB, simply assigning the minimum flash point value of the individual constituents as the mixture flash point does not guarantee a conservative approach and underestimates the risk assessed (Liaw et al., 2003). On the contrary, it has also been demonstrated that a mixture flash point can be more than the maximum flash point (Liaw et al., 2002; Liaw and Lin, 2007) of its individual components, thereby hinting at possibilities of overly conservative estimation of flash point. To overcome these inconsistencies, more rigorous mathematical models have been developed (Liaw et al., 2002, 2004), such as one that can also predict MFPB. Figure 3.8 (Liaw et al., 2002), illustrates an example, where the developed method in the same article can successfully predict MFPB. One can observe that the method proposed by Liaw et al. is in very good agreement with experimental results as opposed to differences shown by some other methods.

Computational model based on molecular structures, as demonstrated in the next section, if constructed and implemented appropriately, has a great potential of predicting flash points of various mixtures in much more efficient manner. The key, however, is to have a validated robust and reliable model for an approach to predict flash points with reasonable accuracy.

### 3.3.1 MOLECULAR MODELING FLASH POINT

QSPR-based estimation methods of flash points of pure compounds, mainly organic compounds include multiple linear regression (MLR), group contribution method, and ANN methods. Description of molecular structure in these studies has often been through use of molecular fragments, topological indices, and quantum mechanical calculation–based descriptors (Liu, 2010). While its necessity has been debated (Puzyn et al., 2008) in favor of semiempirical methods, QSPR-based model can also have its roots based on first principle calculations such as DFT. QSPR approaches find quantitative mathematical mapping between measureable properties of chemical compounds and its intrinsic molecular structure. Molecular descriptors, as required to complete the model, can be calculated by implementation of quantum chemical calculations (Katritzky et al., 2007) and thus have the potential to provide a reliable way of estimating flash point through first principle–based QSPR approaches. We have already discussed briefly how quantum mechanical calculations can be implemented in estimating molecular descriptor values for chemicals in Section 3.1.4. As demonstrated in Section 3.1.4, molecular structure–based methods such as QSPR and QSAR, provides an excellent way to estimate properties and activities of a chemical in relation to its structure. Having

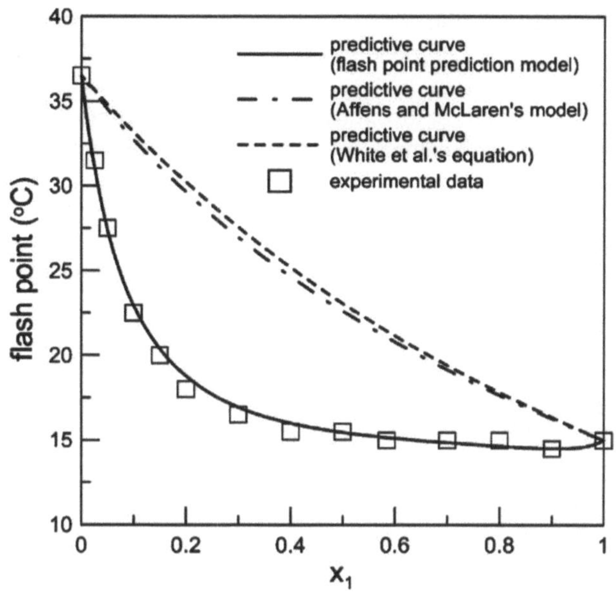


**FIGURE 3.8**

Comparison of the flash point–prediction curves with experimentally derived data for octane (1) 1-butanol (2) solution.

*Reprinted with permission from Liaw et al. (2002).*

the capability of formulating a model with multiple response output and reliable prediction performance, neural network–based QSPR models has made some key breakthroughs in flash point predictions (Liu, 2010). Back propagation (Pan et al., 2007), radial basis function, and group contribution method (Gharagheizi et al., 2008)—based neural network have also been used with QSPR to estimate flash points (Tetteh et al., 1999; Li et al., 2004). Besides estimating flash point for pure components, reliably predicting flash point of mixtures as a function of its composition is also important. To illustrate the importance of incorporating the effect of molecular in flash point estimation of a chemical, let's take an example from the literature (Rogers and Mannan, 2006). Molecular interaction affects vapor pressure of a mixture, which in turn affects flash point of a chemical. A mixture of similar-sized alkanes such as heptane and octane, when mixed in various proportions, will have properties that can be predicted with reasonable accuracy by simple mixing rules. Simple mixing rules do not incorporate effect of nonbonded interactions between individual components. Such an assumption holds reasonably true for these type of system of interest (mixture of alkanes). However, for example, addition of an alcohol in the system changes the interaction dynamics owing to polarity differences. Properties estimated through simple mixing rules in such cases may mislead assessment of hazards present. A repulsive action among the atoms owing to nonbonded forces will consequently result in higher vapor pressure leading to lower flash point of the mixture. Therefore, in a mixture, effect of intermolecular interaction could be significant in determining its flash point. Such interactions, in the absence of experimental data, these are best estimated at the atomistic level.

#### *3.3.1.1 First principle–based approaches*

The need for accurate thermodynamic models is especially evident in flash calculations (Chen and Mathias, 2006) as correctly pointed out by Chen et al. In light of that, however, a survey (Hendriks et al., 2010) found minimal awareness of industry on advanced computational techniques such as quantum mechanics and molecular dynamics. Industry rarely updates or reinvents its current modeling approach unless a clear advantage is proven ahead of implementation (Chen and Mathias, 2006).

Figure 3.9 (Hendriks et al., 2010), which is based on a survey on industrial requirements for thermodynamic and transport properties by the Working Party on Thermodynamic and Transport Properties (http://www.wp-ttp.dk/) of the European Federation of Chemical Engineering, EFCE (http://www.efce.info/) demonstrates the current industrial need for accurate thermodynamic data and highlights advanced scientific tools such as molecular dynamics simulation and quantum mechanics calculations as possible solutions.

We have already demonstrated the significance of predicting MFPB in the previous section. MFPB has been predicted (Rogers and Mannan, 2006) very effectively for binary mixtures using first principle calculations. The intermolecular and intramolecular interactions in presence of other components that have an effect on a mixture's flash point are neglected when using mixing rules but are appropriately captured in calculations through first principles.

Quantum mechanics, ab initio– or first principle–based approaches, has the potential to provide specific input information to help estimate flash point of various chemicals. Being of fundamental nature, such an approach is an efficient and reliable alternative to performing numerous set of experiments for estimating flash point. Before we illustrate these implementations, we must also bear in mind, as seen before, that there are two parts toward estimation of flash point of a chemical, namely that of individual components and that of the mixture. In order to do that, we need to estimate following:

1. Flash point of a pure component
2. Flash point of a mixture

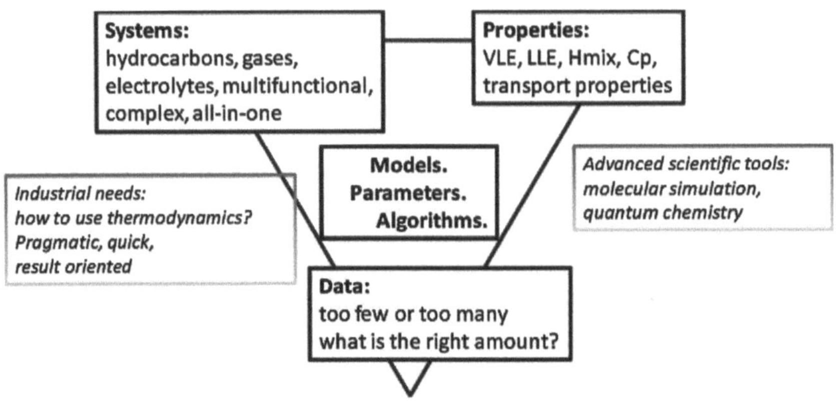


**FIGURE 3.9**

Schematic illustration of the position of thermodynamic tools for the oil and gas, chemical, and process industries (Hendriks et al., 2010).

*Reprinted with permission from American Chemical Society.*

a. Activity coefficient
b. LFL

While estimation of flash point is significant importance, screening chemicals for potential hazardous MFPB is another area where ab initio calculations can help us significantly. COSMO-RS–based estimation of $\gamma^{\infty}$ can help us to identify mixtures exhibiting MFPB. ADF-COSMO-RS (te Velde and Bickelhaupt, 2001), a variant of COSMO-RS, which have been demonstrated in the latter parts of this chapter, uses results from quantum mechanical calculations and Le Chatelier's rule of mixtures to estimate flash point of mixtures.

In the section below, we discuss how approaches based on quantum mechanics or first principles can help us in estimating flash point, by addressing concerns related to individual components and their mixture. For individual component–related approaches we focus on first principle calculations input–based QSPR approaches, whereas for mixtures we focus on using quantum mechanics to estimate activity coefficient through accurately estimating binary parameters. In addition, we also take a look into usage of COSMO-RS–based approach in screening MFPB chemicals. First principle approaches such as DFT can be useful in reliably estimating data such as binary interaction parameters among variety of chemicals, where prior experimental data are not available (Rogers and Mannan, 2006). Binary interaction parameter is an input to activity coefficient model that is required to effectively calculate flash point of a mixture. In this work, the authors did not use first principle approach solely to estimate the flash point of a binary mixture. Instead, quantum mechanics, which accurately calculates energy of a system based on spatial distribution of electronic particles, was used to estimate binary interaction parameter values as input to activity coefficient models where experimental data were lacking. The work is based on an approach (Amadeu, 1999) where ab initio method was used to calculate phase equilibrium of a system. Activity coefficient methods such as UNIFAC, UNIQUAC, and Wilson models are very much dependent on regressed experimental data for interaction energies. Sum et al. (Amadeu, 1999) calculated interaction energies between molecules using ab initio methods and used the same as an input interaction energy parameters to these models, replacing experimental inputs. While brief overview of various quantum mechanical methods such as HF method or DFT can be found in the introductory section, here we will provide the specific approaches taken by the authors to estimate the parameters:

1. Define a cluster of molecules to represent the system fluid. The cluster size (eight molecules or four each of two different kinds in this case) needs to be iteratively determined, by considering consistent energetics and insensitive to initial random structural configuration, reduced end effects, and computational resource usage.
2. Minimize the cluster energy by a less expensive semiempirical method. The authors in this case have used PM3 semiempirical method (Stewart, 1989). The use of a carefully chosen semiempirical method is to develop a reasonably well starting geometry and lessen the burden on subsequent ab initio steps during further refinement. Since parameters in semiempirical methods are based on experimental data, care must be taken in selecting such a method that is appropriate to the set of molecules in question. As an example, in this particular study, PM3 semiempirical method was chosen as it has the ability to qualitatively describe hydrogen bonding. Other semiempirical methods or approach such as molecular mechanics can also be used at this stage.

3. Refine the cluster geometry by minimizing the structure even further by using a higher-level theory, HF method in this particular case with a 6-31G** basis set. To achieve minimization in a relatively less expensive and gradual manner, the authors gradually increased the rigorousness of the basis set from 6-31-G to 6-31G**. The authors chose HF method, as the system studied was weakly interacting system.
4. Select directly interacting (in close proximity) molecular pairs (like and unlike) and record separation distance and relative orientation
5. Calculate interaction energy of each molecular pair (like and unlike) with the HF method and detailed basis set. The authors used 6-311++G (3d,2p) at the separation and orientation obtained in the previous step. The pair energy is computed as

$$E_{AB}^{Int} = E_{AB}[AB] - \mathrm{E}_A[AB] - E_B[AB]$$

where $E$ is the energy of molecule(s) $A$ and/or $B$ and $[AB]$ refers to combined basis set for $A$ and $B$.

6. Average among several interaction energies obtained for the same set of molecular pairs to be used as interaction energy parameters in activity coefficient models.

The study, within a pressure domain between 0.01 bar and 100 bars, showed excellent agreement between experimental observations and ab initio method findings, validating the usefulness and soundness of such approaches for thermodynamic property estimation. Broadly speaking, this method can be divided into two steps. In the first step, minimum energy configuration is determined for a defined cluster of molecules. In the next step, from the minimum energy configuration, interaction energies between like and unlike molecules are calculated and provided as input energy parameters for various activity coefficient models.

Quantum mechanics–based method, conductor-like screening model for realistic solvation or COSMO-RS, has been established as a novel and reliable way to predict thermophysical data for liquid systems (Klamt et al., 2010). It is as pointed out previously (Rogers and Mannan, 2006):

> a theory where the interactions in a fluid is described as local contact interaction of molecular surfaces and the interaction energies are quantified by the values of two screening charge densities $\sigma$ and $\sigma^2$, which form a molecular contact. The screening charge densities can be described as molecular descriptors and provide information about the polarity of the molecules.

As opposed to deriving binary interaction parameters based on functional groups interaction, as done in UNIFAC, COSMO-RS uses the screening charge densities at the local contact area for the same, which is more representative of nonbonded interaction among atoms. Additionally, COSMOS-RS bears the advantage of its ability on screening a large number of compounds from a database (Putnam et al., 2003). Alkanes, alcohols, alkyl in alkanes, ketones in alkanes, and alkenes in alkyl halides are some of the systems best treated through COSMO-RS (Putnam et al., 2003). Lin et al. (Lin and Sandler, 2006) estimated infinite dilution activity coefficients through a group contribution solvation model using ab initio solvation calculations. They found a 7% deviation for their model based on HF method as opposed to 47% and 52% for UNIFAC and modified UNIFAC model while estimating infinite dilution activity coefficient ($\gamma^{\infty}$) for a system with water, *n*-hexane, acetonitrile, and *n*-octanol.

Quantum mechanics has also been used in deriving QSPR parameters to estimate flash points of various chemical components and mixtures. We will deal with quantum mechanics–based QSPR studies in a later section. While it could be resource intensive computationally, generally speaking it is significantly efficient and cheaper in comparison to carrying comprehensive closed cup tests to determine flash point for every mixture of interest.

### 3.3.2 PRACTICAL CONSIDERATIONS

As discussed in the prior section, flash point of materials is an input to fire modeling problems. It defines the ease of generation of a flammable cloud based on environmental conditions. Even though flash point values have been determined experimentally for several pure components, mixture flash point data are scarce (Rogers and Mannan, 2006). In reality, most process fluid in a plant is a mixture of several chemicals with varying compositions. To make matter worse, these compositions can vary widely during upset conditions in a plant or owing to other unwanted disturbances. Estimation of flash point within acceptable accuracy, of such nonideal liquid mixtures with varying composition through conducting experiments is a daunting task. As Jim Olsen from Forum 2000 correctly pointed out (Hendriks et al., 2010):

> We can never conduct all the necessary experiments - but many data measurements are missing. For instance, multi-functional chemicals such as alkanolamines, glycol ethers, excess properties like excess volumes at various temperatures, amides, liquid phase heat capacities, environmental phase equilibrium, safety data e.g., flash-point and auto ignition temperatures, simultaneous physical and chemical equilibrium.

The solution to this problem where need for safety data is a subset probably lies in the following statement from Prausnitz (1999):

> Pioneering molecular thermodynamics is concerned not with improvements where primary understanding has already been achieved, but with shedding light on situations where as yet, we know little.

As a consequence, extrapolating flash point data for mixtures from pure component data is not only misleading in some cases but could also result in implementation of unsafe practices in a process facility. To meet the significant demand for flash point data and in lieu of existing expensive tests for flash points (Liaw et al., 2003), alternative theoretical models are needed (Liu, 2010). While models have been developed to predict flash point for the material of interest, those have limited domain of applicability, largely empirical in nature and needs improvement for complex mixtures.

Typically, determination of flash point for mixtures needs the following:

- Flash points of the individual constituents
- Vapor pressure data of the individual constituents
- Activity coefficients of each component
- A reliable approach to predict the flash point of a mixture from these information

In most cases, experimental data of flash point for individual components of a mixture are available (Smallwood, 1996; Yaws, 2003; Lewis and Lewis, 2008). Vapor pressure data of pure components are similarly available in many places (Perry and Green, 1984; Yaws, 1995; Zwolinski, 1971; Mackay et al., 1982). However, activity coefficients, which capture the binary interactions between components of a mixture, can often be challenging to find and is often estimated through generalized theories (Vidal et al., 2004). Along with a dearth of information on mixture data, the need of additional estimation of flammability limits can be attributed to the technological advances leading to emerging novel materials and new environmental concerns leading to stringent conditions and requirement of more accurate predictions even for the smallest relevant constituent of the mixture. Even with today's computational power, the calculations are nevertheless expensive, but with the advent of increased computational power, computer intensive approaches such as this are becoming more feasible to characterize material and phenomenon more accurately. Quantum mechanical calculations can be used to determine the binary interaction parameter used to determine the activity coefficients in the absence of reliable experimental data (Affens and McLaren, 1972; Amadeu, 1999).

## 3.4 AEROSOL FORMATION

The physical state of liquid or solid particles suspended in air is known as aerosol. Atmospheric aerosol has long been studied and several classic books exist in this area. Even in industry, hazard exhibited by two-phase mixtures, aerosol, is often overlooked (Bowen and Cameron, 1999). It is widely accepted (Lian et al., 2011) that safety in process industry is also shaped by the presence of aerosol. Since aerosol cannot be treated as gas or liquid, assessing hazard associated with it needs a different approach. Pressurized release of high flash point industrial fluids, such as kerosene, has the potential to create a flammable aerosol. Pressure differential as low as few bars (Bowen and Shirvill, 1994) will suffice to initiate such phenomenon. On the contrary, water aerosol has served as diluent in process industries by providing inerting action (Shebeko et al., 1995) to a process stream. In either case, understanding formation mechanism of aerosol has the potential to provide useful insight and add value to modeling approaches related to flammable aerosols. Current understanding related to complexity of aerosol formation and its corresponding combustion process is not complete (Lian et al., 2011; Bowen and Shirvill, 1994) and to accurately determine associated fire and explosion hazard, there is a need to perform more studies in this area.

As addressed above, aerosol modeling approaches or methodologies in the context of process safety can be classified into three areas of interest:

- Formation
- Dispersion of aerosol
- Combustion of aerosol or aerosol particle reactivity

While most used as air quality models, several dispersion models do exist to address the modeling of aerosols. Models such as Air Force Dispersion Assessment Model (ADAM), modified box and Gaussian dispersion model, HGSYSTEM, and the regional modeling system for aerosol and deposition (REMSAD) need information on aerosol present in the atmosphere as prerequisite. Detailed dispersion modeling, which provides further insight into aerosol dispersion is done through the implementation of computational fluid dynamics approaches.

In addition to dispersion modeling, understanding the formation of aerosol droplets, characterization of these aerosol droplets such as their evaporation characteristics, size and movement velocity is important to perform accurate consequence study. Lian et al. (2011) has studied the effects of aerosol properties such as their size and evaporation rate on characteristics of the flame such as their length and propagating speed using heat transfer fluids. These methods, as mentioned above, deal with post aerosol formation and combustion. Approaches or methodologies to obtain insight on the formation of aerosol and its combustion is primarily done through experiments and modeling at the molecular level. Hence the modeling approaches to assess risk associated with flammable aerosol, can be categorized as:

- Macroscale modeling approaches
- Molecular modeling approaches

As briefed in the above section, various modeling approaches exist to address dispersion of aerosol. Sodium fire and combustion is an example where aerosol is formed.

In Yamaguchi and Tajima (2003) the authors have proposed expressions for aerosol dynamics, radiation heat transfer, and evaporation, arising from liquid sodium combustion phenomena. The multiscale nature of liquid sodium combustion phenomena is best depicted by Figures 3.10 and 3.11.

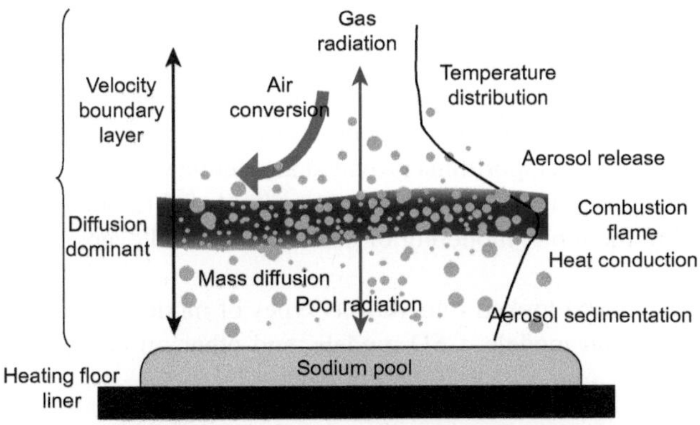


**FIGURE 3.10**

Evaporation from liquid pool (Yamaguchi and Tajima, 2003).

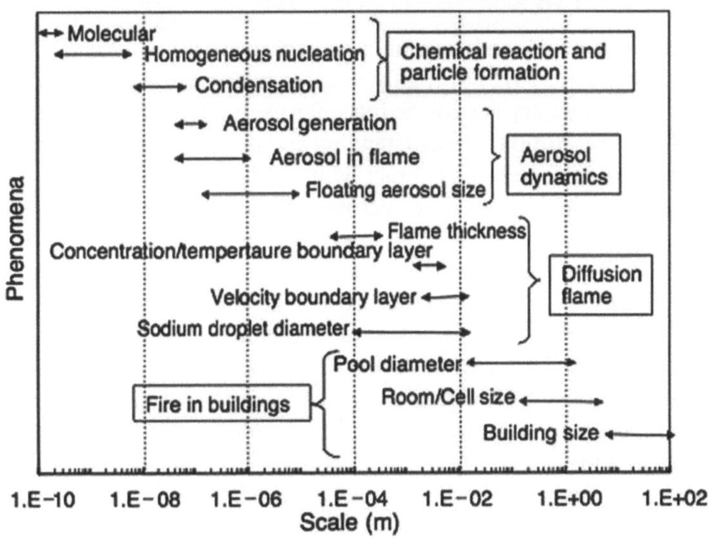


**FIGURE 3.11**

Multiscale nature of liquid sodium combustion (Yamaguchi and Tajima, 2003).

Although the specific work looked into the modeling related to liquid metal fast reactor, the developed model also proposed approaches to capture aerosol generation in the flame zone, coagulation, convection, and deposition. The absorption coefficient, required to calculate the thermal radiation, was adjusted based on aerosol concentration. The authors estimated the evolution of the aerosol particle size distribution by implementing the equation on aerosol density as developed earlier (Wen, 1996). The governing equation, dynamic in nature, helps one to update both the aerosol number density and statistical distribution of particle diameter with time. Incorporation of Brownian movements, relative velocities between aerosol particles including Cunningham correction factor to account noncontinuum effects, and turbulence together builds the model to capture aerosol coagulation (Gelbard and Seinfeld, 1980). Details of aerosol behavior model are not discussed here, and the reader can refer to several references that are available on the subject. Comparative review of various methodologies to model aerosol dynamics, coagulation, growth, nucleation, and gas or particle mass transfer has been done (Zhang et al., 1999). The aerosol modules of CIT (Meng et al., 1998), GATOR (Jacobson, 1997a,b), Models-3 (Binkowski and Shankar, 1995), SAQM-AERO, UAM-AERO (Lurmann et al., 1997), and UAM-AIM (Sun and Wexler, 1998) were reviewed. The reviewers observed that the algorithms available to model coagulation and condensational particle growth can be grouped as sectional or modal, based on particle size distribution. Accurate depiction of coagulation was obtained with a sectional approach, as opposed to modal approach. Within sectional approach, implementation of moving center algorithm as well as hybrid algorithms provided an accurate estimation of condensational growth. A brief comparison of aerosol dynamics is presented in Table 3.6 (Holmes and Morawska, 2006). In all the models referred in the table, modeling of condensation and evaporation processes was included.

The review article (Holmes and Morawska, 2006) looked into various types of models including box models, Gaussian models, Lagrangian/Eulerian models, CFD models, and other miscellaneous models that included aerosol dynamics.

## 3.4.1 MOLECULAR MODELING APPROACHES

In determining the hazards associated with aerosols, understanding aerosol reactivity is very important. For example, direct observation (Savee et al., 2015) of highly reactive hydroperoxyalkyl radical species involved in combustion and aerosol chemistry is expected to improve models of atmospheric and combustion chemistry (Kemsley, 2015). In addition to understanding and predicting the formation of aerosol, another major part is to understand how aerosols interact with volatile compounds, say, for example, formaldehyde. In addition to formation mechanism, aerosol chemistry at the air–water interface is particularly still very poorly understood (Martins-Costa and Anglada, 2012). DFT has been used to investigate the mechanism behind the formation of atmospheric aerosols from sulfuric acid and its hydrates in gas phase (Nadykto et al., 2004). The study focused on the nucleation mechanism behind such phenomena, and used DFT to look into various equilibrium structures of sulfuric acid and its hydrate, up to three water molecules per sulfuric acid molecule. The authors calculated dipole moments, vibrational transitions and values of the total bonding energy for several possible structures. The total bonding energy ($E_{TB}$), difference in energy between the molecule as a whole and its individual constituent atoms, can be equated to:

$$E_{TB} = \Delta E_{electrostatic} + \Delta E_{kinetic} + \Delta E_{coulombic} + \Delta E_{exchange-correlation}$$

**Table 3.6 Comparison of Aerosol Dispersion Models (Holmes and Morawska, 2006)**

| Name Developer | Dispersion Model | Nucleation | Coagulation | Deposition | Particle Size Method | Particle Composition |
|---|---|---|---|---|---|---|
| UHMA University of Helsinki | | B + T | Y | D:Y W:X | Hybrid/Moving center of retacking methods 0.7 nm−2 mm | $H_2SO_4$, Inorganics, Organics |
| MONO32 | Coupled to OSPM | B + T | Y | D:Y W:X | Four size modes, monodisperse approach 7−450 nm | None |
| AERO | Coupled to UAM-IV | Y | | D:Y W:X | 0.01−10 μm | Inorganic, organic, and elemental carbon. Internally mixed |
| GATOR | Eulerian | B | Y | D:Y W:X | Moving size or stationary size | None |
| MADRID | Coupled to CAQM | SOA, B | | D:Y W:X | Multiple size, sectional | |
| AEROFOR | Sectional Box | B, T | Y | D:Y W:Y | 200 groupings | Externally or internally mixed varying with each size group |
| URM | Eulerian | B | X | D:Y W:Y | 4 groups <10 μm | Internally mixed |
| RPM | Incorporated into RADMII | B | Y | D:Y W:Y | 0.01−0.07 μm | Ammonium Sulfate |
| CIT Californian Institute of Technology | | B | X | D:Y W:X | 0.5−10 μm | Ammonium Nitrate Organic Inorganics |

*Y, process included; X, process not included; B, binary; T, ternary; SOA, secondary organic aerosol information; D, dry deposition; W, wet deposition.*

The coulombic energy term here also incorporates the steric and orbital interaction energies. The study estimated properties of optimized molecular structures of gas-phase sulfuric acid and a small cluster through incorporating one, two, or three water molecule associated with one sulfuric acid molecule. Entropies, enthalpies, and Gibbs free energies were estimated using ADF, a DFT-based software package for these molecules. In addition to looking into optimum structures of various sulfuric acid hydrates, the study looked into possible reaction paths leading to the formation of such molecules. Put differently, the study defined end aerosol structures and looked into the relative ease of one forming over others, including the effect of change in temperature and pressure. This method thus implemented not only provide insight on possible aerosol particle size distribution but also favorable pathways of aerosol formation as a function of temperature and pressure. The $\Delta G$ value, change in Gibbs free energy, was calculated to be increasing with increasing pressure and decreasing with increasing temperature. That is, this supports the concern of aerosol formation from pressurized sources of even high flash point liquid. The dipole moment estimation for each structure, which was high in this study for all structures, that is, mono-, di-, and trihydrates of sulfuric acid, provided insight on possibility of nucleation of ultrafine particles in the atmosphere.

The authors used PW91 method to account for exchange correction and correlation parts in conjunction with VWM formal version 5 for local density approximation and the TZP basis set. However, to validate results, the sensitivity of the calculated structural data was investigated against the quality of the basis sets such as DZ, DZP, TZP, TZ2P, ET-pVQZ, and ET-QZ3P-1DIFFUSION.

#### 3.4.1.1 Atomistic simulation–based approaches

Atomistic simulation based methods have been implemented along with quantum mechanical calculations and the Clausius–Clapeyron equation to provide a functional form to temperature-dependent vapor pressure equation (Tong et al., 2004) for oxygenates in atmospheric aerosols. The motivation behind this study was to develop a systematic approach to estimate thermodynamic properties of ambient particulate matter containing multifunctional oxygenates. The system of interest was carboxylic acid with different number of carbon atoms. Low molecular weight carboxylic acids are considered moderate fire hazard (Miller et al., 2010), and hence aerosol formed from carboxylic acids pose as flammable hazard. The approach discussed here looked into estimating liquid vapor pressure of organic compounds through implementation of molecular dynamic simulations.

The vapor pressure of an organic compound can be estimated, as an alternative way to existing empirical equations such as Antoine equation, by intergrating total enthalpy of vaporization over temperature using the Clausius–Clapeyron equation (Baum, 1998), approximating heat of vaporization to be linearly related to temperature and can be expressed as

$$\Delta H_{\text{vap}}(T) = \Delta H_{\text{vap}}(T_b) + \Delta C_p(T_b)(T - T_b)$$

where $\Delta H_{\text{vap}}(T)$ is heat of vaporization at $T$, $\Delta C_p(T)$ is change in specific heat capacity on phase change, and $T_b$ is normal boiling point.

Substituting the above equation in vapor pressure equation of Clausius–Claperyon equation generates vapor pressure equation as a function of heat of vaporization.

**Problem:**

Derive the equation for vapor pressure using the equation for heat of vaporization and Clausius–Clapeyron equation.

The next steps involve accurate estimation of heat of vaporization. It is done through molecular dynamics simulations implemented on the optimized structure including charge distribution, obtained through quantum mechanical calculations, of the molecule of interest. Heat of vaporization can be correlated to cohesive energy density by the following function:

$$\mathrm{CED}(T) = \frac{\Delta H_{\mathrm{vap}}(T) - RT}{V_m}$$

where $V_m$ is the molar volume, and CED($T$) is cohesive energy density.

Cohesive energy density of a pure compound can be calculated using multiple sampling molecular dynamics with periodic boundary conditions. The starting geometry of the molecule along with atomic charges is taken from energy minimized structure as obtained from B3LYP flavor of DFT calculations. The force field used in this particular study was Drieding force field with hydrogen bonding parameters modified based on quantum mechanical calculation results for the dimer using acetic acid dimer. The structural evolution of the system from a reduced density to a target density matching experimentally obtained results was performed in steps, including the use of NVT ensemble in the process of attaining target density and NPT ensemble while equilibrating the system to a final temperature and pressure value after achieving 20% higher than target density. At a specific temperature $T$, the CED is calculated by subtracting the potential energy of the bulk system from the sum of the potential energies of the individual molecules. Potential energies of individual molecules are estimated through nullifying the effect of nonbonded interaction between inter molecules through separating them by large distances. For a more reliable estimation of CED, a set of 10 samples of the same molecular system went through the set of steps to reduce sampling error. Note that the target temperature for obtaining was set such that the molecule is in liquid state, which, in this particular study was 500 K. Using the correlations (Baum, 1998):

$$\Delta C_p(T_b) = -0.8\Delta S_{\mathrm{vap}}(T_b)$$

$$\Delta S_{\mathrm{vap}}(T_b) = \Delta H_{\mathrm{vap}}(T_b)/T_b$$

along with previously stated results, the authors able to reliably predict vapor pressure of carboxylic acids, which are very important while considering atmospheric aerosols.

### 3.4.2 PRACTICAL CONSIDERATIONS

Multiphase chemical reactions in aerosols and heterogeneous processes and formation mechanism of aerosol are still very poorly understood. Despite the plethora of modeling methods available, owing to challenges in accurate estimation, and appropriate approximation of complexities such as particle size distribution, coagulation, formation, and the like, multiscale modeling methods need to be implemented to access both the detail and the larger time and length scales.

In each scale of modeling and simulation, the user must be aware of applicability and limitations of the method. In Nadykto et al. (2004), the authors investigated sensitivity of using six different DFT methods and six different basis sets. To reduce the sampling error as per the $\sqrt{n}$ law, Tong et al. (2004), worked with 10 samples. In the same article, the parameters for hydrogen bonding in Dreiding force field was modified based on DFT results. It was also pointed out that, assumption of linear dependence of heat of vaporization on temperature, can lead to discrepancies in results, as well. However, carefully devised and thought out first principle and atomistic simulation-based fundamental modeling of the system of interest has the potential to provide further insight in understanding flammability hazards arising from aerosol.

## 3.5 REACTIVE HAZARDS

Reactive hazards are of considerable interests to process industries owing to their potential of causing substantial damage. The U.S. Chemical Safety and Hazard Identification Board has defined reactive chemical incidents as: *A sudden event involving an uncontrolled chemical reaction with significant increases in temperature, pressure or gas evolution that has the potential to, or has caused serious harm to people, property or the environment* (Murphy and Holmstrom, 2004). Identification, evaluation, and plan to control reactive hazard are recommended in all phases of a process life cycle to achieve an inherently safer design. All phases of a project typically refer to phases starting from core research and development to industrial scale plant decommissioning. Even for existing processes, understanding of reactive hazards allows one to establish appropriate mitigation techniques including implementing apropos control system. Reactive hazard exists, when changes in chemical structure have the potential to generate heat, energy, and gaseous byproducts beyond that which can be safely absorbed by the immediate surroundings (Urben, 2006). A chemical, which is stable under certain operating conditions, can initiate an undesired set of reactions when subjected to process conditions outside its operating range. The term "outside its operating range" can refer to material incompatibility, change in process-specific factors such as temperature, pressure, material quantity, and catalytic effects. Hence, reactivity hazard is not always an inherent property of an individual chemical substance, and variation in process-specific factors has the ability to generate reactive hazard. Put differently, a chemical identified safe for a specific process may pose the risk of reactive hazard in another process and hence should be reevaluated for every process. Various types of reactive hazards exist in the industry and for the sake of simplicity they can be broadly classified into two categories:

- General reactions such as with water, air, oxidation–reduction
- Self-reactions such as polymerization and decomposition

Reactive chemicals can be attributed to several incidents in the industry and characterization of reactive chemical is of utmost importance in hazard evaluation activity of a process unit. In most reactivity-related incidents, the lack of knowledge on reactive chemistry of the chemical of interest has been found to be the primary root cause (Aldeeb et al., 2003). A report (U.S. Chemical Safety and Hazard Investigation Board, 2003) by CSB suggested that the lack of information on reactivity of chemicals was responsible for over 50% of the studied reactive incidents. The report emphasized on methods and approaches, to better understand and predict reactivity hazard. As mentioned before, a change in process conditions can translate most stable chemical to a reactive material. System contamination resulting in material incompatibility can be a source of triggering a reactive incident. As an example, catalyzed by rust (iron and iron oxide) from an equipment, ethylene hydrogenation has resulted in a runaway reaction (Halle and Vadekar, 1991). Classification of reactive chemicals is subjective and different groups have classified reactive chemicals in their own ways as illustrated in Table 3.7.

However, it is important to find parameters that correlate strongly to the reactive nature of a chemical as a function of process conditions and its surroundings. As is noted later, both qualitative and quantitative methods exist in assessing reactive hazards. Table 3.8, re-created from Johnson et al. (2010), provides examples and definitions of various types of intrinsic reactive hazards.

To assess chemical reactivity hazard, in addition to performing experiments, detailed level modeling with process specific conditions has the potential to provide valuable and usable information. Besides, computational modeling approaches have also the ability to furnish insight inaccessible to its experimental counterpart.

**Table 3.7 Classification of Reactive Chemicals**

| Mannan et al. (Wei et al., 2004b) | CSB Data Analysis for Reactive Chemicals Classification (U.S. Chemical Safety and Hazard Investigation Board, 2003) |
|---|---|
| 1. Pyrophoric<br>2. Peroxide forming chemicals<br>3. Water reactive chemicals<br>4. Oxidizers<br>5. Self-reactive chemicals including decomposition, polymerization, and rearrangement<br>6. Incompatible materials | 1. Incompatibility<br>2. Runaway reaction<br>3. Impact or thermally sensitive materials |

**Table 3.8 Examples of Intrinsic Reactive Hazards (Johnson et al., 2010)**

| Cause of Reactivity Hazard | Definition | Examples |
|---|---|---|
| Unstable | Has the tendency to break down (decompose) over time or when exposed to conditions such as heat, sunlight, shock, friction, or a catalyst with the resulting decomposition products often being toxic or flammable. Decomposition can be rapid enough to give an explosive energy release and can generate enough heat and gases for fires/explosions. | Trinitrotoluene (TNT), dibenzoyl peroxide, ethylene oxide, acetylene, picric acid, hydrogen peroxide (concentrated) |
| Polymerizing | Has the tendency to self-react to form larger molecules, while possibly generating enough heat/gases to burst a container. | Acrylic acid, styrene, 1,3-butadiene |
| Pyrophoric | Will ignite spontaneously when exposed to air. | Phosphorus, silane |
| Peroxide former | Has the tendency to slowly react with oxygen, such as from being exposed to air, to form unstable organic peroxides. | 1,3-Butadiene, isopropyl ether |
| Water reactive | Will react with water or moisture. Some react slowly; others violently. Heat and flammable/toxic gases may be produced. | Sodium, sulfuric acid, acetic anhydride |
| Oxidizer | Will give up oxygen easily or readily oxidize other materials. | Chlorine, nitric acid |

Generally speaking, there is no standard procedure for evaluating chemical reaction hazards (Barton and Rogers, 1997). Two main branch of approaches has been identified by CSB (U.S. Chemical Safety and Hazard Investigation Board, 2003), with companies relying on employing varying degree of both qualitative and quantitative approaches to evaluate reactive hazards for their processes. In qualitative approaches, for example, companies conducting reactions in batch reactors may perform HAZOP studies to identify reactive hazards. Executing "What-If" PHA study belongs to the same category. Although there is little guidance on systematically identifying and evaluating a worst-case scenario for a quantitative reactive hazard assessment, methodologies do exist to estimate helpful parameters, as recommended by CCPS and IChemE, to evaluate reactive hazards.

Figure 3.12 illustrates one of the recommended approaches for chemical hazard evaluation by quantifying chemical reactivity through calculation of seven different parameters.

Among these factors, using 13 sets of calorimeter data, covering polymerization and decomposition reactions, heat of reaction has been found to be the most important factor in assessing the reactivity of a chemical from calorimetric data (Crowl and Elwell, 2004), showing a 77% agreement in the screening test results within the domain of the study. The study set was chosen to represent a very wide range of chemistries and illustrate the complexities arising from several exotherms and changing concentrations. Even though the study could not emerge with a single criterion in assessing the reactivity of chemical mixtures, it suggested following conclusions in regards to reactivity of a chemical mixture:

1. High total heat release or high rates of heat release will represent a greater hazard, as it will produce more heat than can be removed and can potentially lead to a runaway reaction.
2. Generation of noncondensable gases at high rates can lead to high pressure and represent a hazard.

However, systems defined as hazardous due to noncondensable gas generation, are not as common compared to hazardous scenarios arising from heat generation effects (Crowl and Elwell, 2004). This emphasizes the importance of accurate estimation of heat of reaction and heat release rate in assessing reactivity hazards associated with a chemical under defined conditions. Other than having a runaway reaction, in some scenarios, reactivity hazards can result in producing toxic gas, such as $TiCl_4$ that

**FIGURE 3.12**

Chemical hazard evaluation adopted from (Aldeeb et al., 2003).

produces hydrogen chloride gas on in contact with moisture in air (Kapias and Griffiths, 2005). Dealing with such scenarios, that is, generation of toxic gases has been dealt with in earlier here.

### 3.5.1 REACTIVITY HAZARD ASSESSMENT CRITERION

Parameters and combination of parameters, which are used to determine the reactivity of chemical, are summarized (Crowl and Elwell, 2004) and are shown in Table 3.9. It consists of 12 different parameter

**Table 3.9 Criterion Used for Reactivity Hazard Assessment (Crowl and Elwell, 2004)**

| | Criterion | Comments |
|---|---|---|
| 1 | NFPA instability number | As per NFPA 49 (1975),<br>325M (2002), or 704 (2001).<br>The highest NFPA for any of the pure reactants was selected.<br>Based mostly on expert judgment. Limited mostly to a small number of pure components. |
| 2 | Experimental determination of total heat released | Based on temperature versus time data and total reaction mass.<br>Requires heat capacity of reaction mass.<br>Determined using DSC or closed-cell calorimeter.<br>Includes Phi factor for calorimeter. |
| 3 | Experimental heat of reaction | Based on dominant or limiting reactant. Computed from total heat released. Determined using DSC or closed-cell calorimeter. |
| 4 | Instantaneous power density | As described by NFPA 704 (2001)<br>Requires rate determination at 250 °C<br>Determined using closed-cell adiabatic calorimeter.<br>Designed to track NFPA numbers. |
| 5 | Reaction onset temperature | Highly instrument dependent—best to compare values from the same calorimeter.<br>A lower value means a greater hazard, but operating below this value is no guarantee of safety.<br>Determined from temperature–time data<br>Determined from closed-cell adiabatic calorimeter. |
| 6 | Total temperature change | Determined from temperature–time data<br>Determined using closed-cell adiabatic calorimeter. |
| 7 | Total pressure change | Determined from pressure–time data<br>Determined using closed-cell adiabatic calorimeter. |
| 8 | Maximum rate of change of temperature with respect to time | Determined from temperature–time data<br>Determined using closed-cell adiabatic calorimeter. |
| 9 | Maximum rate of pressure change with time | Determined from pressure–time data determined using closed-cell adiabatic calorimeter. |
| 10 | Energy release as calculated by CHETAH | Based on maximum for any stable products. |
| 11 | Calculated heat of reaction | Calculated from heats of formation of reaction and equilibrium products. |
| 12 | Computed adiabatic heat of reaction | Calculated from heat of reaction and heat capacities of products, at equilibrium. Assumes constant enthalpy during reaction. |

values, incorporating both theoretically and experimentally obtained results as separate indicators as shown here:

Note that, the criterions mentioned above are not completely independent of each other. From the above set of criterion, input data requirements for determining reactivity hazard can be lumped into a set of four major different types:

1. NFPA instability number
   a. Expert opinion
   b. Instantaneous power density
2. Heat of reaction/heat released in a reaction
   a. Experimental/calculated
   b. Adiabatic
3. Analysis of time−temperature data
   a. Reaction onset temperature
   b. Change in total temperature
   c. Rate of change of temperature
4. Analysis of time−pressure data
   a. Total pressure change
   b. Rate of change of pressure

Based on screening methods as discussed in the 2003 version of the CCPS concept book (Johnson et al., 2010) on reactivity hazards, Mannan et al. (Wei et al., 2004b) analyzed 167 reactive incidents spanning over 20 years, from the period of June 1980 to January 2001, as reported by the CSB (U.S. Chemical Safety and Hazard Investigation Board, 2003). The study was performed by addressing the questions as posed by the CCPS book through using MSDSs and NOAA reactivity worksheet (NOAA). Put differently, the approach is to estimate various parameters, based on information available from experiments or through calculations, in order to provide a response to the set of questions asked by CCPS concept book on assessing the reactivity of a chemical. Consequently, an important part in evaluating reactive hazard is, estimating answers to the questions presented, such as finding out heat of reaction, time−temperature characteristics and the like, as precisely as possible. So the applicability of different experimental and modeling techniques depends on the factors or parameters that are estimated. As an example, the purpose of employing a calorimeter in reactivity hazard assessment is to use a small quantity of a reactive substance to assess its hazards, through estimating heat of reaction and heat capacity. Nonetheless, a caveat in using experimental techniques is that the data collected is truly an artifact of the apparatus and the procedure (Crowl and Elwell, 2004). Another limitation, in this case, is that the calorimeter conditions, should match with the intended process conditions, which is often not possible in reality, especially in the process design phase.

Unlike estimation of flash point or flammability limits, assessment of reactive hazard is more complicated due to the nature of the hazard, that is, material−process combination specific hazard. Owing to the fundamental nature of data requirements in assessing reactivity hazard, for the estimation methods to be accurate, it needs to incorporate molecular structural information and its corresponding changes during a reaction in a case-by-case basis for different process-specific conditions. This can only be achieved through direct and carefully devised experiments or molecular modeling level techniques. In here, we have focused on various multiscale modeling methodologies that can be utilized to help assessment of reactive hazard. Since fundamentally reactivity hazard is very much a function of the molecules involved and surrounding conditions (temperature, pressure), there is no well-established model at the macroscale to predict reactivity hazard. Based on the types of reactivity hazards as depicted before, two main approaches on assessing reactivity hazards are through experiments and molecular structure based methods.

Before we delve into the specifics of applicability of multiscale modeling approaches to estimate reactive hazards, let us revisit input data requirements which will help us to assess reactive hazards. Keeping requirement for qualitative input information aside (such as input in the form of expert opinion for estimating NFPA instability number), requirement of measurable information includes, as pointed out earlier in Figure 3.12:

- Heat of reaction
- Gibbs free energy of reaction
- Initiation temperature of significant exothermic reaction
- Maximum adiabatic temperature increase
- Arrhenius parameters (activation energy and frequency parameter) for power law kinetics
- Order of reaction
- Adiabatic time to maximum rate

Note that, some parameters, such as instantaneous power density, as required for calculating NFPA instability number can be computed as

$$\text{Instantaneous power density} = \text{Heat of reaction} \times \text{Reaction rate at } 482\ ^\circ\text{F } (250\ ^\circ\text{C})$$

The following section provides illustrations on how molecular structural information can be utilized to address reactivity hazard–related concerns through estimating some of these measurable parameters in the absence of experimental results.

## 3.6 HEAT OF REACTION/HEAT OF FORMATION

Heat of reaction is an important parameter in estimating reactive hazards. If the end product composition is known then heat of reaction can be determined by the difference between chemical energy of products and reactants. Put differently:

$$\Delta H_{rxn} = \Delta H_f(\text{products}) - \Delta H_f(\text{reactants})$$

Typically heat of reaction in a weight basis is more commonly used as opposed to mole basis for reactivity assessment (Medica, 1995).

Present approaches involve estimating heat of reaction using calorimetric techniques (Hsu et al., 2012; Preston et al.). However, assessing reactivity through calorimetric techniques is resource consuming and thus only practical for a limited set of compounds (Saraf et al., 2003). As far as modeling approaches go, Dow Chemical has been a pioneer in using computational chemistry to estimate thermochemical properties, including heat of reaction. Quoting from (Westmoreland, 2002):

> Dave Frurip, a member of the laboratory involved in producing the JANAF tables of thermochemical properties (Dow, 1971), expressed the opinion that computational chemistry for the calculation of change in enthalpy of reaction, a crucial property in safety analysis, has reached "maintenance mode."

Significant cost advantage has already been proven in using computational chemistry to estimate the heat of reaction (Westmoreland, 2002). In 1996, at a conference, a speaker from Dow Chemical showed that the cost of G2 or Gaussian-2 quantum chemical calculation for calculating heat of formation ($20,000) is less than 30% of the cost ($70,000) incurred to obtain the same through calorimetric measurements (Westmoreland, 2002). The advancement of powerful computers, at the beginning of the year 2000, made the comparison looked paler, that is, $2000 for a G3 or Gaussian-3 quantum calculation to $100,000 for a calorimetric measurement. As a side note, Gaussian-n theories are known as quantum chemistry composite methods intending for high accuracy calculations through combining results of several calculations. Several works (Curtiss et al., 1990, 1991, 1998, 2007; Pople et al., 1989; Curtiss and Raghavachari, 1993) exists on Gaussian-n theories and their implementation, details of which is beyond the scope of this book. However, we will look into various methods of estimating heat of reaction. Note that, computational techniques for estimation of heat of reaction are also applicable toward estimation of heat of combustion and heat of formation.

## 3.6.1 MACROSCALE MODELING APPROACHES

CART or calculated adiabatic reaction temperature—based approach to assess reactivity hazards employs Gibbs free energy minimization technique for heat of reaction estimation. The methodology requires the estimation of the adiabatic temperature increase due to the reaction from estimating the heat of reaction and average specific heat information of the reaction mixture. Adiabatic reaction temperature can be calculated as

$$\Delta T_{\text{adiabatic}} = \frac{-\Delta H}{C_p} = \frac{\text{Heat of Reaction}}{\text{Average heat capacity}}$$

A 1400 K value of the temperature rise is defined as the cut-off value for a possible explosion scenario (Saraf et al., 2003). Lower values such as 1200 K have been recommended (Melhem and Shanley, 1996) for a conservative approach.

As mentioned above, minimization of Gibbs free energy can be implemented to estimate adiabatic reaction temperature. Employing Gibbs free minimization of the reactants and possible products, one can find the mixture composition at thermodynamic equilibrium. While both kinetics and thermodynamics of the system is important, for instantaneous or fast reactions such as combustion, a system will achieve thermodynamic equilibrium in a very short span of time, given the reactant with stoichiometric proportions are available. Process Simulation software such as ASPEN Technology has module (RGibbs) built in it to help user find equilibrium composition through minimization of Gibbs free energy for a multicomponent multiphase (solid—liquid—gas) system. Other computer programs, COCO (van Baten, 2010) and APMonitor Optimization Suite, also have the ability to perform same calculations. It has been also illustrated that using Microsoft Excel's solver functionality; such a problem can be solved (Lwin, 2000). Through the process of Gibbs energy minimization, one is able to estimate the heat of reaction and average value of specific heat to leading to calculation of adiabatic reaction temperature.

As an example, let's go through briefly over the steps that are recommended for Gibbs free energy minimization (Lwin, 2000) for a gas phase reaction:

1. Define total Gibbs free energy of the system with $n$ moles as
$nG = \sum_i^N n_i\overline{G_i} = \sum_i^N n_i\overline{G_i^0} + RT\sum_i^N n_i \ln\frac{\hat{f_i}}{f_i^0}$
2. Define fugacity coefficient. For gas phase reaction, it can be defined as: $\widehat{f_i} = \widehat{\varnothing_i} y_i P$
3. Substitute standard state (denoted by superscript "0") property values for each component
4. Substitute fugacity coefficient equation for gas phase reaction into Gibbs free energy equation
5. Minimize the total Gibbs free energy expression subject to the elemental mass balance over the whole system

Implementation of the above mentioned steps are available in the computer program APMonitor Optimization Suite (Hedengren, 2008). Owing to complexities involved, mathematical approaches implemented toward minimization of Gibbs free energy for a multicomponent and multiphase system are various. The set includes linear programming (Rossi et al., 2009), nonlinear programming (Castier et al., 1989; Castillo and Grossmann, 1981; Mather, 1986), global optimization technique (McDonald and Floudas, 1995), interval method (Lin and Stadtherr, 2004), and several other approaches along with variations of the methods mentioned here. Details related to mathematical algorithms of these methods are skipped here.

**Problem:**

Liquid pentane is burned with 20% excess air. The initial temperature of both the reactants is 298 K. Assuming complete combustion, find the adiabatic flame temperature:

1. Assuming constant $C_p$ for each component
2. Assuming $C_p$ takes the form A + BT for each component

The ability to determine multiphase equilibrium allows one to define the final state of a given system. For scenarios with complete combustion.

## 3.6.2 MOLECULAR MODELING APPROACHES

### *3.6.2.1 Molecular group activity*

Amount of energy or heat released from decomposition or reaction correlate strongly to the potential intensity of reactive hazards. CHETAH (Frurip et al., 1989; Davies et al., 1985), an ASTM computational program, which stands for Chemical Thermodynamic and Energy Release Program, has the ability to assess chemical reactivity hazard. It is based on implementation of Benson's method of group activity (Benson et al., 1969; Benson and Buss, 1958). The proposed law can be defined as: "If $\phi$ is a molecular property, then for the disproportional reaction: $RNR + SNS \Leftrightarrow 2RNS$, $\Delta\phi \rightarrow 0$ as the separation between R and S becomes large compared to their dimensions." The basis for the law is analogous to the fact that in infinitely dilute multicomponent systems, molecular properties can be reasonably and accurately estimated by adding contribution from individual atoms toward these properties. This is possible because of the short-range intermolecular forces in the absence of substantial Coulombic interaction, such as in the case of hydrocarbons. However, with the presence of long-range Coulombic forces, such as for polar molecules, such an approach will lead to erroneous molecular property estimation. While CHETAH's principal aim is to predict deflagration/detonation potential from molecular structure, it is as well a useful tool to assess chemical reactivity hazard owing to its ability to estimate heat of reaction and thermodynamic properties of individual substances through the same principles. It identifies and estimates the greatest energy release scenario of reactants decomposing into different product species. Benson group increment theory (Benson and Buss, 1958; Benson et al., 1969), developed additivity rules to be implemented on experimentally obtained heat of formation values for individual groups of atoms to calculate the entire heat of formation of a molecule under investigation. The group additivity relation is generally obeyed within 0.6 kCal/mol for estimation of $\Delta H_f$ (Benson and Buss, 1958).

CHETAH can calculate three major impact sensitivity indicators (Harrison, 2005) to assess chemical reactivity hazard as illustrated below:

1. Maximum enthalpy of decomposition with the assumption that in the worst-case scenario, the chemical decomposes into a set of products releasing maximum amount of energy
2. Modified version of calculating maximum enthalpy of decomposition, taking into account, stoichiometry, number of peroxide bonds, etc.
3. Net plosive density, defined as a group contribution index weighting stabilizing and destabilizing molecular fragments

Accuracy of CHETAH, or put differently, applicability of Benson's group activity method, in assessing chemical reactivity hazard has been tested against experimental data and implementation of all three approaches resulted in more than 80% of correct predictions as illustrated in Figure 3.13 (Harrison, 2005).

A subset of data presented in (Wei et al., 2004b) is presented in Table 3.10 to illustrate effectiveness of CHETAH in estimating heat of formation of various chemical compounds.

While implementation of Benson's group activity method through CHETAH has been very successful, it can fail short of expectation where the values of Benson group properties are missing or erroneous and do not take into account sensitivities of molecules to the different initiation modes or variation in process conditions. To illustrate, let us recreate the example from Saraf et al. (2003) in Figure 3.14. The calculated maximum heat of decomposition (energy released and denoted by the negative sign) for hydrogen peroxide yields a value of −0.743 kcal/g or −3.109 kJ/g, which is higher than the required −2.929 kJ/g criteria used in CHETAH to define high reactivity hazard. Note that,

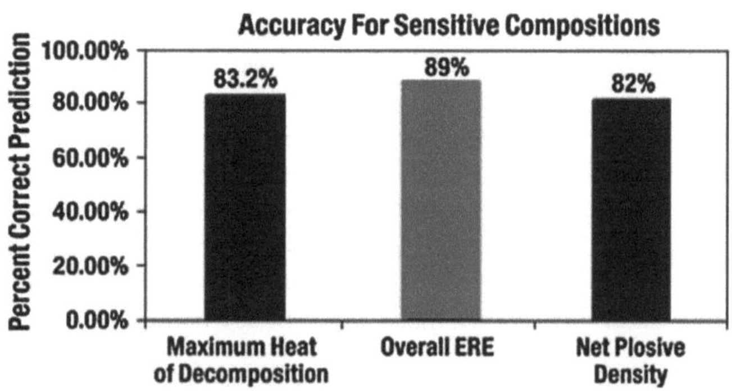


**FIGURE 3.13**

Assessment of CHETAH accuracy against experimental data for predicting the presence of reactive chemical (Harrison, 2005).

**Table 3.10 Comparison of Heat of Formation Values from CHETAH with Experimental Data (Wei et al., 2004b)**

| Chemical | Structure | $\Delta H_f$ (kJ/mol) | $\Delta H_f$ (kJ/mol) |
|---|---|---|---|
| | | CHETAH | Experimental |
| Hydrogen cyanide |  | 130.54 | 135.14 |
| Formaldehyde |  | −125.31 | −115.9 |
| Methanol | 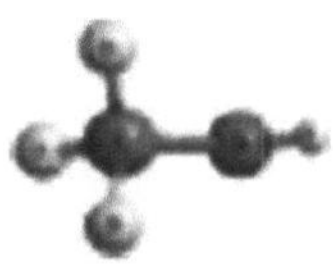 | −200.83 | −201.17 |
| Butadiene | 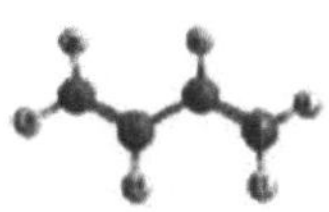 | 110.16 | 108.78 |
| Nitric acid | 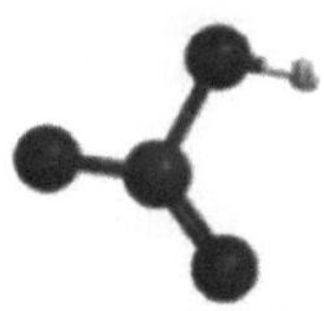 | −134.31 | −134.31 |

such an approach does not address the effect of change in process conditions or estimation of heat of decomposition under various process conditions to guide toward safer process conditions while working with any potential reactivity hazard such as hydrogen peroxide at certain conditions.

In addition, in scenarios where significant long-range molecular interaction is expected, such an approach is not recommended to use as the basic assumption is violated.

Heat of decomposition, or the heat released during the decomposition process, is the energy that is released on the decomposition of the material and is often chosen as a basis for estimating stability of the chemical in question. QSPR-based approaches have been implemented in assessing chemical or material stability (or the lack of) through development of predictive models for estimating heat of decomposition. As discussed before, the basis for QSPR-based approaches is their molecular descriptors. The estimation of these molecular descriptors can be performed in various ways including

**ENERGY RELEASE EVALUTION**

**Compound Name**: Hydrogen Peroxide
**Formula**:$H_2O_2$
**Molecular weight**:34.015
**Amount**: 1 Mole(s)
**Heat of formation at 25 °C:** –32.530 kcal/mol

Plosive Hazard Classification:
**Over-all Energy Release Potential is HIGH** **Value:**–1.375
Contributing Details:

| Criterion | Value | Units | Hazard Classification |
|---|---|---|---|
| Maximum Heat of Decomposition (#2) | –0.743 | kcal/g | HIGH |
| Fuel Value—Heat of Decomposition | 0.000 | kcal/g | HIGH |
| Oxygen Balance (#3) | 47.037 | percent | HIGH |
| CHETAH ERE Criterion 4 | 46.934 | $kcal^2/mol$ | MEDIUM |
| Total Number of Peroxide Bond | 1.000 | | |
| Net Plosive Density (#4) | 0.435 | | PLOSIVE |

Warming: These ratings only apply to hazards assoicated with strong mechanical shock. This does not imply the absence of other hazards.

Notes #1 This evaluation was developed to classify a composition as able or not able to decompose with violence, if subjected to the proper conditions. Information on the interpretation of hazard classification criteria used by CHETAH may be found in the CHETAH documentation. (ASTM publication DS-51A) and in J. Chem. Ed., v66, A137 (1989).

#2 For decomposition products shown.

#3 Experience has shown that the oxygen balance criterion is useful only for compounds composed of the elements C, H, N, and O.

#4 Sum of auxoplosive and plosphoric weights per gram of mixture.

Decomposition Products (chosen to maximize heat of decomposition)

| **Moles** | **State** | **Species** |
|---|---|---|
| 0.500 | ref-gas | $O_2$ Oxygen |
| 1.000 | gas | $H_2O$ Water |

**Heat of Combustion Section**

Fuel value (Net Heat of Combustion)

| **Mass Basis** | **Mole Basis** |
|---|---|
| –0.743 kcal/g | –25.270 kcal/mol |
| –3.108 kJ/g | –105.730 kJ/mol |
| –1337.246 Btu/lb | |

Combustion Products (Chosen for Fuel Value and Net Heat of Combustion)

| **Moles** | **State** | **Species** |
|---|---|---|
| 1.000 | gas | $H_2O$ Water |

**FIGURE 3.14**

Hazard evaluation for hydrogen peroxide using CHETAH 7.2 (Saraf et al., 2003).

first principle calculations and employing experimental data sets. In Fayet et al. (2010a), a QSPR-based approach targeted toward nitroaromatic molecules was developed to assess their thermal stability and electric spark sensitivity. The authors used topological, geometrical, electronic, and quantum chemical molecular descriptors to associate adequate molecular descriptors with the experimental properties of the molecules.

While numerous molecular descriptors have been utilized until date (Karelson, 2000), the work, as shown in Figure 3.15 attempted to build a QSPR model based on a subset of such molecular descriptors. In this particular study, the authors employed the use of Quantum chemical calculations to optimize the molecular structures of the molecules of interest prior to carrying out regression analyses

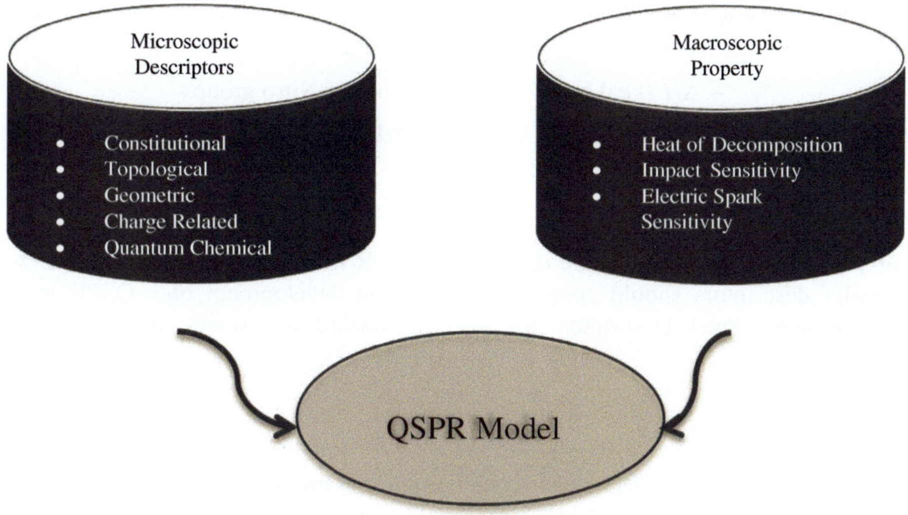


**FIGURE 3.15**

QSPR Model development framework for explosive properties assessment.

for mapping a set of molecular descriptor values of a specific property of interest. The form of the multiple linear regression as implemented in this study was

$$Y = a_0 + \sum_i a_i X_i$$

where $Y$ is predicted property, $X_i$ is $i$th molecular descriptor, $a_i$ is regression constant for $i$th molecular descriptor, and $a_0$ is universal regression constant.

Based on data obtained from differential scanning calorimetry results, in a separate work (Fayet et al., 2009b), the researchers established the following correlation for enthalpy of decomposition:

$$-\Delta H = 33854\eta + 40050\omega - 1030.3\alpha - 33785IP + 25.1DM - 14.0E_{\text{diss}} + 973.2$$

The coefficient of determination, or $R$ squared ($R^2$), was calculated to be 0.91 in this case. The descriptors used are defined as follows:

$$\eta(\text{hardness}) : \text{IP(Ionization Potential)} - \text{EA(Electron Affinity)} = \varepsilon_{\text{LUMO}} - \varepsilon_{\text{HOMO}}$$

$\omega$ (electrophilicity index): $\frac{\mu^2}{2\eta}$ where $\mu$ is chemical potential

$$\alpha(\text{mean polarizability}) : \frac{1}{3}[\alpha_{xx} + \alpha_{yy} + \alpha_{zz}]$$

It must be noted that, while developing a multiple linear regression model for QSPR model development, one may even omit the presence of a universal regression constant, that is a constant, which is not associated with any particular molecular descriptors, and formulate it as

$$Y = \sum_i \mathrm{a}_i X_i$$

As an example, in Saraf et al. (2003), a QSPR model for energy of reaction to nitro compounds was developed as

$$-\Delta H\ (\text{kcal/mol}) = 75 \times \text{number of Nitro groups}$$

The two major sources of estimation of these molecular descriptors typically are:

- Experimental thermokinetic data
- Quantum mechanical calculations

Another very important factor, while developing QSPR models, is to determine which and how many molecular descriptors should contribute toward the development of a QSPR model to the satisfactory confidence level. Descriptors for reactivity hazard assessment can be classified in two different types:

- Local
- Global

Local descriptors are specific to a type of compounds and hence may be undefined for a different kind of compounds. An example of local descriptor could be the number of nitro groups in a compound. Global descriptors, as the name suggests, is applicable to any compound. One such example of global descriptor could be the molecular weight of a compound. Another one, a conventional empirical descriptor used in screening technique for hazard assessment of energetic materials is the oxygen balance as defined in the work by Lothrop and Handrick (1949). Using stoichiometry, it is defined as

$$\text{OB} = \frac{-1600\left(2X + \frac{Y}{2} - Z\right)}{M_w}$$

### *3.6.2.2 Choosing descriptors*

It is recommended that the act of choosing molecular descriptors should be done keeping the following in mind:

- Proven physics-based strong correlation between property and descriptor
- Minimized number of descriptors
- Uncorrelated and unique descriptors
- Invariant with respect to atom indexing, molecular translation, and rotation

Several methods are availed by researchers to help them choose appropriate descriptors such as correlation, F-score, genetic algorithm, and random forest. Note the presence of "physics-based" wording. A good correlation between a descriptor and a property should be explainable physically.

Let us examine the applicability of OB descriptor, which has been defined before. Since in here, fundamentally combustion is taking place, a measure of oxygen balance and attempt to correlate that with explosive properties and intuitive. However, note that there are several factors, which results into a clarion call for in corporation of other molecular descriptors. OB descriptor value does not distinguish between oxygen that is loosely bound and easily available for combustion versus one that is already bound to carbon or hydrogen in a very stable manner. Nor does that addresses reactivity hazard associated oxygen free hazardous compounds such as acetylene, acetylides, azides, and explosive nitrides (Shanley and Melhem, 1995). Sensitivity to reaction initiation can also be not addressed appropriately with oxygen balance approach alone. Thermochemical and kinetic relations based descriptors need to be considered for a better model development.

In addition to choosing molecular descriptors, another important factor to consider is when to stop adding descriptors. A simple technique, called "breaking rule," as employed in several studies (Katritzky et al., 2007; Fayet et al., 2010c), looks into the improvement in statistical quality of the model. It compares the change in validation correlation coefficient $R^2$ (or cross-validation correlation coefficient $R^2_{CV}$), with addition of new descriptors.

Based on the findings as illustrated in Figure 3.16, the authors of the work developed a four parameter–based model for estimating heats of decomposition of nitroaromatic compounds:

$$-\Delta H = -282.3 + 333.5n_{\mathrm{NO_2}} - 1214.5E_{C,\mathrm{avg}} + 7.4\alpha - 275.6\mathrm{IC}_{\mathrm{avg}}$$

where $n_{\mathrm{NO_2}}$ : number of nitro groups, $E_{C,\mathrm{avg}}$ : average one-electron reactivity index for a carbon atom, $\alpha$ : mean polarizability, and $\mathrm{IC}_{\mathrm{avg}}$: average information content, a topological index.

### 3.6.2.3 First principle–based approaches

We have illustrated the importance of accurate estimation of heat of reaction in assessing a chemical reactivity hazard. One of the major challenges in this area lies in the form of unavailable data. If prior

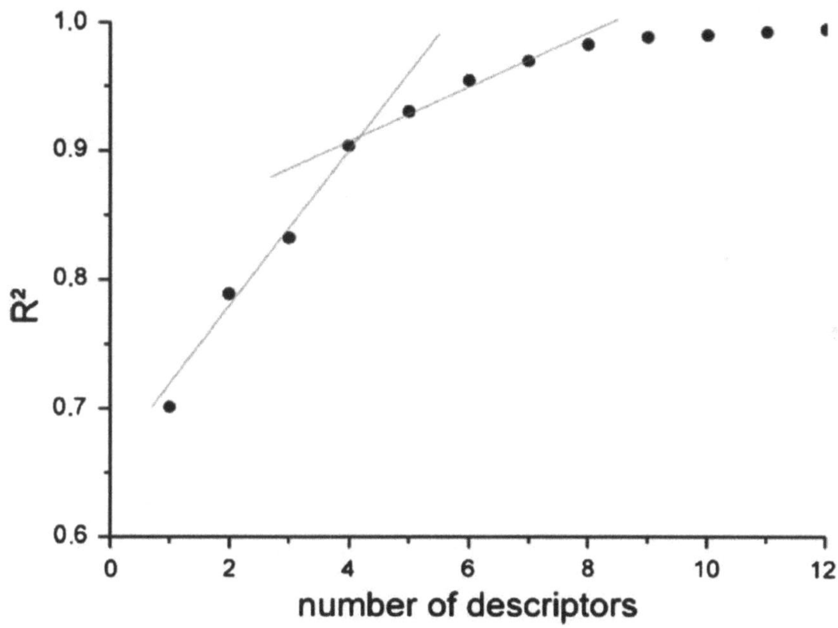


**FIGURE 3.16**

Change in $R^2$ with increment in descriptors for QSPR model development of Heats of decomposition (Fayet et al., 2010c).

data are available, model development as demonstrated before is possible through training and validating a model. In the absence of available data, quantum mechanical calculation based approaches can accurately estimate the heat of reaction, formation or combustion as required.

Let us consider the study done by Miranda et al. (2012). Noting the importance of 2,1,3-benzothiadiazole as organic light emitting diodes, organic photovoltaic and biosensor, the authors pointed out the lack of information on enthalpy of formation for compounds with sulfur–nitrogen bonds, which are not sulfonamides. To address, the authors employed quantum mechanical approach, validated against experiments, in efforts to find out the heat of combustion. In this benzothiadiazole problem, as mentioned above, the authors followed the following steps in order to estimate the gas phase enthalpy formation:

- The structure of 2,1,3-benzothiadiazole molecule is fully optimized with DFT at the B3LYP/6-31G (d) level. The obtained coordinates of the stationary points are characterized as minima through frequency calculation at the same level. It cannot be stressed enough, that obtaining an accurate structure of a molecule is a critical part in this approach.
- Estimate single point energy of the molecule using higher levels of theory such as QCISD (T)/6-31G (d) and MP2/GTMP2Large
- Correct electronic energies through introduction of vibrational, rotational, translational, and the pV terms computed at the B3LYP/6-31G (d) level to obtain final enthalpies at 298.15 K
- Use appropriate reactions demonstrating the formation of 2,1,3-benzothiadiazole and implementing information obtained from above-mentioned steps, estimate standard molar enthalpy of formation at 298.15 K

$$\Delta_r H^0(298\ \text{K}) = \sum_{\text{products}} \Delta_f H^0_{\text{prod}}(298\ \text{K}) - \sum_{\text{reactants}} \Delta_f H^0_{\text{react}}(298\ \text{K})$$

Implementation of the above-mentioned steps for a reaction process would result in estimation of heat of reaction, formation or combustion. However, it is of paramount importance to note that, while quantum mechanical calculations has the potential to provide very accurate information, appropriate selection of model and basis sets are fundamental toward an efficient and accurate set of results. Table 3.11 illustrates how various quantum mechanics models at the semiempirical level compare with respect to their accuracy in energy calculations and their scaling with number of atoms. One can observe the positive correlation between increase in scaling with model accuracy in estimating molecular energetics. Please note that the table is not comprehensive in nature and presented to provide an understanding in cost versus accuracy tradeoff in quantum mechanical calculations.

Let us also revisit the work on evaluation of thermal decomposition hazard associated with hydroxylamine nitrate as done by employing quantum mechanics as a screening tool prior to conducting more expensive experiments (Wei et al., 2006). The authors performed reactivity screening using semiempirical quantum mechanical calculation package, MOPAC (Molecular Orbital Package), and ASTM Chemical Engineering Thermodynamics and Hazard Evaluation (CHETAH). We have already discussed on the usage of CHETAH in the previous section. For the quantum mechanical calculations, the levels of theory used in MOPAC were AM1 and PM3. Similar to the steps described above, the approach applied was as provided in a Gaussian white paper (Ochterski, 2000). Put differently, the methodology consisted of structure optimization and energy minimization, energy estimation, incorporation of energy correction factor through introduction of translational, electronic,

**Table 3.11 Comparison of Various Semiempirical Quantum Mechanical Models**

| Model | Name | Description | Error | Scaling | Cost |
|---|---|---|---|---|---|
| | | | Equil Geom, kcal/mol | ~Basis function | Relative, 1–10 |
| AM1 | Austin Model 1 | Semiempirical | 10–30 | $N^2$ | 1 |
| HF | Hartree Fock | Averaged electron–electron | 10–20 | $N^4$ | 2 |
| B3LYP | Becke, Lee, Yang, Parr | Density functional | 5 | $N^4$ | 3 |
| MP2 | Moller–Plesset | Pertubation, correlated, 2nd order | 10 | $N^5$ | 3 |
| MP3 | Moller–Plesset | Pertubation, correlated, 3rd order | 5 | $N^6$ | 5 |
| CISD | CI singles, doubles | Configuration interaction, correlated | 5.8 | $N^6$ | 5 |
| CCD | CC doubles | Coupled cluster, correlated | 2.4 | $N^6$ | 5 |
| CCSD | CC singles, doubles | Coupled cluster, correlated | 1.9 | $N^6$ | 6 |
| QCISD | QC singles, doubles | Quadratic configuration interaction | 1.7 | $N^6$ | 6 |
| MP4SDQ | Moller–Plesset | Perturb, corr, 4th order, ignore triples | 2.7 | $N^7$ | 6 |
| MP4 | Moller–Plesset | Pertubation, correlated, 4th order | 1.3 | $N^7$ | 7 |
| G2 | G2, multilevel corrections | Composite method | 1.2 | $N^7$ | 6 |
| $G3^+$ | G3, multilevel corrections | Composite method, Truhlar, et al. | 1.0 | $N^7$ | 7 |
| QCISD (T) | QC singles, doubles, (triple) | Quadratic configuration interaction | 0.3 | $N^7$ | 7 |
| MP5 | Moller–Plesset | Pertubation, correlated, 5th order | 0.8 | $N^8$ | 8 |
| MP6 | Moller–Plesset | Pertubation, correlated, 6th order | 0.3 | $N^9$ | 9 |
| CCSD (T) | CC singles, doubles (triple) | Coupled cluster, singles, doubles, triple | 0.3 | $N^8$ | 8 |
| CCSDT | CC singles, doubles triples | Coupled cluster, singles–triple | 0.2 | $N^8$ | 9 |
| MP7 | Moller-Plesset | Pertubation, correlated, 7th order | | $N^{10}$ | 10 |
| CCSDTQ | CC singles, doubles triples, quadruples | Coupled cluster, singles–quadruples | 0.01 | $N^{10}$ | 10 |

*Source: Argonne National Laboratory, US Department of Energy, http://www.cse.anl.gov/*

rotational, and vibrational motion partition function, and, finally, estimation enthalpy of reaction at the desired temperature. A detailed description with a slight variation of the steps is also provided in Foresman and Frisch (1996).

In addition to assisting estimation of heat of formation of a chemical, first principle calculations can also be put to work in places where considerable disagreement between experimental values exists. Consider the case for formaldehyde and acetaldehyde (da Silva et al., 2006). As demonstrated by Silva et al., the disagreement among experimentally estimated values for heat of formation of formaldehyde and acetaldehyde exceeds the allowed standard error margins. The widely reported values for standard enthalpy of formation of formaldehyde are −26.0 and −27.7 kcal/mol at 298 K with acceptable error limits are less than the difference in these values, that is, 1.7 kcal/mol. This illustrates that even with the availability of experimental data, care must be taken in ensuring utilizing accurate information. To address these discrepancies, in da Silva et al. (2006), the authors implemented first principle calculations such as DFT and ab initio—level semiempirical theories to carefully examine the accurate heat of formation for formaldehyde and acetaldehyde. The approach taken by the authors can be broken down in the following set of steps:

1. Identify reactions with known enthalpies from literature involving molecule of interest, that is, formaldehyde in da Silva et al. (2006)
2. Calculate enthalpies of reactions using various levels of quantum mechanical calculations. Interested readers can find the details in the article
3. Compare computationally estimated reaction enthalpies
   a. Among various computational calculations
   b. Among experimentally estimated and computationally estimated values
4. Identify a subset of reactions where reasonable agreement is established among all computational approaches and experimental results
5. Identify a subset of reactions where enthalpies of formation are available for all species participating in the reaction
6. Calculate enthalpy of formation of the molecule of interest using enthalpy of formation of formation of other species from experiments for each reaction
7. Average estimated heat of formation for the molecule of interest over all the reaction as calculated in the previous step.

The work by Wei et al. (2004a) demonstrated the use of semiempirical quantum chemical methods such as AM1 (Dewar and Storch, 1985) and PM3 (Stewart, 1989) to screen reactive chemicals successfully. The low-level quantum mechanical approach was used in this particular study, as the primary objective was to screen the chemicals. The authors also observed better quality prediction of heat of formation with PM1 and AM3 as opposed to MNDO (Dewar and Thiel, 1977) and MINDO/3 (Bingham et al., 1975a,b) methods. This is where understanding the applicability of various methods becomes important.

### 3.6.3 PRACTICAL CONSIDERATIONS

Depending on the approach, calculation of heat of reaction should take into consideration of its associated caveats. At the macro-scale modeling level, reasonably accurate values of parameters such as compressibility factor, acentric factor, and critical constants are required as an input to Gibbs free energy minimization calculations, which results in accurate multiphase calculations of reactive systems. Other concerns relate to choice of appropriate minimization technique with minimal dependence on quality of initial guess on final convergence. While several books exist discussing on minimization and optimization techniques (Bertsekas, 1999; Luenberger and Ye, 2008), for our practical purposes, it is important to be conversant with available computer programs to calculate heat of reaction from thermophysical data.

## 3.7 REACTION RATE

As demonstrated in the previous section, estimation of heat of reaction addresses one of the significant factors related to reactivity of a chemical. However, it is not a sufficient criterion for determination of reactivity hazard. To determine the information related to rate of change of temperature in a chemical reaction, identify the existence and the onset of thermal runaway in a reaction, calculate instant power density and to investigate time–temperature–pressure graph, one also needs to determine the reaction rate as a function of temperature, concentration, and pressure. In other words: $r = f(\zeta, T, P)$ where $\zeta, T,$ and $P$ are concentration, temperature, and pressure, respectively. The effect of temperature and pressure is incorporated within the Arrhenius constant and does not appear separately in a rate equation.

Chemical kinetics in itself is a widely studied field and several books have been written on the quantitative study of the rates and of the factors on which the rates depend. In general, estimating rates of reaction involves estimation of the order of reaction and rate constant. Measurement of reaction rates through experiments can be found in any standard textbook on chemical kinetics such as Laidler (1987). A wide variety of experimental techniques exist to study reactions of various types such homogeneous and heterogeneous reactions. Equally important is the half-life of a reaction. Appropriateness of an experimental technique depends on the half-life of a reaction (Table 3.12).

In the following section, the authors focus on computational modeling methods in determining reaction rates. Although, at a macroscale, the empirical models are developed from experimental data, at the molecular level, as we shall see, a more rigorous computational approach is required for estimation of reaction rates from first principles calculations.

**Table 3.12 Summary of Appropriateness of Experimental Methods Based on Half-lives (Laidler, 1987)**

| Method | Range of Half-lives (s) |
|---|---|
| Conventional | $10^{2}$–$10^{8}$ |
| Flow | $10^{-3}$–$10^{2}$ |
| Relaxation | $10^{-10}$–1 |
| Pressure jump | $10^{-6}$–1 |
| Temperature jump | $10^{-7}$–1 |
| Field tube | $10^{-10}$–$10^{-4}$ |
| Shock tubes | $10^{-9}$–$10^{-3}$ |
| Kinetic spectroscopy | $10^{-15}$–$10^{-10}$ |

### 3.7.1 MACROSCALE MODELING APPROACHES

Quantification of rate of reaction progress can be defined as the rate of reactant consumption of product formation. Even though for simplification purposes, we tend to describe the transformation of reactants to a set of products as a single reaction, it is well known that it is a combination of several independent reactions and species, some of which are intermediate and exist only during the transient phases of the reaction. To illustrate, let us take the following reaction:

$$aA + bB \Leftrightarrow rR + sS + eE + fF$$

In reality, the above-simplified and lumped version of the reaction might well be resulted from a set of reactions, such as

$$aA + bB \rightarrow cC + sS$$

$$cC + bB \rightarrow dD + rR$$

$$dD + bB \rightarrow eE + fF$$

One can easily identify $C$ and $D$ as intermediate species, in this particular set of reactions. Assuming $a$ and $b$ are also the order of the reaction with respect to the reactant $A$ and $B$, respectively, the reaction rate for the above reaction can be equated as

$$R = -k[A]^a[B]^b$$

The order of a reaction is estimated through performing experiments. Same set of experiments performed at different temperatures also helps calculate the Arrhenius parameters for the reaction. Arrhenius equation is widely implemented to estimate the effect of absolute temperature on the reaction rate constant. Two commonly used forms of Arrhenius equation are

$$k = Ae^{-E_a/(RT)}$$

$$k = Ae^{-E_a/(k_bT)}$$

where $A$ is preexponential factor and $E_a$ is activation energy.

Once rate constant ($k$) and reaction orders with respect to reactants are determined, the overall reaction rate can be estimated through the reaction rate equation. The reaction rate equation as discussed above can be derived from collision theory. Any standard text on reaction engineering should cover the derivation of the reaction rate equation from collision theory. Nevertheless, experimental results play a very important role in estimating reaction rates at various temperatures through macroscale modeling.

#### 3.7.1.1 *Fun fact*

Svante Arrhenius, a Swedish chemist, who introduced his famous Arrhenius theory in 1887, received the lowest possible passing grade for his Ph.D. thesis and nearly destroyed his prospects for an academic career. The thesis contained foundation of the work that later earned him the 1903 Nobel Prize in chemistry.

### 3.7.2 MOLECULAR MODELING APPROACHES

In the absence of parameters for Arrhenius equation, which are typically estimated through experiments, one can use first principle calculations in conjunction with transition state theory (TST) to estimate the reaction rate constant. TST explains reaction rates of elementary reactions, reactions having only one transition state. Apart from Born–Oppenheimer approximation, TST relies on two basic assumptions (Henkelman et al., 2000):

1. The reaction rate is slow enough, so that a Boltzmann distribution is accomplished and sustained in the reactant state
2. Existence of only one transition state

The major challenge for successful implementation of TST is the requirement of accurate knowledge of the potential energy surface (PES). Finding the transition state in the multidimensional PES is a challenging computational task. Figure 3.17, shows a two-dimensional figure along the reaction pathway. This is also known as the minimum energy path (MEP).

### *3.7.2.1 First principle–based approaches*

#### 3.7.2.1.1 Transition State Theory

The power of DFT is well established. However, so far we have mostly concentrated on estimating global energy minima of a compound using this method, which refers to calculations performed at 0 K. However, 0 K is excessively cold for reactions to take place. The primary task of estimating reaction rate using first principle approaches is to generate an accurate multidimensional PES constructed from energy minimums. To illustrate, the methodology of estimating reaction rate through TST, we have demonstrated an example borrowed from Sholl and Steckel (2011): Adsorption of Ag atom on Cu ($x$–$y$) surface.

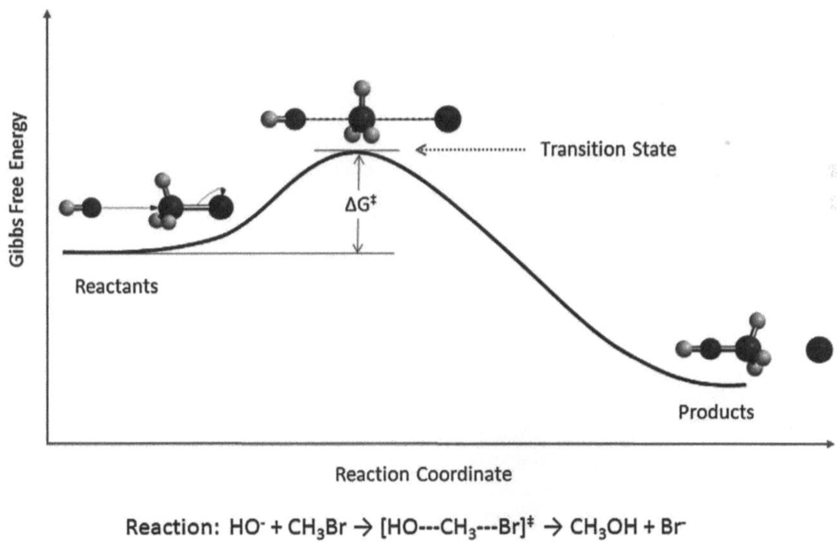


**FIGURE 3.17**

Reaction coordinate diagram for an exothermic reaction.

Let us visualize a set of Cu atoms in the $x-y$ plane. Let us fix the position $(x,y)$ of an Ag atom in the $x-y$ plane and allow it to move along the $z$ plane, while relaxing the Cu atoms in the $x-y$ plane for energy minimization of the Ag atom. In a similar fashion allow the Ag atom to scan the entire $x-y$ plane covered by Cu atoms and perform energy minimization at each point as mentioned above. Such an approach will result into generation of the two-dimensional PES as shown in Figure 3.18. In this figure, the four-fold sites are the only energy minima in the $x-y$ surface. The bridge sites between two four-fold sites are first order saddle points on the energy surface and the on-top site are the second order saddle points. First order saddle points are the bridge between two four-fold sites.

The MEP is defined as the path between two local minima through a first order saddle point. This representation is a very simplified version of a real adsorption process. It does not consider, for example, the exchange of energy between the Ag and Cu atoms. But when one does considers that, the chances that the Ag atom at room temperature with approximately 0.013 eV thermal energy can overcome the energy barrier of roughly 0.36 eV (Figure 3.19) increases significantly. TST is capable of providing the rate of hopping from state A to state B over a given energy barrier. It can be calculated as follows:

$$k_{A \to B} = \frac{1}{2}[\text{average thermal velocity}][\text{probability of finding atom at } x^{\#}]$$

where $x^{\#}$ is the transition state point. The incorporation of the 0.5 factor takes care of the fact that from a transition state activate complex can go toward either forming products or reforming the reactants. Since this book is more application oriented and not about the mathematical fundamentals behind

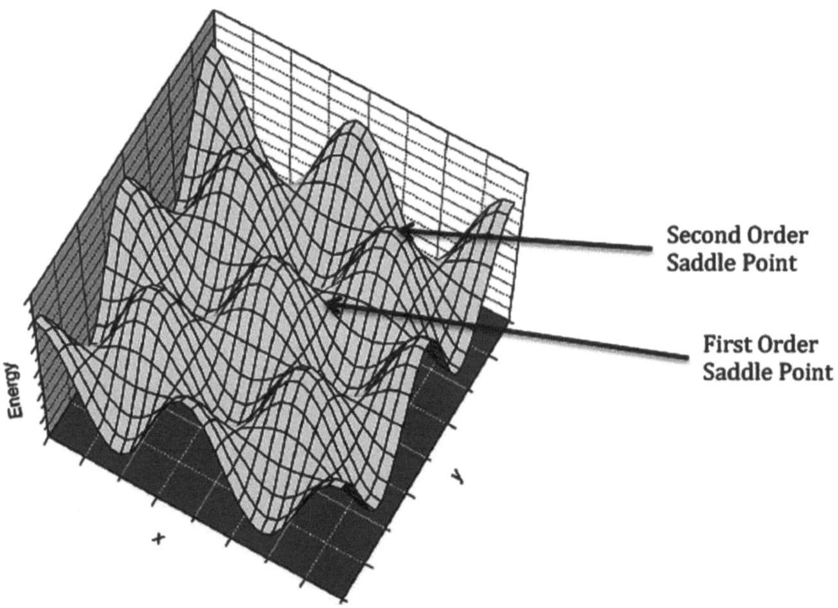


**FIGURE 3.18**

Illustration of two dimensional potential energy surface (Sholl and Steckel, 2011).

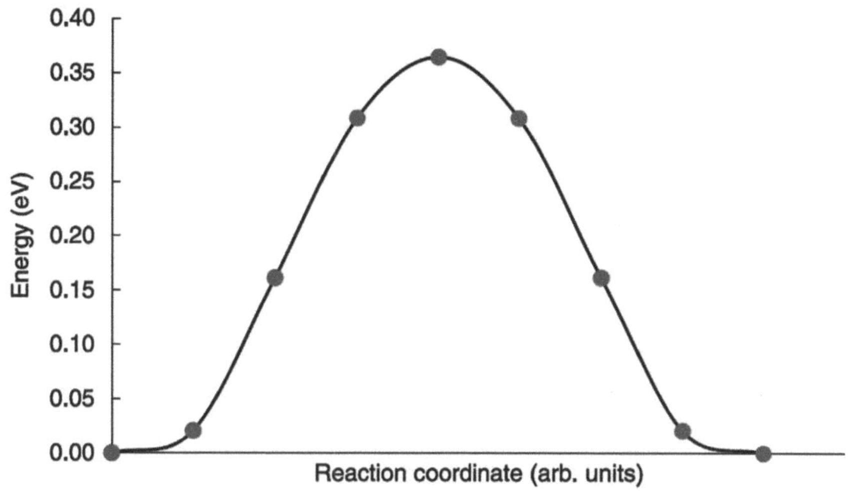


**FIGURE 3.19**

Energies of Ag as calculated by DFT, moving along the minimum energy path between two four-fold sites on Cu (*x*–*y* plane) (Sholl and Steckel, 2011).

various theories, we will skip the part of delving into more details in deriving equations of transition state theory and simply state that it can be shown that

$$k_{A \to B} = \vartheta \exp\left(-\frac{E^{\#} - E^{A}}{k_B T}\right)$$

where $\vartheta$ is the vibrational frequency, $E^{\#}$ is the energy at the transition state #, and $E^{A}$ is the energy minimum at state $A$.

Note that, the equation takes form of the well-known Arrhenius expression for chemical rate. While precise calculation for the preexponential factor, $\vartheta$, can be done by performing DFT calculations, typical values of somewhere between $10^{12}$–$10^{13}\ s^{-1}$ without performing any experiments or calculations at all can safely be used. Please note that the above-derived equation for Ag atom adsorption on Cu ($x$–$y$ plane) surface only considers an one-dimensional MEP. This was naturally done for the ease of illustration it offers. In a real scenario the rate of reaction, derivation of which can be found elsewhere (Sholl and Steckel, 2011), can be stated as

$$k_{A \to B} = \frac{\vartheta_1 X \vartheta_2 X \ldots X \vartheta_N}{\vartheta_1^{\#} X \vartheta_2^{\#} X \ldots X \vartheta_{N-1}^{\#}} \exp\left(-\frac{E^{\#} - E^{A}}{k_B T}\right)$$

This equation builds on the MEP by incorporating contributions from other higher energy trajectories. Specifically, a transition state with $N$ degrees of freedom is characterized by $(N-1)$ real and one imaginary vibrational mode whereas, local minima has $N$ real vibrational modes.

While reaction rates can be estimated reasonably accurately, by implementing transition state theory, the major challenge in such an approach is to correctly locate a transition state in multidimensional PES. It is significantly more complicated than finding a local energy minima (Sholl and Steckel, 2011). Sufficient conditions for transition state point are:

1. Strictly one imaginary frequency of normal vibration mode
2. The configuration is a critical point in the energy surface

However, as brought up earlier, locating transition states are not straightforward and given its significance, several numerical methods have been developed to address this problem. Generally in determining the best suitable method for locating transition state, the following two criteria should be borne in mind:

1. Plane wave DFT calculations do not provide second derivate information from the energy surface.
2. Keeping in mind the computational cost, the chosen method must converge to a single transition point in after small number of iterations.

The several approaches that are typically implemented in finding out transition states are:

1. Elastic band method
2. Nudge elastic band (NEB) method

The methods above are based on transition path sampling. Another approach consisting of a group of methods are the pseudo-Newton–Raphson methods (Cerjan and Miller, 1981; Baker and Gill, 1988; Banerjee et al., 1985), which are all similar to the Cerjan–Miller algorithm (Cerjan and Miller, 1981). The details of the mathematical basis of these methods and algorithms are not much of interest here, given the scope of the book. Instead, let us look for tools available to implement these methods in a real life situation.

In terms of applying transition state theory and its applicability toward both uni- and bimolecular reactions, few programs to the knowledge of authors are available to date. Programs such as POLYRATE, TheRate, and KisTHelP use transition state theory to estimate temperature-dependent rate constant at specific pressures for both unimolecular and bimolecular reactions. KisTHelP (Reinisch et al., 2012) is a kinetic and thermodynamic cross platform open source package that, among other molecular and reaction properties, can also estimate kinetic rate coefficients from transition state theory. KisTHelP is written in Java, making it platform independent, and is designed to perform statistical mechanical calculations of data obtained from ab initio quantum chemical calculations. The program also incorporates one dimensional quantum mechanical effects or tunneling effects on reaction coordinate motion by a multiplicative transmission coefficient. Figure 3.20 shows a captured screen from a KisTHelp calculation on a bimolecular reaction using both transition state theory (TST with and without tunneling effect) and variational transition state theory (VTST with and without tunneling effect).

POLYRATE (Steckler et al., 1995; Zheng et al.), written in FORTRAN, is another useful and powerful program, for the calculation of chemical reaction rates based on variational and conventional transition state theory, multidimensional semiclassical approximations for tunneling and nonclassical reflection. Similar to KisTHelP it is compatible with some of the other well-known application software in quantum chemical calculations such as Gaussian, NWChem, and Gamess. Potential energy surfaces dealt with in POLYRATE can either be analytical functions or relevant information at selected points.

TheRate (Duncan et al., 1998) is another computer application program, which uses information on the PES obtained from ab initio molecular orbital or DFT to calculate reaction rates using variational

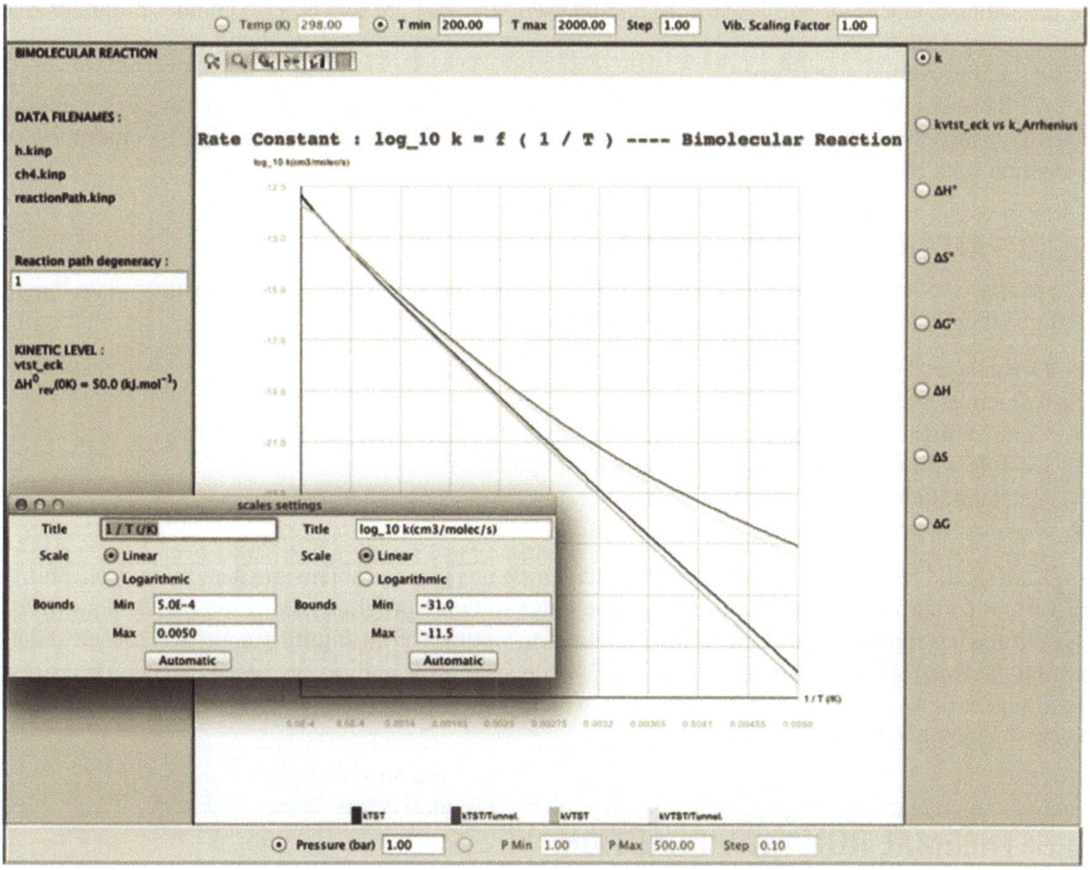


**FIGURE 3.20**

Logarithmic representation of rate constants as calculated using four levels of theory (TST, VTST, TST/tunneling, VTST/tunneling) in KisTHelP (Reinisch et al., 2012).

and conventional transitional state theory. Similar to the programs mentioned above, option for incorporating tunneling corrections is also available in this program. This Java-based program can be interfaced with Gaussian file and request for interfacing with other electronic structure packages is possible by contacting the authors. The program introduces techniques named as focusing technique and energy interpolation procedure. The focusing technique minimizes the number of Hessian calculations by cleverly requiring more around the critical points for transition states and less elsewhere. The energy interpolation technique allows to implement relatively less resource intensive technique, such as DFT, to calculate Hessian and optimum molecular geometries. The article (Duncan et al., 1998) illustrates the implementation of the program by applying it to the reaction:

$$CH_4 + \leftrightarrow HCH_3 + H_2$$

In addition, OpenSOAMS (Reinisch) and GPOP (Miyoshi) are two other computer programs one can use for estimating reaction rates and reaction rate constants based on quantum chemical calculations and transition state theory.

All the above-mentioned programs can also be used for estimating reaction rates through applying RRKM theory (Wieder and Marcus, 2004), a modification of RRK theory through incorporation of transition state theory.

### 3.7.3 PRACTICAL CONSIDERATIONS

The major challenges and concerns in the calculation of reaction rate through transition state theory and quantum calculations are primarily related to:

1. Deriving analytical form of PES
2. Numerical stability and convergence issues
3. Computational challenges in efficiently finding a transition state for a given PES
4. Dealing with systems having multiple reaction pathways
5. Quantum tunneling
6. High temperature/low energy barrier situation

Care should be taken to identify possible challenges in a given problem and determine applicability of various computational methodologies or computer software programs. For example, dealing with a system having multiple transition states, rather than opting for an intuitive approach toward determining the possible reaction pathway, in Peters et al. (2003) the authors proposed biasing a transition state search in a systematic and intuition-free way to identify multiple reaction pathways originating from a common reactant.

## 3.8 THERMAL RUNAWAY REACTIONS

Thermal runaway reactions and self-accelerated decompositions carry with them the potential for inherent risk of loss of control of the process with possible devastating effects on life and the environment. Self-decomposition, for example, is a very likely scenario with organic peroxides, characterized by a single O—O bond with low dissociation energy, in the order of 35—50 kcal/mol (Bach et al., 1996). Thermal runaway reactions are caused from uncontrolled increase in reaction rate resulting from rising temperature in an exothermic reaction. As long as the system has the ability to remove heat at a faster rate than the heat being generated from an exothermic reaction, the reaction is under control. Otherwise, the heat generated from an exothermic reaction accelerates the reaction rate through increased temperature resulting in an uncontrollable circumstance. Poor understanding of reaction chemistry and kinetics can result into undesirable runaway reactions (Etchells, 1997). Thermal runaway reactions has resulted in several accidents in chemical facilities (Westerterp and Molga, 2006; Hsu et al., 2012; Etchells, 1997), most notably the release of methyl isocyanate gas from a Union Carbide facility in Bhopal (Gupta, 2002; Eckerman, 2005; Dhara and Dhara, 1995; Shaluf and Said, 2003), India in 1984 leading to fatalities in the order of several thousands. Another example, as illustrated in Etchells (1997), an assumption of the decomposition temperature in the range of 270—299 °C, based on laboratory scale results as reported in the literature resulted in an explosion from decomposition while scaling up the same setup to plant level. At the plant, it was later found that the decomposition temperature is around 153 °C, at least 100 °C less than the temperature for which safeguards were installed. Factors such as surface area per unit mass, mixing dynamics, heat losses from equipment; a change resulting from scale-up in any or all of these factors has the potential to underestimate the risk of resulting in a thermal runaway reaction.

Among others, the effect of contaminants has emerged as the primary cause of incidents related to runaway reactions in chemical industry. Even the presence of trace impurities can lead to reduction of thermal stability of a chemical resulting in a thermal runaway reaction (Wei et al., 2004a). In most processes, nitro compounds are mixed with other chemicals or are contaminated by impurities which lower their thermal stability (Wei et al., 2004a). Hydroxylamine with acid or base contaminants is another example that has resulted in industrial incidents due to thermal runaway reactions.

The combination of following three line of defense can generally achieve prevention of runaway reaction:

1. Choice of right operating condition
2. Early event detection system
3. Installed safeguards to handle runway reactions such as design of emergency relief systems as per DIERS (Fisher et al., 2010) methodologies

In understanding and determining the right operating conditions, reaction calorimeter has become a widely accepted tool as have been briefed later in this section. Inaccurate understanding of reaction kinetics often leads to choice of overly conservative operating conditions leading to poor productivity. As discussed in Section 2.1, modeling approaches at different scales can be implemented to determine the heat of reaction and reaction rates to better understand reaction kinetics. Even for studying purposes using experimental techniques, reactions with significant exothermic reaction enthalpy is sufficiently hazardous and expensive and computational modeling can provide significant advantages.

In addition to have the ability to understand and estimate reaction kinetics for a particular set of parameters, it is of practical and significant importance to understand the effect of interaction among impurities, reactants and solvents on the thermal stability of the compound and define boundaries on process conditions, outside which this should be considered as hazardous. Put differently, a study should also be recommended to address and model various credible scenarios to help define boundary values of various parameters, which in turn will help design safeguards for risk mitigation. Returning to our earlier discussion on estimating the window for operating conditions, assessment of several parameters and their sensitivities to outer conditions becomes important. Studies of parametric sensitivity on chemical kinetics have been addressed in detail elsewhere (Varma et al., 2005). Instead we will focus here on the parameters themselves. The parameters typically assessed to characterize a runaway reaction are as follows:

- Reaction onset temperature
- Maximum temperature
- Adiabatic temperature rise
- Maximum pressure
- Self-heat rate at onset temperature
- Maximum self-heat rate
- Maximum pressure rate
- Noncondensable gas pressure

Understanding the mechanism behind a runaway reaction is very important from process safety perspective. Accurate understanding of such mechanism can help in processing scale-up situation where process conditions might change and enhance understanding of the associated potential hazards. Assessing thermal runaway reaction hazard is primarily done through experimental calorimetric techniques. The principles of reaction calorimetry are based on the overall mass and energy balance carried out over the reaction vessel. Reactive system screening tool (RSST) and automatic pressure tracking adiabatic calorimeter (APTAC) are two of those calorimetric techniques used for this purpose. In RSST, the chemical of interest is placed in a small container inside a containment cell pressurized with nitrogen. The cell is then heated at a constant temperature rate. Heating at constant temperature is the only heating mode in RSST. RSST, however, is not very sensitive and can only detect temperature rise as low as about 1 °C/min.

APTAC is an analytical instrument that measures dynamic response of a reactive system in the form of adiabatic pressure and temperature. In APTAC, the chemical is placed in a test cell that is then placed in an apparatus. Unlike RSST, APTAC has two heating modes: heat and heat and search. Details on experimental techniques are not discussed here and we instead focus on applicability of computational modeling techniques at various scales to study thermal runaway reactions.

While the final products can be analyzed through various experimental techniques, most of the intermediates and transition states are short lived. In the absence of such information it is difficult to understand and study mechanism behind thermal runaway reaction of the chemical of interest. In order to understand a reaction to a substantial level, knowing the structure of reactants and products can only be considered as a starting point. Effects of some parameter can be predicted using the minimum knowledge of reaction, such as the way one implements Le Chatelier principle to chemical reaction to study effects of changes in state variables of the system and the like. However in a complex or novel scenario, many questions remain unanswered until new experimental studies are performed to capture the study objective. How, for example, one might accurately estimate the effect of a foreign particle in a reaction. Theoretical quantum chemical calculation with assistance from experimental runaway behavior data is a powerful tool to study the most likely reaction mechanism associated with a runaway reaction (Wei et al., 2004a). In other cases, an apparently repeated occurrence of innocuous trivial phenomena can result into building up a runaway reaction. Consider the Seveso incident (Hay, 1978, 1979; Cardillo et al., 1984; Hugo and Steinbach, 1986). It was found out that (Hugo and Steinbach, 1986) small accumulation of reactant over time in a low reactor temperature environment resulted in the runaway reaction. While reaction calorimetric techniques are widely used on assessing thermal runaway reaction potential and predict chemical instability, computational approaches are also implemented for similar purposes.

## 3.8.1 MACROSCALE MODELING APPROACHES

### *3.8.1.1 Parametric sensitivity*

As noted earlier, the contest between the heat generated by an exothermic reaction and heat removed through thermal conduction and convection methods determines the potential of a reaction becoming a runaway. The contest is interesting as the heat generation in a reaction is exponentially dependent on the temperature as opposed to a linear dependence exhibited while heat is removed through conduction and convection. Incorporation of effects such as nonuniform distribution of reactant, effect of small perturbation in the model equations allows one to assess more realistic cases of thermal runaway reactions. Lyapunov exponent–based calculations have been suggested (Strozzi and Zaldívar, 1994), to assess the thermal stability of batch chemical reactors. Note that in batch reactors, one has lesser control in controlling runaway reactions as the amount of reactant into the reactor and cannot be changed once the process has started. On the contrary, in a semibatch reactor or a continuous reactor, manipulating the flow of the reactants to the reactor is another factor in controlling the reaction rate. In addition, in the absence of steady state, control of batch reactors can often get very complex owing to its time-varying sensitive parameters.

The authors (Strozzi and Zaldívar, 1994) demonstrated the calculation of the Lyapunov spectrum (Wolf et al., 1985) consisting of all Lyapunov exponents, in a reactive system with a known set of differential equations and initial conditions describing an exothermic irreversible *n*th order reaction. Application of initial conditions and integrating it for a specified time until almost full conversion of the reactant is achieved on the set of differential equations for mass and energy balance provides information on the evolution of the corresponding reactive dynamical system. In particular, the exponents capture the evolution of the variables with initial conditions in the phase space. The *j*th Lyapunov exponent is defined as

$$\lambda_j = \lim_{t \to \infty} \frac{1}{t} \log_2 \frac{p_j(t)}{p_j(0)}, \quad j = 1, \ldots, n \text{ inital conditions}$$

Lyapunov exponents monitor two closely monitoring points as they evolve over time. While the points on the main trajectory, also known as the fiducial trajectory is defined as the evolution of the differential equations over time, the surface of the sphere are defined by the linearized differential equations on points infinitesimally separated from the fiducial trajectory. Figure 3.21 illustrates the concept of fiducial trajectory (the center line) and Lyapunov exponents.

We will not go into the details of mathematical foundation behind Lyapunov exponents and instead focus briefly on how this method, borrowed from typical chaos theory applications, can be implemented toward batch chemical reactors for assessment of thermal stability. The objective is to estimate parametric sensitivity of reactors to help define operating conditions in low parametric sensitivity region. For a potential thermal runaway reaction, parametric sensitivity is important, as it will describe variation of maximum process temperature with small changes in process conditions. In order to identify such plausible scenario, first one needs to define sensitivity. For thermal runaway reactions, sensitivity has been defined as (Morbidelli and Varma, 1988)

$$s_{\varnothing} = \frac{\partial T^*}{\partial \varnothing},$$

where $T^*$ stands for the maximum temperature.

As per Morbidelli and Varma (1988), a generalized criterion to identify reactions susceptible to runaway can be defined based maximum value of the sensitivity number in the phase space.

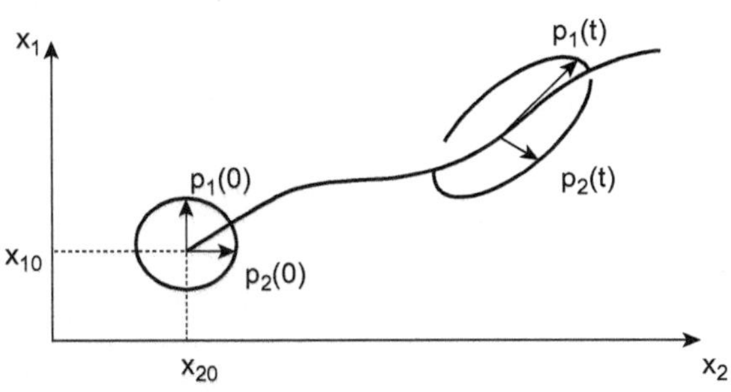


**FIGURE 3.21**

A Schematic representation of the evolution of a dynamical system in the phase space (Strozzi and Zaldívar, 1994).

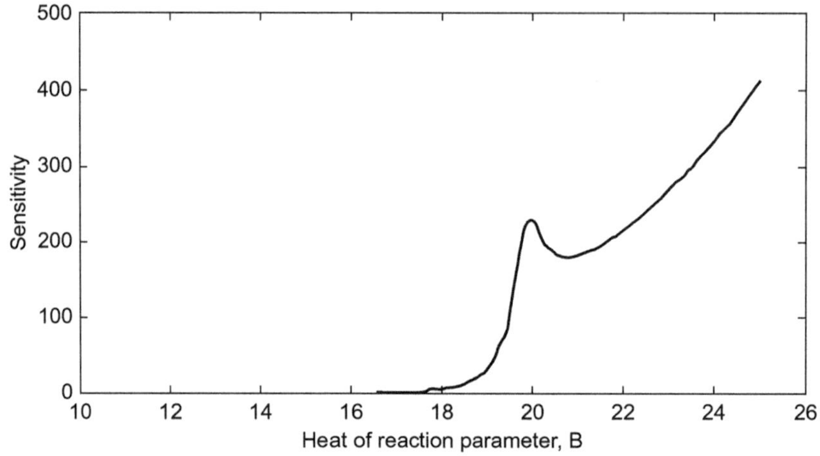


**FIGURE 3.22**

Reaction sensitivity as a function of heat of reaction parameter (Strozzi and Zaldívar, 1994).

The authors also concluded this as a generalized approach as the result is expected to be independent of the choice of parameter for the sensitivity study.

The study by Strozzi and Zaldívar (1994) has defined sensitivity as a function of Lyapunov exponents as

$$s_{\varnothing} = \frac{\Delta \max_i 2^{\mu}}{\Delta \varnothing}$$

where $\mu = \sum_i \lambda_i(t)t$, a function of Lyapunov spectrum of the system.

Figure 3.22 illustrates the sensitivity to thermal runaway with respect to heat of reaction in the approach suggested by Strozzi and Zaldivar (1994). As a side note, the figure also emphasizes the importance of accurate estimation of heat of reaction for the reaction of interest. See Section 2.2 for more information on estimating heat of reaction in the absence of experimental data.

Several other different runaway criteria have been proposed in the literature and have been reviewed (Szeifert et al.; Zaldívar et al., 2003). Parameters such as Semenov number for thermal explosion analysis have also been used to study parameter sensitivity in a reactive system. Parameter sensitivity and Lyapunov exponents are important topics by their own merit and are only briefly discussed here in the context of thermal runaway reactions. Several studies (Luo et al., 1997; Szeifert et al.; Razon and Schmitz, 1987; Varma et al., 2005; Morbidelli and Varma, 1988; Zaldívar et al., 2003; Strozzi and Zaldívar, 1994; Alós et al., 1996) exist on combining parametric sensitivity with potential runaway reactions and treat the matter in greater depth than it is presented here. Much of that are deliberately left out here, as the primary objective is to introduce the reader on availability of advanced methods that can help in process safety applications.

**Problem** (Szeifert et al.):

Consider the following set of equations describing a reactor model:

$$\frac{dc}{dt} = -r$$

$$\frac{dT}{dt} = \beta r - \alpha(T - T_w)$$

$$r(c, T) = \exp\left(\gamma - \frac{\delta}{T}\right) \cdot \varphi(c)$$

Initial condition:
At $t = 0,\ c = c_0,\ T = T_0$
Given:
$\alpha = 5\ 1/\text{h}; \beta = 180\ \text{m}^3\text{K/kmol}; \gamma = 20;\ \delta = 6000\ \text{K};\ C_0 = 1\ \text{kmol/m}^3;\ T_0 = 298\ \text{K}$

1. Find the sensitivity of the temperature ($T$) with respect to the reactor wall temperature ($T_w$), in the range of 250–290 K in 10 K increments.
2. Identify a possible runaway reaction scenario.

## 3.8.2 MOLECULAR MODELING APPROACHES

### *3.8.2.1 First principle calculations*

We know from the previous section that with an accurate know how of the equations describing a reactive system, such as mass balance, energy balance, and reaction information, it is possible to identify reactions susceptible to runaway. However, if accurate understanding of reaction dynamics is absent, then the above-mentioned approach is expected to fall short and may produce misleading information. To overcome such scenarios, and gain more insight, one needs to perform new experiments or perform computational modeling at the electronic level. Ab initio studies have been used to study decomposition reactions (Yu and Yu, 1996). Gaining insight on decomposition of hydroxylamine is one such example where first principle calculations can help.

Hydroxylamine plays an important role in various industries including chemical, pharmaceutical, and semiconductor (Catalysis in Chemistry and Enzymology, 1987). Since it has been reported to be involved in several incidents (Wang et al., 2010; The Explosion at Concept Sciences: Hazards of Hydroxylamine, 2004) (Figure 3.23), further insight on hydroxylamine thermal decomposition is important to understand and mitigate its potential hazard.

**FIGURE 3.23**

Concept Sciences Hydroxylamine Explosion, CSB Investigation Report, February 1, 2002, www.csb.gov

*Image taken from CSB investigation report on the incident.*

The study by Wei et al. (2004a) looked into thermal runaway reactions of hydroxylamine under various concentrations of KOH (representing base) and HCl (representing acid). Majority of this study used RSST and APTAC to determine thermokinetic data of the reaction such as onset temperature, heat of reaction, and others as mentioned in Section 2.1. In that section we have looked into approaches to estimate heat of reaction and calculate reaction rates. One of the interesting and significant parts of this study is that the authors used the power of quantum chemical calculations along with experimental data and knowledge from literature to propose decomposition mechanism of hydroxylamine in the presence of acid and base. The advantage of using first principle calculations is that it can be used to validate proposal based solely on experimental results where information on transition state is minimal.

To understand the mechanism involved on hydroxylamine decomposition in basic solution, a set of elementary reactions, including the ones involving transition state elements was proposed based on prior studies on hydroxylamine. It also involved ruling out reactions as hypothesized in one study based on latest research results available on the subject. The set of five elementary reactions were used to define the decomposition of hydroxylamine are as follows:

$$NH_2OH^- + OH^- \rightarrow NH_2O^- + H_2O \qquad \Delta H_{rxn} = -109.6\ \text{kJ/mol}$$

$$NH_2O^- + NH_2OH \rightarrow NH_2NHOH + OH^- \qquad \Delta H_{rxn} = -14.6\ \text{kJ/mol}$$

$$NH_2NHOH \rightarrow HNO + NH_3 \qquad \Delta H_{rxn} = 29.7\ \text{kJ/mol}$$

$$HNO + NH_2OH \rightarrow NH_2O^* + NH_2O^* \qquad \Delta H_{rxn} = -3.8\ \text{kJ/mol}$$

$$2NH_2O^* \rightarrow N_2 + 2H_2O \qquad \Delta H_{rxn} = -477.8\ \text{kJ/mol}$$

Using B3P86/cc-pVDZ level of theory in Gaussian 98, the heat of reaction was calculated for each of these reactions by using

$$\Delta H_{rxn} = H_{\text{products}} - H_{\text{reactants}}$$

From the observed set of reactions and its corresponding calculated heat of reaction, one can see that the explosive behavior of hydroxylamine decomposition is related to the amount of intermediates produced. It also demonstrates that the pH of the solution affects the decomposition of hydroxylamine. Increased concentration of hydroxide ions resulting from decomposition of hydroxylamine reduces the onset temperature owing to its exothermic nature of the reaction. A similar approach was taken to analyze the hydroxylamine decomposition in acidic solution. The decomposition mechanism proposed based on earlier studies is more complicated in the acidic solution. Treating all species in gas phase, the heat of reaction was calculated using B3P86/cc-pVDZ level of theory in Gaussian 98. The elementary set of reactions for the decomposition of hydroxylamine in acidic solution is as proposed:

$$NH_3OH^+ \rightarrow NH(\text{triplet}) + H_3O^+ \qquad \Delta H_{rxn} = 301.7\ \text{kJ/mol}$$

$$H_3O^+ + NH_2OH \rightarrow NH_3OH^+ + H_2O \qquad \Delta H_{rxn} = -134.3\ \text{kJ/mol}$$

$$NH(\text{triplet}) + NH_2OH \rightarrow {}^*NH_2 + NH_2O^* \qquad \Delta H_{rxn} = -107.1\ \text{kJ/mol}$$

$$NH_2O^* + NH_2OH \rightarrow NH_2NHOH + {}^*OH \qquad \Delta H_{rxn} = 73.2\ \text{kJ/mol}$$

$${}^*OH + NH_2OH \rightarrow H_2O + NH_2O^* \qquad \Delta H_{rxn} = -197.5\ \text{kJ/mol}$$

$$NH_2NHOH \rightarrow HNO + NH_3 \qquad \Delta H_{rxn} = 29.7\ kJ/mol$$

$$NH_2O^* + {}^*NH_2 \rightarrow HNO + NH_3 \qquad \Delta H_{rxn} = -154.8\ kJ/mol$$

$$HNO + HNO \rightarrow N_2O + H_2O \qquad \Delta H_{rxn} = -366.9\ kJ/mol$$

Analyzing the reaction network in both cases using quantum mechanical calculations illustrated among others that NH (triplet), one of the intermediates in the acidic reaction network, is relatively more reactive with hydroxylamine when being compared with the reaction of HNO and hydroxylamine in the basic environment. Based on the calculated heat of reactions in the branching reactions of the two scenarios, one can also see that it explains the fact that the maximum self-heat rate and pressure rate is higher in acid solution. The branching reaction in the acidic solution scenario is more exothermic. The study results also illustrated that based on the type of contamination (e.g., acidic or basic in this case) the nature of thermal runaway reaction can be different and the detailed look through first principle calculations showed that clearly.

Overall the study showed that, while experimental calorimetric techniques helped estimate various parameters that define a thermal runaway reaction, quantum mechanical calculations helped understand the physical reasoning underlying it and demonstrated an approach by which one can reveal the physical phenomenon of thermal runaway reactions explaining the experimental results. Additionally, it is also a useful tool to quantitatively determine, among a set of process conditions, one with relatively safer for a potential thermal runaway reaction hazard.

In a related similar study (Wang et al., 2010), researchers from the same group, studied in detail, through first principle calculations, initial steps involved in the decomposition mechanism of hydroxylamine. The study, as the topic suggests, was focused on identifying initial decomposition pathways for hydroxylamine thermal decomposition using quantum mechanical calculations and analyzing various proposed schemes for initial decomposition pathways of hydroxylamine both in gas and aqueous phase. Based on set of reactions in each proposed scheme, geometry was optimized for each reactant, product, and intermediates. Energy and frequency were calculated for each optimized structure. Effect of solvent was incorporated and calculated by two different approaches, namely, cluster methods and polarizable continuum model, where the solvent is treated a polarizable continuum instead of treating its individual constituents separately to save computational cost. The study helped the researchers to provide information and insight in a number of areas in regard to thermal decomposition of hydroxylamine such as

- Molecular Structure of Hydroxylamine: Obtaining structure of the molecule, in this case, worked more as a validation of the first principle calculations implemented for the whole study. As part of the structure one also gets information on various bond lengths, angles.
- Bond Strength and Bond Dissociation Information: The stability and reactivity of a chemical compound is related to its bond strength and bond dissociation energy. Based on these, one can assess, say for a very reactive system, if and which proposed bond breaking mechanism is more likely to make the system so much reactive. For example, in this study by Wang et al. (2010), it was shown that simple bond dissociations and unimolecular reactions are unlikely to occur in the initial steps of hydroxylamine decomposition.
- Feasible Reaction Pathways: Based on first principle calculations one can assess the various reaction activation barrier of different proposed pathways as done in this study. This, in turn, helps one to determine the most feasible reaction pathway for a given reaction. For example, in this

study, the authors found that, in aqueous solution, bimolecular isomerization step is the most favorable pathway for hydroxylamine decomposition.
- Solvent Effect: For studying hydroxylamine decomposition, studying effect of solvent is important, as they are stored in aqueous solution. In this particular study, the authors found that, activation energy reduces with increasing water molecules as solvent. Both cluster and continuum solvent models were developed to incorporate and illustrate the effect of solvent.

**Problem:**

Identify and calculate the heat of reactions for the two branching reactions using at least two different levels of theories and compare results.

#### *3.8.2.2 Molecular dynamics simulations*

Studying thermal decomposition of materials at the molecular dynamics level has to be performed using force field, which is capable of incorporating reactive nature of the chemical of interest. ReaxFF (van Duin et al., 2001) force field and reactive empirical bond-order (REBO) force fields are well-known force fields that can account for reactive nature of a chemical such as that for decomposition events. Reactive molecular dynamic (RMD) (Stoliarov et al., 2003) simulation that uses a program called MD_REACT is another approach useful for reactive systems by extending classical force field approach.

ReaxFF (van Duin et al., 2001; Chenoweth et al., 2008), a bond order—based reactive force field, initially incorporating hydrocarbons, has been developed to perform large-scale molecular dynamic simulations of reactive systems. It uses general relationship between bond length and bond order on the one hand and between bond order and bond energy on the other to determine feasibility of bond dissociation and bond formation phenomenon. Other bonded interaction such as angles and torsion are also defined as bond order so that on dissociation of a bond, these terms goes to zero in a continuous and smooth fashion. The parameters for ReaxFF force field are derived directly from quantum mechanical calculations. ReaxFF is based on a fundamental assumption that bond order between two atoms in a chemical system can be determined from interatomic distance between the two atoms. Each type of bond (sigma, first pi bond, or second pi bond) has a value based on interatomic distance between a pair of atoms. The sigma bond is valued as unity below ~1.5 Å but negligible above ~2.5 Å. Similarly, the first pi bond is unity below ~1.2 Å and negligible above ~1.75 Å and the second pi bond is unity below ~1.0 Å and negligible above ~1.4 Å. While currently parameters have already been developed for various atom types, should there be a need to develop parameters for new atom type to be used with existing ReaxFF force field, the following set of steps should be followed:

- Define Hamiltonian of the molecular system similar to that of defined by a classical force field system. The Hamiltonian of the system, in addition to incorporating bond, angle, torsion, nonbonded interactions (vdW and Coulombic), takes into account energetics arising from over or under coordination of an atom, conjugation effects, and energy penalty to reproduce stability for atoms with under- or overcoordination. For Coulombic interactions, atomic charges are calculated using electron equilibration method (van Duin et al., 2001; Chenoweth et al., 2008). Since the focus of this book is more on implementation of multiscale modeling on process safety, detailed breakup of the functional form of each of these energetics is not provided and can be found in Shankar et al. (1986).

- Optimize the force field parameters through one-parameter search technique as illustrated in the work by van Duin et al. (2001). The process includes reproducing heats of formation, bond length, and bond angle within an acceptable range of their available literature values. To do that, the structures obtained from quantum chemical calculations were added to the force field training set.
- The quantum chemical calculations in van Duin et al. (1994) were performed using DFT at B3LYP level using the 6-31G** basis set, which uses the GGA (general gradient approximation) and exact exchange energies.
- The bond dissociation calculations in van Duin et al. (2001) were performed using restricted HF method on singlet spin system.

ReaxFF force field has been used in various studies to capture the effect of reaction in a system. In some cases, such as that of combustion of coal char (van Duin et al., 2001), reaction is the primary phenomena defining the system of interest. In others, such as thermal decomposition of an otherwise thermally stable polymer (Castro-Marcano et al., 2012), the event is undesired and may only happen at elevated pressure and temperatures, and is investigated to understand it with a preventive mindset for such a scenario. In either study, attempt has been made to systematically implement power of ReaxFF and investigate the reaction mechanism involved in each case.

In Chenoweth et al. (2005), decomposition of polydimethylsiloxane (PDMS) at high temperature and pressure was investigated using ReaxFF. The rate of decomposition was studied through following the production rate of methane, which is produced as a result of the decomposition of PDMS. The temperature dependency of the methane production was tracked to estimate Arrhenius parameters for the thermal decomposition of PDMS.

- Similar to the earlier referred study, the first step in this study was force field parameterization by training the ReaxFF for Si/SiO system using results from quantum mechanical calculations. Since the parameters in ReaxFF are transferable, as noted earlier, in cases where the parameters are available, this step is not required. As an example, the authors, through this study established parameters for ReaxFF so that it is capable of modeling any system containing Si, O, C, and H. Even though not explicitly mentioned, the Hamiltonian of the system was constructed similar to the earlier example. For optimization of the silicon−carbon bond energy, the dissociation pathway for the Si−C single bond in Si $(CH_3)_2$e $(OCH_3)_2$, $H_3SieCH_3$, and the Si=C double bond in $H_2Si{=}CH_2$ was used. DFT results of the singlet case around equilibrium bond distance were used to optimize the ReaxFF parameters, while the dissociation limit was fitted against the results of the triplet state.
- For optimization of valence angle parameters, the angle bending energy was matched to results from DFT calculations.

RMD, or reactive molecular dynamics as coded in MD_REACT, uses traditional molecular dynamics where one makes use of analytical expression (force field) defining interactions between atoms. In Stoliarov et al. (2004) the authors used consistence valence force field (CVFF) for performing such molecular dynamics simulations to estimate atomic structures near their equilibrium geometries. The trajectories of the system of atoms as obtained through time integration of their classical equation of motion as resulted from set of forces acting on each of the atoms, characterizes the system under study. The RMD technique, in addition, can also model chemical reactions, by incorporating reactive PES. The force field is dynamically modified to model potential energy surfaces associated with chemical reactions. Complex processes such as that takes place in condensed phase systems has been investigated using RMD techniques. Experimental approaches such as thermogravimetric analyses (TGA) along with mass spectrometry can provide reaction rate constants and yield information of the final product. Reactive molecular dynamics approach can however provide more insight and is capable of handling complex and novel systems (Stoliarov et al., 2003). Stoliarov et al. (2003, 2004) and Nyden and Stoliarov (2008) provide a good review of development in this area of extending traditional classical force field–based atomistic simulation to reactive molecular dynamics simulation using reactive force field.

Stoliarov et al. (2003) used reactive molecular dynamics method to study thermal decomposition of polymethyl methacrylate (PMMA). While thermal decomposition of PMMA does not lead to a thermal runaway reaction, the approach used in this study can be implemented while studying thermal runaway reactions for complex processes. In addition to capturing molecular structural details through monitoring information of bond, angle, torsion, and covalent bonds, RMD approach allows a covalent bond to break and form as the system evolves with time.

RMD approach as illustrated (Stoliarov et al., 2003) uses the following set of steps to capture a reactive system evolving with time:

- Define Hamiltonian similar to regular molecular dynamics calculations as shown below:
  - $H = \sum_j^N \frac{p_j^2}{2m_j} + V(q_1, q_2, \ldots, q_{3N})$

    where $V = \sum_{i=1}^{n_{\text{bonds}}} V_{i_{\text{bond}}} + \sum_{j=1}^{n_{\text{angles}}} V_{j_{\text{angle}}} + \sum_{k=1}^{n_{\text{torsions}}} V_{k_{\text{torsion}}} + \sum_{l=1}^{n_{\text{nonbond}}} V_{l_{\text{nonbond}}}$
- Define a function that will help decide formation and deletion of covalent bonds based on fractional bond order as shown below:
  - $\text{SF} = \begin{cases} 1 & r \le r_e \\ 1 - \dfrac{V_{\text{bond}}}{\text{D}} & r > r_e \end{cases}$

where $V_{\text{bond}}$ is the sum of potential energy due to bonded forces and D is the bond dissociation energy $r_e$ is the equilibrium bond length.

- Define bond dissociation and formation criterion. Specific to this study, the authors dissociated or eliminated a bond when the fractional bond order "S" became less than or equal to the bond dissociation criteria or BDC. For bond formation, bonds between active atoms with largest value of $[\text{D} - V_{\text{bond}}]$ were retained. The two atoms who are connected by the eliminated bond, based on bond dissociation criteria, are flagged as active atoms to signify that based on valency, these atoms are available to create another bond. As an example in this study (Stoliarov et al., 2003, 2004), the authors chose a value of 0.7 for BDC. This implied that to be chemically active or a dissociation of a bond, the value of "S" needs to be less than 0.7, or the value of $V_{\text{bond}}$ should be higher than 0.3D.
- After every time step, calculate the fractional bond order for every bond and compare it with its bond dissociation criteria. Eliminate a bond if it's bond dissociation criterion is satisfied.

- Simultaneously with bond dissociation assessment, after each time step, generate a set of possible covalent interactions between all currently defined active atoms.
- If for an active atom, the total number of bonds is equal to its valency and the fractional bond order of each of its bonds is higher than BDC, or in other words if the $V_{bond}$ value is less than a specific value, then the chemically active label is removed from the atom.
- For an atom with chemically active status, it is allowed to form a hypervalent bond to describe reactions taking place through hypervalent states.
- Update atom type based on changes in nature of bonds for chemically active atoms.
- Adjust partial charges of chemically active atoms.
- Simulate next time step, and repeat the above-mentioned steps to identify and define bond dissociation and formation.

In the above approach, it must be noted that for bond dissociation, the bond is not actually eliminated until a more energetically favorable bond replaces it. It must also be noted that the BDC value has been chosen by the authors and not mathematically derived. In such a scenario, experimental validation of results becomes important. However, while exact results may require validation because of chosen parameters, the approach can definitely tell us changes in reaction trends with changes in process conditions. The study also did not look into possibilities of ion formation through reaction. For reactions involving novel materials or otherwise, first principle calculations can also be used for force field parameterization. This type of approach can also be used to investigate generation of flammable materials as a side product of thermal decomposition. Since the system evolves with time and also accounts for its reactive characteristics, the approach is also capable of estimating various kinetic parameters such as rate constant, preexponential factor and activation energy. In addition, the approach is also capable of capturing reaction mechanism of a reactive system.

### 3.8.3 PRACTICAL CONSIDERATIONS

At a macroscale level, sets of differential equations describing the system accurately and accurate initial conditions are important to know in order to assess the potential of thermal runaway reaction. In the absence of that, errors and uncertainties can multiply and provide mislead calculations.

In the absence of experimental data, accuracy of first principle calculations can be a concern. As such, the two components of accuracy, physical accuracy and numerical accuracy need to be differentiated. In a DFT calculation, for example, a problem is approximated for making it numerically solvable, which might lead to introduction of inaccuracies. Physical inaccuracies are introduced for generalization of the problem at a different level. Based on the objective of a problem, accuracy to different extent is acceptable. Screening analysis, not the final study by definition, can tolerate certain degree of inaccuracy. The trade-off between the accuracy and utilization of computational resources is determined by the context of the study and should be appropriately considered.

While implementing force field methods, care should be taken to ensure that the force field parameter estimation and validation is based on similar type of molecules. For implementing ReaxFF, parameters for a specific system might not be available, in which case they need to be generated based on quantum mechanical calculations and validated against experimental data. The effort in such an endeavor should be taken into account while deciding the appropriate methodology.

## 3.9 DUST EXPLOSION

Dust explosions, initiated by rapid combustion of flammable particulates, pose serious explosion hazard in process industry. While to basics are reasonably understood, fires and explosions from solid particulates and liquid particulates continue to contribute toward considerable material damage and loss of life (Lemkowitz and Pasman, 2014). Next to vapor cloud and boiling liquid expanding vapor explosions, dust explosions can be attributed to cause for several injuries, fatalities, and significant financial loss (Abbasi and Abbasi, 2007; Eckhoff, 2003). Given the right conditions, most unoxidized material can burn. While there does not seem to be any unanimous opinion (Abbasi and Abbasi, 2007; Mannan and Lees, 2005; NFPA 68 Guide for Venting of Deflagrations, 2002) on classifying dust particles based on maximum particle size, ranging from less than 420 μm (NFPA 68 Guide for Venting of Deflagrations, 2002) to less than 1000 μm (Palmer, 1973; Mannan and Lees, 2005), severity of dust explosion has been proven as a function of its particle size and distribution (Di Benedetto et al., 2010; Huang et al., 2009; Pritchard, 2004). At the other end it is reported that the maximum explosion pressure does not increase below 50 μm (Eckhoff, 2003). As explained (Pritchard, 2004), the reasoning behind limiting particle size can be explained through understanding of reaction mechanism leading to dust explosion phenomena. In the homogeneous gas phase, the event is initiated by pyrolysis or devolatization, followed by gas phase mixing and combustion. The slowest among devolatization, mixing, and combustion as determined by their time constants will be the rate-determining step in the whole phenomena and thus affecting the maximum explosion pressure. With diminishing particle size, as the surface area is increased resulting in increased devolatisation rate, if it does not remain the rate-determining step of the overall phenomena, the effect of particle size on maximum pressure will cease to exist below a certain particle size.

Modeling dust explosion involves estimation of properties related to the chemical of interest, calculation of sensitivity of the explosion to the flow properties and heat transfer of the dust cloud (Di Benedetto and Russo, 2007). In general, the severity of dust explosion can be quantified by the following parameters (Lemkowitz and Pasman, 2014) as illustrated in the following table:

| Severity Parameters | Description |
|---|---|
| Tendency toward detonation | Increases with increased reactivity and decreasing particle size |
| Maximum explosion pressure or $P_{max}$ | Largely defined by the maximum explosion temperature. The maximum explosion overpressure, $P_{max}$, is the difference between the pressure at the time of ignition and that at the highest pressure attained in the dynamic pressure curve during a dust explosion. |
| Maximum rate of pressure rise or $\left(\frac{dP}{dt}\right)_{max}$ | Generally measured from initial condition. |
| Maximum volume normalized rate of pressure rise or $K_{st}$ | Value of maximum rate of pressure rise, normalized to a volume of 1-$m^3$ volume sphere. |
| Flame speed ($S_f$) or laminar burning velocity ($S_u$) | Speed of flame relative to stationary observer |

Among these, a typical approach to estimate dust explosion severity is to estimate parameters of importance through laboratory scale experiments, such as deflagration index ($K_{st}$) and maximum pressure ($P_{max}$) in 1-m$^3$ laboratory test vessels. The obvious concern in directly using these results in an industrial setting is the clear omission of taking into account the effect of change in scale, which is significant. The cube-root law attempts to address the scale-up concern through the following equation:

$$\left(\frac{dP}{dt}\right)_{max} V^{1/3} = K_{st}$$

The assumption, which has been found generally valid, is that $P_{max}$ will occur at the same time when $\left(\frac{dP}{dt}\right)_{max}$ happens. Even with the cube-root law, however, the applicability of the model is subjected to the following set of restrictions (Dahoe et al., 1996; Eckhoff, 2003):

- Geometrically similar vessels
- Negligible flame thickness
- Same burning velocity
- Point ignition occurs at the center of the vessels

The effect of flame thickness has been dealt with more precisely in Dahoe et al. (1996), specifically for confined dust explosions in spherical vessels as the article title suggests. The three-zone model developed in this work requires the input of laboratory scale experimental results to be scaled up more accurately to the industrial level. The three-zone model was developed to assess the sensitivity of flame thickness on the pressure development and deflagration index for a confined dust explosion in a spherical vessel. It divided, in each scenario (different flame thickness), in three various phases, from initiation of the dust explosion to fully developed expanding flame zone and finally the flame zone reaching to the vessel wall. Based on this analysis, the modified model by the authors can be represented as

$$\left(\frac{dP}{dt}\right)_{max} V^{1/3} = K_{st} f\left(\partial/\mathrm{R}_{vessel}\right)$$

The effect of variation in burning velocity and flame thickness is incorporated into the model through the function $f\left(\partial/\mathrm{R}_{vessel}\right)$. Details of calculations are provided in the article and are not provided here.

One of the generic approaches in modeling dust explosions at the macroscale level involves TNT equivalency modeling with 7–10% explosion efficiency. Vent sizing procedure for dust explosion range from the Nomograph method, the Swedish method, the Norwegian method, the Radandt scaling law, the K-factor method, and the equivalence coefficient method have been discussed and compared with experimental results in several articles (Eckhoff, 2003; Tascón et al., 2009).

### 3.9.1 MOLECULAR MODELING OF DUST EXPLOSION

In the previous section, we have mentioned the cube-root law, utilized to scale-up laboratory scale results related to severity of dust explosion. The factors that affect the deflagration index ($K_{st}$) and maximum pressure ($P_{max}$), such as particle size, moisture content, humidity, oxygen availability, shape, concentration, ignition source, and others (Pritchard, 2004), are not explicitly incorporated in the model and are implicitly embedded within the estimated parameter values. The molecular modeling based approach through QSPR, based on experimentally obtained data set, can be implemented as an alternative route for estimating the deflagration index ($K_{st}$) and maximum pressure ($P_{max}$), for similar chemicals. Such an approach has the potential to obviate or atleast greatly reduce the need for expensive testing (Lemkowitz and Pasman, 2014). Appropriately done, this also translates into having the ability estimate dust explosion characteristics in a consistent manner, independent of influences of experimental methods and equipments, a major caveat in gaining insights from experimental data sets. Gao et al. (2013) demonstrated sensitivity of explosion to various testing procedures. Application of QSPR has been proven to be successful in process safety studies (Saraf et al., 2003, 2004; Katritzky et al., 2007; Pan et al., 2011) from estimating reactivity hazards, flash points, and flammability limits among others. In a similar manner, studies have been performed to characterize dust explosibility through QSPR.

### 3.9.2 PRACTICAL CONSIDERATIONS

The generic concerns for development of QSPR model from experimental data are also applicable here. In addition, for estimation of $K_{st}$ and $P_{max}$, the QSPR models developed based on a representative particle size, for estimation of $K_{st}$ and $P_{max}$ parameters, are good approaches for assessing the level of hazard at a screening level when experimental data are not available. More importantly, as pointed out (Reyes et al., 2011), care should be taken if estimated values are close to category demarcation values ($K_{st} = 200$ and 300). In such a scenario, experimental data should be sought after or a conservative approach should be taken.

## 3.10 WATER REACTIVE CHEMICALS

Water reactive chemicals are generally aggressive materials and can produce chemicals that are hazardous to human health and the environment. It is defined as a reactive chemical, which in contact with water leads to a spontaneous reaction. The exact nature, kinetics, and thermodynamics of this reaction govern the subsequent consequences of this type of release (Fernie et al., 2007). As a result, consequence analysis and dispersion modeling of water reactive chemicals, such as the ones incorporated in facility siting studies, are very challenging. The little experimental data available to validate mathematical models make those models susceptible to discrepancies on of its consequence estimates. As referred to before, some of these chemicals are pyrophoric and have been discussed in a previous section. Common water reactive chemicals include but are not limited to oleum, sulfurtrioxide, trichlorosilane, thionyl chloride, chlorosulfonic acid, phosphorus trichloride, oxalylchloride, methyltrichlorosilane, acetylchloride, and titanium tetrachloride. Some of these chemicals, as might be apparent from their names, produce HCl, a very toxic gas, on reacting with water. Water reactive chemicals have been involved in several industrial incidents (Fernie et al., 2007). In particular, titanium tetrachloride, which is widely used in process industries, has been mapped with several incidents in process industries (Kapias et al., 2001a). It is demonstrated as more hazardous than most chemicals in all of eight ranking systems (Kapias and Griffiths, 2005, and references therein). Modeling methods do exist to model spills of water reactive chemicals (Cruse et al., 2011; Kapias and

Griffiths, 2001a,b; 2005; Kapias et al., 2001a,b). Some of these approaches are simpler (Roy et al., 2003) than others (Kapias and Griffiths, 2005).

## 3.10.1 PRESENT APPROACHES

Fairly accurate dispersion modeling of toxic chemicals owing to spills, accidents, or leaks is of major importance for hazard analysis of process industries. It not only helps in being at par with legal requirements but also aids a site's risk mitigation and emergency planning. Several modeling approaches have been developed, with various assumptions on the underlying reaction mechanism. Most of these attempts are of simpler scope and has made some controversial statements (Kapias and Griffiths, 2005). REACTPOOL (Kapias et al., 2001a,b; Kapias and Griffiths, 2001a,b, 2005), a program specifically designed for water reactive chemicals, attempted to improve on these methods by determining appropriate set of hydrolysis reactions as applicable, with careful examination of existing experimental results and detailed scanning through numerous reaction mechanisms using thermodynamic calculations and various criteria. With regard to dispersion modeling of the toxic cloud, the program also models effect of availability of free ground water, wind speed, and surface roughness. STAWaRS (Cruse et al., 2011) is a latest effort by ESR Technologies to model water reactive chemicals, and it attempts to demonstrate the importance of experimentally validated values as input parameters.

## 3.10.2 CHALLENGES

For water reactive chemicals, in many cases, such as that for $TiCl_4$, the liquid phase hydrolysis reaction is not well understood. So estimation of toxic cloud arising from water reactive chemicals has been particularly difficult as it introduces more complexity to the problem. As examined in Kapias and Griffiths, (2005), controversies exist on the hydrolysis mechanism of $TiCl_4$. In order to derive the appropriate set of hydrolysis reaction, in Kapias and Griffiths (2005), a judgmental approach was taken to define different sets of criteria. However, as pointed out in Cruse et al. (2011) and references therein, the match between theoretical values illustrating hydrolysis of $TiCl_4$ and its corresponding experimental study is rather poor. So, while such an approach might work, a model based directly from either experimental data or first principle calculations removes the uncertainty part associated with this modeling approach involving judgment based criteria selection.

Regardless of being connected to various accidents, dearth of experiments on water reactive nature of chlorosulfonic acid ($HSO_3Cl$) makes a strong case for the need of existing model validation through reliable data (Kapias and Griffiths, 2001a). The same concern is shared for silicon tetrachloride ($SiCl_4$) (Kapias et al., 2001b), phosphorus chloride, and oxychloride (Kapias and Griffiths, 2001b). Mathematical models with inappropriate incorporation of factors, such as temperature effect on the reaction owing to solubility changes, differences in reactant interaction with different phases of water, that is, the chances of the reaction taking place with liquid water or water vapor, change in reaction mechanism with extreme proportions of the reactants such as excess $TiCl_4$ or excess water, has the potential of generating misleading consequence analysis results by undesirable margin. Existing modeling approaches have the ability to do a reasonable job of estimating toxic cloud dynamics arising from water reactive chemicals exposed to atmospheric moisture, if it is equipped with accurate knowledge of reaction mechanism and its sensitivity to various reaction parameters along with its heat of hydrolysis is essential. As illustrated above, lack of experimental data and corresponding validation demonstrate the need for further experiments or first principle—based knowledge extraction for robust modeling. The developers of several of these methods are rightly careful to mention the importance of validation of data.

Additionally, these models do not look into possibility of other products such as aerosol formation, like that from sulfuric acid and their hydrates in gas phase (Temelso et al., 2012). While these may always not be toxic, the products can still have undesirable environmental implications. Predicting formation of such aerosol structures and their properties is challenging and can be addressed with higher-level modeling as has been illustrated below in a separate aerosol-specific section. Narrowing down, owing to their importance, accurate estimation of the following parameters becomes important for reactive chemicals, which in turn becomes input to the overall dispersion model:

Reaction mechanism.

1. In many cases, the reaction mechanism is little or not known at all. As has been illustrated before, determining the hydrolysis reaction is challenging by itself. In addition, it varies with the degree of extent the other factors are present, such as relative availability or abundance of the reactants. In Kapias and Griffiths (2005), it is shown that hydrolysis of $TiCl_4$ changes if excess $TiCl_4$ is present in the system. The lack of experimental data does not help the cause either.

Heat of hydrolysis.

2. It is apparent that the heat of hydrolysis, which has a significant role to play in deterring the consequence, will vary based on the reaction mechanism that takes place. The heat of hydrolysis results in accelerating the vaporization of liquid pool of spilled or leaked chemicals, aggravating the situation. The information is important and needs to be validated experimentally through calorimetric studies prior to using it in the model.

## 3.10.3 MULTISCALE MODELING TECHNIQUES

### *3.10.3.1 First principle calculations*

Lack of satisfactory amount of data to validate and improve on existing models has demonstrated the need for either reliable experimental data or results from first principle calculations to accurately model consequence of accidental release of water reactive chemicals.

Because of the time involved in a chemical reaction, which often times is very short, observing mechanism through experiments is a daunting if not an impractical task. Quantum mechanics or first principle—based approach has the potential to help determine reaction mechanism through transition state theory (Laidler and King, 1983; Eyring, 1935, 1938; Wigner, 1938) approach and access detailed molecular level information that is otherwise inaccessible to regular experiments. Since transition state of a reaction involves large changes in electronic correlation energy, it is important to note that even with first principle calculations use of large basis sets and calculation of electronic correlation must be accounted for meaningful results (Dewar, 1988). In addition, the PES must be rigorously searched for transition states. So predicting reaction mechanism through transition state theory can be computationally demanding and resource intensive. Even then quantum mechanics has been in use by organic chemists, in academia as a practical alternative to experimental work (Dewar, 1988). However, the author argues in favor of the necessity of using higher-level models at all stages of calculations as opposed to a less-intensive approach such as single-point geometrical optimization of transition state structure. The study illustrates the use and the effect of various level of ab initio approach on understanding the mechanism of Cope rearrangement. The same study also shows that transition state of a reaction need not be unique, making a case for rigorous search of reaction pathway in the PES.

While with increase in computational power, it is now possible to afford incorporation of higher level of details in an ab initio calculation, to aid finding transition state in an efficient manner, especially for complex reaction systems, it is important to have an approach insensitive to the starting geometry of the transition state. Approaches such as nudged elastic band and string methods can be used for calculating potential energy surfaces and transition states within but success of such calculation is not independent of the accuracy of the initial guess of the reaction pathway. This means that with little or no prior knowledge of the reaction pathway will lead to a slow convergence or prediction of an incorrect reaction pathway.

Within harmonic approximation to TST, the nudged elastic band method is an efficient tool for finding minimum energy reaction pathway and saddle points (Henkelman et al., 2000; Mills and Jónsson, 1994). In this approach, an elastic band is formed through creating set of images of the system between the initial and final state. Spring like interaction is introduced between adjacent images to build an elastic band and its subsequent minimization overlaps its path with the minimum energy reaction pathway. The NEB method is capable of project forces to ensure no interaction between spring forces and convergence of the band to the minimum energy reaction pathway. If the points along the band are denoted by $X_1$, $X_2$ to $X_n$, then $X_1$ and $X_n$ are fixed as they represent the minimized initial and final states.

In the regular NEB method, the forces on each image are calculated as the sum of true force acting on the chain perpendicularly and spring force parallel to the chain acting on it. While the parallel component, the spring force prevents the images to run downhill, away from transition state, the perpendicular forces are responsible for pulling the images toward MEP or minimum energy path. A minimization of the energy is then done to optimize the image positions and shift the elastic band toward MEP. The climbing image NEB method (Henkelman et al., 2000) slightly modifies this method as it calculates the maximum force in the saddle point independently those of the spring forces along the band. The saddle point image energy is then maximized along the band and minimized in other directions to converge to a relatively accurate saddle point.

The string method is based on the path from reactant to product configuration in the PES.

However, approaches such as these use quantum-level calculations while calculating the energetics of a system. With approaches such as growing string method (GSM) (Peters et al., 2004), this is possible, even though a slightly off initial estimate would need repeated interaction with quantum mechanical calculations leading to computationally intensive set of calculations.

Goodrow et al. (2009) improved on the efficiency of searching transition state method by using three different approaches: a hybrid strategy, energy weighted strategy, and a substring strategy. The idea behind the hybrid strategy is to use a low-level theory and then refine the approach to a higher level of theory for the string calculation. The next approach, energy-weighted strategy divides the PES according to energy density, leading to higher number of points near a transition state, leading to a finer resolution of the transition state. The third approach, the substring strategy, is a slight variation of the hybrid strategy. It differs in the fact that only a portion of the string is refined using higher-level theory as opposed to the whole string.

## 3.11 NANOTOXICITY

Nanoparticles are commonly being employed in a variety of industries including electronics, solar, pharmacy, catalysis, space engineering, and cosmetic. However, with great advancement in this area, as illustrated in the last couple of decades, understanding of nanotoxicity has become a matter of significant importance. As an example, for carbon-based nanomaterials, such as carbon nanotube (CNT), it has been shown that the manufacturing, use, and disposal of these poses the highest risk of exposure to skin, eyes, and lungs (Kayat et al., 2011; Hoet et al., 2004). In addition to being exposed to nanoparticles, emerging nanomedicine—based research requires one to ensure the safety of the same. As an example, it has been shown (Hu et al., 2010) that graphene oxide is very effective against bacteria with potential application in hygiene products and food packaging for lengthening its shelf life. Hence, the benefits have been proven; studies are required to ensure that the possible side effects are within acceptable limits. To a great extent, the characteristic of nanomaterials, which significantly influences the toxicity of a nanoparticle, is still unknown. Since the interaction mechanism of nanoparticles and biological molecules is not well understood, atomistic- and electronic-level computational studies are expected to throw some much needed insight in this area (Gajewicz et al., 2012). Factors such as elemental composition of nanoparticles, surface properties, and their ability to stay uniformly distributed inside a matrix all play a vital role in their interaction with environment and living organisms. Through numerous in vivo (experimentation on living organism) and in vitro (experimentation on dead organism) experimental studies, it has been demonstrated that (Gajewicz et al., 2012)

Upon contact with cell surfaces:

- Nanomaterials form reactive oxygen species (ROS), resulting in oxidative stress and inflammation (Nel, 2005; Nel et al., 2006; Oberdörster and Oberdörster, 2005).
- Lipid peroxidation of cellular membranes, resulting in cell and DNA damage (Nel et al., 2006; Oberdörster and Oberdörster, 2005; Bhabra et al., 2009).
- Abnormalities in cell adhesion (Cui et al., 2005).

It is still mostly unknown how various material properties influence the toxicity of nanoparticles. Owing to differences in the characteristics between bulk material and nano counterpart of the same material, the risk characterization of a bulk chemical cannot be extrapolated to the nanoscale level. Figure 3.24 illustrates the interdisciplinary nature of the problem for assessment of toxicity of CNT.

### 3.11.1 CHALLENGES

With increasing applications of nanotechnology, it has become important for us to understand its associated risk. Owing to their large surface-to-volume ratio, quantum effects such as tunneling, they exhibit significantly different physiochemical properties than demonstrated by their bulk counterpart. With increasing surface-area-to-volume-ratio as one moves from a bulk material to its nano counterpart, parameters that have insignificant consequence at the classical level become more dominant resulting in significantly different set of properties. It is needless to say that to understand the risk associated with nanoparticles of all dimensions, we need to first know and understand their properties, how those are affected and how the nanoparticles interact. Since the properties of the corresponding bulk material is very different from its nano counterpart, lack of available data and low adequacy of

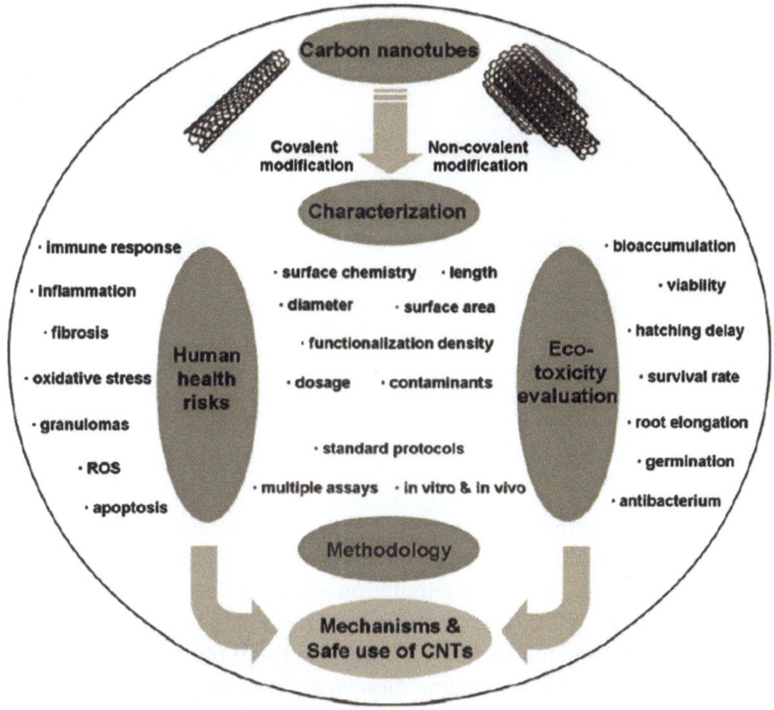


**FIGURE 3.24**

Illustrtation of toxicology of carbon nanotube as an emerging interdisciplinary field (Gajewicz et al., 2012).

experimental protocols make it harder to perform a reliable comprehensive risk assessment of such systems. Furthermore, any experimental approaches to assess the effects of nanoparticles on human live cells will involve rat or mice studies and posess the risk of giving a very different outcome, which might overplay or underplay the risk. In the absence of a reliable comprehensive understanding of the associated risk, in a worse case scenario, the risk would be underestimated to be treated as an overly conservative way in the best case scenario. To combat this, a parallel approach taken by the researchers in this area is computational modeling with greater detail.

## 3.11.2 MOLECULAR MODELING TECHNIQUES

### *3.11.2.1 Quantum mechanics and molecular dynamics*

Current molecular modeling approaches have shown great promise in estimating material properties, reactivity, and reaction mechanism for various molecular systems. Qunatum mechanics– and molecular dynamics–based approaches can be very useful in estimating potential risk of nanomaterials (Gajewicz et al., 2012). Molecular modeling studies determine the low-energy configuration of a molecular system for estimating interaction energies with other molecular systems. In similar lines, the study (Zuo et al., 2011) looked into interaction between nanomaterial and biomolecules using large-scale molecular dynamics simulation. Both experimental and theoretical studies about adsorption of

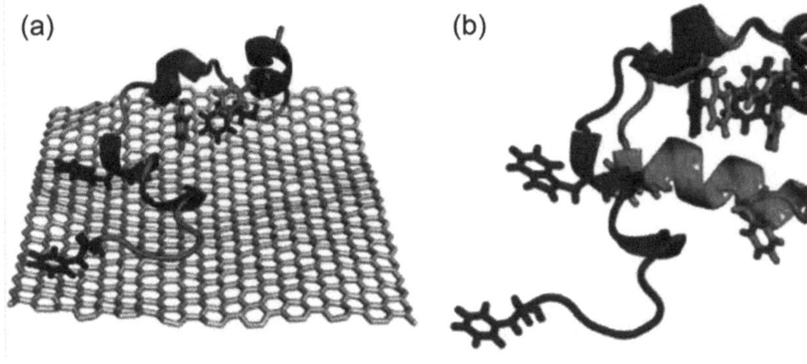


**FIGURE 3.25**

(a) A typical structure of protein villin headpiece (HP35) adsorbed on graphene surface. (b) The superposition of the adsorbed HP35 on graphene (Zuo et al., 2011).

proteins on nanomaterials have shown that these interactions can affect both the structure and function of the protein. Figure 3.25 shows the molecular phenomena of adsorption of protein villin headpiece (HP35) onto graphene structure. The MD simulation showed that the graphene sheet bents (see Figure 3.25(b)) itself to fit better with the aromatic residues in HP35 to form strong $\pi - \pi$ interaction, which is a less dominant phenomenon in single-walled CNT and protein interaction. The study (Zuo et al., 2011) found that despite having the same molecular formula, owing to a difference of surface curvature between graphene and CNT, different interaction mechanisms result with the same protein structure. It can then be deduced that care should be taken in extrapolating nanotoxicity information of graphene to CNT and vice versa.

In their review study (Gajewicz et al., 2012), the authors pointed out the agreement between quantum mechanics based theoretical calculations and experimental results on the fact that graphene or CNT based nanomaterial can be functionalized to reduce toxicity as has been shown previously (Sayes et al., 2006; Mananghaya et al., 2011; Sasidharan et al., 2011). Comprehensive computational study at the atomistic level has found that toxicity of CNT is sensitive to surface properties of the complexes of CNT-surfactants rather than intrinsic surface properties of CNT (Obata and Honda, 2011). While molecular modeling has not yet translated into directly getting ERPG/IDLH or probit values, it certainly has the ability to help in designing inherently safer molecules, assessing relative toxicity between two different molecular systems, and providing insight into underlying mechanism behind nanotoxicity (Gajewicz et al., 2012; Kavlock et al., 2007; Merlot, 2010). While several (Çağın et al., 1999; Chakrabarty and Cagin, 2008) have illustrated the use of atomistic modeling in estimating properties of nanomaterials, computational studies on properties relating to nanotoxicity are still relatively fewer. However, while a multiscale-based method is more appropriate in transferring the insight gained at the atomistic scale to macroscale level simulations, insight gained at the molecular level has been proven to be a powerful tool in assessing nanotoxicity.

#### *3.11.2.2 QSAR/QSPR*

Several studies have implemented QSAR/QSPR approaches for getting better understanding of nanotoxicologic concerns. Figure 3.26 illustrates how the different components such as training data

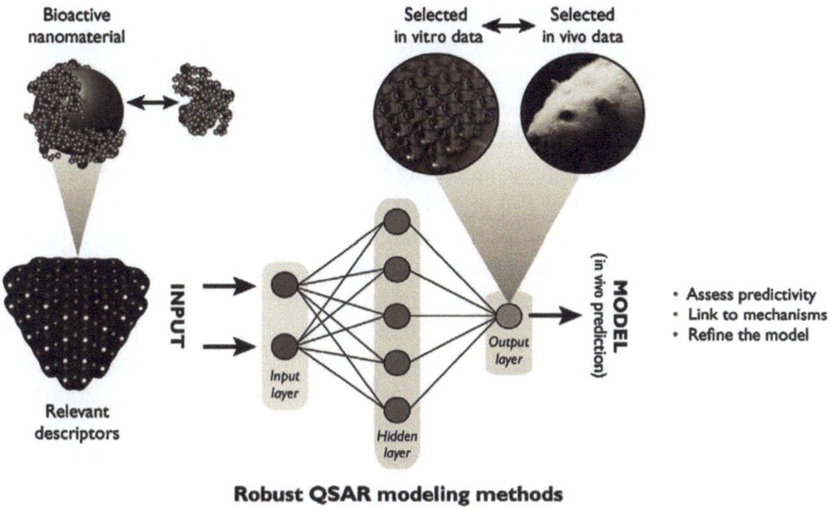


**FIGURE 3.26**

Experimental data based QSAR modeling approach for nanotoxicology (Winkler et al., 2012).

from selected set of in vivo and in vitro data and defining molecular descriptors come together to help develop a robust QSAR model (Winkler et al., 2012). Several studies have been carried out where QSAR/QSPR approaches help provide information on toxicity information.

For regulatory purposes, in vivo effects of nanomaterial are more important (Winkler et al., 2012), although they are more resource intensive and difficult in nature. Combining QSPR/QSAR studies based on in vitro results has the potential to derive in vivo study results. In addition to prediction of in vivo toxicities of industrial chemicals in humans (Lessigiarska et al., 2006), useful models of the in vitro biological effects of nanomaterials have also shown great potential (Puzyn et al., 2009; Fourches et al., 2010). Given the objective of estimating nanotoxicity with QSPR/QSAR-based approaches, the methodologies are also known as quantitative nanostructure toxicity relationships, or QNTR.

The major roadblocks in QNTR approaches include the lack of nanoparticle specific descriptors for estimating nanotoxicity and lack of experimental data. Toxicity information of bulk chemical cannot be used at the nano-scale. While bulk silver is nontoxic, silver nanoparticles are capable of killing bacteria on contact (Choi and Hu, 2008). Since molecular modeling approaches have been proved to be a powerful tool in accurately predicting several molecular properties, there is a great promise in implementing the same to improve on existing and develop robust QNTR models.

## 3.12 CONCLUSION

Molecular modeling has been proven to be a very powerful tool in characterizing materials with various purposes. Similar to other areas of current research interest, such as in designing future generation smart materials, molecular modeling has shown to be a promising and powerful tool in assessing hazardous nature of chemicals under various process conditions. The topics covered does not span over all areas related to process safety where molecular modeling approaches can be implemented are not discussed here. Modeling evaporation (Consolini and Aggarwal, 2003) and its characteristics is one such example. Accidental spills of hazardous liquid may require detailed understanding of evaporation characteristics and characteristics of the cloud formed from it. In addition, the possibility of using molecular modeling in investigating fire incidents is not covered here. Since color of a flame is a function of the materials involved in a combustion, quantum mechanical approaches have the potential to reveal important insight in combustion phenomena of novel materials. Although, beyond the scope of typical process safety applications, the possibility to investigate toxicity effects on biological macromolecules, keeping in mind the pros and cons of the molecular modeling approaches, is also promising (Friedman et al., 2013). In an attempt to study factors affecting risk through long-term exposure, researchers have also studied interaction between aerosol and phenol particles as emitted from automobile and industrial exhaust (Falsig et al., 2006). Examples on implementation of molecular dynamics simulation approaches from a completely different perspective such as finding escape dynamics during emergency evacuation (Rassia and Siettos, 2010) also exist. Some other applications such as corrosion inhibitor studies using quantum mechanical calculations are illustrated in a later chapter relating to equipment failure and are not covered here. Overall, molecular modeling techniques have emerged as a standard tool in a diverse set of material modeling problems including addressing specific concerns related to process safety. The primary aim in this chapter was to introduce the readers to concepts of molecular modeling approaches that might be considered in dealing with assessment of hazardous nature of emerging technologies involving novel processes and materials.

## REFERENCES

Aldeeb, A.A., Rogers, W.J., Mannan, M.S., 2003. New method estimates the parameters for evaluating process reactivity hazards. Oil & Gas Journal 101 (24), 66–70. PennWell Publishing Co. Energy Group, Tulsa, USA.

Abbasi, T., Abbasi, S.A., 2007. Dust explosions—cases, causes, consequences, and control. Journal of Hazardous Materials 140 (1–2), 7–44. http://dx.doi.org/10.1016/j.jhazmat.2006.11.007.

Affens, W.A., 1966. Flammability properties of hydrocarbon fuels – interrelations of flammability properties of n-alkanes in air. Journal of Chemical & Engineering Data 11 (2), 197–202.

Affens, W.A., McLaren, G.W., 1972. Flammability properties of hydrocarbon solutions in air. Journal of Chemical & Engineering Data 17 (4), 482–488. http://dx.doi.org/10.1021/je60055a040. American Chemical Society.

Alder, B.J., Wainwright, T.E., 1957. Phase transition for a hard sphere system. The Journal of Chemical Physics 27 (5), 1208–1209. http://dx.doi.org/10.1063/1.1743957.

Alós, M.A., Strozzi, F., Zaldívar, J.M., 1996. A new method for assessing the thermal stability of semibatch processes based on lyapunov exponents. Chemical Engineering Science.

Amadeu, K., 1999. A novel approach to phase equilibria predictions using ab initio methods. Industrial & Engineering Chemistry Research 36 (7), 2849–2855.

American Society for Testing and Materials (ASTM), 2000. Selection and use of ASTM standards for the determination of flash point of chemicals by closed cup methods, E 502-84. ASTM International.

American Society for Testing and Materials (ASTM), 2002. Standard test method for temperature limit of flammability of chemicals, E 1232-02. ASTM International.

Accelrys Materials Studio for Materials Modeling & Simulation, 2014. http://accelrys.com/products/materials-studio/index.html (accessed March 26).

Baatia, N., Nanchenb, A., Stoesselb, F., Meyera, T., 2013. Quantitative structure-property relationships for thermal stability and explosive properties of chemicals. Chemical Engineering.

Bach, R.D., Ayala, P.Y., Schlegel, H.B., 1996. A reassessment of the bond dissociation energies of peroxides. An ab initio study. Journal of the American Chemical Society 118 (50), 12758–12765.

Baggott, J.E., 2011. The Quantum Story. Oxford University Press.

Bagheri, M., Borhani, T.N.G., Zahedi, G., 2012. Estimation of flash point and autoignition temperature of organic sulfur chemicals. Energy Conversion and Management 58, 185–196. http://dx.doi.org/10.1016/j.enconman.2012.01.014.

Baker, J., Gill, P.M.W., 1988. An algorithm for the location of branching points on reaction paths. Journal of Computational Chemistry 9 (5), 465–475.

Banerjee, A., Adams, N., Simons, J., 1985. Search for stationary points on surfaces. The Journal of Physical Chemistry 89 (1), 52–57.

Barton, K., Rogers, R., 1997. Chemical Reaction Hazards.

van Baten, J., 2010. Using Chemsep, COCO and Other Modeling Tools for Versatility in Custom Process Modeling, AIChE Annual Meeting 2010, Salt Lake City, Utah.

Baum, E.J., 1998. Chemical Property Estimation: Theory and Application.

Benson, S.W., Cruicksh, F., Golden, D.M., Haugen, G.R., Oneal, H.E., Rodgers, A.S., Shaw, R., Walsh, R., 1969. Additivity rules for estimation of thermochemical properties. Chemical Reviews 69 (3), 279–324.

Benson, S.W., Buss, J.H., 1958. Additivity rules for the estimation of molecular properties. Thermodynamic properties. Journal of Chemical Physics 29 (3), 546–572. http://dx.doi.org/10.1063/1.1744539. AIP Publishing.

Bertsekas, D.P., 1999. Nonlinear Programming.

Bhabra, G., Sood, A., Fisher, B., Cartwright, L., Saunders, M., Howard Evans, W., Surprenant, A., et al., 2009. Nanoparticles can cause DNA damage across a cellular barrier. Nature Nanotechnology 4 (1), 876–883. http://dx.doi.org/10.1038/nnano.2009.313.

Bingham, R.C., Dewar, M.J.S., Lo, D.H., 1975a. Ground states of molecules. XXV. MINDO/3. Improved version of the MINDO semiempirical SCF-MO method. Journal of the American Chemical Society 97 (6), 1285–1293.

Bingham, R.C., Dewar, M.J.S., Lo, D.H., 1975b. Ground states of molecules. XXVI. MINDO/3 calculations for hydrocarbons. Journal of the American Chemical Society 97, 1294–1301.

Binkowski, F.S., Shankar, U., 1995. The regional particulate matter model: 1. Model description and preliminary results. Journal of Geophysical Research 100, 26191–26209.

Bodurtha, F.T., 1980. Industrial Explosion Prevention and Protection.

van den Bosch, C.J.H., Weterings, R.A.P.M., 2005. Methods for the Calculation of Physical Effect.

Bowen, P.J., Shirvill, L.C., 1994. Combustion hazards posed by the pressurized atomization of high-flashpoint liquids. Journal of Loss Prevention in the Process Industries 7 (3), 233–241. http://dx.doi.org/10.1016/0950-4230(94)80071-5.

Bowen, P.J., Cameron, L.R.J., 1999. Hydrocarbon Aerosol Explosion Hazards: A Review. Process Safety and Environmental Protection 77 (1), 22–30.

Brazdova, V., Bowler, D.R., 2013. Atomistic Computer Simulations: A Practical Guide. Wiley, Hoboken, NJ.

Britton, L.G., 2002. Using heats of oxidation to evaluate flammability hazards. Process Safety Progress 21 (1), 31–54. http://dx.doi.org/10.1002/prs.680210108.

Britton, L.G., 2006. Two hundred years of flammable limits. Process Safety Progress 21 (1), 1–11. http://dx.doi.org/10.1002/prs.680210104.

Butler, R., Cooke, G., Lukk, G., Jameson, B., 1956. Prediction of flash points of middle distillates. Industrial & Engineering Chemistry 48 (4), 808–812.

Cardillo, P., Girelli, A., Ferraiolo, G., 1984. The Seveso case and the safety problem in the production of 2,4,5-trichlorophenol. Journal of Hazardous Materials 9 (2), 221–234.

Castier, M., Rasmussen, P., Fredenslund, A., 1989. Calculation of simultaneous chemical and phase equilibria in nonideal systems. Chemical Engineering Science 44 (2), 237–248.

Castillo, J., Grossmann, I.E., 1981. Computation of phase and chemical equilibria. Computers & Chemical Engineering 5, 99–108.

Castro-Marcano, F., Kamat, A.M., Russo Jr., M.F., van Duin, A.C.T., Mathews, J.P., 2012. Combustion of an Illinois No. 6 coal char simulated using an atomistic char representation and the ReaxFF reactive force field. Combustion and Flame 159 (3), 1272–1285. http://dx.doi.org/10.1016/j.combustflame.2011.10.022.

Catoire, L., Naudet, V., 2004. A unique equation to estimate flash points of selected pure liquids application to the correction of probably erroneous flash point values. Journal of Physical and Chemical Reference Data 33 (4), 1083–1111. http://dx.doi.org/10.1063/1.1835321.

Cerjan, C.J., Miller, W.H., 1981. On finding transition states. The Journal of Chemical Physics 75 (6), 2800–2801.

Chakrabarty, A., Cagin, T., 2008. Computational studies on mechanical and thermal properties of carbon nanotube based nanostructures. Edited by undefined author Computers Materials & Continua 7 (3), 167–189.

Chen, C.-C., Mathias, P.M., 2006. Applied thermodynamics for process modeling. Aiche Journal 48 (2), 194–200. http://dx.doi.org/10.1002/aic.690480202.

Chenoweth, K., van Duin, A.C.T., Goddard, W.A., 2008. ReaxFF reactive force field for molecular dynamics simulations of hydrocarbon oxidation. Journal of Physical Chemistry A 112 (5), 1040–1053. http://dx.doi.org/10.1021/jp709896w. American Chemical Society.

Chenoweth, K., Cheung, S., van Duin, A.C.T., Goddard, W.A., Kober, E.M., 2005. Simulations on the thermal decomposition of a poly(dimethylsiloxane) polymer using the ReaxFF reactive force field. Journal of the American Chemical Society 127 (19), 7192–7202. http://dx.doi.org/10.1021/ja050980t.

Choi, O., Hu, Z., 2008. Size dependent and reactive oxygen species related nanosilver toxicity to nitrifying bacteria. Environmental Science & Technology 42 (12), 4583–4588. http://dx.doi.org/10.1021/es703238h.

Consolini, L., Aggarwal, S.K., 2003. A molecular dynamics simulation of droplet evaporation. International Journal of Heat and Mass Transfer 42, 4107–4116.

Consultants, Marsh Risk, 2012. The 100 Largest Losses 1972–2011, twenty-second ed. Marsh Energy Practice.

Coward, H.F., Jones, G.W., 1952. Limits of Flammability of Gases and Vapors.

Crowl, D.A., 2012. Minimize the risks of flammable materials. Chemical Engineering Progress, 28–33.

Crowl, D.A., Louvar, J.F., 2011. Chemical Process Safety: Fundamentals with Applications (Prentice Hall International Series in the Physical and Chemical Engineering Sciences), third ed. Prentice Hall.

Crowl, D.A., Elwell, T.I., 2004. Identifying criteria to classify chemical mixtures as "highly hazardous" due to chemical reactivity. Journal of Loss Prevention in the Process Industries 17 (4), 279–289. http://dx.doi.org/10.1016/j.jlp.2004.05.002.

Cruse, H.A., Buston, J.E.H., Vechot, L.N., Tickle, G.A., Rowlands, R., 2011. Modelling spills of water-reactive chemicals. Process Safety and Environmental Protection 89 (6), 415–423. http://dx.doi.org/10.1016/j.psep.2011.09.001.

Cui, D., Tian, F., Ozkan, C.S., Wang, M., Gao, H., 2005. Effect of single wall carbon nanotubes on human HEK293 cells. Toxicology Letters 155, 75–85.

Curtiss, L.A., Raghavachari, K., 1993. Gaussian-2 theory using reduced Mo/Ller–Plesset orders. The Journal of Chemical Physics 98 (2), 1293–1298.

Curtiss, L.A., Jones, C., Trucks, G.W., 1990. Gaussian-1 theory of molecular energies for second-row compounds. The Journal of Chemical Physics 93, 2537.

Curtiss, L.A., Raghavachari, K., W Trucks, G., 1991. Gaussian-2 theory for molecular energies of first- and second-row compounds. The Journal of Chemical Physics 94, 7221–7230.

Curtiss, L.A., Raghavachari, K., Redfern, P.C., Rassolov, V., Pople, J.A., 1998. Gaussian-3 (G3) theory for molecules containing first and second-row atoms. Journal of Chemical Physics 109 (18), 7764. http://dx.doi.org/10.1063/1.477422.

Curtiss, L.A., Redfern, P.C., Raghavachari, K., 2007. Gaussian-4 theory. Journal of Chemical Physics 126 (8), 084108. http://dx.doi.org/10.1063/1.2436888.

Çağın, T., Che, J., Qi, Y., Zhou, Y., Demiralp, E., Gao, G., Goddard, W.A., 1999. Computational materials chemistry at the nanoscale. Journal of Nanoparticle Research 1 (1), 51–69. http://dx.doi.org/10.1023/A:1010009630519.

Catalysis in Chemistry and Enzymology, 1987. http://books.google.com/books?hl=en&lr=&id=LLkSzP-ct0wC&oi=fnd&pg=PA1&dq=Catalysis+in+chemistry+and+Enzymology&ots=YK8V1K_mq4&sig=dbyeC6nrhQMkwpCEQyuAzkB0YjA.

Concept Sciences Hydroxylamine Explosion, 2002. Final Report: Concept Sciences Case Study, CSB. Allentown, PA.

Dahoe, A.E., Zevenbergen, J.F., Lemkowitz, S.M., 1996. Dust explosions in spherical vessels: the role of flame thickness in the validity of the "cube-root law". Journal of Loss Prevention in the Process Industries 9 (1), 33–44.

Davies, C.A., Kipnis, I.M., Chase, M.W., Treweek, D.N., 1985. The Thermochemical and Hazard Data of Chemicals.

Devillers, J., Balaban, A.T., 2000. Topological indices and related descriptors in QSAR and QSPAR.

Dewar, M.J.S., Storch, D.M., 1985. Development and use of quantum molecular models. 75. Comparative tests of theoretical procedures for studying chemical reactions. Journal of the American Chemical Society 107 (13), 3898–3902. http://dx.doi.org/10.1021/ja00299a023.

Dewar, M.J.S., 1988. Use of quantum mechanical models in studies of reaction mechanisms. International Journal of Quantum Chemistry 34 (Suppl. 22), 557–566.

Dewar, M.J.S., Thiel, W., 1977. Ground states of molecules. 38. The MNDO method. Approximations and parameters. Journal of the American Chemical Society 99 (15), 4899–4907.

Dhara, R., Dhara, V.R., 1995. Bhopal—a case study of international disaster. International Journal of Occupational and Environmental Health 1 (1), 58–69.

Di Benedetto, A., Russo, P., 2007. Thermo-kinetic modelling of dust explosions. Journal of Loss Prevention in the Process Industries 20 (4–6), 303–309. http://dx.doi.org/10.1016/j.jlp.2007.04.001.

Di Benedetto, A., Russo, P., Amyotte, P., Marchand, N., 2010. Modelling the effect of particle size on dust explosions. Chemical Engineering Science 65 (2), 772–779. http://dx.doi.org/10.1016/j.ces.2009.09.029.

Diedenhofen, M., Klamt, A., 2010. COSMO-RS as a tool for property prediction of IL mixtures—a review. Fluid Phase Equilibria 294 (1–2), 31–38. http://dx.doi.org/10.1016/j.fluid.2010.02.002.

Drysdale, D., 2011. An Introduction to Fire Dynamics, third ed. Wiley.

Duncan, W.T., Bell, R.L., Truong, T.N., 1998. TheRate: program for ab initio direct dynamics calculations of thermal and vibrational-state-selected rate constants. Journal of Computational Chemistry 19 (9), 1039–1052.

van Duin, A.C.T., Baas, J.M.A., van de Graaf, B., 1994. Delft molecular mechanics: a new approach to hydrocarbon force fields: Inclusion of a geometry-dependent charge calculation. Journal of the Chemical Society. Faraday Transactions 90 (19), 2881–2895. Royal Society of Chemistry.

van Duin, A.C.T., Dasgupta, S., Lorant, F., Goddard, W.A., 2001. ReaxFF: a reactive force field for hydrocarbons. Journal of Physical Chemistry A 105 (41), 9396–9409. http://dx.doi.org/10.1021/jp004368u.

Engineers. Center for Chemical Process Safety, 1993. Guidelines for Engineering Design for Process Safety.

Eckerman, I., 2005. The Bhopal Saga: Causes and Consequences of the World's Largest Industrial Disaster.

Eckhoff, R., 2003. Dust Explosions in the Process Industries: Identification, Assessment and Control of Dust Hazards.

Elliott, J., Richard, S., Suresh, J., Donohue, M.D., 1990. A simple equation of state for non-spherical and associating molecules. Industrial & Engineering Chemistry Research 29 (7), 1476–1485. http://dx.doi.org/10.1021/ie00103a057. American Chemical Society.

Etchells, J.C., 1997. Why reactions run away (Reprinted from IChemE symposium series 141, pp. 361–366, 1997). Organic Process Research & Development 1.

Eyring, H., 1935. The activated complex and the absolute rate of chemical reactions. Chemical Reviews 17 (1), 65–77.

Eyring, H., 1938. The theory of absolute reaction rates. Transactions of the Faraday Society 34, 41–48.

Falsig, H., Gross, A., Kongsted, J., Osted, A., Sloth, M., Mikkelsen, K.V., Christiansen, O., 2006. Uptake of phenol on aerosol particles. Journal of Physical Chemistry A 110 (2), 660–670. http://dx.doi.org/10.1021/jp0536201. American Chemical Society.

Fayet, G., Joubert, L., Rotureau, P., Adamo, C., 2009a. A theoretical study of the decomposition mechanisms in substituted o-nitrotoluenes. Journal of Physical Chemistry A 113 (48), 13621–13627. http://dx.doi.org/10.1021/jp905979w.

Fayet, G., Rotureau, P., Joubert, L., Adamo, C., 2009b. On the prediction of thermal stability of nitroaromatic compounds using quantum chemical calculations. Journal of Hazardous Materials 171 (1–3), 845–850. http://dx.doi.org/10.1016/j.jhazmat.2009.06.088.

Fayet, G., Rotureau, P., Joubert, L., Adamo, C., 2010a. Predicting explosibility properties of chemicals from quantitative structure-property relationships. Process Safety Progress 29 (4), 359–371. http://dx.doi.org/10.1002/prs.10379.

Fayet, G., Rotureau, P., Joubert, L., Adamo, C., 2010b. QSPR modeling of thermal stability of nitroaromatic compounds: DFT vs. AM1 calculated descriptors. Journal of Molecular Modeling 16 (4), 805–812. http://dx.doi.org/10.1007/s00894-009-0634-7.

Fayet, G., Rotureau, P., Joubert, L., Adamo, C., 2010c. Development of a QSPR model for predicting thermal stabilities of nitroaromatic compounds taking into account their decomposition mechanisms. Journal of Molecular Modeling 17 (10), 2443–2453. http://dx.doi.org/10.1007/s00894-010-0908-0.

Fernie, L., Wright, P., Kapias, T., 2007. Water reactive materials – incorporation into safety and environmental risk assessments. Process Safety and Environmental Protection 85, 162–168. http://dx.doi.org/10.1205/psep06033.

Fisher, H.G., Forrest, H.S., Grossel, S.S., Huff, J.E., Muller, A.R., 2010. Emergency Relief System Design Using DIERS Technology: The Design Institute for Emergency Relief Systems (DIERS) Project Manual.

Foresman, J.B., Frisch, Æ., 1996. Exploring Chemistry with Electronic Structure Methods.

Fourches, D., Pu, D., Tassa, C., Weissleder, R., Shaw, S.Y., Mumper, R.J., Tropsha, A., 2010. Quantitative nanostructure–activity relationship modeling. ACS Nano 4 (10), 5703–5712. http://dx.doi.org/10.1021/nn1013484.

Frenkel, D., Smit, B., 2002. Understanding Molecular Simulation: From Algorithms to Applications [Electronic Resource]. Elsevier (formerly published by Academic Press).

Friedman, R., Boye, K., Flatmark, K., 2013. Molecular modelling and simulations in cancer research. Biochimica et Biophysica Acta 1836 (1), 1–14. http://dx.doi.org/10.1016/j.bbcan.2013.02.001.

Frurip, D.J., Freedman, E., Hertel, G.R., 1989. A new release of the ASTM CHETAH program for hazard evaluation: versions for mainframe and personal computer. Plant/Operations Progress 8 (2), 100–104. http://dx.doi.org/10.1002/prsb.720080215.

Gajewicz, A., Rasulev, B., Dinadayalane, T.C., Urbaszek, P., Puzyn, T., Leszczynska, D., Leszczynski, J., June 2012. Advancing risk assessment of engineered nanomaterials: application of computational approaches. Advanced Drug Delivery Reviews. http://dx.doi.org/10.1016/j.addr.2012.05.014.

Gao, W., Zhong, S., Miao, N., Liu, H., 2013. Effect of ignition on the explosion behavior of 1-octadecanol/air mixtures. Powder Technology 241, 105–114.

Gelbard, F., Seinfeld, J.H., 1980. Simulation of multicomponent aerosol dynamics. Journal of Colloid and Interface Science 78 (2), 485–501.

Gharagheizi, F., 2009. A QSPR model for estimation of lower flammability limit temperature of pure compounds based on molecular structure. Journal of Hazardous Materials 169, 217–220. http://dx.doi.org/10.1016/j.jhazmat.2008.02.010.

Gharagheizi, F., Alamdari, R.F., Angaji, M.T., 2008. A new neural network-group contribution method for estimation of flash point temperature of pure components. Energy & Fuels 22 (3), 1628–1635. http://dx.doi.org/10.1021/ef700753t.

Goodrow, A., Bell, A.T., Head-Gordon, M., 2009. Transition state-finding strategies for use with the growing string method. Journal of Chemical Physics 130 (24). http://dx.doi.org/10.1063/1.3156312. American Institute of Physics: 244108–244108–14.

Gupta, J.P., 2002. The bhopal gas tragedy: could it have happened in a developed country? Journal of Loss Prevention in the Process Industries 15 (1), 1–4. http://dx.doi.org/10.1016/S0950-4230 (01)00025-0.

Gustin, J.L., 1998. Runaway reaction hazards in processing organic Nitro compounds. Organic Process Research & Development 2 (1), 27–33.

Haessler, W.M., 1989. Fire: Fundamentals and Control.

Hager, T., 1995. Force of Nature.

Halle, R.T., Vadekar, M., June 1991. Rust catalyzed ethylene hydrogenation causes temperature runaway. Oil and Gas Journal (United States) 89, 24.

Hansel, J.G., Mitchell, J.W., Klotz, H.C., 1992. Predicting and controlling flammability of multiple fuel and multiple inert mixtures. Plant/Operations Progress.

Harrison, K.B., 2005. Using the ASTM Computer Program for the Prevention of Reactive and Flammability Hazards. http://www.astm.org/SNEWS/OCTOBER_2005/harrison_oct05.html.

Hay, A., 1979. Seveso: the crucial question of reactor safety. Nature 281, 521.

Hay, D.A., 1978. Seveso: no answers yet. Disasters.

Hedengren, J.D., 2008. A nonlinear model library for dynamics and control. Computer Aids for Chemical Engineering (CACHE) News. http://www.hedengren.net/research/models.htm.

Heisenberg, W., December 1925. Über quantentheoretische umdeutung kinematischer und mechanischer beziehungen. Zeitschrift für Physik 33, 879–893. http://dx.doi.org/10.1007/BF01328377.

Hendriks, E., Kontogeorgis, G.M., Dohrn, R., de Hemptinne, J.-C., Economou, I.G., Zilnik, L.F., Vesovic, V., 2010. Industrial requirements for thermodynamics and transport properties. Industrial & Engineering Chemistry Research 49 (22), 11131–11141. http://dx.doi.org/10.1021/ie101231b.

Henkelman, G., Uberuaga, B.P., Jónsson, H., 2000. A climbing image nudged elastic band method for finding saddle points and minimum energy paths. Journal of Chemical Physics 113 (22). http://dx.doi.org/10.1063/1.1329672. American Institute of Physics: 9901–4.

Hill, T.L., 2012. An Introduction to Statistical Thermodynamics. Courier Corporation.

Hoet, P.H.M., Brüske-Hohlfeld, I., Salata, O.V., 2004. Nanoparticles – known and unknown health risks. Journal of Nanobiotechnology 2 (1). http://dx.doi.org/10.1186/1477-3155-2-12. BioMed Central Ltd: 12.

Holmes, N.S., Morawska, L., 2006. A review of dispersion modelling and its application to the dispersion of particles: an overview of different dispersion models available. Atmospheric Environment 40 (30), 5902–5928. http://dx.doi.org/10.1016/j.atmosenv.2006.06.003.

Hshieh, F.Y., 1997. Correlation of closed-cup flash points with normal boiling points for silicone and general organic compounds. Fire and Materials 21 (6), 277–282.

Hsu, J.-M., Su, M.-S., Huang, C.-Y., Duh, Y.-S., May 2012. Calorimetric studies and lessons on fires and explosions of a chemical plant producing CHP and DCPO. Journal of Hazardous Materials 217–218, 19–28. http://dx.doi.org/10.1016/j.jhazmat.2011.12.064.

Hu, W., Peng, C., Luo, W., Lv, M., Li, X., Li, D., Huang, Q., Fan, C., 2010. Graphene-based antibacterial paper. Acs Nano 4 (7), 4317–4323. http://dx.doi.org/10.1021/nn101097v.

Huang, Y., Risha, G.A., Yang, V., Yetter, R.A., 2009. Effect of particle size on combustion of aluminum particle dust in air. Combustion and Flame 156 (1), 5–13.
Huggett, C., 1980. Estimation of rate of heat release by means of oxygen consumption measurements. Fire and Materials 4 (2), 61–65. http://dx.doi.org/10.1002/fam.810040202.
Hugo, P., Steinbach, J., 1986. A comparison of the limits of safe operation of a SBR and a CSTR. Chemical Engineering Science 41 (4), 1081–1087. http://dx.doi.org/10.1016/0009-2509(86)87195-0.
Ifa, Gestis-dust-ex, 2014. http://staubex.ifa.dguv.de/explosuche.aspx?lang=e (accessed March 26).
Jacobson, M.Z., 1997a. Development and application of a new air pollution modeling system—II. Aerosol module structure and design. Atmospheric Environment 31A, 131–144.
Jacobson, M.Z., 1997b. Development and application of a new air pollution modeling system—Part III. Aerosol-phase simulations. Atmospheric Environment 31, 587–608.
Jensen, P., Bunker, P.R., 2006. Molecular Symmetry and Spectroscopy.
Johnson, R.W., W Rudy, S., Unwin, S.D., 2010. Essential Practices for Managing Chemical Reactivity Hazards.
Kapias, T., Griffiths, R.F., 2001a. Spill behaviour using REACTPOOL Part I. Results for accidental releases of chlorosulphonic acid ($HSO_3Cl$). Journal of Hazardous Materials 81, 19–30.
Kapias, T., Griffiths, R.F., 2001b. Spill behaviour using REACTPOOL Part III. Results for accidental releases of phosphorus trichloride ($PCl_3$) and oxychloride ($POCl_3$) and general discussion. Journal of Hazardous Materials 81 (3), 223–249.
Kapias, T., Griffiths, R.F., 2005. Accidental releases of titanium tetrachloride ($TiCl_4$) in the context of major hazards – spill behaviour using REACTPOOL. Journal of Hazardous Materials 119 (1–3), 41–52. http://dx.doi.org/10.1016/j.jhazmat.2004.12.001.
Kapias, T., Griffiths, R.F., Stefanidis, C., 2001a. REACTPOOL: a code implementing a new multi-compound pool model that accounts for chemical reactions and changing composition for spills of water reactive chemicals. Journal of Hazardous Materials 81, 1–18.
Kapias, T., Griffiths, R.F., Stefanidis, C., 2001b. Spill behaviour using REACTPOOL Part II. Results for accidental releases of silicon tetrachloride ($SiCl_4$). Journal of Hazardous Materials 81 (3), 209–222.
Karelson, M., 2000. Molecular Descriptors in QSAR/QSPR.
Karelson, M., Lobanov, V.S., 1996. Quantum-chemical descriptors in QSAR/QSPR studies. Chemical Reviews.
Kaszniak, M., 2010. Oversights and omissions in process Hazard analyses: lessons learned from CSB investigations. Process Safety Progress 29 (3), 264–269.
Katritzky, A.R., Stoyanova-Slavova, I.B., Dobchev, D.A., Karelson, M., 2007. QSPR modeling of flash points: an update. Journal of Molecular Graphics and Modelling 26 (2), 529–536. http://dx.doi.org/10.1016/j.jmgm.2007.03.006.
Katritzky, A.R., Petrukhin, R., Jain, R., Karelson, M., 2001. QSPR analysis of flash points. Journal of Chemical Information and Computer Sciences 41 (6), 1521–1530.
Kavlock, R.J., Ankley, G., Blancato, J., Breen, M., Conolly, R., Dix, D., Houck, K., et al., 2007. Computational toxicology – a state of the science mini review. Toxicological Sciences 103 (1), 14–27. http://dx.doi.org/10.1093/toxsci/kfm297.
Kayat, J., Gajbhiye, V., Tekade, R.K., 2011. Pulmonary Toxicity of Carbon Nanotubes: A Systematic Report.
Kemsley, J., 2015. Scientists Observe Elusive Radical Involved in Combustion and Aerosol Chemistry. http://cen.acs.org/articles/93/i6/Scientists-Observe-Elusive-Radical-Involved.html.
Klamt, A., 1995. Conductor-like screening model for real solvents: a new approach to the quantitative calculation of solvation phenomena. The Journal of Physical Chemistry 99 (7), 2224–2235. http://dx.doi.org/10.1021/j100007a062. American Chemical Society.
Klamt, A., Eckert, F., Arlt, W., 2010. COSMO-rs: an alternative to simulation for calculating thermodynamic properties of liquid mixtures. Annual Review of Chemical and Biomolecular Engineering 1 (1), 101–122. http://dx.doi.org/10.1146/annurev-chembioeng-073009-100903.

Klamt, A., Volker, J., Bürger, T., Lohrenz, J.C.W., 1998. Refinement and parametrization of COSMO-RS. Journal of Physical Chemistry A 102 (26), 5074–5085. http://dx.doi.org/10.1021/jp980017s. American Chemical Society.

Knotts, T.A., Wilding, W.V., Oscarson, J.L., Rowley, R.L., 2001. Use of the DIPPR database for development of QSPR correlations: surface tension. Journal of Chemical and Engineering Data 46 (5), 1007–1012.

Kohn, W., Sham, L.J., 1965. Self-consistent equations including exchange and correlation effects. Physical Review 140 (4), 1133–1138. http://dx.doi.org/10.1103/PhysRev.140.A1133.

Kondo, S., Takizawa, K., Takahashi, A., Tokuhashi, K., 2007. Flammability limits of isobutane and its mixtures with various gases. Journal of Hazardous Materials 148 (3), 640–647.

Kondo, S., Takizawa, K., Takahashi, A., 2006. Extended Le Chatelier's formula for carbon dioxide dilution effect on flammability limits. Journal of Hazardous Materials 138 (1), 1–8.

Kuchta, J.M., 1985. Investigation of Fire and Explosion Accidents in the Chemical, Mining, and Fuel-Related Industries: A Manual.

Kuchta, J.M., 1986. Investigation of fire and explosion accidents in the chemical, mining, and fuel-related industries – a manual. Bulletin.

Laidler, 1987. Chemical Kinetics. Pearson Education India.

Le Chatelier, H., 1891. Estimationof firedampby flammability limits. Annales des Mines.

Laidler, K.J., King, M.C., 1983. Development of transition-state theory. The Journal of Physical Chemistry 87 (15), 2657–2664.

Leardi, R., Lupiáñez González, A., 1998. Genetic algorithms applied to feature selection in PLS regression: how and when to use them. Chemometrics and Intelligent Laboratory Systems 41 (2), 195–207.

Leardi, R., Boggia, R., Terrile, M., 1992. Genetic algorithms as a strategy for feature selection. Journal of Chemometrics 6, 267–281.

Lee, J.G., 2011. Computational Materials Science. CRC Press.

Lee, K.-H., Sandler, S.I., 1987. The generalized van der waals partition function—IV. Local composition models for mixtures of unequal-size molecules. Fluid Phase Equilibria 34 (2–3), 113–147. http://dx.doi.org/10.1016/0378-3812 (87)80028-6.

Lees, F.P., 2005. Lee's Loss Prevention in the Process Industries: Hazard Identification, Assessment, and Control.

Lemkowitz, S.M., Pasman, H.J., 2014. A review of the fire and explosion hazards of particulates. KONA Powder and Particle Journal 31, 53–81.

Lessigiarska, I., Worth, A.P., Netzeva, T.I., Dearden, J.C., Cronin, M.T.D., 2006. Quantitative structure-activity-activity and quantitative structure-activity investigations of human and rodent toxicity. Chemosphere 65 (10), 1878–1887. http://dx.doi.org/10.1016/j.chemosphere.2006.03.067.

Lewars, E., 2010. Computational Chemistry. Springer Science & Business Media.

Lewis, R.J., Lewis Sr, R.J., 2008. Hazardous Chemicals Desk Reference. Wiley-Interscience.

Li, Q.F., Chen, X.G., Hu, Z.D., 2004. Quantitative structure-property relationship studies for estimating boiling points of alcohols using calculated molecular descriptors with radial basis function neural networks. Chemometrics and Intelligent Laboratory Systems 72 (1), 93–100. http://dx.doi.org/10.1016/j.chemolab.2004.03.002.

Lian, P., Ng, D., Mejia, A.F., Cheng, Z., Mannan, M.S., 2011. Study on flame characteristics in aerosols by industrial heat transfer fluids. Industrial & Engineering Chemistry Research 50 (12), 7644–7652. http://dx.doi.org/10.1021/ie200179y. American Chemical Society.

Liaw, H.J., Lee, T.P., Tsai, J.S., Hsiao, W.H., Chen, M.H., Hsu, T.T., 2003. Binary liquid solutions exhibiting minimum flash-point behavior. Journal of Loss Prevention in the Process Industries 16 (3), 173–186. http://dx.doi.org/10.1016/S0950-4230 (03)00004-4.

Liaw, H.J., Tang, C.L., Lai, J.S., 2004. A model for predicting the flash point of ternary flammable solutions of liquid. Combustion and Flame 138 (4), 308–319. http://dx.doi.org/10.1016/j.combustflame.2004.06.002.

Liaw, H.J., Lee, Y.H., Tang, C.L., Hsu, H.H., Liu, J.H., 2002. A mathematical model for predicting the flash Point of binary solutions. Journal of Loss Prevention in the Process Industries 15 (6), 429–438.

Liaw, H.-J., Lin, S.-C., 2007. Binary mixtures exhibiting maximum flash-point behavior. Journal of Hazardous Materials 140 (1–2), 155–164. http://dx.doi.org/10.1016/j.jhazmat.2006.06.050.

Liaw, H.-J., Chen, C.-C., Chang, C.-H., Lin, N.-K., Shu, C.-M., 2012. Model to estimate the flammability limits of fuel–air–diluent mixtures tested in a constant pressure vessel. Industrial & Engineering Chemistry Research 51 (6), 2747–2761. http://dx.doi.org/10.1021/ie2011824.

Liekhus, K.J., Zlochower, I.A., Cashdollar, K.L., Djordjevic, S.M., Loehr, C.A., 2000. Flammability of gas mixtures containing volatile organic compounds and hydrogen. Journal of Loss Prevention in the Process Industries 13 (3–5), 377–384. http://dx.doi.org/10.1016/S0950-4230 (99)00034-0.

Lin, S.-T., Sandler, S.I., 2006. Infinite dilution activity coefficients from ab initio solvation calculations. Aiche Journal 45 (12), 2606–2618. http://dx.doi.org/10.1002/aic.690451217.

Lin, S.-T., Chang, J., Wang, S., Goddard, W.A., Sandler, S.I., 2004. Prediction of vapor pressures and enthalpies of vaporization using a COSMO solvation model. Journal of Physical Chemistry A 108 (36), 7429–7439. http://dx.doi.org/10.1021/jp048813n. American Chemical Society.

Lin, Y., Stadtherr, M.A., 2004. LP strategy for the interval-Newton method in deterministic global optimization. Industrial & Engineering Chemistry Research 43 (14), 3741–3749. http://dx.doi.org/10.1021/ie034073d.

Liu, X., 2010. Research Progress on Flash Point Prediction. Journal of Chemical & Engineering Data 55 (9), 2943–2950. http://dx.doi.org/10.1021/je1003143.

Lothrop, W.C., Richard Handrick, G., 1949. The relationship between performance and constitution of pure organic explosive compounds. Chemical Reviews 44 (3), 419–445. http://dx.doi.org/10.1021/cr60139a001.

Luenberger, D.G., Ye, Y., 2008. Linear and nonlinear programming.

Luo, K.-M., Lu, K.-T., Hu, K.-H., 1997. The critical condition and stability of exothermic chemical reaction in a non-isothermal reactor. Journal of Loss Prevention in the Process Industries 10 (3), 141–150. http://dx.doi.org/10.1016/S0950-4230 (97)00003-X.

Lurmann, F.W., Wexler, A.S., Pandis, S.N., Musarra, S., 1997. Modelling urban and regional aerosols—II. Application to California's south coast air basin. Atmospheric Environment 31, 2695–2715.

Lwin, Y., 2000. Chemical Equilibrium by Gibbs Energy Minimization on Spreadsheets. International Journal of Engineering Education 16 (4), 335–339.

Ma, T., 2011. A Thermal theory for estimating the flammability limits of a mixture. Fire Safety Journal 46 (8), 558–567. http://dx.doi.org/10.1016/j.firesaf.2011.09.002.

Mack, E., Boord, C.E., Barham, H.N., 1923. Calculation of flash points for pure organic substances. Industrial & Engineering Chemistry 15, 963–965.

Mackay, D., Bobra, A., Chan, D.W., Shiu, W.Y., 1982. Vapor-pressure correlations for low-volatility environmental chemicals. Environmental Science & Technology 16 (10), 645–649. http://dx.doi.org/10.1021/es00104a004.

Mananghaya, M., Rodulfo, E., Santos, G.N., Villagracia, A.R., 2011. Theoretical investigation on the solubilization in water of functionalized single-wall carbon nanotubes. Journal of Nanotechnology 2012. Hindawi Publishing Corporation.

Mannan, M.S., Aldeeb, A.A., Rogers, W.J., 2006. Understanding the role of process chemistry in fires and explosions. Process Safety Progress 21 (4), 323–328. http://dx.doi.org/10.1002/prs.680210409.

Mannan, S., Lees, F.P., 2005. Lee's Loss Prevention in the Process Industries. Elsevier.

Markowski, A.S., Sam Mannan, M., Kotynia Bigoszewska, A., Siuta, D., 2010. Uncertainty aspects in process safety analysis. Journal of Loss Prevention in the Process Industries 23 (3), 446–454. http://dx.doi.org/10.1016/j.jlp.2010.02.005.

Martins-Costa, M.T.C., Anglada, J.M., 2012. Reactivity of volatile organic compounds at the surface of a water droplet. Journal of the American Chemical Society 134 (28), 11821–11827.

Mashuga, C.V., Crowl, D.A., 2000. Derivation of le Chatelier's mixing rule for flammable limits. Process Safety Progress 19 (2), 112–117.

Mather, A.E., 1986. Phase equilibria and chemical reaction. Fluid Phase Equilibria 30, 83–100.
McDonald, C.M., Floudas, C.A., 1995. Global optimization and analysis for the gibbs free energy function using the UNIFAC, Wilson, and ASOG equations. Industrial & Engineering Chemistry Research 34 (5), 1674–1687. http://dx.doi.org/10.1021/ie00044a020.
McQuarrie, D.A., 2000. Statistical Mechanics. Sterling Publishing Company.
Medica, L.A., 1995. Guidelines for safe storage and handling of reactive materials. Storage and Handling of Reactive Materials 364.
Melhem, G.A., 1997. A detailed method for estimating mixture flammability limits using chemical equilibrium. Process Safety Progress 16 (4), 203–218. http://dx.doi.org/10.1002/prs.680160403.
Melhem, G.A., 2004. A detailed method for estimating mixture flammability limits using chemical equilibrium. John Wiley and Sons, New York, pp. 203–218.
Melhem, G.A., Shanley, E.S., 1996. On the estimation of hazard potential for chemical substances. Process Safety Progress 15 (3), 168–172. http://dx.doi.org/10.1002/prs.680150311.
Meng, Z., Dabdub, D., Seinfeld, J.H., 1998. Size-resolved and chemically resolved model of atmospheric aerosol dynamics. Journal of Geophysical Research 103, 3419–3435.
Merlot, C., 2010. Computational toxicology-a tool for early safety evaluation. Drug Discovery Today 15 (1–2), 16–22. http://dx.doi.org/10.1016/j.drudis.2009.09.010.
Metcalfe, E., Metcalfe, A.E.M., February 1992. Communication: On the Correlation of Flash Points, vol. 16, 153–54.
Metropolis, N., Rosenbluth, A.W., Rosenbluth, M.N., Teller, A.H., Teller, E., 1953. Equation of state calculations by fast computing machines. Journal of Chemical Physics 21 (6). http://dx.doi.org/10.1063/1.1699114. abstract. copyright_name.value: 1087–92.
Mewis, J.J., Pasman, H.J., 1995. Loss Prevention and Safety Promotion in the Process Industries.
Miller, M.W., Wesley, D., Connolly, B., 2010. User Centered Design in Chemical Emergency Software Development.
Mills, G., Jónsson, H., 1994. Quantum and thermal effects in $H_2$ dissociative adsorption: evaluation of free energy barriers in multidimensional quantum systems. Physical Review Letters 72 (7), 1124–1127. http://dx.doi.org/10.1103/PhysRevLett.72.1124.
Miranda, M.S., Matos, M.A.R., Morais, V.M.F., 2012. 2, 1, 3–Benzothiadiazole: study of its structure, energetics and aromaticity. The Journal of Chemical Thermodynamics 50, 30.
Miyoshi, A. GPOP Software. http://www.frad.t.u-tokyo.ac.jp/~miyoshi/gpop/.
Moghaddam, A.Z., Rafiei, A., Khalili, T., 2012. Assessing prediction models on calculating the flash point of organic acid, ketone and alcohol mixtures. Fluid Phase Equilibria 316, 117–121. http://dx.doi.org/10.1016/j.fluid.2011.12.014.
Morbidelli, M., Varma, A., 1988. A generalized criterion for parametric sensitivity: application to thermal explosion theory. Chemical Engineering Science 43 (1), 91–102. http://dx.doi.org/10.1016/0009-2509(88)87129-X.
Murphy, J.F., Holmstrom, D., 2004. Understanding reactive chemical incidents. Chemical Engineering Progress 100 (3), 31–33.
Nadykto, A.B., Mikkelsen, K.V., Yu, F.Q., Ruuskanen, J., 2004. Sulfuric acid and sulfuric acid hydrates in the Gas phase: a DFT investigation. Journal of Physical Chemistry A 108 (41), 8914–8929.
Nel, A., 2005. Air pollution-related illness: effects of particles. Science 308 (5723), 804–806.
Nel, A., Xia, T., Mädler, L., Li, N., 2006. Toxic potential of materials at the nanolevel. Science 311 (5761), 622–627.
NOAA. Chemical Reactivity Worksheet. http://response.restoration.noaa.gov/reactivityworksheet.
Nyden, M.R., Stoliarov, S.I., 2008. Modeling the Thermal decomposition of large molecules and nanostructures. Multiscale Simulation Methods for Nanomaterials.
NFPA 68 Guide for Venting of Deflagrations, 2002. National Fire Protection Association.

Obata, S., Honda, K., 2011. Dynamic behavior of carbon nanotube and bio-/artificial surfactants complexes in an aqueous environment. The Journal of Physical Chemistry C 115 (40), 19659–19667. ACS Publications.

Oberdörster, G., Oberdörster, E., 2005. Nanotoxicology: an emerging discipline evolving from studies of ultrafine particles. Environmental Health Perspectives 113 (7), 823–839.

Ochterski, J.W., 2000. Thermochemistry in Gaussian. Gaussian Inc.

Palmer, K.N., 1973. Dust Explosions and Fires. Halsted Press.

Palucis, M., Glowienka, T., Van Brunt, V., Ervin, R., Chastain, W., Kline, R., Lodal, P., 2007. Prediction of flammability speciation for the lower alkanes, carboxylic acids, and esters. Process Safety Progress 26 (1), 4–9. http://dx.doi.org/10.1002/prs.10148.

Pan, Y., C Jiang, J., Wang, R., J Jiang, J., 2011. Predicting the net heat of combustion of organic compounds from molecular structures based on ant colony optimization. Journal of Loss Prevention in the Process Industries 24, 85–89.

Pan, Y., Jiang, J., Wang, Z., 2007. Quantitative structure-property relationship studies for predicting flash points of alkanes using group bond contribution method with back-propagation neural network. Journal of Hazardous Materials 147, 424–430. http://dx.doi.org/10.1016/j.jhazmat.2007.01.025.

Pan, Y., Jiang, J., Wang, R., Cao, H., Cui, Y., 2009. A novel QSPR model for prediction of lower flammability limits of organic compounds based on support vector machine. Journal of Hazardous Materials 169 (2–3), 217–220. http://dx.doi.org/10.1016/j.jhazmat.2009.03.083.

Pan, Y., Jiang, J., Ding, X., Wang, R., Jiang, J., 2010. Prediction of flammability characteristics of pure hydrocarbons from molecular structures. Aiche Journal 56 (3), 690–701. http://dx.doi.org/10.1002/aic.12007.

Panayiotou, C., 2003. Equation-of-state models and quantum mechanics calculations. Industrial & Engineering Chemistry Research.

Panayiotou, C., Sanchez, I.C., 1991. Statistical thermodynamics of associated polymer solutions. Macromolecules 24 (23), 6231–6237.

Panayiotou, C., Vera, J.H., 1980. The quasi-chemical approach for non-randomness in liquid mixtures. Expressions for local surfaces and local compositions with an application to polymer solutions. Fluid Phase Equilibria 5 (1–2), 55–80. http://dx.doi.org/10.1016/0378-3812 (80)80043-4.

Patil, G.S., 1988. Estimation of flash-point. Fire and Materials 12 (3), 127–131.

Pauling, L., Wilson, E.B., 1985. Introduction to Quantum Mechanics: With Applications to Chemistry. Dover Publications, New York, NY.

Perry, R.H., Green, D.W., 1984. Perry's Chemical Engineers' Handbook.

Peters, B., Liang, W.Z., Bell, A.T., 2003. Biasing a transition state search to locate multiple reaction pathways. The Journal of Chemical Physics 118, 21.

Peters, B., Heyden, A., Bell, A.T., Chakraborty, A., 2004. A growing string method for determining transition states: comparison to the nudged elastic band and string methods. Journal of Chemical Physics 120 (17), 7877. http://dx.doi.org/10.1063/1.1691018.

Pople, J.A., Head Gordon, M., Fox, D.J., 1989. Gaussian-1 theory: a general procedure for prediction of molecular energies. The Journal of Chemical Physics 110, 7650–7657.

Prausnitz, J.M., 1999. Thermodynamics and the Other Chemical Engineering Sciences: Old Models for New Chemical Products and Processes, vol. 158. pp. 95–111.

Preston, J., Leger, M., Teng, Y. Bomb Calorimetry—Heats of combustion. oocities.com.

Pritchard, D.K., 2004. Literature Review: Explosion Hazards Associated with Nanopowders.

Prugh, R.W., 1973. Estimation of flash point temperature. Journal of Chemical Education 50 (2), A85.

Putnam, R., Taylor, R., Klamt, A., Eckert, F., 2003. Prediction of infinite dilution activity coefficients using COSMO-RS. Industrial & Engineering Chemistry Research 42 (15), 3635–3641.

Puzyn, T.T., Suzuki, N.N., Haranczyk, M.M., Rak, J.J., 2008. Calculation of quantum-mechanical descriptors for QSPR at the DFT level: is it necessary? Journal of Chemical Information and Modeling 48 (6), 1174–1180. http://dx.doi.org/10.1021/ci800021p.

Puzyn, T., Leszczynska, D., Leszczynski, J., 2009. Toward the development of "nano-QSARs": advances and challenges. Small 5 (22), 2494–2509. http://dx.doi.org/10.1002/smll.200900179.

Rahman, A., 1964. Correlations in the motion of atoms in liquid argon. Physical Review 136 (2), 405–411. http://dx.doi.org/10.1103/PhysRev.136.A405.

Rassia, S.T., Siettos, C.I., 2010. Escape dynamics in office buildings: using molecular dynamics to quantify the impact of certain aspects of human behavior during emergency evacuation. Environmental Modeling & Assessment 15 (5), 411–418.

Razon, L.F., Schmitz, R.A., 1987. Multiplicities and instabilities in chemically reacting systems—a review. Chemical Engineering Science 42, 1005–1047.

Reinisch, G. OpenSOAMS. http://opensoams.google-code.com/svn-history/r2/doc/html/index.html.

Reinisch, G., Miki, K., Vignoles, G.L., Wong, B.M., Simmons, C.S., 2012. An efficient and accurate formalism for the treatment of large amplitude intramolecular motion. Journal of Chemical Theory and Computation 8 (8), 2713–2724. http://dx.doi.org/10.1021/ct300278x.

Reyes, O.J., Patel, S.J., Sam Mannan, M., 2011. Quantitative structure property relationship studies for predicting dust explosibility characteristics (kst, pmax) of organic chemical dusts. Industrial & Engineering Chemistry Research 50 (4), 2373–2379. http://dx.doi.org/10.1021/ie1013663.

Rogers, W.J., Mannan, M.S., 2006. Prediction of minimum flash Point behaviour for binary mixtures. Process Safety and Environmental Protection 84 (B1), 1–9.

Rossi, C., Cardozo-Filho, L., Guirardello, R., 2009. Gibbs free energy minimization for the calculation of chemical and phase equilibrium using linear programming. Fluid Phase Equilibria 278, 117–128.

Rowley, J.R., Freeman, D.K., Rowley, R.L., 2010. Flash point: evaluation, experimentation and estimation. International Journal of Thermophysics 31, 875–887.

Roy, P.K., Bhatt, A., Rajagopal, C., 2003. Quantitative risk assessment for accidental release of titanium tetrachloride in a titanium sponge production plant. Journal of Hazardous Materials 102, 167–186. http://dx.doi.org/10.1016/S0304-3894 (03)00220-6.

da Silva, G., Bozzelli, J.W., Sebbar, N., Bockhorn, H., 2006. Thermodynamic and ab initio analysis of the controversial enthalpy of formation of formaldehyde. ChemPhysChem 7 (5), 1119–1126. http://dx.doi.org/10.1002/cphc.200500667.

Sandler, S.I., 2010. An Introduction to Applied Statistical Thermodynamics. John Wiley & Sons.

Santon, R. Mist explosions incident survey. ukelg.ps.ic.ac.uk.

Saraf, S.R., Rogers, W.J., Ford, D.M., Mannan, M.S., August 2004. Integrating molecular modeling and process safety research. Fluid Phase Equilibria 222-223, 205–211. http://dx.doi.org/10.1016/j.fluid.2004.06.048.

Saraf, S.R., Rogers, W.J., Mannan, M.S., 2003. Prediction of reactive hazards based on molecular structure. Journal of Hazardous Materials 98, 15–29.

Sasidharan, A., Panchakarla, L.S., Chandran, P., Menon, D., Shantikumar, N., Rao, C.N.R., Koyakutty, M., 2011. Differential nano-bio interactions and toxicity effects of pristine versus functionalized graphene. Nanoscale 3 (6), 2461–2464. http://dx.doi.org/10.1039/c1nr10172b.

Satyanarayana, K., Katati, M.C., 1991. Note: correlation of flash points. Fire and Materials 15 (2), 97–100.

Satyanarayana, K., Rao, P.G., 1992. Improved equation to estimate flash points of organic-compounds. Journal of Hazardous Materials 32 (1), 81–85.

Savee, J.D., Papajak, E., Rotavera, B., Huang, H., Eskola, A.J., Welz, O., Sheps, L., Taatjes, C.A., Zádor, J., Osborn, D.L., 2015. Direct observation and kinetics of a hydroperoxyalkyl radical (QOOH). Science 347 (6222), 643–646. American Association for the Advancement of Science.

Sayes, C.M., Liang, F., Hudson, J.L., Mendez, J., Guo, W., Beach, J.M., Moore, V.C., Doyle, C.D., West, J.L., Billups, W.E., 2006. Functionalization density dependence of single-walled carbon nanotubes cytotoxicity in vitro. Toxicology Letters 161 (2), 135–142. Elsevier.

Schrodinger, E., 1926. An undulatory theory of the mechanics of atoms and molecules. Physical Review 28 (6), 1049–1070.

Shaluf, I.M., Said, A.M., 2003. A review of disaster and crisis. Disaster Prevention and Management 12, 24–32.

Shankar, S., Mortier, W.J., Ghosh, S.K., 1986. Calculation of atomic charges in large molecules. Annals of the New York Academy of Sciences 482, 82–84. http://dx.doi.org/10.1111/j.1749-6632.1986.tb20939.x.

Shanley, E.S., Melhem, G.A., 1995. The oxygen balance criterion for Thermal hazards assessment. Process Safety Progress 14 (1), 29–31. http://dx.doi.org/10.1002/prs.680140105.

Shebeko, Yu.N., Tsarichenko, S.G., Korolchenko, A.Y., Trunev, A.V., Navzenya, V.Y., Papkov, S.N., Zaitzev, A.A., 1995. Burning velocities and flammability limits of gaseous mixtures at elevated temperatures and pressures. Combustion and Flame 102 (4), 427–437. http://dx.doi.org/10.1016/0010-2180 (95)00002-N.

Sholl, D., Steckel, J.A., 2011. Density Functional Theory: A Practical Introduction.

Smallwood, I.M.N., 1996. Handbook of Organic Solvent Properties. Halsted Press.

Steckler, R., Hu, W.P., Liu, Y.P., Lynch, G.C., 1995. POLYRATE 6.5: a new version of a computer program for the calculation of chemical reaction rates for polyatomics. Computer Physics Communication.

Stewart, J.J.P., 1989. Optimization of parameters for semiempirical methods I. Method. Journal of Computational Chemistry 10 (2), 209–220. http://dx.doi.org/10.1002/jcc.540100208.

Stoliarov, S.I., Lyon, R.E., Nyden, M.R., 2004. A reactive molecular dynamics model of Thermal decomposition in polymers. II. Polyisobutylene. Polymer 45, 8613–8621.

Stoliarov, S.I., Westmoreland, P.R., Nyden, M.R., 2003. A reactive molecular dynamics model of Thermal decomposition in polymers: I. Poly(methyl methacrylate). Polymer 44, 883–894.

Strozzi, F., Zaldívar, J.M., 1994. A general method for assessing the thermal stability of batch chemical reactors by sensitivity calculation based on lyapunov exponents. Chemical Engineering Science 49 (16), 2681–2688.

Stølen, S., Grande, T., 2004. Chemical Thermodynamics of Materials. John Wiley & Sons.

Sun, Q., Wexler, A.S., 1998. Modeling urban and regional aerosols near acid neutrality—application to the 24–25 June SCAQS episode. Atmospheric Environment.

Szabo, A., Ostlund, N.S., 2012. Modern Quantum Chemistry. Courier Corporation.

Szeifert, F., Chován, T., Nagy, L., Abonyi, J., Árva, P. Runaway of chemical reactors: parametric sensitivity and stability. konyvtar.uni-pannon.hu.

Tascón, A., Aguado, P.J., Ramírez, A., 2009. Dust explosion venting in silos: a comparison of standards NFPA 68 and EN 14491. Journal of Loss Prevention in the Process Industries 22 (2), 204–209.

Temelso, B., Morrell, T.E., Shields, R.M., Allodi, M.A., Wood, E.K., Kirschner, K.N., Castonguay, T.C., Archer, K.A., Shields, G.C., 2012. Quantum mechanical study of sulfuric acid hydration: atmospheric implications. Journal of Physical Chemistry A 116 (9), 2209–2224. http://dx.doi.org/10.1021/jp2119026. American Chemical Society.

Tetko, I.V., 2003. The www as a tool to obtain molecular parameters. Mini Reviews in Medicinal Chemistry 3 (8), 809–820. http://dx.doi.org/10.2174/1389557033487638. Bentham Science Publishers.

Tetteh, J., Suzuki, T., Metcalfe, E., Howells, S., 1999. Quantitative structure-property relationships for the estimation of boiling point and flash point using a radial basis function neural network. Journal of Chemical Information and Computer Sciences 39 (3), 491–507.

The Explosion at Concept Sciences: Hazards of Hydroxylamine, 2004. http://onlinelibrary.wiley.com/doi/10.1002/prs.10013/full.

Thornton, W.M., 1917. The relation of oxygen to the heat of combustion of organic compounds. Philosophical Magazine 33, 196–203.

Todeschini, R., Consonni, V., 2008. Handbook of Molecular Descriptors. Wiley-VCH.

Todeschini, R., Consonni, V., 2009. Molecular Descriptors for Chemoinformatics, vol. 1. Alphabetical Listing; vol. 2. Appendices, References.

Todeschini, R., Consonni, V., Mauri, A., 2006. Characterization of DNA primary sequences by a new similarity/diversity measure based on the partial ordering. Journal of Chemical Information and Modeling 46 (5), 1905–1911.

Tong, C., Blanco, M., Goddard, W.A., 2004. Thermodynamic properties of multifunctional oxygenates in atmospheric aerosols from quantum mechanics and molecular dynamics: dicarboxylic acids. Environmental Science & Technology 38 (14), 3941–3949.

U.S. Chemical Safety and Hazard Investigation Board, 2003. Hazard Investigation: Improving Reactive Hazard Management. Chemical Safety and Hazard Investigation.

Urben, P., 2006. Bretherick's Handbook of Reactive Chemical Hazards: 2-Volume Set.

Varma, A., Morbidelli, M., Wu, H., 2005. Parametric Sensitivity in Chemical Systems.

te Velde, G., Bickelhaupt, F.M., 2001. Chemistry with ADF. Journal of Computational Chemistry 22 (9), 931–967.

Vidal, M., Rogers, W.J., Holste, J.C., Mannan, M.S., 2004. A review of estimation methods for flash points and flammability limits. Process Safety Progress 23 (1), 47–55. http://dx.doi.org/10.1002/prs.10004.

Voutsas, E., Lampadariou, M., Magoulas, K., Tassios, D., 2002. Prediction of vapor pressures of pure compounds from knowledge of the normal boiling point temperature. Fluid Phase Equilibria 198 (1), 81–93(13).

Wang, Q., Wei, C., Pérez, L.M., Rogers, W.J., Hall, M.B., Mannan, M.S., 2010. Thermal decomposition pathways of hydroxylamine: theoretical investigation on the initial steps. Journal of Physical Chemistry A 114 (34), 9262–9269. http://dx.doi.org/10.1021/jp104144x.

Wang, S., Sandler, S.I., Chen, C.C., 2007. Refinement of COSMO-SAC and the applications. Industrial & Engineering Chemistry Research 46, 7275–7288.

Wei, C., Rogers, W.J., Mannan, M.S., 2006. Thermal decomposition hazard evaluation of hydroxylamine nitrate. Journal of Hazardous Materials 130 (1–2), 163–168.

Wei, C., Saraf, S.R., Rogers, W.J., Mannan, M.S., 2004a. Thermal runaway reaction hazards and mechanisms of hydroxylamine with acid/base contaminants. Thermochimica Acta 421 (1–2), 1–9. http://dx.doi.org/10.1016/j.tca.2004.02.012.

Wei, C., Rogers, W.J., Mannan, M.S., 2004b. Application of screening tools in the prevention of reactive chemical incidents. Journal of Loss Prevention in the Process Industries 17 (4), 261–269. http://dx.doi.org/10.1016/j.jlp.2004.04.003.

Wen, C.S., 1996. The Fundamentals of Aerosol Dynamics.

Westerterp, K.R., Molga, E.J., 2006. Safety and runaway prevention in batch and semibatch reactors - a review. Chemical Engineering Research & Design 84 (A7), 543–552. http://dx.doi.org/10.1205/cherd.05221.

Westmoreland, P.R., 2002. Applying Molecular and Materials Modeling.

White, D., Beyler, C.L., Fulper, C., 1997. Flame spread on aviation fuels. Fire Safety Journal 28, 1–31.

Wieder, G.M., Marcus, R.A., 2004. Dissociation and isomerization of vibrationally excited species. II. Unimolecular reaction rate theory and its application. The Journal of Chemical Physics.

Wigner, E., 1938. The transition state method. Transactions of the Faraday Society 34, 29–41.

Winkler, D.A., Mombelli, E., Pietroiusti, A., Tran, L., Worth, A., Fadeel, B., McCall, M.J., November 2012. Applying quantitative structure–activity relationship approaches to nanotoxicology: current status and future potential. Toxicology. http://dx.doi.org/10.1016/j.tox.2012.11.005.

Wolf, A., Swift, J.B., Swinney, H.L., Vastano, J.A., 1985. Determining lyapunov exponents from a time series. Physica D: Nonlinear Phenomena 16 (3), 285–317.

Yamaguchi, A., Tajima, Y., 2003. Numerical simulation of non-premixed diffusion flame and reaction product aerosol behavior in liquid metal pool combustion. Journal of Nuclear Science and Technology 40 (2), 93–103. http://dx.doi.org/10.1080/18811248.2003.9715338. Taylor & Francis Group.

Yao, X.J., Panaye, A., Doucet, J.P., Zhang, R.S., Chen, H.F., Liu, M.C., Hu, Z.D., Fan, B.T., 2004. Comparative study of QSAR/QSPR correlations using support vector machines, radial basis function neural networks, and multiple linear regression. Journal of Chemical Information and Modeling 44 (4), 1257–1266. http://dx.doi.org/10.1021/ci049965i. American Chemical Society.

Yaws, C.L., 2003. Yaws' Handbook of Thermodynamic and Physical Properties of Chemical Compounds.

Yaws, C.L., 1995. Handbook of Vapor Pressure: Inorganic Compounds and Elements.

Yeoh, G.H., Yuen, K.K., 2009. Computational Fluid Dynamics in Fire Engineering: Theory, Modelling and Practice.

Young, D., 2004. Computational Chemistry: A Practical Guide for Applying Techniques to Real World Problems.

Yu, J.-S.K., Yu, C., 1996. Ab initio study of the decomposition of formaldehyde. Chemical Physics Letters 271 (4), 259–265. http://dx.doi.org/10.1016/S0009-2614 (97)00456-9.

Zabetakis, M.G., 1965. Flammability Characteristics of Combustible Gases and Vapors.

Zaldívar, J.M., Cano, J., Alós, M.A., Sempere, J., Nomen, R., Lister, D., Maschio, G., et al., 2003. A general criterion to define runaway limits in chemical reactors. Journal of Loss Prevention in the Process Industries 16 (3), 187–200. http://dx.doi.org/10.1016/S0950-4230 (03)00003-2.

Zhang, Y., Seigneur, C., Seinfeld, J.H., Jacobson, M.Z., Binkowski, F.S., 1999. Simulation of aerosol dynamics: a comparative review of algorithms used in air quality models. Aerosol Science and Technology 31 (6), 487–514. http://dx.doi.org/10.1080/027868299304039.

Zheng, J., Zhang, S., Lynch, B.J., Corchado, J.C., Chuang, Y.Y., Fast, P.L., Hu, W.P., et al., POLYRATE version 2010-A. Minneapolis. http://t1.chem.umn.edu/polyrate/.

Zuo, G., Zhou, X., Huang, Q., Fang, H., 2011. Adsorption of villin headpiece onto graphene, carbon nanotube, and C60: effect of contacting surface curvatures on binding affinity. The Journal of Physical Chemistry C 115 (47), 23323–23328.

Zwolinski, B.J., 1971. Handbook of Vapor Pressures and Heats of Vaporization of Hydrocarbons and Related Compounds.

CHAPTER

# COMPUTATIONAL FLUID DYNAMICS SIMULATION IN PROCESS SAFETY

# 4

## 4.1 INTRODUCTION

Computational fluid dynamics (CFD) simulation has been applied extensively in process safety to assess possible catastrophic consequences such as fire, explosion, and dispersion of flammable and toxic materials from accidental release due to loss of primary containment events. If conducted properly, CFD simulations provide not only accurate hazard assessments such as radiation level, overpressure contour, and distribution of toxic cloud, but also detailed information about the spatial and temporal evolution of accidental events. A typical CFD simulation for a consequence modeling requires an extensive geometric setup, multiple input parameters, and related physical models representing physical phenomena behind the event. The overall setup of a CFD simulation consists of four main elements:

- Preprocessor or geometry representation and grid setup
- Transport equation and submodel parameter setup
- Solver setting to solve the transport and related equation2s
- Postprocessor in which the simulation results are analyzed and useful information is extracted and visually displayed in accessible manners.

These four elements of a CFD simulation setup are shown in Figure 4.1. This setup procedure is applicable for all types of consequence assessment using CFD modeling in process safety such as fire, explosion, and dispersions of flammable and toxic materials.

**Preprocessor**: This element mainly focuses on geometry representation, the grid, and boundary condition setups. One of the main advantages and the most notable characteristic of CFD simulation over simpler models, such as integral or phenomenal types, is its capability to correctly represent complex geometries. This capability is especially critical in the typical type of accidental events in process safety; it is well known that the interactions between fluid flow and obstacles have significant effects on the final outcome of consequences, such as vapor cloud explosion (VCE) or dispersions of flammable or toxic gases.

Other parameters to be considered in this section are the computational domain size, the grid size, the boundary conditions, the simulation time step, and the total duration of a simulation. The selection of these parameters is controlled by the temporal and spatial scales (i.e., the time and space scales) of the modeled events. Table 4.1 lists the typical temporal and spatial scales of three types of process safety consequences.

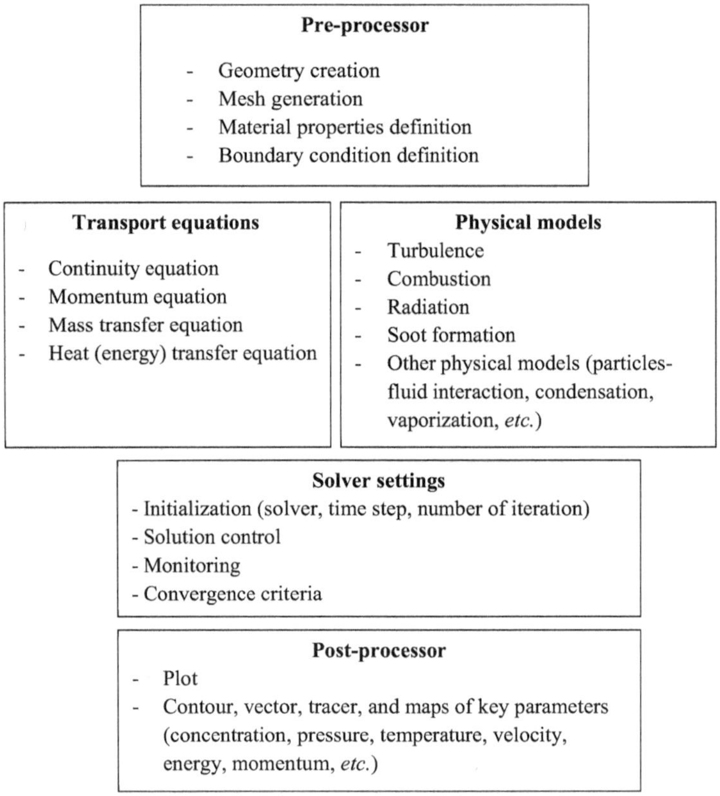


**FIGURE 4.1**

Main elements of CFD simulations in process safety consequence assessment.

*Adapted from Tu et al. (2008).*

**Table 4.1 Temporal and Spatial Scales of Process Safety Incidents**

| Scales | Fire | Explosion | Dispersion |
|---|---|---|---|
| Temporal | Medium (seconds to minutes) | Short (seconds) | Long (hours) |
| Spatial | Small (0–100 m) | Medium (0–1000 m) | Large (0–10,000 m) |

**Transport equation and physical models**: These elements provide physical description of the fluid flow, such as the Navier-Stokes equation of the fluid flow, and other related phenomena such as combustion, radiation, and reaction in a modeled event. Table 4.2 lists a set of transport equations in their general derivative form.

**Table 4.2 Set of Transport Equations (Bird et al., 2002)**

| Equations | Derivative Form |
|---|---|
| Continuity equation | $\frac{\partial \rho}{\partial t} + \nabla(\rho v) = 0$ |
| Momentum transfer equation | $\frac{\rho D v}{Dt} = -\nabla p - [\nabla \tau] + pg$ |
| Mass (scalar) transfer equation | $\frac{\partial C_p DT}{Dt} = -(\nabla q) - \left(\frac{\partial \ln \rho}{\partial \ln T}\right)_p \frac{Dp}{Dt} - (\tau: \nabla v)$ |
| Heat (energy) transfer equation | $\frac{\rho D w_\alpha}{Dt} = -(\nabla j_\alpha) + r_\alpha$ |

The physical models used in CFD simulation to describe certain phenomena in modeled events have not been covered in the transport equations. In the context of process safety, the use of CFD simulation is to model catastrophic consequences that could occur in industry such as:

- *Fire*: Typical focus on large-scale jet and pool fire
- *Explosion*: Focus on VCE, boiling liquid expanding vapor explosion (BLEVE), combustible dust cloud explosion, and deflagration to detonation (DDT) transition of VCE
- *Dispersion*: Focus on flammable and toxic materials' dispersion

Thus, there are five types of concerned physical models often used in these types of CFD simulations:

- Flow field with turbulent models
- Combustion models
- Radiation models
- Source models
- Soot formation models

The use of these physical models in process safety consequence modeling is listed in Table 4.3.

All of those physical models are currently active research subjects with focuses on development of models, validation, and incorporation and implementation into CFD modeling.

**Table 4.3 Use of Physical Models in Process Safety CFD Simulations**

| Models | Fire | Explosion | Dispersion |
|---|---|---|---|
| 1. Turbulent model | Yes | Yes | Yes |
| 2. Source model | Yes | Yes | Yes |
| 3. Combustion model | Yes | Yes | No |
| 4. Radiation model | Yes | Often ignore | No |
| 5. Soot formation model | Yes | No | No |

The research on CFD simulation on consequence analysis in process safety could be roughly divided into two main groups:

1. Model development and validation
2. Application studies

In the first group, efforts are focused on developing new or improved existing mathematical-physical models for better descriptions of certain aspects of the modeled phenomena. For example, research of this type includes the development of new models to describe the combustion phenomenon associated with combustible dust explosion or new models to describe flame propagation in deflagration to detonation. Implementation of these new models is also a subject of these studies; for example, how to incorporate new models effectively into current CFD packages or how to implement large eddy simulation (LES) or direct numerical simulation (DNS) into large-scale simulations (Blazek, 2007).

In the second group, efforts are focused on the implementation of the current CFD model on consequence studies to exploit the most important advantage of CFD simulation, which is its capability to represent and simulate complex geometries. For example, typical studies of this type are often done for the investigation of the possible consequence from VCEs in complex offshore platforms or the distribution of toxic gas during accidental releases in complex urban environments. Although process safety CFD modeling focuses on both types of studies, process safety professionals often apply CFD in application studies using available commercial or self-developed CFD packages.

## 4.2 FIRE MODELING

CFD simulation of fire incidents (e.g., large-scale jet fire and pool fire) will provide information about the flow field and characteristics of the incident fires, especially in terms of hazards associated with fire events. In the context of process safety, these two incidents are associated with large-scale releases of flammable materials in gas or liquid phases because of loss of primary containment events in industrial settings. Other types of fires such as compartmental fires (e.g., in residential or office buildings or industrial facilities), large-scale fires (e.g., forest fires), and deliberated fire activities (e.g., flare fires or fires in combustion chambers) have been covered extensively in published literature (Novozhilov, 2001). This also limits the boundary of hazards associated with large-scale jet fire and pool fires in process safety. Generally, only radiation hazard and flame impingement on nearby objects, which could set up a sequence of incident events due to the domino effect, are considered in process safety consequence modeling using CFD (Cozzania et al., 2009; Darbra et al., 2012). Other types of fire hazards such as combustion products dispersion, toxic gases dispersion, and fire spread are covered elsewhere.

### 4.2.1 CFD MODELING OF FIRE

CFD simulation of fire in the context of process safety is greatly benefitted from CFD studies in combustion engineering and fire safety engineering studies (Baukal et al., 2001; Novozhilov, 2001; Rasbash, 2004). Combustion models, soot formation models, and radiation models, developed in these two disciplines, are used directly in process safety large-scale jet fire and pool fire simulations. Essentially, in CFD simulation of large-scale jet fires and pool fires, there are three important elements: (1) geometry; (2) the mechanism for energy, momentum, and mass transportation provided by the CFD models; and (3) the detailed descriptions of the mechanisms and the kinetics of the fire combustion provided by the fire models (listed in Table 4.3). Extensive reviews on CFD simulation of fire and related physical models have been published (Novozhilov, 2001). Guidelines for using CFD for fire and smoke modeling are listed in a publication provided by the Canadian National Research Council (Hadjisophocleous and McCartney, 2005).

#### 4.2.1.1 Geometry representation

There are typically three factors to be considered in CFD geometry for fire modeling: (1) computational domain, (2) grid size, and (3) boundary conditions. The computational domain of an entire event is the control volume encompassing the affected areas and the surrounding areas. The size of the computational domain for fire incidents is dependent on the size of the fire. The other parameters of the computational domain, including thermal and hydrostatic boundary conditions, are then defined to complete the computational domain. The grid size is determined based on the computational domain sizes and computing resources available to provide grid independent, accurate, and affordable solutions. The smaller the grid size, the more accurate is the solution but the higher is the cost of computing resources. Thus, each scenario is unique and can have its own optimum grid size.

In large-scale jet fire and pool fire modeling, relatively simple geometries are used. It is often rectangular or cylindrical volumes surrounding the incident fire. The grid size is required to be very fine near the fire sources but may be set larger away from the sources. Symmetry rule may be applied to reduce the number of grid cells, thus reducing the computing time.

#### 4.2.1.2 Turbulent model

The use of Reynolds-averaged or Farve-averaged (density averaged) Navier–Stokes (RANS) with two-equation closure for a turbulence k–ε model is the primary type of turbulent model used in fire modeling due to its robustness, balance between accuracy and computational cost (Blazek, 2007; Novozhilov, 2001). Many modifications have been made to the original k–ε turbulence model either as additional terms or additional equations to address phenomena such as buoyancy, stratification, near wall treatment, and fire shapes. The basic assumption for the k–ε model is the eddy viscosity model where k is the turbulence kinetic energy and ε is the energy dissipation rate. The two parameters k and ε arecalculated from the following two equations (Novozhilov, 2001):

$$\frac{\partial(\rho k)}{\partial t} + \nabla(\rho \vec{U} k) = \nabla\left(\frac{\mu_\varepsilon}{\sigma_k}\nabla k\right) + P + G - \rho\grave{o} \tag{4.1}$$

$$\frac{\partial(\rho\grave{o})}{\partial t} + \nabla(\rho \vec{U} \grave{o}) = \nabla\left(\frac{\mu_e}{\sigma_{\grave{o}}}\nabla \grave{o}\right) + \frac{\grave{o}}{k}(c_1[P + c_3\max(G,0)] - c_s\rho\grave{o}) \tag{4.2}$$

where

$$P = \mu_e\left\{2\left[\left(\frac{\partial u}{\partial x}\right)^2 + \left(\frac{\partial v}{\partial y}\right)^2 + \left(\frac{\partial w}{\partial z}\right)^2\right] + \left(\frac{\partial u}{\partial y} + \frac{\partial v}{\partial x}\right)^2 + \left(\frac{\partial v}{\partial z} + \frac{\partial w}{\partial y}\right)^2 + \left(\frac{\partial u}{\partial z} + \frac{\partial w}{\partial x}\right)^2\right\} \tag{4.3}$$

$$G = g\left(\frac{\mu_e}{\sigma_h}\right)\left(\frac{1}{\rho}\right)\left(\frac{\partial p}{\partial z}\right) \tag{4.4}$$

$$\mu_e = \mu_{lam} + \mu_t \tag{4.5}$$

$$\mu_t = \rho c_\mu \frac{k^2}{\grave{o}} \tag{4.6}$$

where $\mu_e$ is the eddy viscosity, $\mu_{lam}$ is the laminar viscosity, $\mu_t$ is the turbulence viscosity, and the constants: $\sigma_h = 0.7$, $c_1 = 1.44$, $c_2 = 1.92$, and $c_3 = 1.0$.

Other turbulent models available to be used with RANS, such as the Reynolds normalization group (RNG) k–ε and the realizable k–ε model, or more sophisticated turbulent models, such as the Reynolds stress model, have been used in combustion modeling but have not been used in large-scale fire simulations (Papageorgakis and Assanis, 1999; Novozhilov, 2001; Blazek, 2007). LES and the DNS, because of their high computing power demand, have been only used in small-scale combustion studies (Novozhilov, 2001; Calhoun et al., 2005; Tugnoli et al., 2008; Brennan et al., 2009).

#### 4.2.1.3 Fire models

The fire models, which are used in both jet fire and pool fire modeling, provide the detailed description of the mechanisms and kinetics of the fire combustion, including combustion model, soot formation model, and radiation model.

##### Combustion model

The combustion chemistry is very complex. Depending on fuels, it could involve hundreds of elementary reactions; thus it requires sophisticated models and extensive computing resources for a complete description of the combustion phenomenon (Baukal et al., 2001; Novozhilov, 2001). In order to simulate fire, especially large-scale fires such as jet fire and pool fire, simplified combustion models are often used. The conserved scalar model with fast chemistry assumption and the eddy break-up model are two simple combustion models, which describe the combustion reaction between fuel and oxygen. The main benefit of these two models is simplicity in implementation, but the main drawback is the inability to predict the formation of species, especially the toxic combustion products such as carbon monoxide.

- *The conserved scalar model with fast chemistry* (Novozhilov, 2001): The conserved scalar model uses an assumption that the reaction time is significantly shorter than the mixing time and the combustion reaction is a complete, nonreversible reaction. This assumption is somewhat reasonable in certain types of fire modeling such as hydrogen jet fire. The conserved scalar is the quality that does not change during combustion reaction, and using the conserved scalar greatly simplifies the turbulent combustion rate calculation. For example:

$$F = nO \rightarrow (n+1)P \tag{4.7}$$

where $F$ is the fuel; $O$ is the oxidant; $P$ is the combustion products; and $n$ is the stoichiometric number for the combustion reaction.

The scalars in this case are mixture fractions $\xi$ and $\beta$:

$$\beta = Y_F - \frac{Y_O}{n} \tag{4.8}$$

$$\xi = \frac{\beta - \beta_F}{\beta_F - \beta_O} \tag{4.9}$$

where the indexes $F$ and $O$ refer to the fuel and oxidant streams, respectively.

- *The eddy break-up model* (Novozhilov et al., 1996; Novozhilov, 2001): The eddy break-up combustion model and its derivatives are very popular models in fire study, especially for large-scale fire modeling. This model uses two variant approaches assuming that the chemical reaction rate is controlled by the rate of molecular mixing of reactants, which in turn is related to the rate of dissipation of turbulent eddies. The reaction rate, *R*, is calculated by using the following equation:

$$R = -C_R \rho \frac{\grave{o}}{k} \min\{\overline{Y}_F, \overline{Y}_O/r\} \tag{4.10}$$

where $\min\{\overline{Y}_F, \overline{Y}_O/r\}$ is the lesser rate in fuel rich stream or in oxygen rich stream and $C_R$ is the eddy break-up constant.

The constant $C_R$ is calculated from:

$$C_R = 23.6\left(\frac{\nu \grave{o}}{k^2}\right)^{1/4} \tag{4.11}$$

To limit the rate at low temperatures, a controlling term is used. For example, the Arrhenius equation is used to calculate the reaction rate at low temperatures:

$$R = \overline{Y}_F \overline{Y}_O A_p \exp\left(\frac{-E}{R\overline{T}}\right) \tag{4.12}$$

where $A_p$ is the Arrhenius constant, *E* is the activated energy, *R* is the gas constant, and $\overline{T}$ is the average temperature.

For a complete description of fire, the detailed chemistry models are needed. They allow accurate predictions of toxic gases, such as CO, $NO_x$, and $SO_x$, generated in a fire. These toxic gases are also one of the major fire hazards, especially for fire in confined spaces. For example, models available for this type of detailed chemistry include the flamelet model which offers detailed chemistry, based on laminar flamelet considerations. The flamelet approach considers the diffusion flame as a statistical ensemble of thin laminar flames, called flamelets, and assume that the changes in the local mixing environment is much smaller than the mixing time scale (Novozhilov, 2001). The flamelet model is able to describe a complex combustion reaction and is therefore able to predict combustion species. Other more advanced models are the Monte Carlo joint PDF (Pope, 1985) and the conditional moment closure models (Klimenko and Bilger, 1999). These two sophisticated models, although more expensive in terms of computing resources, are becoming more popular in fire modeling to describe the detailed combustion chemistry (Yeoh and Yuen, 2009).

### Radiation model

Radiation exchange plays a very important role in fires and is the major hazard in the context of process safety (2005). The radiation model is also a very important part of a realistic CFD fire modeling because, in many situations, radiation is a dominant mode of heat transfer in the vicinity of a fire source, especially in fires of very large scale (Novozhilov, 2001). In large-scale accidental jet fire and pool fire modeling, the radiation model is used to calculate the surface emissivity power (SEP), which in turn is used in a Probit equation to calculate the limit exposure time and safe distances for personnel and equipment. The radiation model calculates the spectral intensity $I_\lambda$, defined as the radiant energy (per unit time and per unit wave length interval) passing per unit surface area normal to the direction $\Omega$, into a unit solid angle centered around $\Omega$ (Novozhilov, 2001). The spectral intensity is calculated as follows (Novozhilov, 2001):

$$\frac{1}{c}\frac{\partial I_\lambda}{\partial t} + \frac{\partial I_\lambda}{\partial s} = W_\lambda \tag{4.13}$$

where $s$ is the distance along a ray in the direction $\Omega$, $W_\lambda$ is the source term, and $c$ is the characteristic velocity for the radiation exchange.

The source term $W_\lambda$ is from emission adsorption and scattering of species presented in fires:

$$W_\lambda = W_{emission} - W_{absorption} + W_{scattering-in} - W_{scattering-out} \tag{4.14}$$

The radiative characteristics of gas mixtures in a fire involve many species and are very complicated; thus, certain simplifications need to be made so that the calculation is justifiable with reasonable computing resources. It is observed that in a fire, soot, $CO_2$, and $H_2O$ account for more than 95% of the radiation absorption and radiation. This allows the radiation model to be divided into two groups: (1) a gray gas model, where soot presence is significant in the fire; and (2) a nongray gas model, where soot presence is less significant. Hydrocarbon fire, either in a large-scale jet fire or pool fire, often results in gray flame because of a high level of soot formation. The following assumptions are often used in radiation calculation:

- *Soot-band emission assumption*: A simple model to calculate radiation in a fire, assuming that the fuel-side soot layer of the flame zone is the only source of radiation. This simplification assumption may lead to underestimation of the radiation (Novozhilov, 2001).
- *Gray gas assumption*: This model assumes that the adsorption and emission coefficients are not dependent on the radiation wavelength, simplifying the radiation calculations. Using this model, the absorption coefficients of the simulated fire can be split into two separated parts: gas and soot. Furthermore, scattering parts of the equation may also be neglected (Novozhilov, 2001).

Using these two assumptions, two methods are available to calculate the radiation of a gray gas fire:

- *Flux method* (Adiga et al., 1989, 1990; Novozhilov, 2001): The flux method used to calculate fire radiation is popular because it also uses the same numerical techniques used to solve the main hydrodynamic equations (the set of transport equations). In the flux method, radiation equations are solved to calculate the radiating fluxes for three forward and three backward directions in $x$, $y$, $z$ coordinates. Several improvements have been made in order for this model to take into account angle dependence in Cartesian coordinates.
- *Discrete transfer method* (Lockwood and Shah, 1981; Novozhilov, 2001): This model calculates the radiation fluxes by tracing "rays" of radiation through computational domains and solving appropriate equations for the radiation density along each ray. A separated radiation computational grid, besides the main computational grid for the transport equations, is needed; and the accuracy of the method can be adjusted by changing the input parameters to control the numbers and the directions of each ray. This model is widely used because it provides good accuracy with a reasonable computing resource.

Nongray fire is fire having a relatively small amount of soot, like jet fire of non-hydrocarbon fire such as hydrogen jet fire or much diluted hydrocarbon fire. The radiative properties of the non-gray fires can be calculated from methods such as the narrow-band statistical model or the exponential wide-band model (Yan and Holmstedt, 1999; Novozhilov, 2001). The exponential wide-band model takes into account that the infrared radiation of each species in a fire is concentrated into six wide-band regions associated with the principal vibrational transitions of the combustion species. The narrow-band statistical model calculates the radiation by dividing the spectrum into small intervals for each radiating species (i.e., combustion products such as $CO_2$ and $H_2O$). These intervals are assumed to follow an exponential probability distribution and have random locations. This method requires a large number of tabulated parameters and has not been used widely in large-scale jet fire or pool fire simulations. Simplified models based on this method have been developed.

#### Soot formation model

Soot formation is a very complex process in combustion. Soot is very important in fire simulation because of its effect on the radiation through the absorption and scattering process. Whereas in certain combustion processes, such as hydrogen combustion or methane combustion, soot formation may be ignored, in other hydrocarbon combustion processes such as ethylene and acetylene, soot presence is significant to the combustion products.

A number of soot formation models have been introduced (Novozhilov, 2001). For example, a soot formation model can be as simple as a constant factor related to a local value of mixture fraction for each fuel (Bilger, 1977; Orloff et al., 1987), or using a similar eddy break-up approach for combustion process, or more comprehensive using a set of equations for soot fraction and particulate number density (Novozhilov, 2001). Another aspect to be considered of a soot formation model is the coupling between soot production and radiative heat loss from flame because soot is often the primary radiating species (Brookes and Moss, 1999; Novozhilov, 2001).

#### Fuel source

The fuel source models are used to determine the amount of fuel going into the fire such as the release rate in jet fires or the evaporating rate in pool fires. These parameters are calculated using experimental data or other empirical model programs then incorporated as input into the CFD fire modeling setup.

### 4.2.2 APPLICATIONS OF CFD MODELING DATA INTO A RADIATIVE PROBIT FUNCTION

Results from CFD simulations can be used to calculate the surface emission power of the accidental fires, which then are used to determine the safe distances for personnel and equipment. The results can also be used in the Probit equation to calculate the maximum allowance time when exposed to radiation (Rew and Hulbert, 1996; Crowl and Louvar, 2002; Mannan and Lees, 2005; Pontiggia et al., 2010; Raj, 2005, 2007a–c; Pontiggia et al., 2010).

$$\mathrm{Pr} = a + b\ln(D) \tag{4.15}$$

where $D$ is the radiation dose, and $a$ and $b$ are specific parameters for each situation.

The radiation dose is calculated from:

$$D = \int_{0}^{\infty} SEP^{n} dt \tag{4.16}$$

where *SEP* is the surface emission power, *t* is time, and *n* is specific parameters for each situation.

The relationship between the death probability, *P*, and the corresponding Probit value, Pr, is:

$$P = 0.5\left[1 + erf\left(\frac{\text{Pr} - 5}{\sqrt{2}}\right)\right] \tag{4.17}$$

### 4.2.3 CFD MODELING CODES

There are several CFD fire modeling packages available, including commercial codes, government codes, and research codes (Vaagsaether et al., 2007; Skjold, 2010). They can be categorized into two groups: (1) fire modeling codes, which are designed only for modeling fire such as JASMINE (Cox and Kumar, 1987), KAMELON, SMARTFIRE, and SOFIE; and (2) general purpose CFD codes that can be used for fire modeling such as PHOENICS, CFX, and FLUENT. These CFD codes and their developers are listed in Table 4.4.

In a fire event, typically there are five stages involved, including ignition, growth, flashover, fully developed, and decay (Novozhilov, 2001). These five stages have been modeled and studied using CFD, particularly in fire safety engineering. In process safety, CFD simulation of fire is limited to open fields, allowing non-premixed fully developed fires, which results from loss of primary containment events, to release flammable materials. The following sections will discuss CFD simulations in each type of accidental fire (e.g., large-scale jet fire and pool fire).

**Table 4.4 Several Commercial Available CFD Packages for Fire Simulations**

| Name | Developer |
|---|---|
| CFX | ANSYS |
| FLUENT | ANSYS |
| PHOENICS | FRS |
| JASMINE | FRS |
| SMARTFIRE | UoG |
| KAMELON | SINTEF/NTH |
| SOFIE | Cranfield/FRS |
| ALOFT-FT | NIST |
| FDS | NIST |
| KOBRA-3D | Technik GmbH |
| SOLVENT | Massachusetts Highway Department; Innovative Research, Inc.; Parsons Brinckerhoff Quade & Douglas, Inc. |
| STAR-CD | Computational Dynamics Ltd |

FIGURE 4.2

A large-scale jet fire resulted from the release of flammable gas.

### 4.2.4 JET FIRE

Jet fires are formed when the high-speed releases of various fuel types from pressurized high momentum sources ignite to form flames with trajectories determined by the orientation of the leak. Figure 4.2 illustrates the simplest case of jet fire in which the source is a pressurized flammable gas release, such as compressed natural gas, hydrogen, or other flammable gases. This jet fire was set up as part of a training exercise for firefighters in realistic industrial settings at the Brayton Fire Training Field in College Station, Texas.

Gas phase jet fires are the most encountered jet fire incidents in industry. If material sources contain pressurized gas–liquid mixtures, the ignited flammable streams will be two-phase jet fire. Two-phase jet fires can also form due to the release of highly pressurized liquid if storing conditions allow the liquids to flash evaporation upon release into atmospheric conditions such as liquid natural gas or liquid petroleum gas (LPG).

The main hazards of jet fire are (Lowesmith et al., 2007):

1. Radiation on nearby equipment and personnel, and
2. Possible impingement of the jet flames on nearby objects.

If the impinged objects are storage vessels, this type of incident could lead to vessel failure and other larger incidents such as VCE or boiling liquid expanding vapor explosion (BLEVE) (Crowl and Louvar, 2005; Lowesmith et al., 2007). The radiation is expressed as radiative heat flux iso-surface, which is used as an input parameter in a Probit equation to calculate the maximum allowance exposure time for personnel and equipment. For example, the maximum allowance time for personnel when exposed to radiative heat flux of 4.7 kW/m$^2$ is 3 min (Houf et al., 2011). For liquefied natural gas (LNG) facilities (i.e., production, storage, and handling sites), the National Fire and Protection Association (NFPA) specifies 5 kW/m$^2$ as a safe level of exposure at a property line (Raj, 2005).

CFD simulations have been used for evaluating the effect of barriers on mitigating the hazards of accidental hydrogen jet release and jet fire at the Sandia National Laboratory in Albuquerque, NM

(Houf et al., 2009, 2010, 2011). The purpose of these simulation studies is to assess the effectiveness of barriers in mitigating the consequences including deflecting the jet flame and reducing the extent of the flammable cloud, the magnitude of the radiative heat flux, and the overpressure due to explosion. Several different arrangements of those physical barriers were investigated. The CFD simulations were conducted using Sandia self-developed FUEGO code; the turbulence of the hydrogen flow was modeled using the Farve-averaged RANS RNG k–ε turbulence model; the combustion of hydrogen was modeled as a nonreversible one reaction between hydrogen and oxygen using the Eddy Dissipation Concept model; the radiative heat loss of the hydrogen flames was computed using Sandia-developed gray discrete ordinates SYRINX code with only the spectral band of water vapor radiation absorption in the Lecker model included (Houf et al., 2009, 2010, 2011). The hydrogen jet source was modeled as a subsonic jet inflow boundary condition using the Mach disk model (Houf et al., 2009, 2010, 2011). The simulation results indicate that having appropriate barriers surrounding hydrogen sources could effectively reduce the total emission surface, which is in agreement with experimental data (see Figures 4.3 and 4.4). Figure 4.3 shows graphical rendering of the radiative heat flux

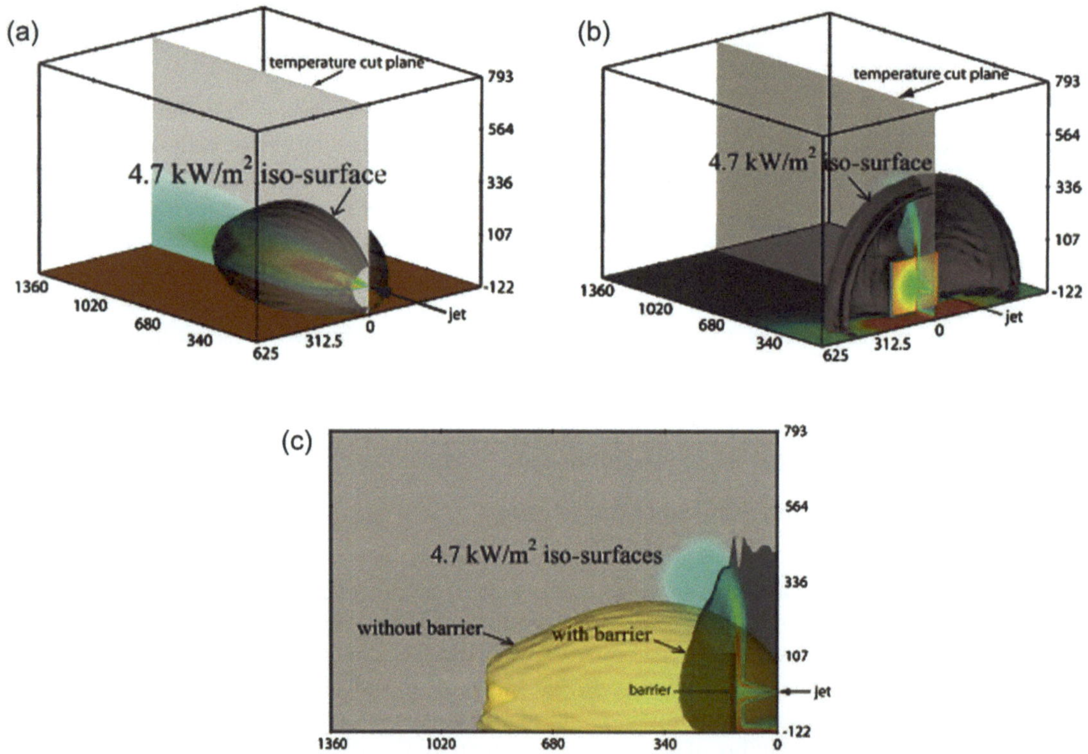


**FIGURE 4.3**

Radiative heat flux iso-surface from the CFD simulation results: (a) jet flame with no barrier; (b) jet flame directed towards a barrier; (c) side view of iso-surfaces shown in (a) and (b).

*Reprinted with permission from Houf et al. (2011).*

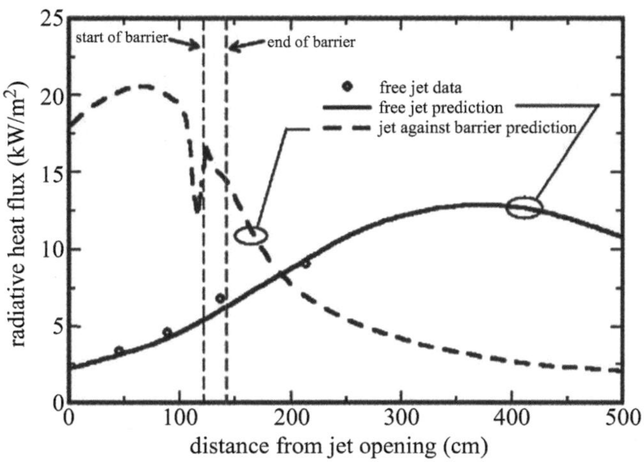


**FIGURE 4.4**

The comparison between radiative heat flux from experimental (circles) and CFD simulations (line) from a hydrogen jet fire with barriers (dash line) and without barrier (solid line).

*Reprinted with permission from Houf et al. (2010).*

iso-surface and Figure 4.5 shows jet deflection from barriers from the CFD simulation results. The study was performed at Sandia National Laboratory to evaluate the effectiveness of a mitigating barrier in accidental hydrogen jet release and jet fire. The iso-surfaces of 4.7 $kW/m^2$ are shown in Figure 4.3 including (a) without physical barrier; (b) (top view) with one vertical barrier located 3 feet in front of the hydrogen jet flame; and (c) (side view) with one vertical barrier located 3 feet in front of the hydrogen jet flame (gray color), and the iso-surface of the simulation without barrier (light yellow color). The simulation result clearly shows the reduction of total affected areas when a physical barrier was used.

Figure 4.5 shows the effects of mitigation barriers including (a) vertical barrier with the hydrogen jet flame directed at the center of the wall; (b) vertical barrier with the hydrogen jet flame directed at the top of the barrier; (c) 60° tilted barrier with the hydrogen jet flame directed at the center of the barrier; and (d) three vertical barriers for a complete enclosure of the hydrogen jet flame.

Because of the possible impingement, jet fire flame length is one of the important characteristics that could help to determine appropriate separation distances between equipment. Cumber and Spearpoint proposed a CFD methodology that mimics how a camera or human eye senses a fire to calculate the propane jet fire length (Cumber and Spearpoint, 2006). Generally, in CFD simulation, the flame length of a jet fire must be inferred in certain way (i.e., temperature profile of the flame), which may not be accurate in the sense of reality. In this study, the authors calculate the flame length based on the senses of the receiver on the radiation intensity field given off by the incandescent soot particles and participating species. The two-equation k–ε turbulence model was used with modifications to account for the round jet/plane jet anomaly and buoyancy-induced turbulence; a probability density function (PDF) with infinitely fast reaction assumption was used to model propane combustion; and soot formation was modeled using Lindstedt's two-equation soot model (Cumber and Spearpoint, 2006).

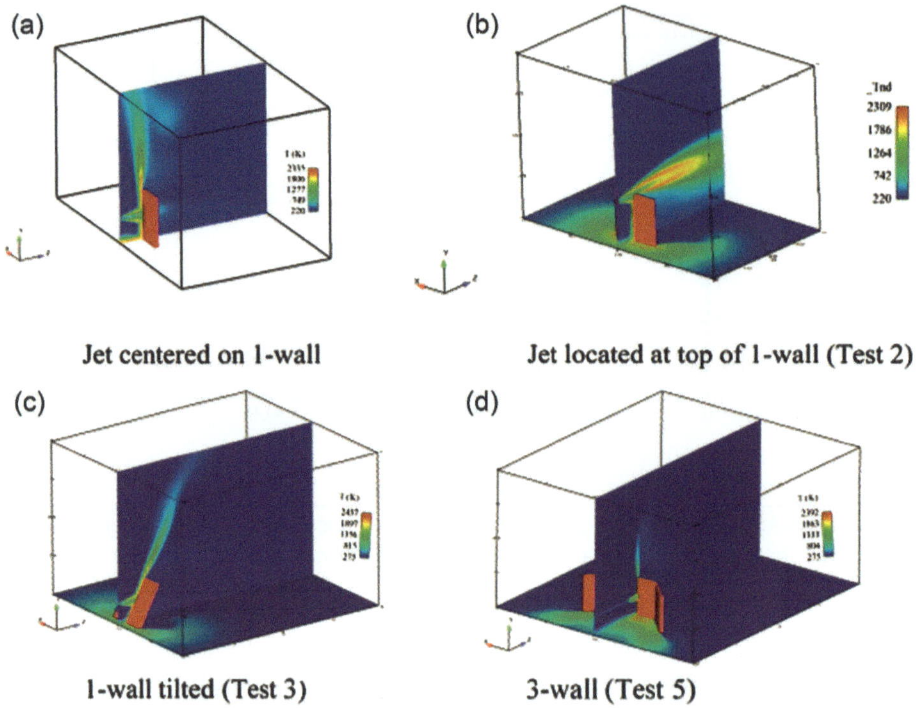


**FIGURE 4.5**

Graphical interpretation of jet deflection from barriers from the CFD simulation results.

*Reprinted with permission from Houf et al. (2010).*

Radiation heat loss was calculated via the specific enthalpy perturbation transport equation (Cumber and Spearpoint, 2006). The model is calibrated using a selected set of jet fire experiments and then validated against a wider range of data.

Kim et al. (2010) used CFD to evaluate the heat load characteristics of steel and concrete tubular members under jet fire in terms of temperature and heat flux. The author used ANSYS, CFX, and KFX codes to conduct their simulations (see Figures 4.6 and 4.7). The jet fire was directed into the bottom of the horizontal bars. They found that CFD simulations were a useful tool to predict jet fire loads on nearby objects; the predicted temperature distribution was similar to those observed in the experiment, although they also noted that the CFD simulation parameters needed to be carefully selected to avoid anomalous results.

Yang et al. (2011) conducted CFD simulations to evaluate the dangerous areas by calculating the shape and size of jet fires on gas pipelines. In this study, the authors used the k–ε turbulence model to describe the jet flow; the turbulent non-premixed combustion process was modeled using the PDF/laminar flamelet model. Radiation heat transfer is described using an adaptive version of the discrete transfer method. The model is compared with experiments on a horizontal jet fire in a wind tunnel in the literature with success. The influence of wind and jet velocity on the fire shape has been

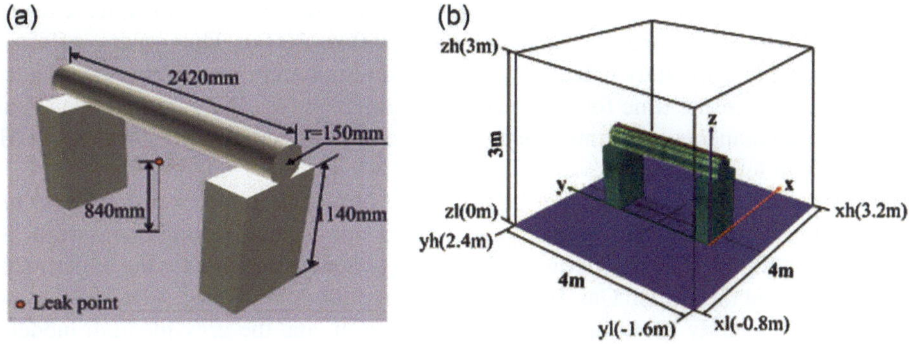


**FIGURE 4.6**

CFD simulation setup for the heat loads of jet fire on steel and concrete tubular bars: (a) dimension of individual objects; (b) the simulation domain.

*Reprinted with permission from Kim et al. (2010).*

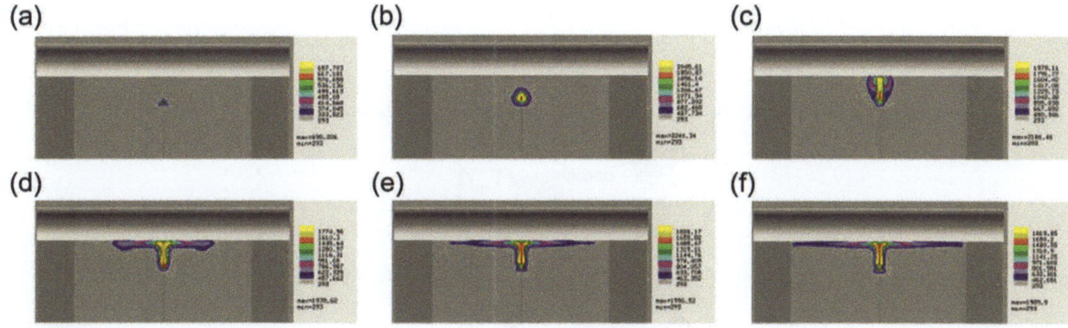


**FIGURE 4.7**

Evolution of the jet impingement on a pipe structure: (a) 0.02 s; (b) 0.04 s; (c) 0.10 s; (d) 0.50 s; (e) 1.00 s; (f) 5.00 s.

*Reprinted with permission from Kim et al. (2010).*

investigated, and a correlation based on numerical results for predicting the stoichiometric flame length is proposed.

Aloqaily and Chakrabarty (2010) through implementation of CFD have showed that such approaches are extremely useful in estimating thermal hazards resulting from jet flames to personnel working in process facilities. As illustrated by the same group in a follow-up article (Chakrabarty et al., 2011), upon narrowing down high-risk jet fire scenarios from QRA studies, CFD-based approaches can assist facilities comply with thermal hazard regulation in an more accurate and efficient manner.

## 4.2.5 POOL FIRE

CFD simulation has been used to evaluate many hazards associated with large-scale pool fire events in process safety. Most important among them is the radiation hazard due to the very large size of accidental pool fires often encountered due to loss of primary containment events. Other types of hazards associated with pool fires, such as toxic gases in confined environments, have been covered extensively in fire safety engineering literature (Novozhilov, 2001). Thus, many efforts in CFD modeling are directed at accurately calculating the SEP, an important parameter used to calculate the maximum allowance exposure time for personnel and equipment (see Figure 4.8). CFD simulation of pool fires allows instantaneous and time-averaged temperature fields and iso-surfaces for SEP to be predicted with a high level of accuracy.

CFD modeling was used to investigate the instability of an unconfined kerosene pool fire with a diameter of 20 m in a stagnant, stable atmosphere and in atmosphere having cross wind (Sinai and Owens, 1995; Sinai, 2000). These modeling studies were conducted parallel to the experimental work by Shell Research Ltd at the British Gas Test at Site Spadeadam. A buoyancy-modified k–ε model was used for turbulence, the eddy break-up model for combustion, and the gray-medium model for radiation. The pool was modeled as a stable pool without any extraneous turbulence sources at the pool or within the fire. The modeling results show realistic behavior where large toroidal vortices were generated and rose within the modeled fire, similar to those observed experimentally.

Lin et al. (2010) conducted a CFD simulation to investigate the characteristics of radiative heat transfer in liquid pool fire to predict the thermal radiation and flame characteristics numerically in small to medium-sized heptane pool fires (14–38 cm). This work used an LES model for turbulence, an

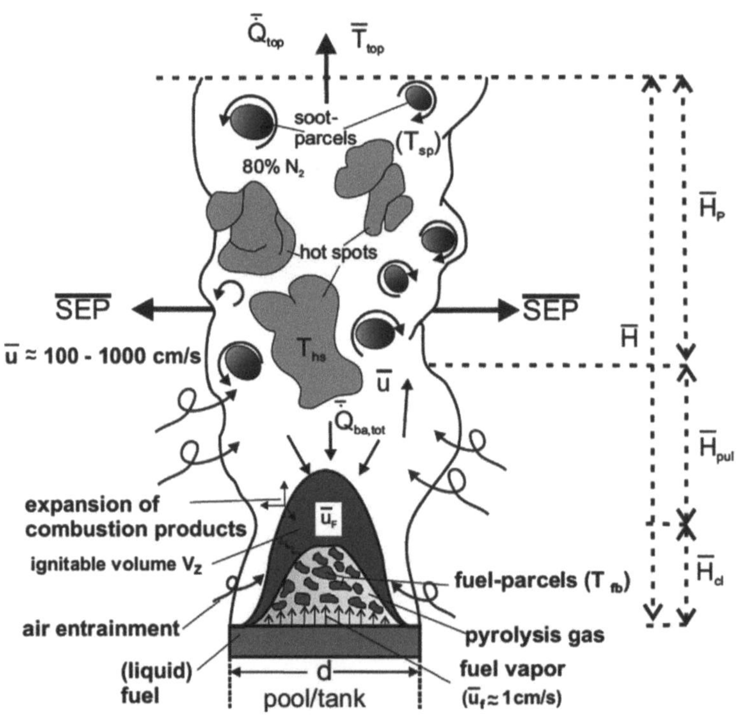


**FIGURE 4.8**

A pool fire diagram and its associated physical processes.

*Reproduced by permission with Schalike et al. (2011).*

eddy dissipation concept with extremely fast chemistry for combustion, and a gray gas narrow-band model for radiation. The modeling results are in agreement with experimental data.

Schalike et al. (2011) conducted CFD simulations to predict the thermal radiation hazards of large LNG pool fires having diameters of 1, 6.1, and 30 m whose radiations were also measured in experimental study (see Figure 4.9). Flow turbulence was modeled by LES, combustion process was modeled using the PDF approach with laminar flamelet for 21 species and elementary reactions, radiation was modeled using discrete ordinates, and soot formation was modeled using the Moos Brookes model. It can be seen from Figure 4.8 that the model pool fire has three rather distinctive surface emission zones (represented as different colors in the figure). The simulation results show that for LNG pool fires having a diameter of 1, 6.1, and 30 m, the averaged flame temperatures of T = 1320, 1298, and 1281 K were observed, and the SEPs of 55, 130, and 230 kW/m$^2$ were predicted, respectively. The predicted SEPs as well as their increases with the increase of pool fire diameter agree well with the experimental data. This result indicates that there is a maximum SEP for large-scale pools; that value may be observed for pool fires with diameters greater than 35 m, which have not been simulated. They also found that the higher mass burning rate and the higher flame temperatures lead to the higher SEP values for larger LNG pool fires in comparison with other hydrocarbon pool fires. This modeling result shows that accurate surface emission power can be calculated using a CFD model.

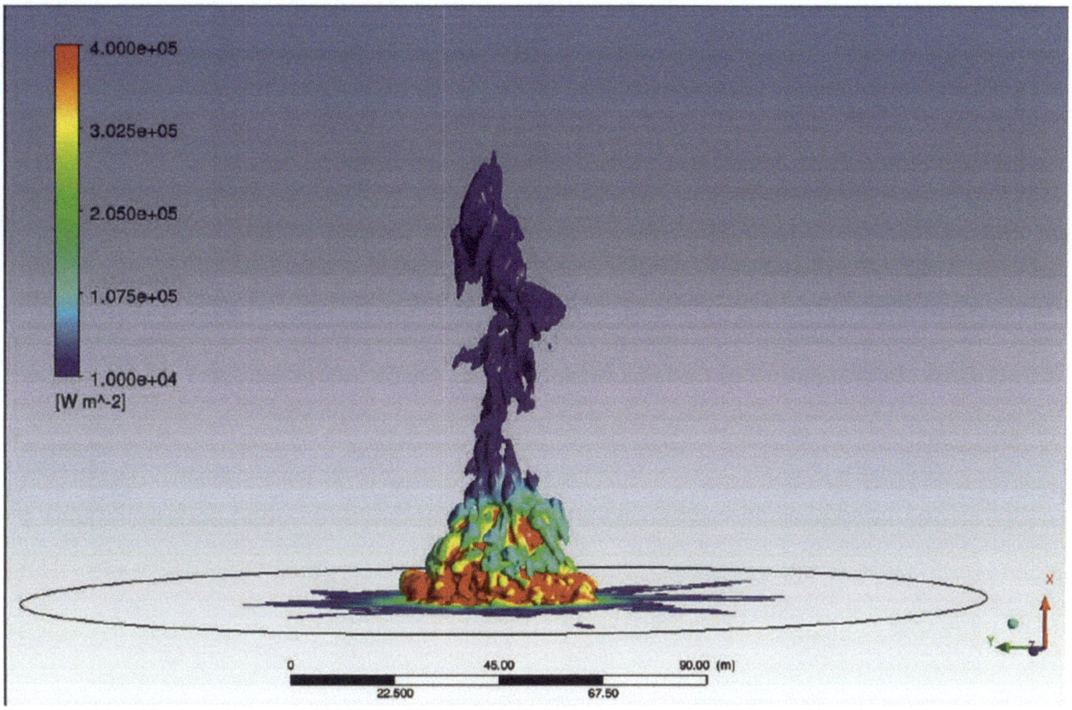


**FIGURE 4.9**

Graphic rendering of the surface emission power using a CFD simulation as a result of a large-scale LNG pool fire.

*Reproduced by permission with Schalike et al. (2011).*

## 4.3 DISPERSION OF FLAMMABLE GASES

When using CFD simulation for an accidental flammable material release, the gas cloud distribution can be accurately calculated; specifically, the upper flammability limit (UFL) and the lower flammability limit (LFL) fronts are predicted (see Figure 4.10). The UFL and LFL are the important parameters needed to calculate the exclusion distance, which is determined by one half of the LFL, to protect the public from being reached by flammable vapor and other possible consequences such as flash fire and VCE during an accidental release. This information is also valuable regarding detection (i.e., distribution and arrangement of gas detectors); mitigation (i.e., the use of water curtains); and in emergency response (i.e., evacuation). CFD simulations of flammable gas dispersion also shed light on the past incidents as lessons learned for the future.

It is well known that the interaction between obstacles and the fluid flow could result in certain trap zones where the concentration of flammable gas is higher than the surrounding areas. Figure 4.9 shows the effect of building on the delaying of the dispersion, which leads to the formation of a local cloud in an underground garage (the building is in the middle of the figure), where the concentration of flammable gas is larger than the LFL/2 limit, while the LFL front is near. Incident report shows that a partially confined explosion was observed at this site.

The accuracy of prediction from CFD models of flammable gas dispersion is critical; thus, CFD dispersion models, as well as other simpler models are needed to satisfy certain criteria listed in a report published by the UK Health and Safety Laboratory as follows (Ivings et al., 2007):

- Model accepts a credible source term
- Model accounts for the effects of wind speed
- Model accounts for the effects of surface roughness on dispersion
- Atmospheric stability

**FIGURE 4.10**

Graphical rendering of the distribution of UFL and LFL of flammable gas from a CFD simulation study.

*Reproduced by permission with Pontiggia et al. (2011).*

- Passive dispersion
- Gravity-driven spreading
- Buoyancy on dilution
- Finally, the numerical methods are based on accepted/published good practice

Those criteria need to be carefully examined when conducting CFD simulations on dispersion of flammable materials in order to achieve realistic prediction results.

## 4.3.1 DISPERSION PARAMETERS

Generally, CFD simulation setup on a dispersion of flammable material study requires multiple input parameters including suitable models for turbulence, representation of complex geometries, and other physical models (Qiao and Zhang, 2010). Figure 4.11 lists the typical parameters needed to be considered for an LNG vapor dispersion study by Cormier et al. (2009); these parameters are also applicable for CFD simulation setup for other types of flammable materials such as natural gas or hydrogen. Note that these parameters follow the overall CFD simulation framework presented in Figure 4.1.

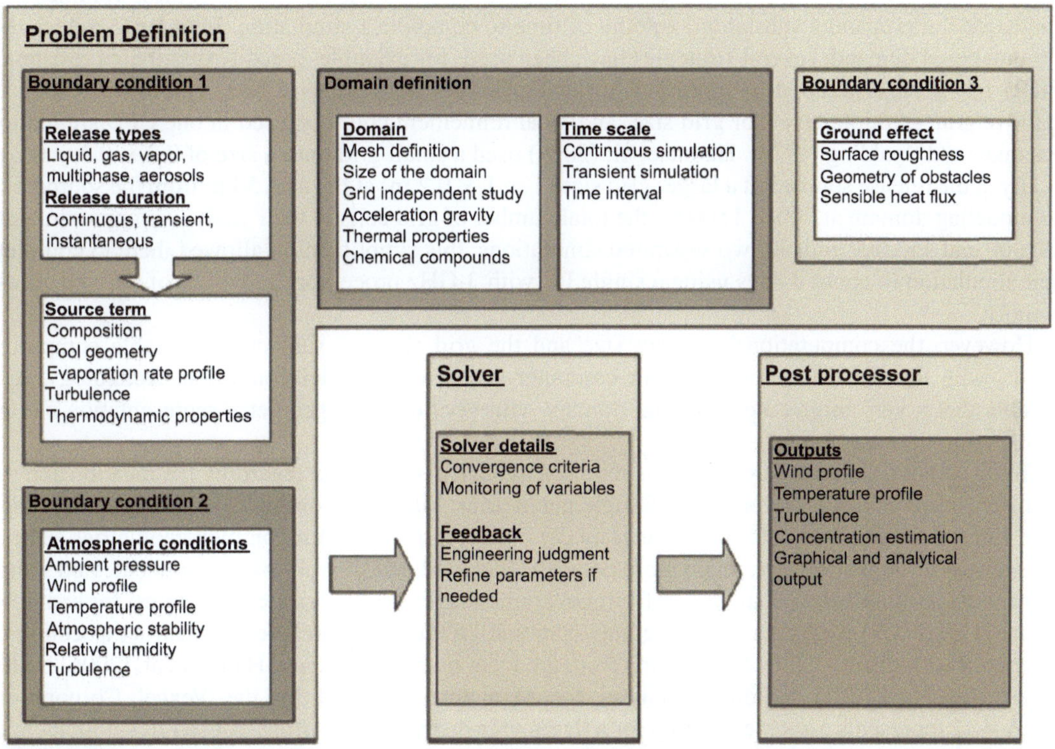


**FIGURE 4.11**

Representative problem setup for CFD simulation study on the dispersion of LNG.

*Reproduced by permission with Cormier et al. (2009).*

### 4.3.2 GEOMETRY REPRESENTATION

Generally, CFD simulation of dispersion gas is conducted for scenarios in which the interaction of the flow with obstacles is very important, such as in scenarios involving flammable gas releases in complex geometries (i.e., urban areas or complex landscape). For example, CFD simulations have been conducted to evaluate the consequences of LNG releases in urban areas (Pontiggia et al., 2011) or tracer gas in the Manhattan area (Hanna et al., 2006).

Geometry is considered first among the required setup because it is the primary reason why CFD needs to be used and is the most challenging part of the simulation setup. The challenges in geometry setup for a CFD dispersion study are:

1. Computational domain size and the grid size; and
2. Availability of data to construct such geometry.

Realistic geometries such as sections of a city represented in a dispersion study require a simulation domain ranging from several 100 feet to several miles. For example, Pontiggia et al. (2011) constructed a 350 × 500 × 45 m computational domain to study the dispersion of LNG in an urban setting. Further, obstacles disperse widely in sizes ranging from meter scale to smaller, requiring very fine grid setup to correctly present them; this leads to a significant number of control volumes in a simulation domain and a substantial amount of time to complete a simulation. In order to reduce this computational demand, several strategies have been used; for example, porosity distributed resistance (PDR) has been developed to represent obstacles smaller than the grid size. Approaches such as adaptive grid, multiple levels of grid size, and local refinement could be used in one CFD simulation (Patankar and Spalding, 1974). Hanna et al. (2009) used a small grid with a size of 0.2 × 0.2 × 0.2 m locally at the release source but a larger grid size (1 × 1 × 0.2 m) at the area 50 m from the source. For the modeling domain at 500 × 1500 m, the total number of grid cells in their study was approximately 900,000 and 450,000 cells in two separated simulations; this simplification allowed them to complete their simulation in about 3 days using a single PC with 3-GHz processors and 4 GB of random access memory.

However, the computational domain size and the grid size are subject to available computing power; with the current development in computer speed and parallel computing technology, it is possible that a very large computational domain with very detailed geometry representation can be simulated in the near future.

To construct a simulated domain, it is required to have detailed data regarding the location and the geometry of each and every obstacle. Simple geometries with a few obstacles can be generated and built with ease using available commercial model builders; however, for complex geometries such as city sections or industrial sites, much more complete, detailed data as well as more sophisticated setups are needed. Available data could come from government agencies or private companies; for example, Hanna et al. (2009) used data from Argonne National Laboratory's archive to construct a section of Chicago; Pontiggia et al. (2011) used data from the local municipality; and Hanna et al. (2006) built a three-dimensional (3D) building database for Manhattan licensed by the Vexcel Corporation. Figure 4.12 represents a geometry setup for a flammable dispersion resulting from a rail car incident in an urban environment in Italy, demonstrating the typical scale and complexity in geometry setup. The size of this computational domain is 350 × 500 × 45 m, representing a part of the city of Viareggio, Italy. This type of study is often conducted based on real or hypothetical large-scale accidental events

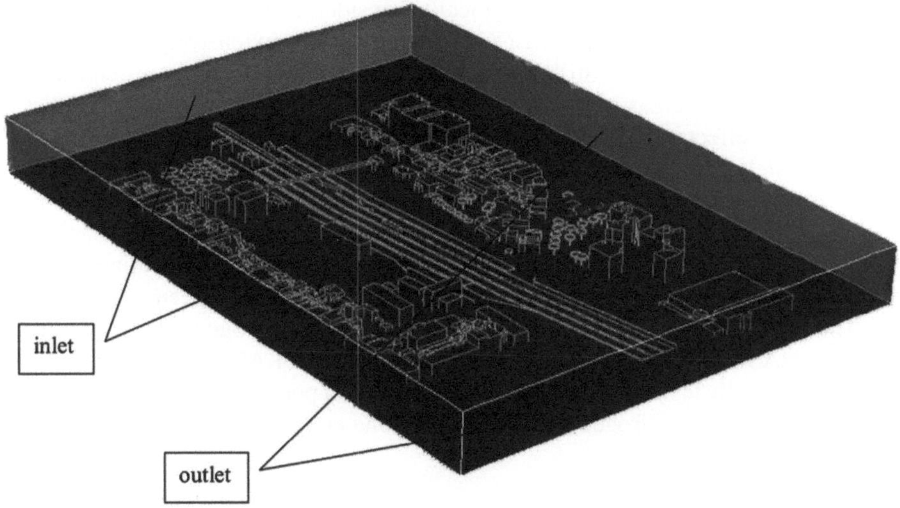


**FIGURE 4.12**

A geometry setup for CFD simulation study of the dispersion from an LNG release.

*Reproduced by permission with Pontiggia et al. (2011).*

to collect information for future emergency preparation and planning. It is also used to validate the performance and reliability of CFD simulation codes in predicting such large-scale events, but this is limited. The constraint in geometry setup further poses a limitation in using CFD simulation on real-time dispersion monitoring, because significant time is required for a comprehensive setup of high-quality, up-to-date geometry and weather data (city landscapes and weather conditions are constantly changing).

For industrial sites such as refineries or chemical plants, data may readily be available through CAD systems; many efforts have been made to develop interface programs to transfer the data from CAD systems into a CFD grid. Many of such interfaces are now available in commercial CFD codes such as FLACS and EXSIM.

In small-scale simulation studies, efforts have been made for the investigation of the effects of certain types of obstacles and environmental factors in the fluid flow. For example, the effect of vehicle wake or the effects of certain type of buildings on the fluid flow have been investigated; the results from these studies could be used for further refinement in large-scale CFD simulations (Hanna et al., 2002; Richards, 2002). There has been a study focusing on the incorporation of CFD into meso-scale meteorological model domains in order to obtain better boundary conditions able to link the flows outside and inside of a building using information on the total building exchange rate (Tewari et al., 2010).

## 4.3.3 TYPE OF SIMULATED FLAMMABLE MATERIALS

Generally, CFD can simulate any type of flammable material, either one-phase fluid, gas or liquid, or two-phase fluid. In the context of process safety, dense gas dispersion (e.g., LNG) and other gas dispersion (e.g., methane or hydrogen) are typically considered for flammable material dispersion because they are widely used in industry. LNG has been studied extensively using CFD simulation because the infrastructure for LNG distribution systems is often found near or inside highly populated areas. A release inside these areas will result in complex flow patterns and hard to predict gas plume distribution due to the effects of obstacles on the flow (Pontiggia et al., 2010, 2011).

Similar CFD models have also been used to simulate other types of flammable gases, for example, Middha et al. (2009) conducted studies on the validation of the FLACS CFD tool on the dispersion of hydrogen. In this study, FLACS was used to simulate the dispersion of $H_2$ from subsonic and sonic jet release; the effect of jet impingement on obstacles was also modeled. Experimental data from a parallel study indicated that the modeling predictions were in good agreement with experimental observation. The FLACS was also used to study the dispersion of hydrogen from pool liquid hydrogen using data from a NASA experiment (Middha et al., 2011). The FLACS new pool model, which was developed and successfully tested for simulation of LNG liquid pool, was used to model the formation, spread, and evaporation of a hydrogen liquid pool (Middha et al., 2009, 2011).

### 4.3.4 SOURCE MODELS

There are multiple different sources of flammable gases depending on the scenarios. The scenarios most often encountered in process safety incidents are: (1) single-phase releases such as a flammable gas from a jet release resulting from the rupture of high-pressure containments and (2) two-phase (both gas and liquid phases) releases such as a release resulting from the rupture of an LNG tank. The source term may include the geometry (hole), area (size), gas phase velocity profile, turbulence (jet, constant evaporating), and location (on water, concrete, soil, etc.). For simplification during a jet release simulation, a steady-state jet release rate is often assumed or data for jet release from other models such as the FLASH model in the FLACS code are used (Calay and Holdo, 2008; Middha et al., 2009, 2011). For two-phase releases, flammable gases disperse starting from the evaporation process of jet sources or liquid pools formed on ground surfaces at the incident sites. The ground surfaces on which releases occur also have certain effects on the evaporation rate because of the heat transfer; for example, the evaporation rate of flammable gases from an LNG pool on a concrete surface is different from that of an LNG pool on soil or on water (Cormier et al., 2009; Qi et al., 2010).

Hydrogen dispersion from a jet release source at free subsonic, sonic, and impingement jets have been reported by Middha et al. (2009). In this study, steady-state jet release sources were used (see Figure 4.13).

A new FLACS pool model for a cryogenic pool has been reported to have good agreement with large-scale test using LNG and liquid hydrogen (see Figure 4.14) (Middha et al., 2011). Figure 4.14 shows the volume fraction contours of hydrogen evaporated from the liquid pool at 20 s after the spill started. The top figure shows the results for weather condition with stable (F) stability class (night with moderate clouds and light/moderate wind); and the bottom figure shows the results for weather condition with neutral (D) stability class (little sun and high wind or overcast/windy night).

In several studies, the evaporating rate of the LNG pool is estimated from the concentration profile experimentally measured onsite (Cormier et al., 2009; Qi et al., 2010). Pontiggia et al. (2011) used the jet release source constructed using integral model PHAST—which in turn used empirical model to determine the jet release (Pontiggia et al., 2011).

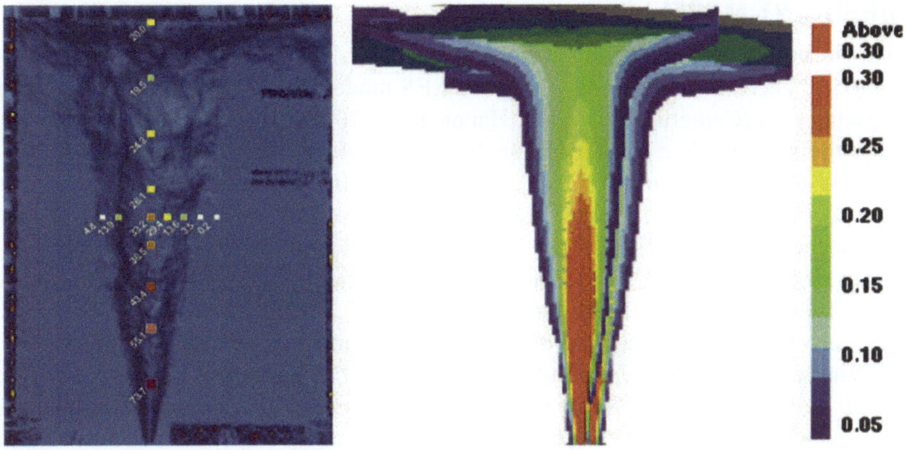


**FIGURE 4.13**

The experimental hydrogen jet release (left) and the graphic rendering of the same release using CFD simulation.

*Reproduced by permission with Middha et al. (2009).*

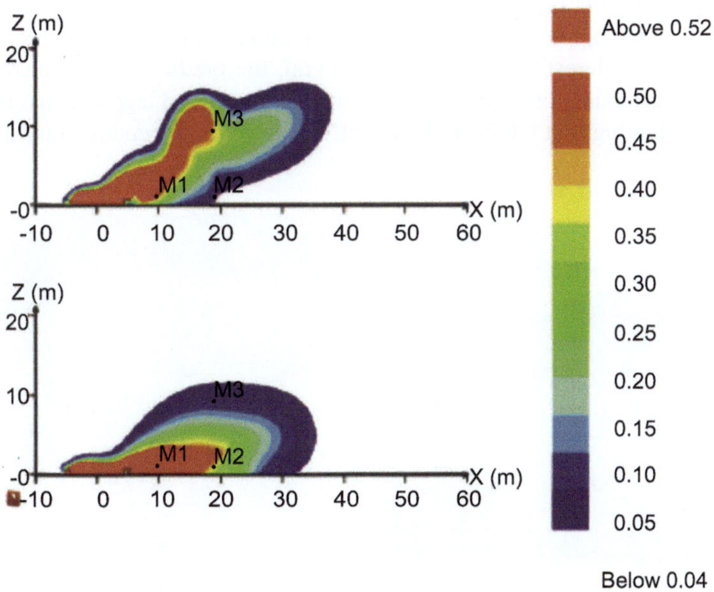


**FIGURE 4.14**

Results from CFD simulation of the dispersion of hydrogen from a pool of liquid hydrogen.

*Reproduced by permission with Middha et al. (2011).*

### 4.3.5 TURBULENT MODEL

Most CFD models use Reynolds-averaged Navier–Stokes (RANS) equations with a two-equation k–ε turbulence model. There are CFD codes using the LES model available, but they come with a much higher computing cost (Camelli et al., 2004; Hanna et al., 2006).

Generally, the standard k–ε turbulence model gives good results for neutral buoyancy gas dispersions. But for dense gas dispersion, such as LNG, certain modifications are needed to take into account the effect of buoyancy and stratification of the dense gas layers. Cormier et al. (2009) used three k–ε turbulence models, including the RNG k–ε, SSG Reynolds stress turbulence, and standard k–ε turbulence models in a CFD simulation study of a medium-scale LNG spill at the Brayton Fire Training Field (see Figure 4.15). In this medium-scale test, 10,000 gallons of LNG were spilled in several consecutive dispersion and fire tests. The CFD simulation was performed using ANSYS FLUENT for the same LNG test. The results show a reasonable agreement in the vapor concentration contours for all three RNG k–ε, SSG Reynolds stress turbulence, and standard k–ε turbulence models. The results using RNG k–ε and SSG Reynolds stress turbulence models show a wider distribution than those from the simulation using standard k–ε turbulence models, but the difference is considered insignificant.

Sklavounos and Rigas (2004) conducted a study to evaluate turbulence models used in a dense gas dispersion study. The four models investigated in their study were the standard k–ε, standard k–ω, SST, and SSG Reynolds stress turbulence models. They constructed a relatively simple geometry setup so that a parallel experiment can be conducted to validate the CFD predictions. The simulation results show that the k–ε and SST models showed improved robustness during solver processing. The SSG model entailed increased CPU time without significant enhancement of the accuracy of results. The SSG, k–ε, and SST models appeared to overestimate maximal concentrations recorded in the trials, whereas the k–ω model underestimated them. In general, the predictions obtained from CFD simulation study on the dispersion of flammable materials show good agreement compared with the experimental data, indicating that CFD can be effectively used in dispersion consequence assessment procedures.

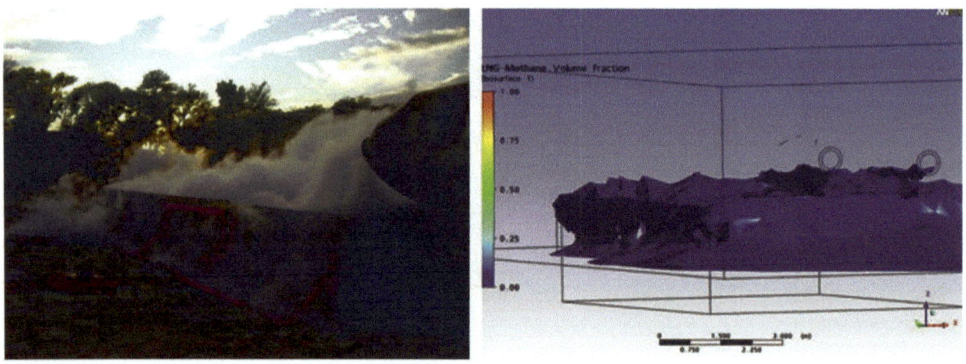

**FIGURE 4.15**

A medium scale LNG spill test by the Mary Kay O'Connor Process Safety Center at the Brayton Fire Training Field, College Station, Texas (left), and the graphic representation from the results of a CFD simulation (right).

*Reproduced by permission with Cormier et al. (2009).*

More advanced models, such as LES, have been used in simulating the dispersion of flammable gases but are only applicable in small-scale and with relatively simple geometries. For example, Koutsourakis et al. (2012) applied LES to investigate the dispersion of hydrogen within a non-ventilated garage at ambient pressure using the ADREA-HF CFD code. This type of research indicates the possibility of applying advanced CFD models for realistic, large-scale dispersion studies is achievable in the near future, with the current rate of computing power development.

### 4.3.6 PUBLISHED VALIDATION DISPERSION STUDIES

A large number of validation studies on available CFD codes have been conducted based on available experimental data in literature. Data available are large-scale Maplin Sands, the Burro and Coyote experiments using LNG (Sklavounos and Rigas, 2006; Middha et al., 2011), Brayton Fire Training Field medium-scale release using LNG (Cormier et al., 2009; Qi et al., 2010), Kit Fox large-scale release using dense $CO_2$, large-scale Mock Urban Setting Test (MUST experiment) using neutral buoyant gas tracer, Prairie Grass using neutral buoyant gas tracer and Experimental Measurement Uncertainty (EMU) using gas tracer (Hanna et al., 2004). Small-scale testing in simple obstacle settings (Hanna et al., 2002; Baik et al., 2003) with different geometry setups has also been used in CFD code validation. Various CFD simulation studies were conducted on the hydrogen dispersion in an urban environment (Venetsanos et al., 2003) and on LNG dispersion (Luketa-Hanlin, 2006; Qi et al., 2010, 2012; Olewski et al., 2011). CFD simulations were also conducted to further understand and assess the evolution of the dispersion of a flammable cloud in past incidents such as the release of LPG from a derailed freight train having 14 tank wagons, each carrying approximately 46.7 tons of LPG in Viareggio, Italy, in 2009 (Brambilla and Manca, 2010; Pontiggia et al., 2011), the release of gasoline due to a storage tank overfilling at Buncefield Oil Storage Depot near Hemel Hempstead, UK, in 2005 (Markatos et al., 2009; Argyropoulos et al., 2010; Gant and Atkinson, 2011) (see Figure 4.16), and the release of cyclohexane due to connecting pipe failure at the Nypro plant in Flixborough, UK, in 1974 (Hoiset et al., 2000; Venart, 2004a,b, 2007a,b) (see Figure 4.17). In the Buncefield incident, hundreds of tons of gasoline were released due to an overfilling of one of the large storage tanks at the Buncefield Oil Storage Depot near Hemel Hempstead, UK, in 2005. The gasoline overflowed at the top of the tank and fell 14 m to the ground. The resulting vapor cloud was estimated to be 2 m deep and covered an area 500 × 400 m. This CFD simulation result shows the spread of the cloud starting from the wall of Hertfordshire Oil Storage Limited (HOSL) Bund A. The number in each contour indicates the minutes from the time the mist appeared over the HOSL Bund A. ANSYS-CFX code was used in this simulation study.

In the Flixborough incident, approximately 30 tons of cyclohexane was released due to a connecting pipe failure. Figure 4.16 presents the flammable cloud at the lower flammability limit boundary at 20.4 s from the southwest direction. EXSIM CFD (Lea, 2002) code was used in this simulation study.

The Mary Kay O'Connor Process Safety Center at Texas A&M University has conducted a series of studies on the medium-scale dispersion of LNG in the Brayton Fire Training Field in College Station, Texas (Cormier et al., 2009; Suardin et al., 2009; Qi et al., 2010, 2012). In this project, approximately 10,000 gallons of LNG were used in several release tests to assess multiple factors involved in the dispersion as well as to evaluate mitigation methods for LNG dispersion. Parallel to these medium-scale tests, CFD simulations were carried out using FLUENT and CFX to evaluate the

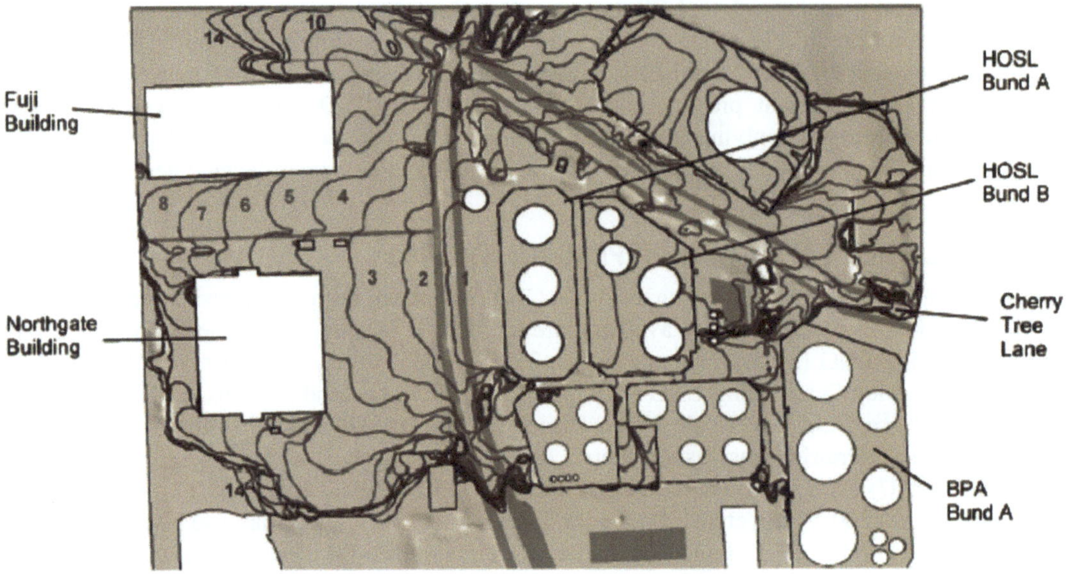


**FIGURE 4.16**

Graphic representation of the CFD prediction on the evolution of a flammable vapor cloud in the Buncefield incident.

*Reproduced by permission with Gant and Atkinson (2011).*

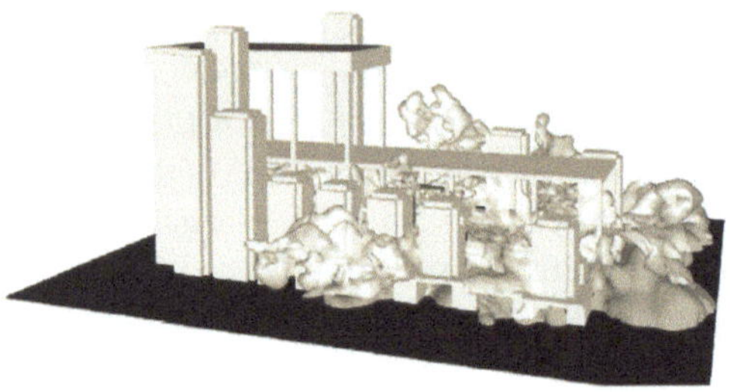

**FIGURE 4.17**

Graphical rendering from CFD simulation results of the cyclohexane cloud formed in the Flixborough incident.

*Reproduced by permission with Venart (2007b).*

CFD codes and affecting parameters (Cormier et al., 2009; Qi et al., 2010). Multiple modeling parameters involving the dispersion were investigated, including wind velocity, sensible heat flux, obstacles effects, source-term model, and turbulence models (Cormier et al., 2009). The CFD predictions were validated against experimental data of medium-scale testing of LNG dispersion at the Brayton Fire Training Field.

Subsequent studies on LNG dispersion and mitigation methods were continued at Texas A&M University at Qatar. Olewski et al. (2011) used Gexcon's FLACS CFD code to evaluate the

**FIGURE 4.18**

Graphic representation of CFD simulation on the effectiveness of a water curtain to mitigate LNG release.

*Reproduced by permission with Olewski et al. (2011).*

effectiveness of a water curtain as the primary mitigation method for LNG dispersion. Water curtains have been recognized as a promising mitigation method to reduce hazards in process safety consequences, especially in terms of flammable material dispersion and radiation hazards. The 2-dimensional (2D) water curtain model in this work was simulated as a porous disk, while the LNG release source model was modeled as an air jet model. The general effects of a water curtain such as "hold back" and "lift up" were observed in CFD simulations, similar to those found in experimental work and presented in Figure 4.18. Although it is clear that further development is needed, these results indicate that CFD could be an effective tool in evaluating water curtains as protective barriers.

## 4.3.7 CFD CODES AVAILABLE

There are many available CFD codes for flammable material dispersion study, either commercially or available for free from government agencies (Hanna et al., 2006; Ivings et al., 2007). Table 4.5 lists several popular CFD codes and their developers.

**Table 4.5 Lists of CFD Codes and Their Developers for Flammable Material Dispersion Study (Hanna et al., 2006; Ivings et al., 2007)**

| Software | Developer |
|---|---|
| FEM3A | University of Arkansas, US |
| CFX | ANSYS |
| FLUENT | ANSYS |
| FLACS | Gexcon AS, Norway |
| STAR-CD | CD-Adapco |
| FEM3MP | Lawrence Livermore National Laboratory |
| CFD-Urban | CFD research corporation |
| FLUENT-EPA | NOAA—National Oceanic and Atmospheric Administration |
| FEFLO | George Manson University |

## 4.4 EXPLOSION MODELING

CFD explosion modeling is focused on the understanding of possible consequences from large-scale industrial accidental events such as VCE, BLEVE, DDT, and dust explosion. Besides data regarding overpressures and impulse generated from these types of explosion, which is also the main hazards of explosion in context of process safety, CFD simulation results provide extensive information regarding the development of flammable clouds, the movement of flame and pressure waves, and the interaction between the pressure wave and the obstacles. These data are critical for the design, layout, control, and mitigation of explosion hazards. Even though the four phenomena described in this section are all related to explosion and pose the same type of hazards, their explosion kinetics and mechanism, temporal and spatial scales, and involved geometries are significantly different in terms of CFD simulation; thus, they will be covered in separated sub-sections below.

### 4.4.1 VAPOR CLOUD EXPLOSION

The main advantage of CFD in simulating VCEs is its capability to accurately represent complex geometries to a detailed level. This capability is critical in studying VCE, because it is well known that obstacles have significant effects on the development of flammable cloud and VCE strength (e.g., the overpressure magnitude and affected radii). The interaction of gas flow with obstacles generates turbulence, which increases the energy release rate and the flow velocity of the combustion process and further intensifies the turbulence ahead of the flame. This feedback cycle increases the flame acceleration and strengthens the overpressure generated from VCE. This requires very accurate representation of all geometrical features, making CFD a clear advantage over simpler models in the capability of simulating VCEs in complex geometries.

Besides providing predictions regarding the development of flammable cloud, overpressure magnitude, and affected radii, CFD simulations are conducted to gather comprehensive information on the effects of overpressures on structural members, the possible arrangement of explosion vents, the effectiveness of protective barriers, and other mitigation measures.

#### *4.4.1.1 Geometry representation*

To take full advantage of a CFD simulation in VCE, there needs to be a detailed and accurate geometry representation. However, practically for large-scale simulation in a typical process facility (e.g., a refinery, a chemical plant, or an offshore platform), there are multiple obstacles including tube pipe bundles, structural trusses, and pieces of process plant equipment, and surrounding environmental objects such as trenches and trees (Bakke et al., 2010), whose numbers can be up to several thousand with dimensions ranging in several orders of magnitude (millimeters to meters). This reality makes geometry representation one of the most challenging aspects of using CFD in a VCE study for an actual industrial facility. A relatively simple approach using a Cartesian grid with porosity distributed resistance (PDR) (Patankar and Spalding, 1974) is a popular concept used in many CFD codes such as FLACS, AutoReaGas, and EXSIM (see Figures 4.19 and 4.20) (Lea, 2002; Cant et al., 2004; Baraldia et al., 2009).

Detailed and accurate geometry representations of large-scale industrial facilities as presented in Figure 4.19 are critical in CFD simulation of VCE. The Cartesian/PDR approach was invented in order to present large-scale complex geometries by dividing them approximately into resolved geometrical

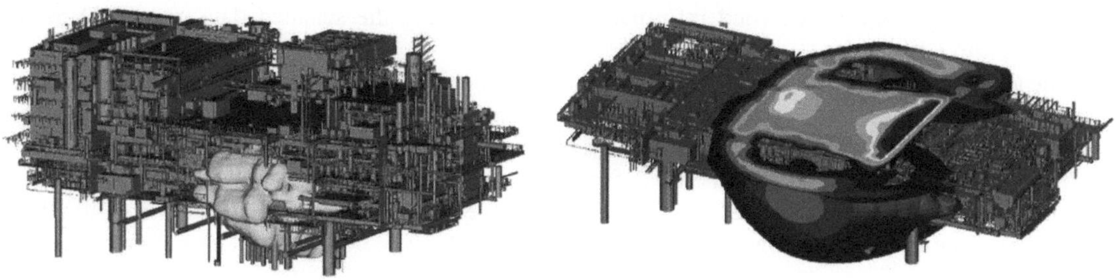

**FIGURE 4.19**

Rendering explosion modeling results from a CFD simulation for an offshore platform using FLACS.

*© FAO, 1988. In Vitro Toxicology. pp. 1–14. http://agris.fao.org/agris-search/search.do?recordID=US19890077884 (02.02.15.).*

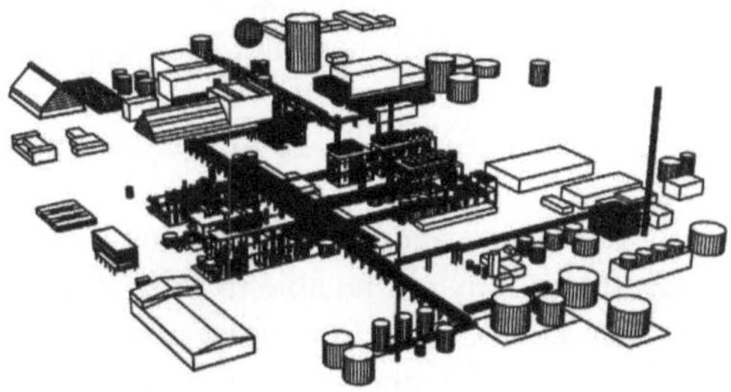

**FIGURE 4.20**

Graphic representation of the Flixborough plant for CFD simulation using EXSIM CFD code.

*Reproduced by permission with Hoiset et al. (2000).*

objects, often large objects whose dimensions could be represented in the grid, and unresolved geometrical objects, often smaller objects whose dimensions are smaller than the grid size. By that, the resolved geometries (e.g., large objects) are treated explicitly on a computational grid, whereas the unresolved geometries (e.g., small objects) are represented using the PDR method as area porosities on the control volume faces and as volume porosities within the control volume faces. The porosity is defined as the fraction of a face or volume available for fluid flow, as opposed to the blockage, which is the fraction blocked from access by the fluid. The flow resistance, turbulence, source-term, and the flame speed enhancement are then calculated using these defined porosities. In this way, complex geometries in oil and gas or chemical facilities such as pipe systems or skid mounted equipment can be divided into multiple parts and represented by PDR "sub-modules", simple, primitive geometries such as isolated cylinders or square bars, fully developed tube-bundle flows, simple planar grids, and uniform porous surfaces or regions (Cant et al., 2004).

Computational geometry can be prepared either by manually defining a set of primitives (boxes, cylinders, and planes) representing corresponding obstacles or automatically by importing from CAD drawings. Many, but not all, commercial CFD packages such as FLACS or AutoReaGas provide certain types of interface to convert CAD data into CFD 3D models (Lea, 2002; Cant et al., 2004). The process of defining all objects in large facilities such as refineries or offshore platforms requires an extremely large amount of time. This process can be sped up by using the current technology in laser

3D scanning to obtain the detailed information and converting the scanned data into CAD before importing into CFD codes.

The geometry representation on the CFD grid needs major improvement over the current Cartesian/PDR approach, which has been criticized for being unable to accurately and completely represent complex geometries (Lea, 2002). More recently, other more advanced methods have been explored to fully represent the interested geometries; for example, the unstructured body-fitting grid or the adaptive meshing has been successfully implemented in certain CFD codes (Fairweather et al., 1999; Lea, 2002; Cant et al., 2004; Baraldia et al., 2009).

#### 4.4.1.2 Turbulence model

The two-equation k–ε turbulence model is the model of choice for all CFD codes for VCE simulations because it provides a reasonable balance between accuracy and computing cost (Lea, 2002). This model is based on the gradient transport assumption and the Boussinesq eddy dissipation turbulence model (Blazek, 2007). Other more advanced derivatives of the k–ε turbulence model such as the Reynolds normalization group (RNG) and the realizable k–ε turbulence model, as well as newer models such as Reynolds stress turbulence models have not been implemented in CFD codes for VCE modeling (Lea, 2002). The advanced LES methods such as the eddy viscosity LES and RNG-LES have been applied in hydrogen explosion study (Colin et al., 2000; Durand and Polifke, 2007; Baraldia et al., 2009; Lecocq et al., 2011).

The k–ε model was originally developed and validated for only certain fundamental nonreacting constant flow and has been shown that it may not be able to predict the large-scale turbulent reacting flow in a complex 3D geometry (Lea, 2002). Therefore, more research is needed on the application of turbulence models, such as exploring more advanced versions of the k–ε model and other turbulence model such as Reynolds stress model implementation in CFD codes for VCE studies.

#### 4.4.1.3 Combustion model

Combustion model is used to calculate the combustion reaction rate in VCE events. Generally, it is reasonable to use a fast reaction assumption for premixed fuel–air mixtures in explosion modeling, thus allowing some simplification in combustion models needed. Available CFD codes use either prescribed reaction rates based on empirical correlations of burning velocity or the eddy break-up model (Magnussen and Hjertager, 1977), which requires a certain resolution of the flame front to work (Lea, 2002; Cant et al., 2004). The latter is used in commercial CFD codes such as EXSIM and CFX-4 models, while the former is used in AutoReaGas (Lea, 2002). Newer versions of FLACS and AutoReaGas use an eddy break-up–based combustion model called the "β transformed" gradient method for flame tracking with Bray's correlation for turbulent burning velocity (Bray, 1990; Arntzen, 1998; Durand and Polifke, 2007; Baraldia et al., 2009). These two simple approaches for modeling combustion are also used in newer and more sophisticated CFD codes such as COBRA (empirical), NEWT (eddy break-up), and REACFLOW (eddy break-up). The relatively new CREBCOM combustion model is implemented in CFD codes COM3D-3.4 and b0b (Efimenko and Dorofeev, 2001; Baraldia et al., 2009).

Other more combustion advanced models available are the laminar flamelet, probability PDF transport, and conditional moment closure; even though they have been developed and used in combustion studies for a long time, they have not been used in VCE CFD modeling (Vianna and Cant, 2010; Lecocq et al., 2011). Studies have been done to incorporate more detailed chemistry and advanced models into combustion models but only on a very small experimental scale (Vianna and Cant, 2010; Lecocq et al., 2011). However, the possibility of using those models in a real large-scale facility is still open for discussion.

#### 4.4.1.4 Published validation studies

Many studies have been published to validate the capabilities of CFD code in a VCE. In here, the focus is on independent validation projects that allow unbiased evaluation of the accuracy of available CFD explosion models. Projects of this type include the Modeling and Experimental Research into Gas Explosions (MERGE) (Popat et al., 1996) and Extended Modeling and Experimental Research into Gas Explosions projects funded by EU, the Joint Industry Project on blast and fire engineering for top side structures phases 2 (JIP-2) (Lea, 2002), and the European Commission–funded Network of Excellence, "Hydrogen Safety as Energy Carrier" (Baraldia et al., 2009).

In the MERGE project sponsored by the EU, experiments were designed to have an extensive evaluation of important components of CFD codes for VCE modeling. This project has three distinctive phases (Popat et al., 1996; Lea, 2002):

- Phase 1: Evaluation of the submodels such as combustion, turbulence, etc. incorporated into CFD codes.
- Phase 2: Verification of the CFD explosion prediction against small- and medium-scale geometries.
- Phase 3: "Blind" prediction form the CFD code developed for the explosion overpressures in the large-scale MERGE geometries, which consisted of a regular cuboidal pipe array filled with a combustible gas mixture. The blind prediction is considered the most stringent and precise test because the modeling predictions have to be submitted before the actual experiments are conducted.

Four CFD codes were used in the MERGE project: COBRA, EXISM, FLACS, and AutoReaGas. Among them, EXISM, FLACS, and AutoReaGas all use Cartesian/PDR for geometry representation; only COBRA, which is considered an advanced CFD model, is able to handle both Cartesian and cylindrical polar/arbitrary hexahedral meshes. All four CFD codes use the two-equation k–ε turbulence model for the turbulent flow. Although the blind predictions from all four models were considered scattered in comparison to the experimental data, they were within the accuracy range generally expected from simple CFD codes.

The MERGE project was extended into the EMERGE project in which the four aforementioned CFD codes were further evaluated in many more experimental tests. The results indicate the participating CFD codes can give acceptable results for practical application, but further improvement is required.

The JIP-2 projects had both experimental and modeling parts. The experimental portion was aimed at providing a better understanding of the mechanism and affecting factors behind VCE and establish a large database for CFD model calibration. Similar to the MERGE and EMERGE projects, the modeling portion of the JIP-2 project has three phases but with different focuses (Lea, 2002):

- CFD code blind predictions on a given geometry (8 m wide).
- Using the experimental results from the test in geometry in phase 1 to tune the models and repredict the experimental result.
- CFD code blind predictions on a larger geometry (12 m wide).

**Table 4.6 List of CFD Codes and Participants in the Exercise**

| CFD Codes | Participants |
|---|---|
| FLACS | DNV, Det Norske Veritas AS, Norway<br>GXC, GexCon AS, Norway |
| COM3D-3.4 (CREBCOM) | FZK, GmbH, Germany |
| AutoReaGas | HSL, Health And Safety Laboratory, UK |
| ReacFLOW | JRC, European Commission Joint Research Center, Institute for Energy, the Netherlands |
| B0b | KI, Research Center Kurchatov Institute, Russia |
| FLUENT | UU, University of Ulster, UK |

Results from this JIP-2 project indicate that certain models were sensitive to geometrical parameters as well as input conditions; small changes in the geometry can lead to large changes in their predictions. CFD models performed better after being retuned using available experimental data, suggesting that simple CFD models were valid only when used in geometries similar to those used in calibration.

In the hydrogen safety project, seven participants used six different CFD codes. DNV and GexCon both used FLACS for their modeling prediction and joined an intercomparison exercise on CFD model capabilities to predict a hydrogen explosion in a simulated vehicle refueling environment (Baraldia et al., 2009). The CFD codes and participants are listed in Table 4.6. This project was designed to test blind predictions from a wide range of CFD models available; however, only two submitted results were considered truly blind, the other prediction results came after the experimental data were available and were listed as "post prediction". The turbulence models used ranges from the nonviscous Eulerian model, two-equation k–ε models, and LESs; the combustion models include the eddy dissipation concept, constant burning velocity model and other models depended on local turbulence characteristics. In general, all CFD predictions in this project had reasonable agreement with experimental data, even predictions using simple combustion models with constant burning velocity given that a correct burning velocity was chosen.

One important note from all the independent validation projects is that achieving reasonable predictions using CFD for VCE requires certain expertise and experience in both CFD methodology and VCE knowledge.

#### 4.4.1.5 Available CFD codes

CFD codes available for VCE modeling can be categorized into two groups, simple and advanced, based on the level of complete description of the physical and chemical processes involved, the geometry representation capability, and the accuracy of numerical schemes (Lea, 2002). For example, the simple CFD models make use of the Cartesian/PDR approach for representing smaller-scale obstacles, while the advanced CFD codes use adaptive mesh refinement to resolve all obstacles. Table 4.7 lists several CFD models available for VCE.

**Table 4.7 CFD Package Available for VCE (Lea, 2002; American Institute of Chemical Engineers Center for Chemical Process Safety, 2010)**

| CFD | Type | Developer |
|---|---|---|
| Exsim | Simple CFD | Telemark Technological R&D Center, Norway and Shell Global Solutions, UK |
| AutoReaGas | Simple CFD | Century Dynamics Ltd and TNO |
| Flacs | Simple CFD | CMR-Gexcon, Norway |
| BWTI | Simple CFD | (Geng and Thomas, April, 2007) |
| CEBAM | Simple CFD | (Clutter and Mathis, 2002) |
| CFX-4 | Advanced CFD | AEA Technology Engineering Software at Harwell, UK |
| Cobra | Advanced CFD | Mantis Numerics Ltd and Advantica Technologies Ltd |
| NewT | Advanced CFD | Cambridge University |
| Reacflow | Advance CFD | Joint Research Center of the European Union, Ispra, Italy |
| Research code | Advance CFD | Imperial College, UK |

## 4.4.2 BOILING LIQUID EXPANDING VAPOR EXPLOSION

Boiling liquid expanding vapor explosion (BLEVE) is one of the major consequences in process safety incidents, which involve not only the overpressure and radiation hazards from the fire ball but also the missile hazard due to vessel disintegration in the explosion process. BLEVE events have been observed mostly with LPG vessels but are also possible for LNG, hydrogen, and other liquefied gases if the containers lose their cooling capabilities. The mechanism behind a BLEVE event is rather complicated, involving multiple phase interactions. Even in the simplest BLEVE case, there are two fluid phases involved, including liquid and gas phases, whose interactions and transformation pose formidable challenges in simulating a BLEVE event using CFD. Currently, published CFD simulation studies only focus on certain aspects of BLEVE events, such as two-phase release, without considering container disintegration. Because of this, simpler integral models offering reasonable predictions of the blast wave and the radiation generated in BLEVE events are often used. For example, Makhviladze and Yakush (2005) conducted a study to investigate the formation and combustion of accidentally released fuel clouds. In this study, they present a "start-to-finish" modeling of a BLEVE event, which incorporated both blast effects of expanding volumes of superheated liquid and fireballs following ignition of the fuel cloud.

Generally, two types of BLEVEs have been observed (Luther and Muller, 2009):

- *"Cold"* BLEVE: In this type of BLEVE, there is typically only a small flash fire, followed by a large spill fire because either the initial momentum of the release is too low or oxygen supply is insufficient;
- *"Hot"* BLEVE: In this type of BLEVE, the hemispherical expansion of the BLEVE fire ball is followed by a buoyancy-induced upward flow, and the fireball forms into the well-known

mushroom structure. It normally occurs if the initial momentum of the release is large and enough oxygen is sufficient to start turbulent combustion. The resulting expanding fireball is self-sustaining at a relatively constant rate by absorbing surrounding fresh air to continue burning.

Tan et al. (2011) used CFD to investigate the small-scale instantaneous flashing release of pressurized freon from small laboratory spherical flasks. This instantaneous release is considered an example of a "cold" BLEVE event. In the experiment, the event started with an instantaneously vaporizing mass of liquid that rapidly increased the volume leading to the total disintegration of the container. Upon release, the liquid phase came in contact with the ambient conditions of nucleates and moved into a superheated phase, then broke into small droplets violently. In the CFD simulation study, the authors used the Eulerian–Lagrangian approach to simulate the discrete phase model, in which the particles were assumed to be smooth and spherical; the drag on particles was modeled by a spherical drag submodel. The simulations were run in FLUENT commercial CFD codes, and started at the moment when shattered droplets were already on the surface of the flask (see Figure 4.21). Their modeling results show good agreement with experimental data.

Luther and Muller (2009) used a fire dynamic simulator (FDS) from NIST to conduct simulations of the fuel fireball from a hypothetical commercial airplane crash on a generic nuclear power plant to explore the potential external hazard threatening the nuclear power plant. This type of BLEVE event is considered an example of a "hot" BLEVE. The "hot" BLEVE develops through three phases (Luther and Muller, 2009):

1. *The initial phase*: The liquid–vapor mixture is blown out at a high momentum from the container and rapidly expands.
2. *The expansion phase*: Momentum-controlled phase in which the expanding cloud forms a hemispherical fireball. In this phase, the hot, rich in fuel fireball expands in the cold ambient air forming an unstable surface. The fireball could stick to the ground.
3. *The uplift phase*: Hot gas buoyancy controlled phase. The hot fireball rises into a spherical mushroom-shaped plume, and the vortices formed in the fire ball continue sucking in ambient air to sustain the fireball. The remaining fuel is burned almost completely.

The model was validated using the experimental data of the well-documented fireball from the BAM 1999 BLEVE experiment (see Figure 4.22) (Droste et al., 1999). They noted that phenomena governing the evolution of small fireballs are the same as those for large fireballs so that it is justifiable to apply CFD codes to simulate large fireballs for which no experimental data are available as an extrapolation of small-scale behavior to larger size. The simulation was conducted to answer three key questions:

- How does the presence of the nuclear power plant buildings affect the evolution of the fireball?
- Which safety-related openings of the buildings are reached by the fireball?
- Which temperature levels and duration, velocities, and fuel mixtures are reached at these openings?

They used the Smagorinski model for turbulent flow, with an assumption that the fireball started with no turbulence at all and turbulence was developed during the fireball evolution. Near wall turbulence treatment and radiation were ignored. Buoyancy was taken into account, and the Z model was used for combustion, assuming that all fuel was immediately burnt if oxygen were available. Three approaches were used to introduce fuel into the fire ball: (1) prescribed HRR (heat-released rate), (2) initial fuel droplet cloud, and (3) dynamic fuel spray. The evolution of the BLEVE event in the

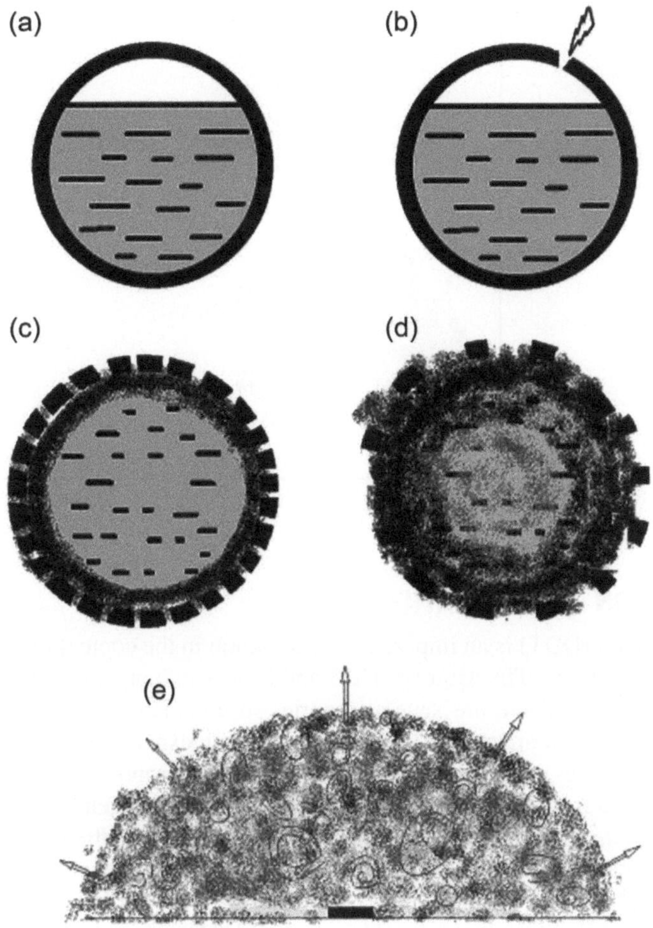


**FIGURE 4.21**

The simulation sequence of the instantaneous two-phase release, starting from (a) the initial condition; (b–d) the evolution of the release; and (e) the complete release resulting in the total integration of the flask.

*Reproduced by permission with Tan et al. (2011).*

simulation agreed well with observations, providing more detail for further understanding of this hypothetical event. Key findings in this study include:

- The building structures could have a significant impact on the fireball evolution, as the local turbulence caused by interactions between the fluid with obstacles and corners may result in "hot spots."
- In the early phase (the initial phase), the fireball is cool, rich in fuel, and lean in oxygen. The highest temperature was only reached after the fireball had risen above the facility corresponding to the uplift phase.

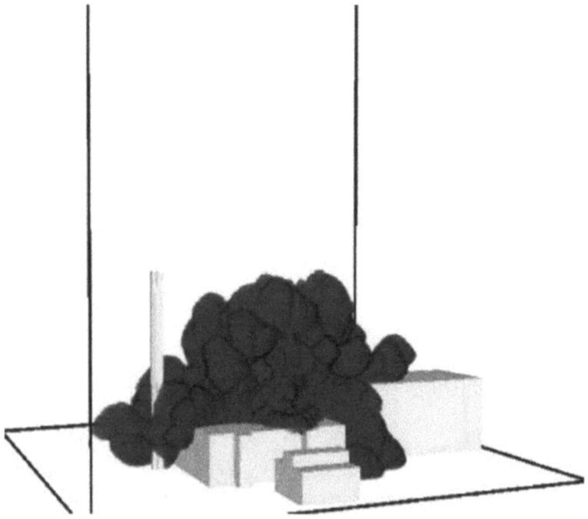

FIGURE 4.22

Graphical rendering of the fireball resulted from the impact of an airplane on a nuclear plant.

*Reproduced by permission with Luther and Muller (2009).*

### 4.4.3 DEFLAGRATION TO DETONATION

Deflagration to detonation (DDT) is an important phenomenon in the context of process safety because of its devastating consequences. The nature of DDT and factors influencing DDT are currently subjects of ongoing research efforts. There are several theories to elucidate the mechanism behind a DDT phenomenon. For example, Sherman et al. (1989), by analyzing experimental data from the large-scale tests of hydrogen–air mixtures at the Sandia National Laboratory, proposed that DDT could occur when the shock fronts generated from the expanding unburned gas and those generated by the autoignition of preheat-unreacted mixture merge together (Sherman et al., 1989; Middha and Hansen, 2008). Other theories have been developed such as "local explosion" proposed by Oppenheim (Meyer et al., 1970), the "Shock Wave Amplification by Coherent Energy Release" (SWACER) concept proposed by Lee et al. (1978), or the instabilities caused by sudden venting by Dorofeev et al. (1995) and Middha and Hansen (2008). However, those theories have not been able to provide a full picture of what really happens or led to a practical mechanism behind the observations in DDT events in experimental works. Similarly, at this state, because of the complexity and the lack of clear understanding of DDT mechanisms, CFD modeling works on DDT only to try to shed light on the question of whether CFD as it is today is capable of describing DDT events. Current CFD studies on DDT could be divided into two different approaches: one is trying to precisely simulate a DDT event using available computational resources, and the other is trying to predict the possibility of DDT event in an indirect way.

The first approach is pursued by Oran's research group and other researchers efforts in simulating DDT from first principles have been published (Khokhlov and Oran, 1999; Khokhlov et al., 1999; Gamezo et al., 2001, 2005, 2008; Tegner and Sjogreen, 2002; Oran and Gamezo, 2007; Vaagsaether et al., 2007). An extensive list of several notable CFD simulation studies conducted to understand and predict DDT is also included in their review (Oran and Gamezo, 2007). The main objective of their current research is to provide an insight into DDT and to answer how, when, and where DDT occurs, mainly by numerically recreating and analyzing laboratory experiments. Even though their simulation systems were typically simple in a macroscopic view (i.e., shock-tube experiments), they pose formidable challenges because the simulation system has the capability to resolve up to 12 orders of magnitude ranging from the size of the system to the flame thickness (Oran and Gamezo, 2007). Through a series of sophisticated simulations, they have identified that shock–flame interactions play a central role in creating the environments leading to DDT. In these positive feedback interactions, the turbulent flames enhance the shock and vice versa, generating conditions leading to the formation of "hot spots," or the ignition center of unreacted fuel mixtures that ultimately produce a detonation. They also have shown, through their simulation results, that the obstacles and boundary layers are critical in creating the suitable conditions for "hot spots" formation.

Even though these efforts could ultimately bring forward better understanding of the nature of DDT, the current computing resources only allows CFD simulation to be conducted in relatively small 2D or 3D systems; thus, predicting the possibility of DDT in realistic large-scale geometries such as offshore platforms or refineries by conducting first-principle CFD simulations may not be possible in the near future.

In the latter effort, notably works from Gexcon with their FLACS simulation packages, instead of directly simulating DDT, their efforts are on evaluating the likelihood of a DDT event (e.g., identifying whether DDT is likely in a given scenario and indicating the regions where it might occur) (Middha and Hansen, 2008). They proposed using a key indicator such as a parameter proportional to the spatial pressure gradient across the flame front or the normalized spatial pressure gradient (Middha and Hansen, 2008). This parameter provides visualized information on when the flame front captures the pressure front, which is the case in situations when fast deflagrations transition to detonation. Reasonable agreement was obtained with experimental observations in terms of explosion pressures, transition times, and flame speeds. By calculating this indicator for a given system, the likelihood of DDT could be inferred. For instance, Middha and Hansen (2008) proposed that the magnitude of this indicator around 10 or larger indicate a high possibility of DDT. The normalized spatial pressure gradient is calculated from the following equation (Middha and Hansen, 2008):

$$\left.\frac{dP}{dx}\right|_{normalized} = \left.\frac{dP}{dx}\right|_{actual} \frac{X_{cv}}{P_0} \tag{4.18}$$

where $P_0$ is the initial pressure, and $X_{cv}$ is the grid resolution.

A newer model has been introduced by GexCon to take into account the comparison of detonation cell size with geometrical dimensions, thus, correcting some of the short comings of CFD predictions by using the normalized spatial pressure gradient indicators such as DDT occurrence in very small geometries in which a detonation front may not be possible to propagate or in regions where there is insufficient fuel. The new indicator is introduced as DDT Length Scale, or DDTLS, which is calculated from the relevant geometric length in FLACS, the LSLIM:

$$\text{DDTLS} = \frac{\text{LSLIM}}{\lambda} \tag{4.19}$$

where DDTLS is a key parameter to predict the occurrence of DDT, LSLIM is the relevant geometric length scale in FLACS, and $\lambda$ is the detonation cell size.

Figure 4.23 shows a result from a 2D CFD simulation in which the DDTLS is presented for a detonation tube experiment at two different times. According to GexCon, this new model has been

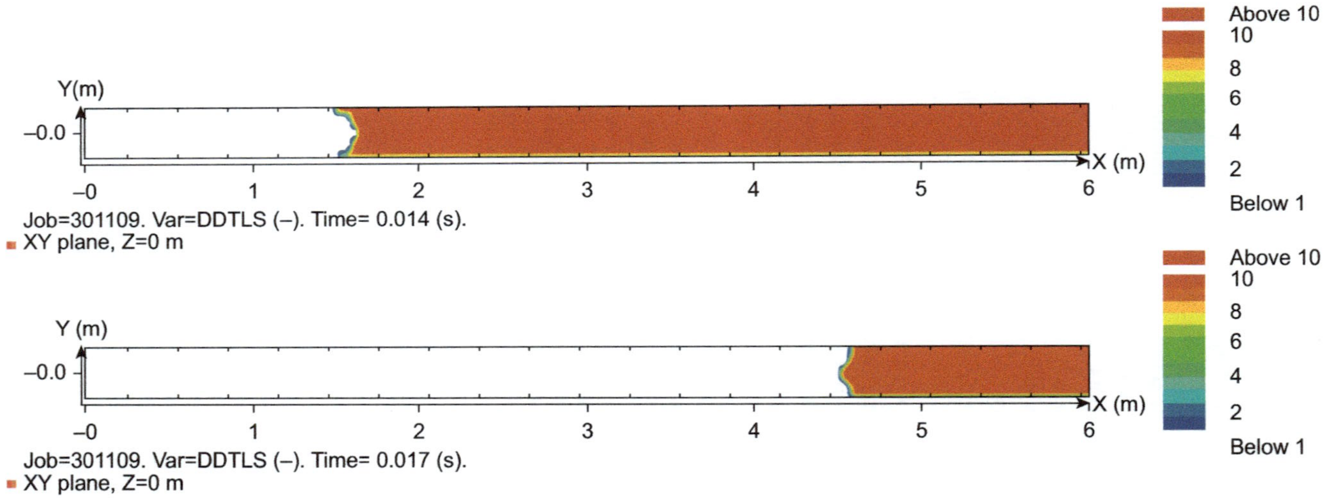


**FIGURE 4.23**

Graphical representation of the DDT length scale (DDTLS) results in the 30 cm diameter detonation tube. The top is the CFD simulation result at 14 ms and the bottom is at 17 ms.

*Reproduced by permission with Middha and Hansen (2008).*

shown to be able to predict with reasonable accuracy in multiple validation experiments. For example, in a detonation experiment using three detonation tubes with diameters D at 5, 15, and 30 cm at McGill University, the new model predicted that the DDT is most likely to occur near the baffle of the 30-cm tubes (DDTLS is around 20), while the possibility of having DDT in a smaller tube is lower (DDTLS is around four for the 5-cm tube). In other validation tests such as the GexCon 1.44 m explosion channel tests, or the Fh-ICT lane experiments, the tests conducted by SRI International and tests in the EU-sponsored HYCOM project, the simulation results were in reasonable agreement with the experimental data.

This new model is still in the development stage, and multiple challenges still need to be solved before it can be applied in industry for DDT prediction. For example, the relevant geometric length scale in FLACS (LSLIM) is an important parameter to obtain the DDTLS, but it is not clear how GexCon calculates this parameter in different scenarios, particularly because FLACS uses the Cartesian/PD approach to present the geometry. Further, the detonation cell, $\lambda$, is currently calculated using a fitting equation against available experimental data; this kind of fitting equation, as noted by Middha and Hansen (2008), is only valid within the range of the experimental data and may not be available for all kinds of fuel. Nevertheless, the results show some promise in its capability to predict the likelihood of DDT, and with further development it could be a very useful tool in process safety.

### 4.4.4 DUST EXPLOSION

Complete understanding and a full description of dust explosion using CFD are still a formidable question in process safety. There are several challenges associated with dust explosion simulation using CFD, namely (1) complex computational domain involving internal geometries at industrial scale; (2) complex transient, turbulent, compressible two-phase flow (solid or particle and gaseous); and (3) complex incomplete combustion model involving rapid phase transitions (solid phase to gas phase) and multiple chemical reactions (Skjold, 2010). Unlike other types of CFD simulation such as dispersion or VCE, the first challenges, complex geometry representation involving internal system at an industrial scale, happens to be the most straightforward obstacle in CFD dust explosion simulation, while for the other two, there is need for better mathematical models describing the turbulent flow of dust clouds, models describing the combustion associated with combustible dust particles to accurately simulate dust explosion. Because of the chaotic and random nature of a combustible dust cloud, the inconsistency of available experimental data, even with data from standardized methods such as 20-liter vessel dust explosion experiments, adds more obstacles to the efforts of establishing a reliable CFD methodology for effective prediction of dust explosion hazards (Skjold, 2007).

Several key factors involving the dust cloud explosion listed by Eckhoff in his book *Dust Explosion in Process Industries* are (Eckhoff, 2003):

- Chemical composition of the dust, including its moisture content.
- Chemical composition and initial pressure and temperature of the gas phase.
- Distribution of particle size and shape in the dust: this distribution determines the specific surface area of the dust in the fully dispersed state.
- Degree of dispersion, or agglomeration, of dust particles: this determines the effective specific surface area available to the combustion process in the dust cloud in an actual industrial situation.

- Distribution of dust concentration in an actual cloud.
- Distribution of initial turbulence in an actual cloud.
- Possibility of generation of explosion-induced turbulence in the still unburned part of the cloud (location of ignition source is an important parameter).
- Possibility of flame front distortion by mechanisms other than turbulence.
- Possibility of significant radiative heat transfer (highly dependent on flame temperature, which in turn depends on particle chemistry).

Therefore, the capability of an ideal CFD model should be sufficient to describe all those factors in order to accurately predict the characteristics of a dust explosion phenomenon. However, with current computing power, certain simplifications must be devised so that CFD simulation is feasible for dust cloud explosion.

#### 4.4.4.1 Turbulence model

Modeling accuracy of a particle-laden turbulent flow in dust explosion simulation is a formidable challenge for many reasons. For example, dust particles, because of their wide distribution in size, density, and shape, never form a perfect, uniform dispersed cloud. This complexity, in CFD simulation, generally, is solved by certain simplifications such as uniform particle size distribution or non-interaction between particle assumptions. In this way, dust–air flow variables can be cataloged into two classes based on multiphase submodels: (1) Eulerian–Eulerian (E–E) model considers particle phase as a fluid phase; and (2) Eulerian–Lagrangian (E–L) model in which the particles are treated as a noncontinuous discrete phase (Zhong and Deng, 2000; Radandt et al., 2001; Zhong et al., 2001; Skjold et al., 2005; Zhang and Chen, 2007).

In the E–E model, the dust particles and air are assumed to be in thermodynamic equilibrium or, more specifically, they have the same velocity and temperature, implying that the size of particles are relatively small with uniform temperature throughout the particles (Skjold, 2007; Shi et al., 2009). Dust particles are further assumed to not contribute to the pressure of the system, and dust–gas mixtures behave like ideal gas. This assumption is reasonable when the size of the dust particle is relatively small, the density of the cloud is low and the thickness of the actual combustion front is much smaller than the characteristic dimension of the problem under consideration.

The second model, the E–L model, presents a more realistic system in which dust particles are treated as separate discrete phases, or the Lagrangian phase. This model allows simulated dust clouds to have different sizes, velocities, and interactions between particles, thus avoiding the thermodynamic equilibrium assumption used in the E–E model. The flow and combustion properties of dust particles (i.e., trajectories, mass changes, motions, and combustion) are calculated in Lagrangian coordinates (Radandt et al., 2001). However, the E–L approach requires much more computing power than the E–E model, indicating that it may not be applicable for large-scale simulations of dust explosions in the near future.

#### 4.4.4.2 Combustion models

The combustion model used in dust explosion simulation is to describe the rate of the combustion process of premixed dust clouds (e.g., rate of conversion from reactant to product) and to define the reaction zone (Skjold, 2007). The combustion process for dust particles is much more complicated to be modeled in comparison to those of gaseous mixtures because of several issues involved such as multiple-step process, incomplete reactions, and complex product formation. To address this issue, Krause and Kasch (2000) divided and categorized the whole combustion process into a number of successive and manageable steps, including (1) devolatilization of the volatile fraction of the solid fuel, (2) mixing of the volatiles with oxygen in the gas phase, (3) combustion of the volatiles, and (4) combustion of the remaining solid fraction, which consists of mostly char. Further simplification is often applied to reduce this multiple-step complex process into solvable combustion problem in modeling. For instance, Collecutt et al. (2009) used the assumption that only $CH_4$ and $H_2$ were extracted from volatile, thus using single-rate Arrhenius equations to model two gas phase combustion reactions:

$$CH_4 + 2O_2 \rightarrow CO_2 + 2H_2O$$

$$2H_2 + O_2 \rightarrow 2H_2O$$

and a reaction on the surface of the char particles:

$$C + O_2 \rightarrow CO_2$$

This combustion model was used in a 3D CFD model using OpenFOAM, an open source CFD code, to study coal dust explosions and their suppression in underground coal mines (Collecutt et al., 2009). This study is part of a larger funded project to develop a practical active explosion barrier, specifically to develop the specifications for a prototype active explosion barrier which was designed as a ring of water injectors at the end of the tunnel. The simulation results so far were validated against a range of explosion conditions in the Simtars Siwek 20-L chamber and the 200-m explosion tunnel.

#### *4.4.4.3 Published CFD studies in dust explosion*

CFD modeling studies published in the literature could be divided into two groups based on the used aforementioned multiphase approaches: E–E or E–L approach. The former has been used in more CFD simulation studies because it is less expensive in terms of computing resources compared to the latter (Radandt et al., 2001; Zalosh, 2008). For example, Krause and Kasch (2000) developed a CFD model for vented dust explosion simulations using an FLUENT commercial CFD package to simulate and analyze time-dependent 2D or 3D local distributions of flow velocities, density, pressure, enthalpy, and species concentrations. In this model, the two-equation k–ε turbulence model with wall and buoyancy correction was used. They used an Arrhenius combustion reaction rate for laminar flames, and a Magnussen-Hjertager model for turbulent flames, and a multiple-step solid phase reaction model including devolatilization, gas phase mixing, and combustion and char combustion. This CFD model was used to calculate the laminar flame propagation through a mixture of lycopodium powder with air in a vertical tube of 2 m in length and 100 mm in diameter. An assumption in which the tested system was a time-dependent 2D laminar flow, particles with negligible inertia, and ignition near the closed end was used. It was further assumed that during devolatilization, mainly propane is released at an amount equivalent to the volatile content of the solid and combusted in the gas phase for further simplification of the combustion model. This simulation had good agreement with the experimentally observed flame speed, which subsequently was used to calculate dust concentration distributions in a 40-$m^3$ silo, focusing on the flow and turbulence fields and the particle distribution. The results from this CFD model can be used for tests in which dust concentration and turbulence parameters have not been measured. It is also useful to predict the homogeneity of a dust cloud in real size, complex geometries, helping establish operating conditions to avoid the formation of dust clouds in the "optimum" range of concentration.

GexCon also applied the E–E approach in their effort to develop a CFD code called Dust Explosion Simulation Code (DESC) in order to simulate dust explosions in large-scale complex industrial facilities (Skjold et al., 2005, 2006; Skjold, 2007; Shi et al., 2009). DESC is part of a larger combinatorial project that uses both modeling and experiments to achieve a more comprehensive understanding and obtain more and better data on important but relatively not well-known aspects of dust explosion such as dust lifting and dust cloud formation. A review of the DESC project including summaries of the various sub-projects has been published (Skjold, 2007). The modeling parts in the DESC project include (1) the development of a combustion model for the turbulent dust cloud and (2) the development and validation of a CFD code. The sub-projects of the DESC include:

1. Turbulent flow measurement to measure the decay of the dispersion induced and to extract the empirical decay formulas for the root mean square of the turbulent velocity fluctuation and the integral turbulent length scale;
2. Measurements of burning velocities and flame speed to obtain the empirical correlation between the laminar and turbulent burning velocities, or the flame propagation velocity relative to the unreacted mixture under laminar and turbulent flow conditions; and
3. Dust dispersion phenomena to understand the mechanisms of the dust cloud formation (e.g., the process of lifting dust layers into dust suspensions) (Zydak and Klemens, 2007).

DESC CFD simulation code is based on the commercial FLACS CFD code developed by GexCon. The geometry is presented by the Cartesian/PDR model, turbulence by the two-equation k–ε model and particle-laden flows by the E–E multiphase approach. The particle-air mixture is assumed as an equilibrium mixture in which the dust particles are in dynamic and thermal equilibrium with the gaseous phase. The turbulent burning velocity is calculated by using the following equation:

$$S_t = 15.1 S_L^{0.784} u'^{0.412}_{rms} l_I^{0.196} \tag{4.20}$$

where $S_L$ is the laminar burning velocity of the dust cloud, $S_t$ is the turbulent burning velocity of the dust cloud, $u'_{rms}$ is the root mean square of the turbulent velocity fluctuation, and $l_I$ is the integral turbulent length scale.

The DESC dust explosion CFD code has been validated against several dust explosion experiments described in literature (Skjold et al., 2005, 2006) and has achieved reasonable agreement between the prediction results and the experimental data. DESC has been used to estimate the flow conditions inside the roller mill during normal operating conditions (van Wingerden et al., 2011) and in a filter system (Shi et al., 2009).

Large-scale comparisons to evaluate the effectiveness of DESC models also have been conducted. For example, DESC has been used to simulate large-scale explosions in 20- and 2-m$^3$ interconnected cylindrical vessels designed by the UK Health and Safety Laboratory (see Figure 4.24). In this test, four different types of dusts including coal, silicon, and two types of potato starch were used as the representative of a wide range of combustible dust materials. One of the notable conclusions from this experiment is that there is poor repeatability of the experimental data, even between identical test

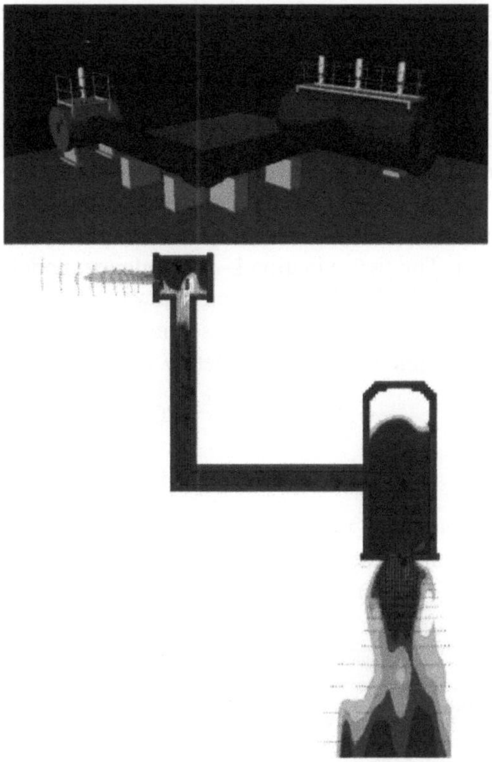

**FIGURE 4.24**

Rendering of the simulation system and the results using DESC. The top figure presents the simulated system and the bottom figure presents the simulation results of the dust explosion.

*Reproduced by permission with Skjold (2007).*

settings, posing a significant challenge to validate the CFD model. Ideally, the necessary data to validate the CFD model would consist of relevant turbulence parameters, dust concentration, flame temperatures, burning velocity, and flow velocity, etc. In practice, it is not easy to obtain those data because of several reasons such as the reliability of the measuring method or the repeatability of the large-scale testing. Currently, only a beta version of the code, the DESC version 1.0 released in 2005, is available. Further development and validation are needed before DESC could be used as a full-fledged dust explosion predictive tool.

The E–L approach, due to its higher cost of computing power, has a limited number of CFD simulation studies reported. Radandt et al. (2001) used the E–L model to conduct 2D CFD simulation for dust explosion in a closed vessel system with different height/diameter ratios. This CFD model used both gas phase and heterogeneous surface combustion of solid particles and took into account the heat loss to the vessel walls. In this model, the two-equation k–ε turbulence model was used for the turbulent flow. This modeling setup achieved good predictions for dust explosions in vessels with varying height/diameter ratios and agreed with experimental data. Although the calculations were limited only to 2D simulations, it can be seen that a similar approach can be extended to a 3D model for a realistic prediction at the industrial scale.

Zhong and Deng (2000) also applied similar E–L approaches in their CFD modeling of maize starch explosions in a large industrial-scale 12-$m^3$ silo. The initial conditions including dust concentration, flow velocity, and turbulent RMS velocity in the silo in their CFD simulation were obtained from published experimental data. In their simulation, they took into account more factors known to have significant effects on the dust dispersion and combustion, including "dust granule" to consider dispersion efficiency, vaporization of water from particles, and radiation in the gas phase and between the gas/particle phases. The two-equation k–ε turbulence model was used to model turbulent flow. For volatile combustion, the eddy break-up model was used with the assumption that only $CH_4$ and CO were extracted from volatile, and CO was generated by surface reaction of solid carbon. The combustion process was modeled as two completely irreversible reactions:

$$2CO + O_2 \rightarrow 2CO_2$$

$$CH_4 + 2O_2 \rightarrow CO_2 + H_2O$$

Although this simulation result has not been compared with experimental data, this promising result suggests that this model could be developed into large-scale realistic 3D geometry simulations of dust explosions with better particle motion and combustion models so that it can be applied in the design process.

## 4.5 TOXIC DISPERSION AND HVAC DESIGN

### 4.5.1 RESEARCH

CFD modeling provides valuable information in terms of safety in the dispersion of toxic gas in complex environments, either in outdoor environment such as in urban areas with multiple obstacles and building structures, or indoor environment with complex flow direction. There are multiple challenges associated with this type of modeling, such as the requirement of accurate presentation of complex geometries, the type of toxic materials, the release sources, the environmental factors, the interaction between toxic materials flow and obstacles, the possible reactions with atmospheric gases or moisture, the possibility of condensation, the dispersion in extreme weather conditions or under the effects of other types of incidents.

### 4.5.2 DISPERSION

Consequences due to the release of hazardous materials into the atmosphere are one of the major concerns in process safety. Important information needed in a toxic dispersion scenario is the hazardous cloud footprint such as the location and magnitude of concentrations, the time evolution of the gas flow, direction of the flow, and the dosages or the amounts of gas absorbed by receivers over time. This information is critical for:

1. Emergency responders in order to prepare in an event of toxic gas release; and
2. Land use planners to mitigate the possible consequence through action such as inventory reduction, transportation patterns, or other mitigation measurements.

One of the most significant toxic dispersion simulations is the range coverage. Because the threshold values for most of the toxic materials, especially toxic gases, are often less than 100 ppm, the affected areas could be up to several miles from the release source. For example, in the Bhopal disaster, toxic gas affected villages and towns several miles away from the plant (2005). The range coverage will dictate the computation domain; the large range will significantly increase the need for more computing power or require a longer run time. Even though the CFD simulation of the dispersion of toxic materials has similar characteristics, especially in terms of modeling parameters, the differences in the possible affected areas between those two types of simulations requires substantial differences in simulation setup. Generally, in flammable gas release scenarios, once the gas concentration in air is lower than its LFL (typical in the range of 1000 ppm), the hazard is not substantial. It is noted that several toxic gases are also flammable gases such as carbon monoxide or hydrogen sulfide; however, their toxicity is far more dangerous than their flammability.

The UK Health and Safety Executive, in its hazardous installations directorate offshore division lists the several important factors needed to be considered in the dispersion of toxic gases and vapor in offshore platform. They emphasize the importance of dispersion and ventilation of toxic and flammable materials on offshore installations, and recommend hazard assessments to evaluate the following:

- Dispersion of flammable or toxic gases or vapors;
- Dispersion within a module, over an installation, over the sea surface, and sub-sea;
- Dispersion of releases that are passive, momentum-dominated, or buoyancy-driven;
- Ventilation of open, sheltered or enclosed areas, primarily as a means of minimizing gas buildup and aiding dispersion;
- Forced or natural ventilation; and
- Ingress of smoke.

The HSE report "Evaluating Vapor Dispersion Models for Safety Analysis of LNG Facilities" lists the following important criteria for CFD dispersion model (Ivings et al., 2007):

- Model accepts a credible source term
- Model accounts for the effects of wind speed
- Model accounts for the effects of surface roughness on dispersion
- Atmospheric stability
- Passive dispersion
- Gravity-driven spreading
- Buoyancy on dilution
- The numerical methods are based on accepted/published good practice

These criteria, even though written for the dispersion of LNG, are applicable for the dispersion of toxic materials, especially dense gases such as chlorine or ammonia.

A typical setup for CFD simulation of toxic gas dispersion is similar to the setup required for flammable gas dispersion as illustrated in Figure 4.11. Among those parameters needed for simulation setup, two most important ones are reviewed here: (1) geometry setup and (2) turbulence model.

### 4.5.3 GEOMETRY REPRESENTATION

The main advantage of CFD simulation over other simpler models on the dispersion of toxic gases, as mentioned earlier, is its ability to accurately present complex geometries and take into account the effect of obstacles such as buildings, tanks, railcars, and trees; however, this is also the main challenge of CFD simulations because it takes much longer to set up this type of geometry and obtain results. Typical running time of a CFD simulation on a sizable geometry ranges from several hours to several days, depending on the available computing power. Thus, CFD dispersion model is yet to be popular in process safety professionally. It is mostly used in research modes for planning and also for providing data for other simplified models designed for quick response. For example, Hanna et al. (2009) conducted CFD simulations to assess consequences from hypothetical chlorine rail car releases in Chicago urban areas. Large portions of the dispersion research have been done using CFD, and data available from experimental work were used to validate the results (Hanna et al., 2009).

It is important to specify that the toxic materials dispersion in the context of process safety only covers large-scale releases of certain types of industrial chemicals that have acute or fatal effects in short time periods on humans. Other types of hazardous material which may pose long-term effects on human health, such as environmental pollution (i.e., $NO_x$, $SO_x$, or dust particles) or contaminations (i.e., pesticide from agriculture usage and radioactive dust from nuclear fallout), are equally important but require a different methodology to study since the interested areas in dispersion studies of those materials could cover hundreds to thousands of miles. Further, the effect of obstacles in complex geometries is less important in those cases, implying that CFD simulations may not have much advantage over simpler integral and phenomenal models. Readers interested in those studies can find information from published literature (2005). There have been studies focused on the incorporation of CFD computational domains into meso-scale meteorological model domains in order to obtain better boundaries, and are able to link the flows outside and inside a building using information on the total building exchanges rate (Tewari et al., 2010); however, they have not been explored outside of their research range.

Geometry is discussed first because it is the primary reason why CFD needs to be used and also the most challenging part during simulation setup. The challenges in geometry setup for a CFD dispersion study are the:

1. Simulation domain size and the grid size and
2. Data available to construct such geometry.

Realistic geometries such as sections of a city represented in a dispersion study require a simulation domain that can range from several 100 feet to several miles. For example, Pontiggia et al. (2010) constructed a large section of Lecco, Italy, for CFD simulations to study the dispersion of ammonia in an urban setting. In order to present obstacles, whose size is in meter scale, the grid setup needs to be very fine; this condition can lead to a significant number of control volume in simulation domains, which in turn require substantial amounts of time to complete. In order to reduce the computational demand, several strategies have been used. For example, porosity distributed resistance concepts to represent obstacles smaller than the grid size (refer to the VCE section for more information about the porosity distributed resistance); adaptive grid, and multiple levels of grid size could be used in one CFD simulation. For example, Hanna et al. (2009) used grid size $0.2 \times 0.2 \times 0.2$ m at the release source and $1 \times 1 \times 0.2$ m at the area within 50 m from the source—the modeling domain is

500 × 1500 m. The total number of grid cells simulated in their study is 900,000 and 450,000 cells, even with this simplification, the time required to complete their simulation was about 3 days using a single PC with 3-GHz processors and 4 GB of random access memory.

However, the domain size and the grid size are only subject to the available computing power. With current developments in computer speed and parallel computing technology, it may not be a hurdle in the future. To construct a simulated domain, it is required to have detailed data regarding the location and geometry of each obstacle. Simple geometries of interest can be generated and built easily; however, for complex geometry such as city sections and industrial sites, available data often come from government agencies. For example, Hanna et al. (2009) used data from Argonne National Laboratory's archive to construct a section of the Chicago city; Pontiggia et al. (2011) used data from the local municipality; Hanna et al. (2006) used a 3D building database for Manhattan licensed by the Vexcel Corporation. This poses a limitation in using CFD simulation on real-time dispersion studies, without high-quality, up-to-date geometry data (city landscapes constantly change); the results may not be significant.

The advantage of accurate representation of complex geometries in a CFD simulation is also tapped to assess the dispersion of toxic materials in indoor environments, which is critical in (1) buildings or enclosed structures designed to maintain natural convection and (2) heating, venting and air conditioning (HVAC) designed for an effective control of the air quality in indoor environments. For example, a CFD simulation application on the HVAC duct design for offshore platforms was discussed by O'Donnell and Lea (2008).

### 4.5.4 TURBULENCE MODEL

Turbulence model is one of the most important aspects in CFD dispersion modeling. The two-equation k–ε turbulence model based on the eddy dissipation concept is the most popular choice (refer to the VCE section for more information on the k–ε turbulence model). In order to take into account important effects, additional terms are added into the model; for example, the model will need to take into account the effects of buoyancy, atmospheric conditions, and reactions with humidity (Pontiggia et al., 2010).

For gas dispersion, the standard k–ε turbulence model is often sufficient, but for dense gas dispersion, such as for the dispersion of pressurized liquefied chlorine, additional terms need to be added into the model for correctly predicting the flow pattern. It is because dense gas, in high concentration regions, is also flowing within itself; this generated another layer of complexity into the models. For example, Hanna et al. (2009) mentioned that in FLACS, the number of equations needed to be solved for each time steps are 10–15, which significantly increases the computing time. The "frozen flow" algorithm, based on the assumption that air pollution is passive in the sense that it does not affect the flow field that is given, was applied in this case to reduce the number of equations (Hanna et al., 2009).

### 4.5.5 SOURCE MODEL

CFD simulation for the dispersion of toxic materials uses mostly steady-state source models such as a steady-state release rate from a jet release or a steady-state evaporating rate from a pool. These data could be inferred from experimental data or using empirical integral models. For example, Hanna et al. (2009) used FLASH, a sub-program of FLACS to estimate the conditions in a gas jet at distances from the source in a CFD simulation of the jet release of pressurized liquefied chlorine gas. Pontiggia et al. (2010) use PHAST to construct the chlorine jet release to obtain information such as liquid fraction after the release, final temperature, two-phase final velocity, and total discharge mass flow rate.

### 4.5.6 TOXIC MATERIALS

Generally, the dispersion of all kinds of toxic materials could be assessed using CFD simulation. The materials of simulation are neutrally buoyant gases, lighter than air gases, and dense gases such as ammonia, chlorine, and $CO_2$ (Hanna et al., 2004).

### 4.5.7 APPLICATIONS OF CFD DATA IN PROBIT FUNCTION

The ability of CFD to present complex geometry and calculate the effect of the geometry on flow and concentration is critical in risk assessment. For example, the Probit functions are used to calculate the probability of death by exposure to a toxic substance for a given time (Crowl and Louvar, 2002, 2005; Pontiggia et al., 2010) is given by equation 4.15, where $D$ is the toxic dose, and $a$ and $b$ are gas-specific parameters.

The value of $D$ is calculated from the following equation:

$$D = \int_0^\infty C^n dt \tag{4.21}$$

where $C$ is the concentration of the toxic gas, $t$ is time, and $n$ is gas-specific parameter.

The relationship between the death probability $P$ and the corresponding Probit value, $Pr$, is identical to, as estimated for radiation doses and is given by equation 4.17.

This equation shows that the toxic dose depends on the concentration of toxic gases at a given location in a given time; therefore, any stagnant zones, often observed in street canyons and behind buildings, can significantly increase the toxic dose. This information is readily converted into Emergency Response Planning Guidelines (ERPGs) values for emergency planning and response. There are three levels of ERPG (ERPG-1, ERPG-2, ERPG-3) describing possible consequences of exposure up to 1 h for a specific substance. The definitions of those three values are given as following:

Up to 1 h of exposure:

- **ERPG-1**: Maximum airborne concentration below which it is believed that nearly all individuals could be exposed without experiencing anything other than mild transient adverse health effects or perceiving a clearly defined, objectionable odor.
- **ERPG-2**: Maximum airborne concentration below which it is believed that nearly all individuals could be exposed without experiencing or developing irreversible or other serious health effects or symptoms that could impair an individual's ability to take protective action.
- **ERPG-3**: Maximum airborne concentration below which it is believed that nearly all individuals could be exposed without experiencing or developing life-threatening health effects.

Figure 4.25a shows an example of the possible information on the distribution of toxic gas concentration and the dosage calculated and represented from CFD simulation results (Pontiggia et al., 2010). It can be seen that using CFD simulation results, the areas having three ERPG levels are clearly

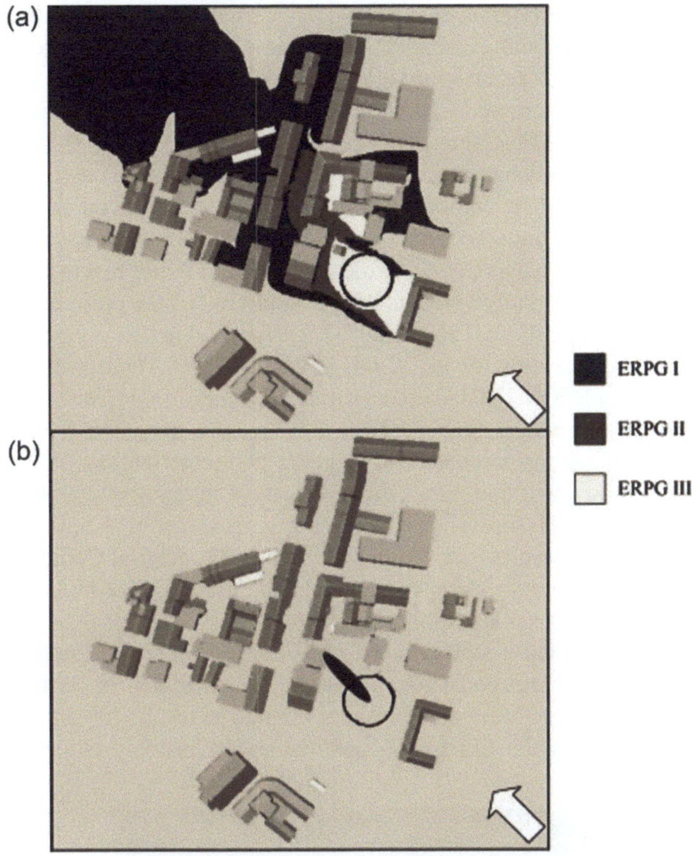


**FIGURE 4.25**

Graphical representation of iso-concentration contours of ammonia using (a) CFD simulation and (b) an integral model.

*Reproduced by permission with Pontiggia et al. (2010).*

presented in the figure. On the other hand, it is difficult to obtain the same results from a simpler integral model (Figure 4.25b) (Pontiggia et al., 2010).

## 4.5.8 PUBLISHED VALIDATION STUDIES

It is necessary to note that due to their toxicity, many validation tests were conducted using inert gases as tracers (such as $CO_2$), or using results from flammable gas (LNG, propylene, etc.) release, but the results should be applicable for other types of toxic gases.

CFD codes have been validated using extensive experimental data based on large-scale testing of dispersion; for example, the Kit Fox with dense $CO_2$ release (Hanna et al., 1993), the Mock Urban Setting Test (MUST) with neutrally buoyant gas tracer release (Yee and Biltoft, 2004), Prairie Grass with neutrally buoyant gas tracer (Hanna et al., 2004), Experimental Measurement Uncertainty (EMU) with tracer (Hanna et al., 2004), and in small-scale testing with simple obstacle setups (Hanna et al., 2002; Baik et al., 2003). A large number of validation studies for CFD codes using data from those tests have been published, for example FLACS (Hanna et al., 2004), ADREA and STAR-CD (Bartzis et al., 2007), and FEFLO-URBAN (Camelli et al., 2004).

There are also multiple published studies using CFD simulation to assess the possible consequences of toxic gases released in complex environments, especially in urban environments. For example, Hanna et al. (2006) conducted simulations of an accidental release into Madison Square Garden, Manhattan, New York, using five CFD models. The modeling result in this study shows that the CFD model predictions reasonably agree on the flow pattern and the recirculation behind buildings and in street canyons (see Figure 4.26). Five CFD codes were used in this project: CFD-Urban (Coirier and Kim, 2006a,b), FLACS (Hanna et al., 2004), FLUENT-EPA (Huber et al., 2005), FEM3MP (Calhoun et al., 2005), and FEFLO-URBAN (Camelli and Lohner, 2004; Camelli et al., 2004). Among those, only FEFLO-URBAN used advanced LES. The results from the CFD simulation provide valuable information for planning purposes and to guide parameterizations in simpler and fast real-time wind flow models; on the other hand, however, they noted that the long time required to obtain results from CFD models prevents it from being used for real-time emergency response. Simpler models based on CFD results such as Quick Urban and Industrial Complex (QUIC) (Williams et al., 2004) are available. They are parameterized using data from CFD simulations and are designed for fast response, allowing for real-time monitoring.

Hanna et al. (2009) conducted a simulation using FLACS studies to simulate a release of pressurized liquefied chlorine from railroad tank cars in an industrial site in Festus, MO, and at a rail junction in a Chicago urban area. Figure 4.26 presents simulation results for chlorine dispersion from a hypothetical release scenario in the Chicago downtown area. Observed chlorine cloud is used as a

**FIGURE 4.26**

Graphical representation of tracer gas flume after 1000 s after a hypothetical toxic gas release.

*Reproduced by permission with Hanna et al. (2006).*

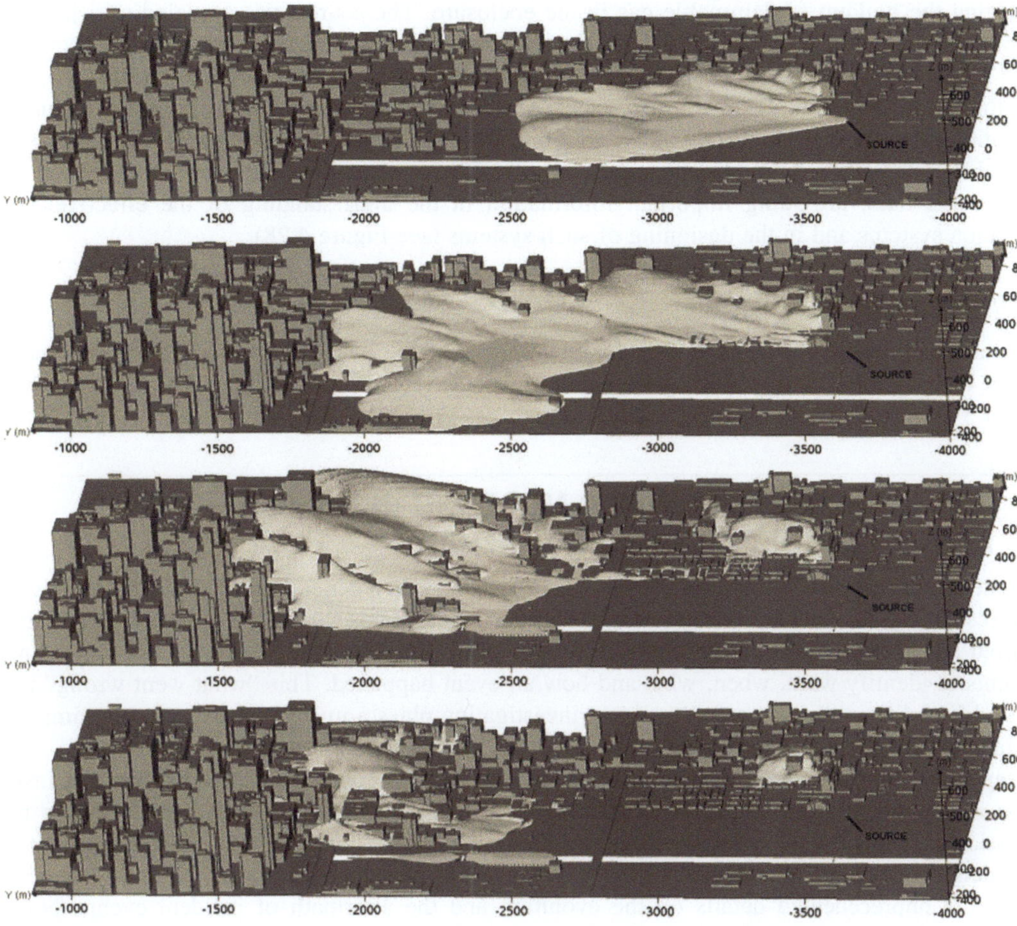

**FIGURE 4.27**

Graphical representation of the 100 ppm chlorine gas flume from CFD simulation.

*Reproduced by permission with Hanna et al. (2009).*

validation for their CFD model. The modeling results show that significant amounts of the hazardous gas, chlorine, may become trapped in obstacle wakes for many minutes after the main cloud has passed. This finding demonstrated the importance of fully understanding the effects of buildings and terrain on the dispersion of toxic gas. The "trapped" gas could increase the doses of those present in the affected areas. They also found that because of the dense gas nature, the presence of obstacles such as buildings or fences near the source further slowed the dispersion and even contained the gas and increased the concentration over time. However, obstacles may have some benefit further down wind as they help generate more turbulence, enhance mixing, and dilute the toxic gas.

For HVAC design, Ivings et al. (2010) used ANSYS CFX to conduct a CFD simulation study of low-pressure jet release in a ventilated enclosure to assess the effectiveness of the ventilation system in preventing the buildup of flammable gas in the enclosure. The source was modeled as a jet with a constant flow rate of 0.86 g/s, equivalent to the release rate through a 2.5-mm$^2$ hole at 1 bar pressure. Three locations of jet sources within the room (corresponding to the three configurations 1, 2, and 3 from the top in Figure 4.28) were investigated. Affecting factors addressed in this study include the ventilation rate, the distribution of the ventilation, the sources and the amount of the gas (toxic or flammable) released, and the obstacles present within the system. The results are in agreement with experimental data, providing important information in the understanding of the effectiveness of ventilation systems and in the designing of such systems (see Figure 4.28).

### 4.5.9 CFD CODES AVAILABLE

There are many available CFD models for toxic gas dispersion study. They are either commercial sources or available for free from government agencies (Hanna et al., 2006; Ivings et al., 2007). Table 4.5 lists several popular CFD codes and their developers.

## 4.6 APPLICATION OF COMPUTER MODELING IN INCIDENT INVESTIGATION AND RECONSTRUCTION

Incident investigation is an important topic in process safety. In a general sense, incident investigation in process safety, like any kind of retrospective investigation, helps shine light into events that already occurred and searches for comprehensive information of combinations of separate events leading to incidents to identify what, when, why, and how an event happened. This "what went wrong" information learned from process safety incident investigation plays a much broader and more important role in shaping future action to prevent the same and similar things from happening.

Depending on the type of process safety incident, methodologies available for incident investigation range widely from relatively simple for local and isolated incidents (Kletz, 2009) to more sophisticated methodologies to study catastrophic events. Among those, the use of computer modeling in incident investigation and reconstruction allows investigators to obtain additional multi-perspective views with unprecedented details on the evolution and the aftermath of incident events. Notable examples in using computer simulation in incident investigation are those by the Chemical Safety and Hazard Investigation Board whose detailed incident reconstructions and incident animations have been used extensively as training tools in process safety.

Historically, there are many catastrophic incidents whose general accounts are widely available but the details are still subject to debate, especially when evidence is scarce. This is where computer simulation could be conducted to fill in the gap to help explain the situation leading to an incident. For example, Khan and Amyotte (2007) investigated the BP Texas City refinery incident by constructing three different scenarios using different simulation tools (Shell Scope 4.1, Shell FRED 4.0, and MAXCRED) to assess the likelihood of what happened during the event including the distribution of the flammable cloud and the magnitude of the overpressure resulting from the explosion. Even though there were significant differences in those three scenarios, similar modeling results were observed, indicating that should computer modeling be used to assess possible consequences, reasonably accurate predictions obtained can help assess the probability and severity of disasters. Further investigation by Manca and Brambilla (2012) showed that the details of how the ISOM tower flooded and how the liquid coming out of the knockout pot have not been agreed on. They argued that three key

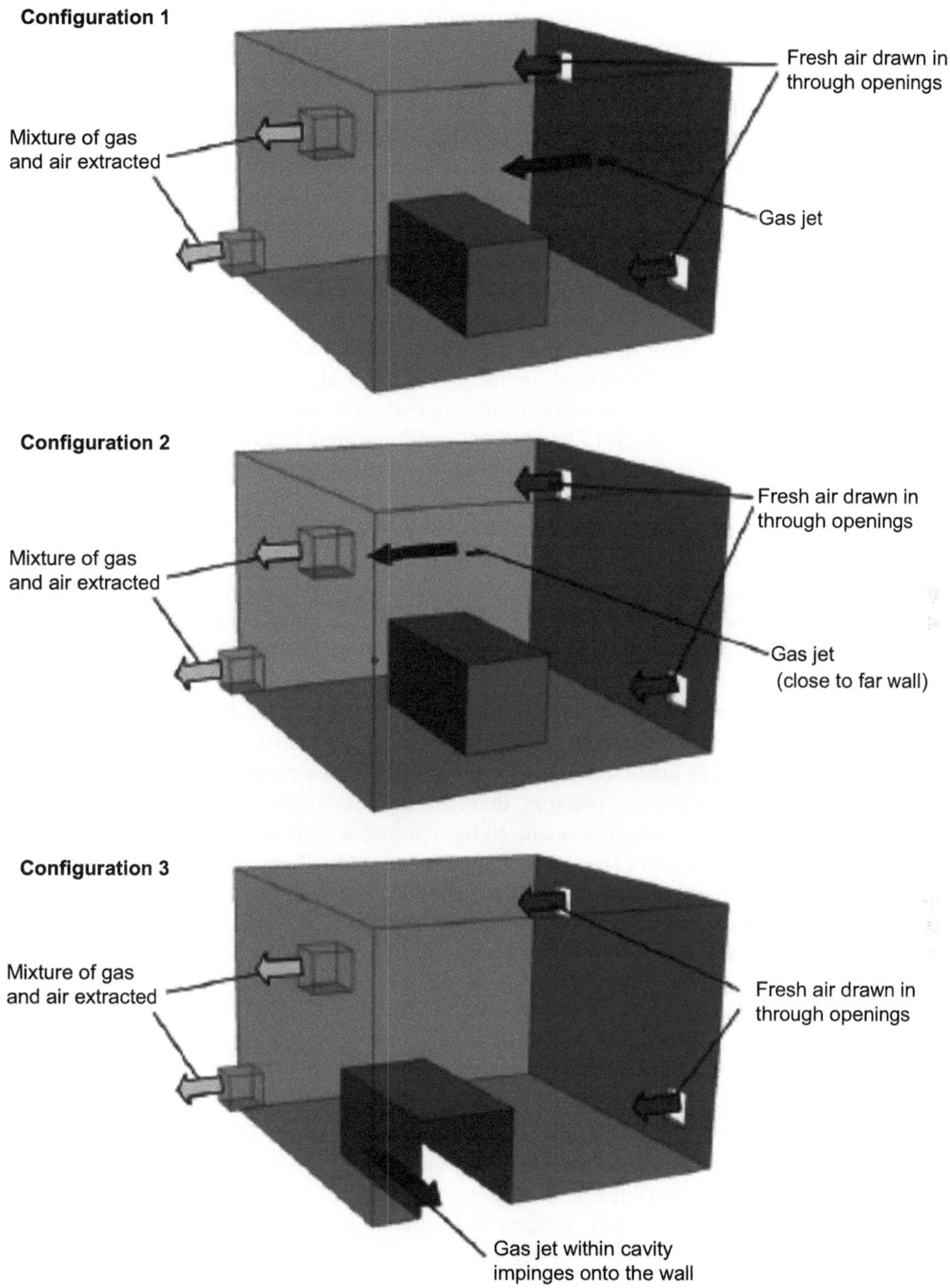


**FIGURE 4.28**

Graphical representation of the CFD simulation system for the investigation of low pressure jet release in a ventilated enclosure.

*Reproduced by permission with Ivings et al. (2010).*

issues related to the cause of the ISOM column overfilling, the two-phase flow in the pipe connected the ISOM column with the blowdown system, and the blowdown drum filling dynamics presented by a CSB report (Investigation Report Refinery Explosion and Fire: Final Investigation Report 2007) and a BP report may not be sufficient. Using dynamic process simulator, they found that the cause of the column overfilling cannot be explained by the mass balance and the liquid thermal expansion as well as the vapor cap formed inside the column. They argued using simulation results that the partial vaporization of the feed stream and the dispersion of vapor bubbles could be more likely to cause the column to overflow. They found that there were significant changes in the thermodynamic conditions along the pipe connecting the ISOM column with the blowdown drum; however, their homogeneous equilibrium model did not provide a good reason to explain the blowdown drum filling and release, suggesting that further studies should be conducted to achieve better understanding of the event.

Computer modeling has been used extensively by authors such as Venart and Hoiset et al. to investigate and reconstruct the Flixborough incident (Hoiset et al., 2000; Venart, 2004a,b, 2007a,b). Hoiset et al. (2000) found that the simulation results agreed well with literature data as it showed that the maximum overpressure may reach 15 bar and the ignition source had relatively insignificant effects on the overpressure value. It is interesting that conducting calculation, or simulation in this case, as found by Hoiset et al., could alter the conclusions of investigators regarding the magnitude of the explosion. For example, those who performed simulations had higher estimations in the maximum explosion pressure than those who only relied on visual inspection of the damage at the incident site. Although the actual data at the incident site (i.e., overpressure values), may never be obtainable in order to judge who is favored; these findings suggest that using well-validated modeling tools based on real physical phenomena with detailed reconstruction of multiple alternative scenarios at the incident scenes could be a better alternative than heuristic judgment based only on experience.

The use of computer modeling to gain further information in order to assess the credibility of multiple incident scenarios is highlighted by the work of Venart also on the Flixborough disaster (Venart, 2004a,b, 2007a,b). In order to explain how the bypass pipe ruptured leading to the release of the flammable material from the two reactors, there were three possibilities: it could be ruptured by (1) internal pressure; (2) a small explosion caused by flammable gas escaping from a small tear in the bellows (or a two-stage event); and (3) a small explosion caused by flammable gas from the 8-inch line rupture. By conducting extensive analysis of the computer simulation results on both bellow failure (finite element method using ANSYS-5.3 code) and the development of the flammable cloud (3D CFD using FloSYS CFD code), Venart showed that among those three scenarios, results for the two-stage event, in which the bypass rupture was in better agreement with a witness's account, indicated that this possibility was more likely to be the right explanation.

Computer modeling can also allow investigators a new look into certain features which could be overlooked or considered negligible. For example, Bakke et al. (2010) applied CFD computer simulation on the dispersion of a gasoline cloud in the Buncefield incident and obtained some interesting findings that help explain why a strong explosion was observed. CFD simulation, as shown in Figure 4.29, suggests that the layer of dense trees surrounding the tanks contributed significantly to the strong explosion observed at the Buncefield incident. The layer of trees could create and channel this strong explosion (Bakke et al., 2010). Other CFD simulation studies on the Buncefield incident indicate that it is possible to reconstruct the release event and the development of the flammable mist with reasonable accuracy (see Figure 4.30), providing more information for investigators to understand the details of the event (Gant and Atkinson, 2011). The flammable cloud is shown as the gray mass

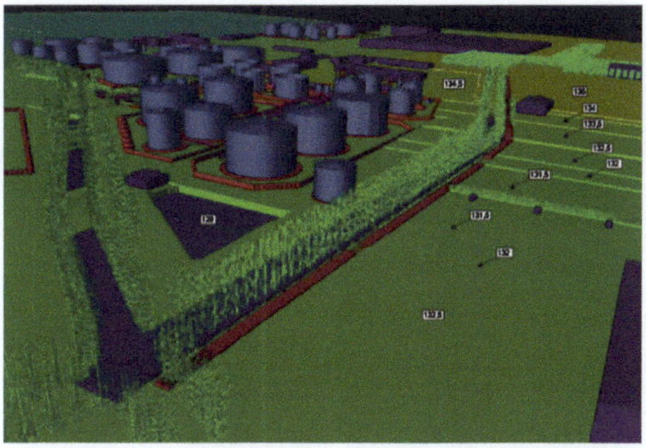

**FIGURE 4.29**

Graphical rendering of the 3D model for CFD simulation using FLACS of the Buncefield tank farm.

*Reproduced by permission with Bakke et al. (2010).*

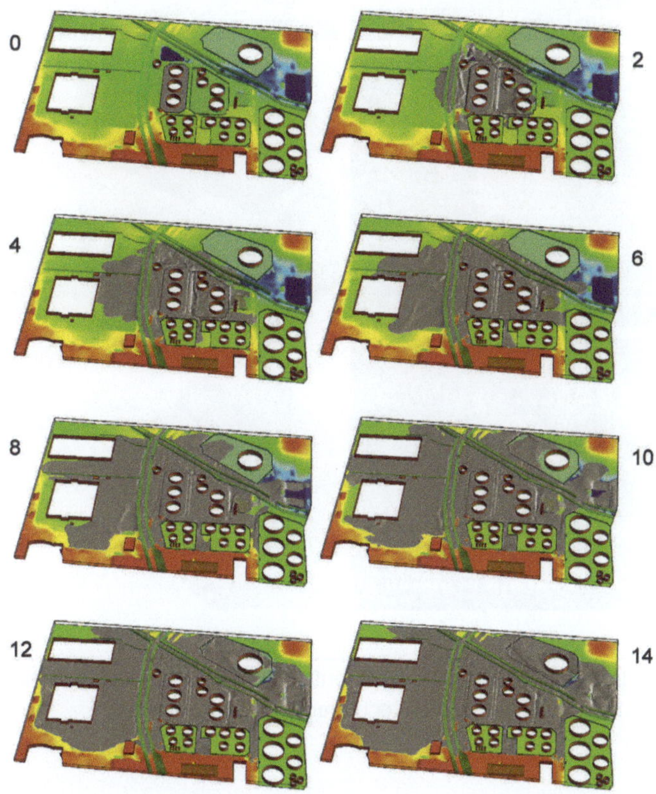


**FIGURE 4.30**

Graphical representation of the dispersion of the flammable vapor cloud over time in the Buncefield incident from CFD simulation results.

*Reproduced by permission with Gant and Atkinson (2011).*

**FIGURE 4.31**

Frames from Fox news videos (left) and graphic representations of CFD simulation using FLACS (right) of the liquefied chlorine release from a rail car incident occurred in Festus, Missouri.

*Reproduced by permission with Hanna et al. (2009).*

developing from the wall of Bund A in sequence starting from time 0 to time 14 min (Figure 4.29). The color code represents the elevation of the ground.

Pontiggia et al. (2011) used CFD modeling for the investigation of an LPG railcar accident in Viareggio, Italy. In this incident, one of the 14 railcars involved ruptured and released approximately 46.7 tons of LPG, forming a flammable cloud traveling into nearby buildings and residential apartments. The subsequent flash fire from the ignition of the cloud caused extended flash fire and confined fires and explosions in several places where the flammable gas was trapped. The authors were able to reconstruct the release of LPG and the evolution of the flammable cloud in terms of LFL and UFL using CFD simulation whose predictions indicated that the large obstacles in simulation (e.g., houses and garages at the incident site) had a significant effect on the cloud distribution. In the simulation, the concentration of the flammable cloud inside the underground garage was greater than LFL/2, while the LFL front was several meters away. The incident report shows that a partially confined explosion was observed at this site, indicating that modeling reconstruction was reasonable in presenting the cloud concentration.

Hanna et al. (2009) conducted a CFD simulation to reconstruct the liquefied chlorine release from a rail car incident that occurred in Festus, Missouri. The authors found that several main features observed in this incident were well simulated using FLACS, including the chlorine jet shape, jet ground impingement, flows under the railcar, horizontal and vertical dimension of the cloud, and the concentration distribution 10 min after the release (see Figure 4.31). This result suggests that using CFD simulation tools such as well-validated FLACS will allow investigators to have a more detailed description of the evolution of incident events.

In summary, computer modeling has been used extensively in incident investigation and reconstruction because of multiple reasons such as its unique ability in capturing and visually presenting detailed information of investigated incidents, as well as its capability in examining alternative scenarios to shine light into unknown factors. To this extent, however, computer modeling, just like any other investigation tools and techniques available, has its own limitations and cannot and should not be relied on as the only source of information. It is well known that computer simulation and its prediction should not be treated as a black box, and the accuracy of a modeling prediction is based on factors such as simulation parameters, scenario definition, and codes calibration, which are in turn dependent on the expertise and experience of the modelers. Those factors need to be carefully considered and results need to be validated against experimental observations in every case.

## REFERENCES


Adiga, K.C., Ramaker, D.E., Tatem, P.A., Williams, F.W., 1989. Modeling pool-like gas flames of propane. Fire Safety Journal 14 (4), 241–250.

Adiga, K.C., Ramaker, D.E., Tatem, P.A., Williams, F.W., 1990. Numerical predictions for a simulated methane fire. Fire Safety Journal 16 (6), 443–458.

American Institute of Chemical Engineers Center for Chemical Process Safety, 2010. Guidelines for Vapor Cloud Explosion, Pressure Vessel Burst, BLEVE, and Flash Fire Hazards. Wiley, Hoboken, NJ.

Aloqaily, A., Chakrabarty, A., 2010. Jet flame length and thermal radiation: evaluation with CFD simulations. In: American Institute of Chemical Engineers 6th Global Congress on Process Safety, 2010 Spring Meeting, San Antonio, TX.

Argyropoulos, C.D., Sideris, G.M., Christolis, M.N., Nivolianitou, Z., Markatos, N.C., 2010. Modelling pollutants dispersion and plume rise from large hydrocarbon tank fires in neutrally stratified atmosphere. Atmospheric Environment 44 (6), 803–813.

Arntzen, B.J., 1998. Modeling of turbulence and combustion for simulation of gas explosion in complex geometries. In: Dr Ing. NTNU.

Baik, J.J., Kim, J.J., Fernando, H.J.S., 2003. A CFD model for simulating urban flow and dispersion. Journal of Applied Meteorology 42 (11), 1636–1648.

Bakke, J.R., van Wingerden, K., Hoorelbeke, P., Brewerton, B., 2010. A study on the effect of trees on gas explosions. Journal of Loss Prevention in the Process Industries 23 (6), 878–884.

Baraldia, D., Kotchourkob, A., Lelyakinb, A., Yanezb, J., Middhac, P., Hansenc, O.R., Gavrikovd, A., Efimenkod, A., Verbeckee, F., Makarove, D., Molkove, V., 2009. An inter-comparisonexercise on CFDmodel capabilities to simulate hydrogen deflagrations in a tunnel. International Journal of Hydrogen Energy 34 (18), 7862–7872.

Bartzis, J.G., Sfetsos, A., Efthimiou, G., Andronopoulos, S., Venetsanos, A., 2007. Validation excercise utilizing adrea and STAR-CD codes in urban scale. In: The Msut Experiment. 11th Conference on Harmonisation within Atmospheris Dispersion Modelling for Regulatory Purposes, Cambirdge, UK.

Baukal, C.E., Schwartz, R., Company, J.Z., 2001. The John Zink Combustion Handbook. CRC Press, Boca Raton, FL.

Bilger, R.W., 1977. Reaction-rates in diffusion flames. Combustion and Flame 30 (3), 277–284.

Bird, R.B., Stewart, W.E., Lightfoot, E.N., 2002. Transport Phenomena. J. Wiley, New York.

Blazek, J., 2007. Computational Fluid Dynamics: Principles and Applications. Elsevier, Oxford, UK.

Brambilla, S., Manca, D., 2010. The Viareggio LPG railway accident: event reconstruction and modeling. Journal of Hazardous Materials 182 (1–3), 346–357.

Bray, K.N.C., 1990. Studies of the turbulent burning velocity. Proceedings: Mathematical and Physical Sciences 431 (1882), 315–335.

Brennan, S.L., Makarov, D.V., Molkov, V., 2009. LES of high pressure hydrogen jet fire. Journal of Loss Prevention in the Process Industries 22 (3), 353–359.

Brookes, S.J., Moss, J.B., 1999. Predictions of soot and thermal radiation properties in confined turbulent jet diffusion flames. Combustion and Flame 116 (4), 486–503.

Calay, R.K., Holdo, A.E., 2008. Modelling the dispersion of flashing jets using CFD. Journal of Hazardous Materials 154 (1–3), 1198–1209.

Calhoun, R., Gouveia, F., Shinn, J., Chan, S., Stevens, D., Lee, R., Leone, J., 2005. Flow around a complex building: experimental and large-eddy simulation comparisons. Journal of Applied Meteorology 44 (5), 571–590.

Camelli, F.E., Lohner, R., 2004. Assessing maximum possible damage for containment release events. Engineering Computations 21 (7), 748–760.

Camelli, F.E., Lohner, R., Hanna, S., 2004. FEFLO-URBAN CFD model evaluation of the MUST experiment. In: Buscaglia, G., Dari, E., Zamonsky, O. (Eds.), Mecanica Computacional, vol. XXIII.

Cant, R.S., Dawes, W.N., Savill, A.M., 2004. Advanced CFD and modeling of accidental explosions. Annual Review of Fluid Mechanics 36, 97–119.

Chakrabarty, A., Raibagkar, A., Aloqaily, A., 2011. Thermal Hazard evaluation for process buildings using CFD analysis techniques. In: Hazards XXII. IChemE, Liverpool, UK.

Clutter, J.K., Mathis, J., 2002. Computational modeling of vapor cloud explosions in offshore rigs using a flame speed based on combustion models. Journal of Loss Prevention in the Process Industries 15, 391–401.

Coirier, W.J., Kim, S., 2006a. Summary of CFD-urban results in support of the Madison Square Garden and urban dispersion program field texts. In: Sixth Symposium on the Urban Environment, Atlanta, GA.

Coirier, W.J., Kim, S., 2006b. CFD modeling for urban area contaminant transport and dispersion: model description and data requirements. In: Sixth Symposium on the Urban Environment, Atlanta, GA.

Colin, O., Ducros, F., Veynante, D., Poinsot, T., 2000. A thickened flame model for large eddy simulations of turbulent premixed combustion. Physics of Fluids 12 (1843).

Collecutt, G., Humphreys, D., Proud, D., 2009. CFD simulation of underground coal dust explosion and active explosion barriers. In: Seventh International Conference on CFD in the Minerals and Process Industries. CSIRO, Melbourne, Australia.

Cormier, B.R., Qi, R.F., Yun, G.W., Zhang, Y.C., Mannan, M.S., 2009. Application of computational fluid dynamics for LNG vapor dispersion modeling: a study of key parameters. Journal of Loss Prevention in the Process Industries 22 (3), 332–352.

Cox, G., Kumar, S., 1987. Field modeling of fire in forced ventilated enclosures. Combustion Science and Technology 52 (1–3), 7–23.

Cozzania, V., Tugnolia, A., Salzanob, E., 2009. The development of an inherent safety approach to the prevention of domino accidents. Accident Analysis & Prevention 41 (6), 1216–1227.

Crowl, D.A., Louvar, J.F., 2002. Chemical Process Safety: Fundamentals with Applications. Prentice Hall PTR, Upper Saddle River, NJ.

Cumber, P.S., Spearpoint, M., 2006. A computational flame length methodology for propane jet fires. Fire Safety Journal 41 (3), 215–228.

Darbra, R.M., Palacios, A., Casal, J., 2012. Domino effect in chemical accidents: main features and accident sequences. Journal of Hazardous Materials 183 (1–3), 565–573.

Dorofeev, S.B., Bezmelnitsin, A.V., Sidorov, V.P., 1995. Transition to detonation in vented hydrogen air explosion. Combustion and Flame 103, 243–246.

Droste, B., Probst, U., Heller, W., 1999. Impact of an exploding LPG rail tank car onto a castor spent fuel cask. RAMTRANS (Nulcear Technology Publishing) 10 (4), 231–240.

Durand, L., Polifke, W., 2007. Implementation of the thickened flame model for large eddy simulation of turbulent premixed combustion in a commercial solver. Proceedings of the American Society of Mechanical Engineers Turbo Expo 2, 869–878.

Eckhoff, R.K., 2003. Dust Explosions in the Process Industries. Elsevier Science, USA.

Efimenko, A.A., Dorofeev, S.B., 2001. CREBCOM code system for description of gaseous combustion. Journal of Loss Prevention in the Process Industries 14 (6), 575–581.

Fairweather, M., Hargrave, G.K., Ibrahim, S.S., Walker, D.G., 1999. Studies of premixed flame propagation in explosion tubes. Combustion and Flame 116 (4), 504–518.

Gamezo, V.N., Khokhlov, A.M., Oran, E.S., 2001. The influence of shock bifurcations on shock-flame interactions and DDT. Combustion and Flame 126 (4), 1810–1826.

Gamezo, V.N., Ogawa, T., Oran, E.S., 2008. Flame acceleration and DDT in channels with obstacles: effect of obstacle spacing. Combustion and Flame 155 (1–2), 302–315.

Gamezo, V.N., Oran, E.S., Khokhlov, A.M., 2005. Three-dimensional reactive shock bifurcations. Proceedings of the Combustion Institute 30, 1841–1847.

Gant, S.E., Atkinson, G.T., 2011. Dispersion of the vapour cloud in the Buncefield Incident. Process Safety and Environmental Protection 89 (6), 391–403.

Geng, J.H., Thomas, J.K., April 2007. Simulation and application of blast wave-target intercation. In: 41st Annual Loss Prevention Symposium, Houston, Texas.

Hadjisophocleous, G.V., McCartney, C., 2005. Guidelines for the use of CFD simulations for fire and smoke modeling. ASHRAE Transactions 111 (Pt 2), 583–594.

Hanna, S.R., Brown, M.J., Camell, F.E., Chan, S.T., Coirier, W.J., Hansen, O.R., Huber, A.H., Kim, S., Reynolds, R.M., 2006. Detailed simulations of atmospheric flow and dispersion in downtown Manhattan:

an application of five computational fluid dynamics models. Bulletin of the American Meteorological Society 87 (12), 1713–1726.

Hanna, S.R., Chang, J.C., Strimaitis, D.G., 1993. Hazardous gas-model evaluation with field observations. Atmospheric Environment Part A-General Topics 27 (15), 2265–2285.

Hanna, S.R., Hansen, O.R., Dharmavaram, S., 2004. FLACS CFD air quality model performance evaluation with Kit Fox, MUST, Prairie Grass, and EMU observations. Atmospheric Environment 38 (28), 4675–4687.

Hanna, S.R., Hansen, O.R., Ichard, M., Strimaitis, D., 2009. CFD model simulation of dispersion from chlorine railcar releases in industrial and urban areas. Atmospheric Environment 43 (2), 262–270.

Hanna, S.R., Tehranian, S., Carissimo, B., Macdonald, R.W., Lohner, R., 2002. Comparisons of model simulations with observations of mean flow and turbulence within simple obstacle arrays. Atmospheric Environment 36 (32), 5067–5079.

Hoiset, S., Hjertager, B.H., Solberg, T., Malo, K.A., 2000. Flixborough revisited - an explosion simulation approach. Journal of Hazardous Materials 77 (1–3), 1–9.

Houf, W., Schefer, R., Evans, G., Merilo, E., Groethe, M., 2010. Evaluation of barrier walls for mitigation of unintended releases of hydrogen. International Journal of Hydrogen Energy 35 (10), 4758–4775.

Houf, W.G., Evans, G.H., Schefer, R.W., 2009. Analysis of jet flames and unignited jets from unintended releases of hydrogen. International Journal of Hydrogen Energy 34 (14), 5961–5969.

Houf, W.G., Evans, G.H., Schefer, R.W., Merilo, E., Groethe, M., 2011. A study of barrier walls for mitigation of unintended releases of hydrogen. International Journal of Hydrogen Energy 36 (3), 2520–2529.

Huber, A., Freeman, M., Spencer, R., Bell, B., Kuehlert, K., Schwarz, W., 2005. Applications of CFD simulations of pollutant transport and dispersion within ambient urban building environments: including homeland security. In: AWMA 98th Annual Conf., Minneapolis, MN.

Investigation Report Refinery Explosion and Fire: Final Investigation Report, 2007. T. C. S. Board, BP Texas City, Texas.

Ivings, M.J., Gant, S.E., Saunders, C.J., Pocock, D.J., 2010. Flammable gas cloud build up in a ventilated enclosure. Journal of Hazardous Materials 184 (1–3), 170–176.

Ivings, M.J., Jagger, S.F., Lea, C.J., Webber, D.M., 2007. Evaluating Vapor Dispersion Models for Safety Analysis of LNG Facilities. U. H. S. Laboratory, Derbyshire, UK.

Khan, F.I., Amyotte, P.R., 2007. Modeling of BP texas city refinery incident. Journal of Loss Prevention in the Process Industries 20 (4–6), 387–395.

Khokhlov, A.M., Oran, E.S., 1999. Numerical simulation of detonation initiation in a flame brush: the role of hot spots. Combustion and Flame 119 (4), 400–416.

Khokhlov, A.M., Oran, E.S., Thomas, G.O., 1999. Numerical simulation of deflagration-to-detonation transition: the role of shock-flame interactions in turbulent flames. Combustion and Flame 117 (1–2), 323–339.

Kim, B.J., Seo, J.K., Park, J.H., Jeong, J.S., Oh, B.K., Kim, S.H., Park, C.H., Paik, J.K., 2010. Load characteristics of steel and concrete tubular members under jet fire: an experimental and numerical study. Ocean Engineering 37 (13), 1159–1168.

Kletz, T.A., 2009. What Went Wrong?: Case Histories of Process Plant Disasters and How They Could Have Been Avoided. Gulf Professional Pub, Burlington, MA.

Klimenko, A.Y., Bilger, R.W., 1999. Conditional moment closure for turbulent combustion. Progress in Energy and Combustion Science 25 (6), 595–687.

Koutsourakis, N., Venetsanos, A.G., Bartzis, J.G., 2012. LES modelling of hydrogen release and accumulation within a non-ventilated ambient pressure garage using the ADREA-HF CFD code. International Journal of Hydrogen Energy 37 (22).

Krause, U., Kasch, T., 2000. The influence of flow and turbulence on flame propagation through dust-air mixtures. Journal of Loss Prevention in the Process Industries 13 (3–5), 291–298.

Lea, C.J., 2002. A Review of the State of the Art in Gas Explosion Modeling. U. H. A. S. Laboratory. Harpur Hill, Buxton, UK.

Lecocq, G., Richard, S., Colin, O., Vervisch, L., 2011. Hybrid presumed pdf and flame surface density approaches for large-eddy simulation of premixed turbulent combustion. Part 2: early flame development after sparking. Combustion and Flame 158 (6), 1215–1226.

Lee, J.H., Knystautas, R., Yoshikawa, N., 1978. Photochemical initiation of gaseous detonations. Acta Astronautica 5 (11–12), 971–982.

Lin, C.-H., Ferng, Y.-M., Hsu, W.-S., Pei, B.-S., 2010. Investigations on the characteristics of radiative heat transfer in liquid pool fires. Fire Technology 46, 321–345.

Lockwood, F.C., Shah, N.G., 1981. A new radiation solution method for incorporation in general combustion prediction procedures. In: 18th Symposium (International) on Combustion. The Combustion Institute, Pittsburgh, PA, pp. 1405–1414.

Lowesmith, B.J., Hankinson, G., Acton, M.R., Chamberlain, G., 2007. An overview of the nature of hydrocarbon jet fire hazards in the oil and gas industry and a simplified approach to assessing the hazards. Process Safety and Environmental Protection 85 (B3), 207–220.

Luketa-Hanlin, A., 2006. A review of large-scale LNG spills: experiments and modeling. Journal of Hazardous Materials 132 (2–3), 119–140.

Luther, W., Muller, W.C., 2009. FDS simulation of the fuel fireball from a hypothetical commercial airliner crash on a generic nuclear power plant. Nuclear Engineering and Design 239 (10), 2056–2069.

Magnussen, B.F., Hjertager, B.H., 1977. On mathematical modeling of turbulent combustion with special emphasis on soot formation and combustion. Symposium (International) on Combustion 16 (1), 719–729.

Makhviladze, G.M., Yakush, S.E., 2005. Modelling of formation and combustion of accidentally released fuel clouds. Process Safety and Environmental Protection 83 (B2), 171–177.

Manca, D., Brambilla, S., 2012. Dynamic simulation of the BP Texas City refinery accident. Journal of Loss Prevention in the Process Industries 25 (6), 950–957.

Mannan, S., 2005. Lees' Loss Prevention in the Process Industries, Third Edition: Hazard Identification, Assessment and Control. Elsevier.

Markatos, N.C., Christolis, C., Argyropoulos, C., 2009. Mathematical modeling of toxic pollutants dispersion from large tank fires and assessment of acute effects for fire fighters. International Journal of Heat and Mass Transfer 52 (17–18), 4021–4030.

Meyer, J.W., Urtiew, P.A., Oppenheim, A.K., 1970. On the inadequacy of gasdynamic processes for triggering the transition to detonation. Combustion and Flame 14 (1), 13–20.

Middha, P., Hansen, O.R., 2008. Predicting deflagration to detonation transition in hydrogen explosions. Process Safety Progress 27 (3), 192–204.

Middha, P., Hansen, O.R., Storvik, I.E., 2009. Validation of CFD-model for hydrogen dispersion. Journal of Loss Prevention in the Process Industries 22 (6), 1034–1038.

Middha, P., Ichard, M., Arntzen, B.J., 2011. Validation of CFD modelling of $LH_2$ spread and evaporation against large-scale spill experiments. International Journal of Hydrogen Energy 36 (3), 2620–2627.

Novozhilov, V., 2001. Computational fluid dynamics modeling of compartment fires. Progress in Energy and Combustion Science 27 (6), 611–666.

Novozhilov, V., Moghtaderi, B., Fletcher, D.F., Kent, J.H., 1996. Computational fluid dynamics modelling of wood combustion. Fire Safety Journal 27 (1), 69–84.

O'Donnell, K., Lea, C., 2008. Assessment of gas detection strategies for offshore HVAC ducts based on CFD modelling. In: 53rd Technical Meeting on "Dispersion Modelling". FABIG, Ascot, Berkshire, UK.

Olewski, T., Nayak, S., Basha, O., Waldram, S., Vechot, L., 2011. Medium scale LNG-related experiments and CFD simulation of water curtain. Journal of Loss Prevention in the Process Industries 24 (6), 798–804.

Oran, E.S., Gamezo, V.N., 2007. Origins of the deflagration-to-detonation transition in gas-phase combustion. Combustion and Flame 148 (1–2), 4–47.

Orloff, L., Deris, J., Delichatsios, M.A., 1987. Chemical effects on molecular-species concentrations in turbulent fires. Combustion and Flame 69 (3), 273–289.

Papageorgakis, G.C., Assanis, D.N., 1999. Comparison of linear and nonlinear RNG-based k-epsilon models for incompressible turbulent flows. Numerical Heat Transfer Part B-Fundamentals 35 (1), 1–22.

Patankar, S.V., Spalding, D.B., 1974. A calculation procedure for the transient and steady-state behavior os shell- and -tube heat exchangers. In: Afgan, N.H., Schlunder, E.V. (Eds.), Heat Exchangers: Design and Theory Sourcebook. McGraw-Hill, New York, pp. 155–176.

Pontiggia, M., Derudi, M., Alba, M., Scaioni, M., Rota, R., 2010. Hazardous gas releases in urban areas: assessment of consequences through CFD modelling. Journal of Hazardous Materials 176 (1–3), 589–596.

Pontiggia, M., Landucci, G., Busini, V., Derudi, M., Alba, M., Scaioni, M., Bonvicini, S., Cozzani, V., Rota, R., 2011. CFD model simulation of LPG dispersion in urban areas. Atmospheric Environment 45 (24), 3913–3923.

Popat, N.R., Catlin, C.A., Arntzen, B.J., Lindstedt, R.P., Hjertager, B.H., Solberg, T., Saeter, O., VandenBerg, A.C., 1996. Investigations to improve and assess the accuracy of computational fluid dynamic based explosion models. Journal of Hazardous Materials 45 (1), 1–25.

Pope, S.B., 1985. Pdf methods for turbulent reactive flows. Progress in Energy and Combustion Science 11 (2), 119–192.

Qi, R.F., Ng, D., Cormier, B.R., Mannan, M.S., 2010. Numerical simulations of LNG vapor dispersion in Brayton Fire Training Field tests with ANSYS CFX. Journal of Hazardous Materials 183 (1–3), 51–61.

Qi, R.F., Prem, K.P., Ng, D., Rana, M.A., Yun, G., Mannan, M.S., 2012. Challenges and needs for process safety in the new millennium. Process Safety and Environmental Protection 90 (2), 91–100.

Qiao, A., Zhang, S., 2010. Advanced CFD modeling on vapor dispersion and vapor cloud explosion. Journal of Loss Prevention in the Process Industries 23 (6), 843–848.

Radandt, S., Shi, J.Y., Vogl, A., Deng, X.F., Zhong, S.J., 2001. Cornstarch explosion experiments and modeling in vessels ranged by height/diameter ratios. Journal of Loss Prevention in the Process Industries 14 (6), 495–502.

Raj, P.K., 2005. Large LNG fire thermal radiation - modeling issues and hazard criteria revisited. Process Safety Progress 24 (3), 192–202.

Raj, P.K., 2007a. LNG fires: a review of experimental results, models and hazard prediction challenges. Journal of Hazardous Materials 140 (3), 444–464.

Raj, P.K., 2007b. LNG pool fire spectral data and calculation of emissive power. Journal of Hazardous Materials 142 (3), 720–729.

Raj, P.K., 2007c. Large hydrocarbon fuel pool fires: physical characteristics and thermal emission variations with height. Journal of Hazardous Materials 140 (1–2), 280–292.

Rasbash, D., 2004. Evaluation of Fire Safety. West Sussex, England; Hoboken, NJ, J. Wiley & Sons.

Rew, P.J., Hulbert, W.J., 1996. Development of Pool Fire Thermal Radiation Model. U. HSE.

Richards, K., 2002. Computational Modelling of Polution Dispersion Near to the Wake of Vehicle (Ph.D. thesis). University of Nottingham.

Schalike, S., Wehrstedt, K.D., Schonbucher, A., 2011. CFD simulation to predict the thermal radiation of large LNG pool fires. In: Proceedings of the European Combustion Meeting.

Sherman, M.P., Tieszen, S.R., Benedick, W.B., 1989. The Effect of Obstacles and Transverse Venting on Flame Acceleration and Transition to Detonation for Hydrogen-air Mixtures at Large Scale. S. N. Laboratories, Albuquerque, NM, USA. SAND85–1264, R3.

Shi, J.Y., Li, X.G., Ping, Z., Radandt, S., 2009. Dust explosion simulations in a filter using DESC. In: 8th International Conference on Measurement and Control of Granular Materials, Proceedings, pp. 353–357.

Sinai, Y.L., 2000. Exploratory CFD modelling of pool fire instabilities without cross-wind. Fire Safety Journal 35, 51–61.

Sinai, Y.L., Owens, M.P., 1995. Validation of CFD modelling of unconfined pool fires with cross-wind: flame geometry. Fire Safety Journal 24, 1–34.

Skjold, T., 2007. Review of the DESC project. Journal of Loss Prevention in the Process Industries 20 (4–6), 291–302.

Skjold, T., 2010. Modelling of industrial dust explosions in complex geometries. In: Forty-sixth UKELG One-day Discussion Meeting on "Causes, Severity and Mitigation of Aerosol and Particulate Explosions". Imperial College, London, UK.

Skjold, T., Arntzen, B.J., Hansen, O.R., Storvik, I.E., Eckhoff, R.K., 2006. Simulation of dust explosions in complex geometries with experimental input from standardized tests. Journal of Loss Prevention in the Process Industries 19 (2–3), 210–217.

Skjold, T., Arntzen, B.J., Hansen, O.R., Taraldset, O.J., Storvik, I.E., Eckhoff, R.K., 2005. Simulating dust explosions with the first version of DESC. Process Safety and Environmental Protection 83 (B2), 151–160.

Sklavounos, S., Rigas, F., 2004. Validation of turbulence models in heavy gas dispersion over obstacles. Journal of Hazardous Materials 108 (1–2), 9–20.

Sklavounos, S., Rigas, F., 2006. Simulation of Coyote series trials—part I: CFD estimation of non-isothermal LNG releases and comparison with box-model predictions. Chemical Engineering Science 61 (5), 1434–1443.

Suardin, J.A., Wang, Y.J., Willson, M., Mannan, M.S., 2009. Field experiments on high expansion (HEX) foam application for controlling LNG pool fire. Journal of Hazardous Materials 165 (1–3), 612–622.

Tan, W., Liu, X., Liu, Y., Bua, Y., Wang, Z., 2011. Modeling the two-phase cloud evolution from instantaneous flashing release using CFD. Journal of Loss Prevention in the Process Industries 24, 420–425.

Tegner, J., Sjogreen, B., 2002. Numerical investigation of deflagration to detonation transitions. Combustion Science and Technology 174 (8), 153–186.

Tewari, M., Kusaka, H., Chen, F., Coirier, W.J., Kim, S., Wyszogrodzki, A.A., Warner, T.T., 2010. Impact of coupling a microscale computational fluid dynamics model with a mesoscale model on urban scale contaminant transport and dispersion. Atmospheric Research 96 (4), 656–664.

Tu, J., Yeoh, G.H., Liu, C., 2008. Computational Fluid Dynamics - A Practical Approach, second ed. Elsevier.

Tugnoli, A., Khan, F., Amyotte, P., Cozzani, V., 2008. Safety assessment in plant layout design using indexing approach: implementing inherent safety perspective: part 2—Domino Hazard Index and case study. Journal of Hazardous Materials 160 (1), 110–121.

Vaagsaether, K., Knudsen, V., Bjerketvedt, D., 2007. Simulation of flame acceleration and DDT in $H_2$-air mixture with a flux limiter centered method. International Journal of Hydrogen Energy 32 (13), 2186–2191.

Venart, J., 2007a. Flixborough triggers. Tce (793), 62.

Venart, J.E.S., 2004a. Flixborough: the explosion and its aftermath. Process Safety and Environmental Protection 82 (B2), 105–127.

Venart, J.E.S., 2004b. Flixborough explosion: case closed? Progress in Safety Science and Technology 4 (Pts A and B 4), 1083–1097.

Venart, J.E.S., 2007b. Flixborough: a final footnote. Journal of Loss Prevention in the Process Industries 20 (4–6), 621–643.

Venetsanos, A.G., Huld, T., Adams, P., Bartzis, J.G., 2003. Source, dispersion and combustion modelling of an accidental release of hydrogen in an urban environment. Journal of Hazardous Materials 105 (1–3), 1–25.

Vianna, S.S.V., Cant, R.S., 2010. Modified porosity approach and laminar flamelet modelling for advanced simulation of accidental explosions. Journal of Loss Prevention in the Process Industries 23 (1), 3–14.

van Wingerden, K., Pedersen, G., Wilkins, B., Berg, M., Nielsen, N.O.F., 2011. A full-scale experimental and modeling investigation of dust explosions in a roller mill. Process Safety Progress 30 (1), 87–96.

Williams, M.D., Brown, M.J., Boswell, D., Singh, B., Pardyjak, E., 2004. Testing of the QUIC-plume model with wind-tunnel measurements for a high-rise building. In: 13th Conf. On Applic. of Air Poll. Meteor. with AWMA, Amer. Meteor. Soc., Vancouver, Canada.

Yan, Z.H., Holmstedt, G., 1999. Three-dimensional computation of heat transfer from flames between vertical parallel walls. Combustion and Flame 117 (3), 574–588.

Yang, Z., Lai, F.B., Liu, L., 2011. Numerical modeling of fires on gas pipelines. Applied Thermal Engineering 31 (6–7), 1347–1351.

Yee, E., Biltoft, C.A., 2004. Concentration fluctuation measurements in a plume dispersing through a regular array of obstacles. Boundary-Layer Meteorology 111 (3), 363–415.

Yeoh, G.H., Yuen, K.K., 2009. Computational Fluid Dynamics in Fire Engineering Theory, Modelling and Practice. Elsevier.

Zalosh, R., 2008. Explosion Venting Data and Modelig Research Project. The Fire Protection Rsearch Foundation, MA, USA.

Zhang, Z., Chen, Q., 2007. Comparison of the Eulerian and Lagrangian methods for predicting particle transport in enclosed spaces. Atmospheric Environment 41 (25), 5236–5248.

Zhong, S., Deng, X., Shi, J., Radandt, S., 2001. Modelling of vented maize starch explosions in a silo. Safe Handling of Combustible Dusts 1601, 423–435.

Zhong, S.J., Deng, X.F., 2000. Modeling of maize starch explosions in a 12 $m^3$ silo. Journal of Loss Prevention in the Process Industries 13 (3–5), 299–309.

Zydak, P., Klemens, R., 2007. Modelling of dust lifting process behind propagating shock wave. Journal of Loss Prevention in the Process Industries 20 (4–6), 417–426.

CHAPTER

# FINITE ELEMENT ANALYSIS IN PROCESS SAFETY APPLICATIONS

5

## 5.1 INTRODUCTION

Finite element analysis is a dominant computational method in science and engineering. It is a numerical procedure that can be applied to obtain solutions to a variety of problems in engineering including steady, transient, linear, or nonlinear problems. Advancement of computational power along with advanced computational methods, applied mathematics, has aided the broad implementation of finite element–based approaches in designing products and processes. Similar to other modeling approaches, the mathematical models describing the physical phenomena form the foundation of finite element–based methods. Typically, representing a real-world problem into a mathematical problem requires one to explain the physical phenomena in the form of or combination of partial differential equations, ordinary differential equations, and algebraic equations. The exact solutions through analytical approaches in solving these are a rare possibility for real-world problems. A finite element method provides a practical alternative for solving these issues, especially for scenarios involving complex geometries and boundary conditions, which is usually the case in representing real-world problem scenarios. Within the context of process safety applications, finite element analysis has been proved to be a useful tool for gaining valuable insights as illustrated in the following sections. While for practical purposes it is not required to have a full understanding of the theoretical aspect of finite element analysis, learning tools only without some basic understanding of underlying mathematics and principles can be an expensive mistake. Barring some brief overview in the next section, this book does not intend to provide fundamentals of finite element analysis. Books are suggested for further reading (Hutton, 2005; Moaveni, 2015; Cook et. al., 2001; Zienkiewicz et al., 2013a; Brebbia, 2013). Here, we have provided few examples demonstrating the potential of finite element analysis in process safety applications. The examples are by no means comprehensive.

## 5.2 FINITE ELEMENT ANALYSIS

Finite element analysis starts with dividing the geometry of the problem. In other words, discretization of the whole domain of interest into a number of finite elements or discrete sub-regions is fundamental in the preprocessing part of finite element analysis. Hence, finite element analysis is also defined as a *general discretization procedure of continuum problems posed by mathematically defined problems* (Zienkiewicz et al., 2013a). The process of discretization is commonly known as meshing. The building block of subdivision or discretization of geometry through meshing can take

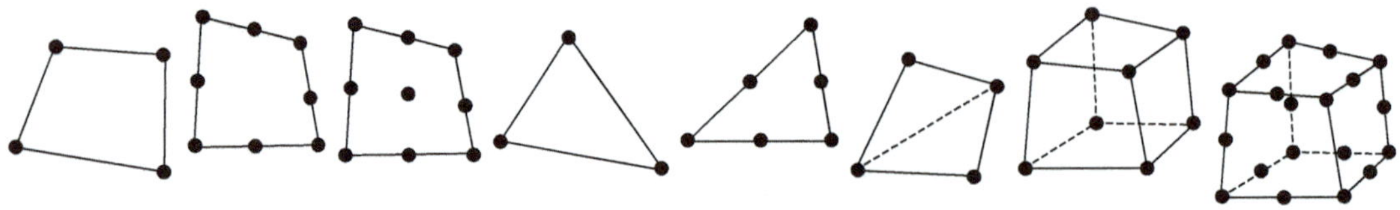

**FIGURE 5.1**

Examples of mesh element and nodes.

*Tabatabaian (2013).*

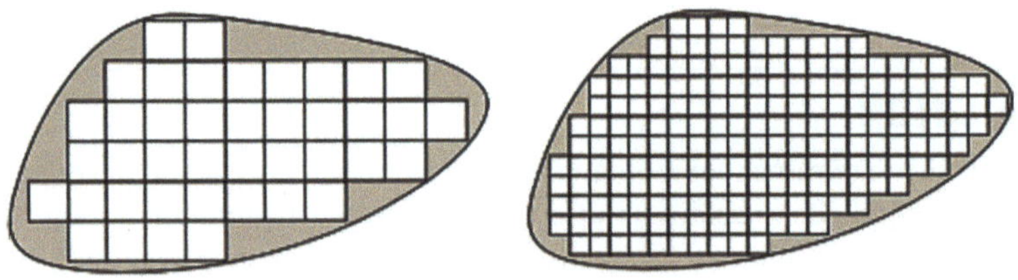

**FIGURE 5.2**

Examples of meshes of two different resolution of the same geometry (Hutton, 2005).

any geometrical shape such as triangle, quadrilateral, tetrahedron, or hexagon, as depicted in Figure 5.1 (Tabatabaian, 2013).

In addition to choosing the type of mesh to be implemented, one also needs to consider the resolution required for its appropriate subdivision. Figure 5.2 demonstrates how the same geometry can be discretized with square elements of different resolutions. Typically, based on the physical understanding of a problem, one needs to have higher-resolution mesh in particular areas of the domain of interest. Commercial codes have large sets of element libraries, enabling one to use very specific types of discretization techniques along with a range of resolutions.

It must be noted that depending on the extent of regularity in problem geometry, the generated mesh can be very irregular (i.e., high-resolution near geometrical discontinuity) as depicted in Figure 5.3.

Upon completion of the meshing exercise, in general one does not need to worry about the underlying problem transformation mechanism to a finite element analysis approach, owing to the fact that the ability is inherent to any commercially available software tool intended to solve finite element analysis problems. The data set incorporating geometry and meshing information along with the problem, physics, initial conditions, boundary conditions, and material properties is then used as input to the finite element code that constructs and solves a system of linear or nonlinear set of algebraic equation, approximated and transformed from the original set of differential equations describing the problem.

In order to transform the governing equations for finite element analysis, the variation of the variables in question, governed by a set of differential equations, is approximated using simple functions with the values of the dependent variables at the nodes of the elements. The transformation of differential equations to a set of algebraic equations is generally done through either of variational principle or weighted-residual method. Figure 5.4 illustrates typical steps involved in transforming the initial set of equations to a set of algebraic equations for finite element analysis. Hence, for each element one obtains a set of algebraic equations and one now has a large set of algebraic equations accounting for the whole domain that can be solved by using standard and advanced solvers.

**FIGURE 5.3**

Illustration of irregular mesh (Hutton, 2005).

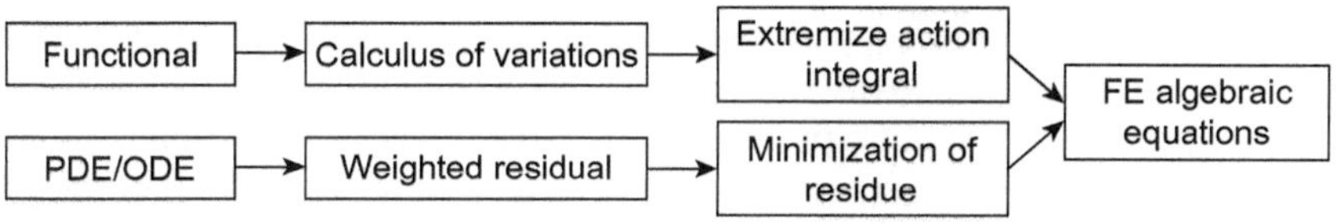


**FIGURE 5.4**

Typical steps in formulating a problem for finite element analysis.

*Tabatabaian (2013).*

Ideally, the results of finite element analysis should converge toward the exact solution in a stable manner in somewhat of a damped oscillation fashion. Hence, monitoring convergence of the simulation in ensuring its approach toward stable *exact* solution is matter of primary importance. Typically, variation of mesh resolution and/or implementing higher order polynomials for shape function offer a better convergence with increased computational cost. The shape function is a function that helps interpolate variable values in between two different nodes with discrete values of the variable.

## 5.3 THERMOMECHANICAL RESPONSE OF STRUCTURES

Fires are major safety concerns to a facility. Especially in offshore platforms, the threat to structural integrity from facility fire is of significant importance. Computational fluid dynamics (CFD) studies inspecting consequence of fire has been explored in Chapter 4. It looked into flow field and characteristics of the incident fires along with hazardous consequences from the same through estimation of associated thermal radiation in an attempt to define safe distances for personnel and equipment. While it also looked into thermal radiation values on the surface of the buildings, it did not look into the structural response of the same in such a scenario. However, estimating the mechanical response of structures subjected to fire is of prime importance in order to incorporate appropriate fire safety measures in structural design of buildings.

Given that thermal stresses can be significant in a fire incident depending upon the vicinity of a structure to the source of incident, assessment of thermal load on a building for consequent analysis of mechanical response of building structures during fire is made. As suggested (Reed and Peterson, 2012), simulation of the mechanical response of structures exposed to thermal radiation provides an opportunity to understand structural member failure path, weight redistribution, structure collapse, and so on. A hybrid analysis combining CFD and finite element analysis can be taken into account during the design phase itself for the structure of interest for designing economical and inherently safer structure for both onsite and offshore facilities.

Flame–wall interactions have also been widely studied for scenarios inside combustion devices (*Revisit Theoretical and Numerical Combustion* by Poinsot). Albeit with a different objective, the approaches developed are equally applicable for process safety applications, as in both scenarios it is a case of flame–wall interaction with an objective of (Revisit).

As an alternative to determination of fire resistance of a building in the laboratory, the study by Kucera (2007) looked into the thermomechanical response of concrete structures under fire. The objective of the study was to get better insight through improving simplified European codes (Eurocode 1: Actions on Structures, 2002; Eurocode 2 - Design of Concrete Structures: Part 1-1-General Rules and Rules for Buildings, 2004) through development of a mathematical model using CFD and finite element analysis approaches. The model offers the possibility of studying the differences in mechanical properties arising from the fact that different parts of structures are heated differently on its exposure to fire. The thermal loading on the building was calculated by application of CFD modeling leading to estimation of a time-dependent temperature profile. To study the effect of the resultant unsteady thermal field on the concrete structure, one also needs to incorporate accurate estimation of the study relevant temperature-dependent material properties such as specific heat and thermal conductivity. As shown in Figure 5.5, the total strain on a structure element was estimated by:

$$\varepsilon = \varepsilon_{\mathrm{th}} + \varepsilon_{\sigma} + \varepsilon_{\mathrm{tr}} + \varepsilon_{\mathrm{creep}}$$

where

$\varepsilon_{\mathrm{th}}$ is the thermal strain (the topmost strain in Figure 5.5)
$\varepsilon_{\sigma}$ is the elastic strain (the second topmost strain in Figure 5.5)
$\varepsilon_{\mathrm{tr}}$ is the creep strain (the third topmost strain in Figure 5.5)
$\varepsilon_{\mathrm{creep}}$ is the transient state strain (the bottommost strain in Figure 5.5)

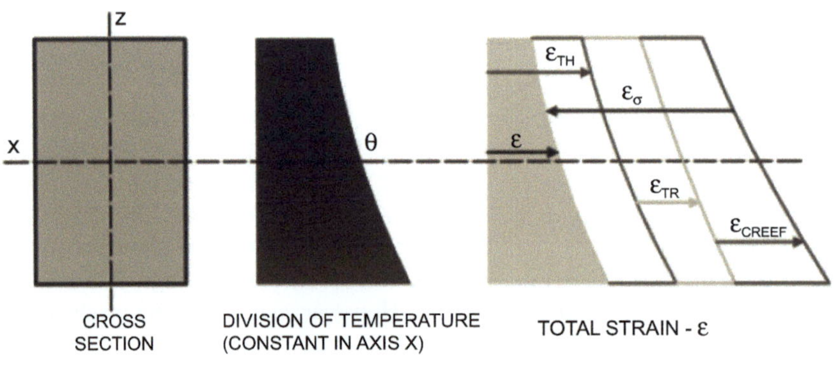


**FIGURE 5.5**

Total strain (not to scale) on a structure (Kucera, 2007).

Internal forces on the structure, a prefabricated space box segmental structure, similar to a garage for personal use, were calculated based on the structural mechanics principle. Avoiding thermal deformation on an intermediate structure results in internal forces such as shearing forces and bending moments. The analytical expression for the shear force, $N_T$, was expressed as the integration of the thermal stress over the whole cross-sectional area:

$$N_T = \int_A \sigma_T dA$$

The corresponding bending moment, $M_T$, was defined as:

$$M_T = \int_A \sigma_T z dA$$

Based on the analysis, the study could predict the time-dependent development of thermal strain in different parts of the structure. It also demonstrated how the bending moments increase due to increasing internal forces as a result of rising temperature inside the structure.

The understanding of the thermomechanical response of composite materials exposed to fire is of significant importance given its increased use in various types of structures. In that context, the modeling approach (Davidy, 2012) on the thermomechanical analysis of composite material investigating the consequence of fire in a polymer laminate is very relevant. The authors assumed the following for modeling the thermomechanical response of the composite material:

- Layered composite material
- Heat on the surface causes degradation and eventually ignition of the material. Degradation produces gas and char material. This is modeled as: P (Initial plastic) $\rightarrow \alpha G\,(gas) + (1 - \alpha)\,C\,(char)$
- Migration of degradation products from each layer to the layers below
- Estimation of pyrolysis reaction was done using a first-order reaction involving polymer thickness. In other words: $\frac{dr_p}{dt} = -kr_p$, where $r_p$ denotes the thickness of the polymer layer
- The rate constant was defined through the Arrhenius relationship: $k_0 = A_0 \exp^{\left(-E_0/RT\right)}$
- Heat transfer through layers of composite material was assumed in single dimension (1 D) and an energy balance equation was written accordingly:

$$\rho C_p \frac{\partial T}{\partial t} = \frac{\partial}{\partial x}\left(k\frac{\partial T}{\partial t}\right) + \dot{Q}$$

- The rate of heat generation consisted of heat of reaction and convective heat transfer

  Figures 5.6 and 5.7 clearly illustrate:

- The diminishing thickness of the composite material with time
- The degradation of material strength with time and location

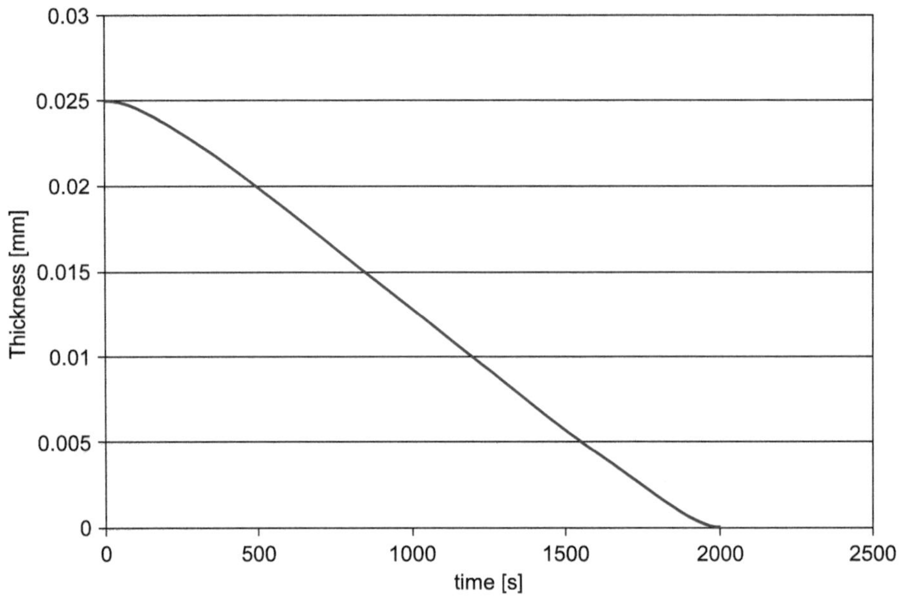


**FIGURE 5.6**

Thickness of the composite material with time (Davidy, 2012).

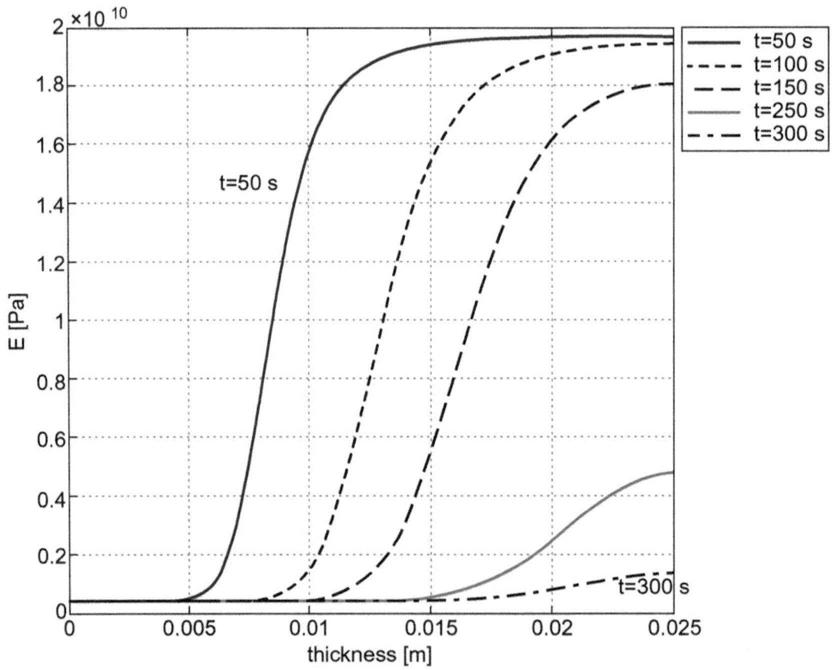


**FIGURE 5.7**

Young's modulus of the composite as a function of time and location (Davidy, 2012).

## 5.4 STORAGE AND TRANSPORTATION

Storage of cylinders containing unstable gases requires ensuring that a number of safety measures are in place. Let us take storage of acetylene, regarded as a dangerous substance and an explosive, as an example (Ferrero et al.). The instability of acetylene lies in the fact that it can decompose explosively into carbon and hydrogen. Accordingly, it is stored at relatively low pressures. As an additional safety measure, such cylinders are filled with porous material that can stop the propagation of the decomposition reaction. Such cylinders, when exposed to external fires, are expected to lead to decomposition of acetylene leading to an increase in internal pressure resulting in further consequence in the form of fireball and explosion.

Several incidents, such as the accident (1999) in the acetylene factory in Brisbane, Australia; accident (2007) in a gas factory in Dallas, USA, demonstrates the need for safer storage and handling practices of acetylene. In an attempt to better understand the risk mitigation strategy in such scenarios, the authors (Ferrero, Beckmann-Kluge, and Holtappels) looked into heating up of such a cylinder and subsequent cooling down with the help of external risk mitigation measures such as external cooling with water. Since the decomposition phenomenon of acetylene can propagate long after fire has been extinguished, it is important to understand how long the external cooling actions should continue in order to ensure the temperature inside the cylinder is reduced below the acetylene decomposition temperature. The mathematical model after implementing a series of valid assumptions leads to a system of equations incorporating heat transfer by conduction in the cylinder interior, heat transfer by conduction and convection in the exterior part of the cylinder, Navier-Stokes equation accounting for momentum balance, and continuity equation for fluid surrounding the cylinder. In an attempt to use a conservative approach, the model assumed a finite cavity in the porous material in the interior of the cylinder. Based on this assumption, mathematical representation of coupled mass, momentum, and heat transfer equation, the study demonstrated the potential to estimate time to explosion, cooling time required for various scenarios.

Similar to the storage of flammable materials, transportation of flammable materials has its associated hazards. Design approval for containers with the purpose of transporting fuel requires it to go through thermal tests for ensuring its response is within acceptable limits in the case of an accidental fire. Since temperature arising from a fire can be significantly high within a closed container, melting of the shielding material is a plausible phenomenon that can take place during a fire test or an actual fire incident. In the study (Goyal et al., 2009) on thermal analysis of a spent fuel transportation cask (Figure 5.8), the shielding material was lead. However, for an accurate consequence analysis finite element analysis also requires one to model the phase change dynamically. In their study (Goyal et al., 2009), the authors did not explicitly demarcate the moving boundary between liquid and solid during the transient phase and instead treated both phases in one continuous region. The melting phenomenon was modeled using the effective specific heat method. In this approach, the latent heat is approximated by a sudden change in the specific heat of the material. The fact that during a phase change temperature remains constant, it will pose mathematical difficulty. Accordingly, an approximation in the form of assuming a small change in temperature during a phase change needs to be assumed. In the particular study (Goyal et al., 2009), the authors assumed $\Delta T = 1$ °C. Accordingly, the specific heat capacity equivalent during melting was calculated by *(Latent Heat)*/$\Delta T$. The resultant slope discontinuity, that is, the discontinuity in the specific heat, needs to be addressed by small integration time steps around the melting point region. Accordingly, the authors noted that the use of a generalized alpha method is more suitable as opposed to the backward differentiation formula in

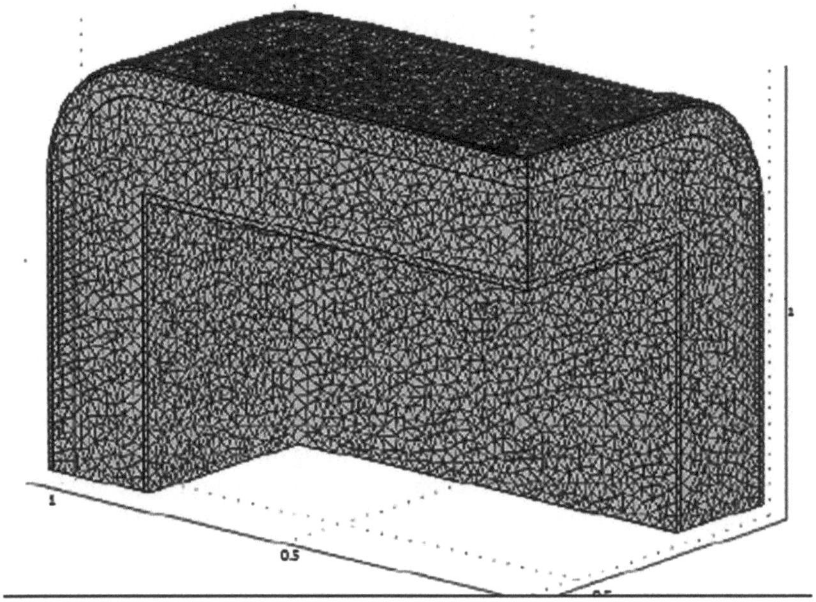

FIGURE 5.8

Mesh element distribution for the study (Goyal et al., 2009).

solving problems involving sharp discontinuity of material characteristics. The differential equation addressing the temperature distribution can be represented as a standard energy balance equation:

$$\rho C_p \frac{\partial T}{\partial t} = \nabla \cdot (k \nabla T) + \dot{Q}$$

where

$\rho$ is density
$C_p$ is specific heat capacity
$T$ is temperature as a function of both spatial coordinates and time
$t$ is time
$k$ is thermal conductivity
$\dot{Q}$ is the rate of heat generation

Boundary conditions, under normal scenarios, for both the inner and outer surfaces of the vessel, can be developed, respectively, as:

$$-k_{\text{inner}} \frac{\partial T_{\text{inner}}}{\partial n} = \frac{Q}{Area_{\text{inner}}}$$

$$-k_{\text{outer}} \frac{\partial T_{\text{outer}}}{\partial n} = h_{\text{outer}}(T_{\text{outer}} - T_{\text{ambient}}) + Q_{\text{radiation}}$$

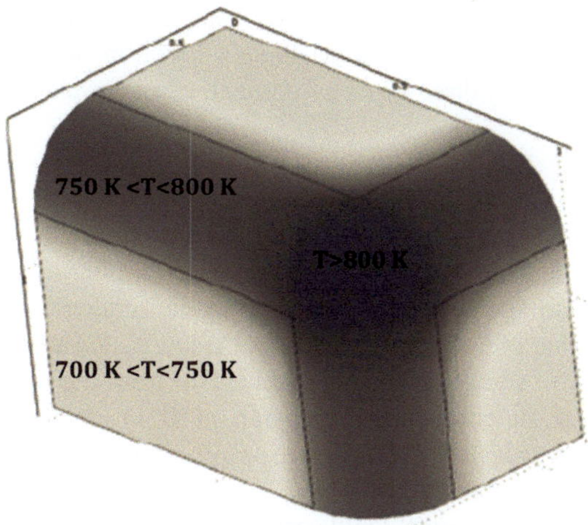


**FIGURE 5.9**

Temperature distribution inside the cask after 30 min of fire (Goyal et al., 2009).

The numerical calculations using finite element analysis showed in the study that under normal conditions, while the cask is heated from solar flux, the temperature is within the acceptable limits. However, during fires, the temperatures inside the cask can reach exceedingly high levels, as high as over 500 °C at specific locations inside the cask after being subjected to fire for 30 min. The finite element analysis very effectively captures the temperature distribution as a function of both space and time and provides insight into consequences without having to perform expensive fire testing. The analysis also predicted the possibility of melting of about 20% (by volume) portion of the shielding material, lead, in this case study. It could also predict the maximum melt penetration area, corner portion in this case study. Figure 5.9 depicts the temperature distribution inside the cask and clearly illustrates the temperature distribution including the above-mentioned findings.

## 5.5 DAMAGE DETECTION

Detecting damage in a timely fashion is of significant importance to process industry. Various techniques exist for the detection of damage depending on the material of interest. In several occasions, these are done through the use of embedded sensors, which in turn has the potential to initiate damage in the material through the very act of embedding. In lieu of that, Thiagarajan (2011) developed an alternative modeling approach, implementing finite element analysis for damage detection in conductive composites. Conductive composites are increasingly used in safety critical aerospace applications. Failures in these composite structures are often initiated from a failure at a much smaller scale. Damage detection capability by monitoring macro-scale properties at the multi-scale level is hence of a significant value.

In the case of damage detection in conductive composites, the authors (Thiagarajan, 2011) examined the electrical response of these materials as a function of typical defects in the material. Typically, in a composite material, three types of damage can take place:

- Fiber damage
- Matrix damage
- Interfacial damage

Fibers are the main load-bearing members in a composite, similar to, say, carbon nanotube or graphene in a nanocomposite. Damage in fibers has a far more consequence in terms of overall damage of the composite as opposed to that of a matrix. Damage in the interfacial area between a fiber and a matrix will translate into an ineffective load-transferring mechanism from the matrix to the fiber, leading to a loss of mechanical strength of the composite. Thiagarajan (2011) specifically looked into the effect of damage on fiber and how that translates into change in resistivity of the composite. Accordingly, the work, implementing finite element analysis using COMSOL, showed that regular monitoring of electrical resistivity of conducting composites could serve as a damage detection technique in conducting composites. Figure 5.10 illustrates similar results obtained with experimental results and finite element analysis–based calculations, both predicting an increase in resistivity with increased damage in fiber. It also found similar sensitivity of electrical resistance of conductive composite for delamination scenarios. Hence, by exploiting this sensitivity, it is possible to monitor electrical resistance of a conductive composite for the development of effective damage detection techniques.

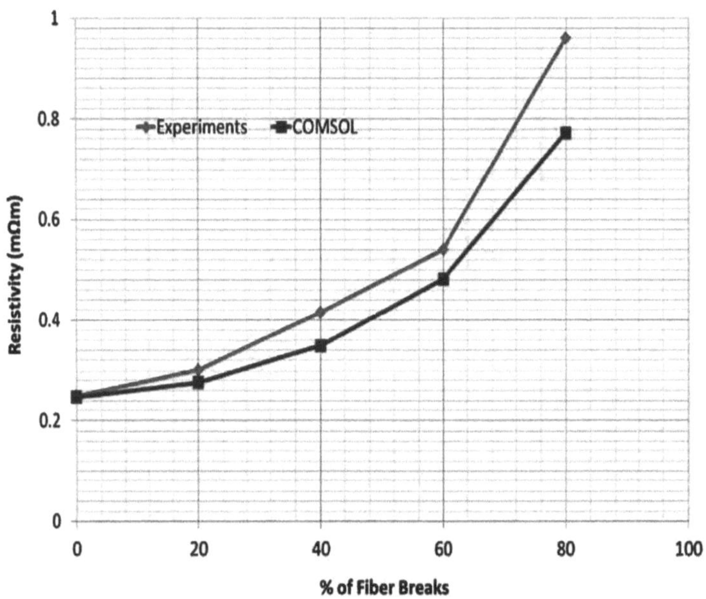


**FIGURE 5.10**

Increase in resistivity of a conductive composite with increased damage in fibers. A comparison study between finite element analysis (COMSOL) and experimental data.

*Thiagarajan (2011).*

## 5.6 HEAT RELEASE RATE

Cone calorimeter is a device used for predicting real-time fire behavior and is able to determine parameters such as ignition time, heat release rate, mass loss, and other properties relevant to fire characteristics. Among these, heat release rate is calculated by knowing the percentage of oxygen consumed during combustion. This is an important parameter in fire modeling, and can dictate the safety aspects of a flammable release or incidents such as thermal runaway reactions. Hence, having the ability to model heat release rate through a finite element analysis approach can be of great benefit, especially for looking into hazardous reactive chemicals and processes.

In their work, Statler and Gupta (2008) developed finite element analysis–based methodologies for determination of heat release rate during pyrolysis and combustion of polymer system. The assumptions in the modeling are similar in lines to the assumptions used while modeling thermo mechanical behavior of composites exposed to fire. The kinetics of the pyrolysis reaction was simplified by assuming that the polymer system breaks into gas and char with a first-order reaction rate. The rate constant was assumed to follow the Arrhenius relationship. Owing to constant consumption of the polymer material, the problem is of moving boundary nature. Performing a mass balance implementing Fick's diffusion law enables one to present the diffusion of the gas produced by pyrolysis as:

$$\frac{\partial C}{\partial t} - D\frac{\partial^2 C}{\partial x^2} = \dot{G}$$

where $C$ is the gas concentration, $D$ is the diffusivity constant, and $\dot{G}$ is the rate of gas generation. In the polymer zone, the rate of gas generation is determined by the rate of the first order pyrolysis reaction as described before. Outside the polymer zone, the rate of gas generation is zero and both the terms, $\frac{\partial C}{\partial t}$ and $D\frac{\partial^2 C}{\partial x^2}$, cancel out each other. In a similar fashion, to mass balance across polymer zone and char formation zone, energy balance or, more specifically, heat transfer is equated differently in polymer and char formation zone. The generalized equation applicable for both zones can be presented as:

$$\rho C_p\frac{\partial T}{\partial t} - k\frac{\partial^2 T}{\partial x^2} = \sum \Delta H$$

where $\rho$ is the material density, $C_p$ is the specific heat, $k$ is the thermal conductivity, and $\Delta H$ is heat generated or absorbed due to various processes such as reaction, decomposition, and volatilization, as applicable. In addition to replacing the parameters by their respective values in the polymer and char formation zone, the difference in the two-energy balance equation in two different regions is the fact that, in the char zone, the heat transfer through a conduction mode is the sole reason for change in temperature with no heat generated or absorbed. On the contrary, in the polymer zone, as the polymer is transferred to gas and char, the heat of volatilization should be considered. The heat transfer equation at the surface of the char needs to incorporate other modes of heat transfer such as convection and radiation.

The heat release rate can be calculated as:

$$\text{HRR} = \Delta H \cdot \left(-\frac{\partial C}{\partial x}\right)_{\text{surface}} \cdot D$$

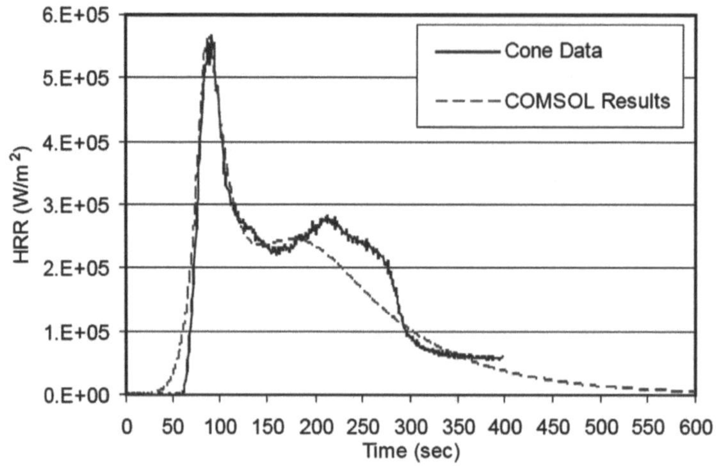


**FIGURE 5.11**

Comparison of heat release data as obtained from experiments versus finite element analysis in COMSOL for char-forming polycarbonate sample (Statler & Gupta, 2008).

where HRR is the heat release rate, $\Delta H$ is the heat of combustion, and $D$ is diffusion coefficient of the gas in the char zone. The problem is complicated by its moving boundary nature that can be attributed to constant char formation. The rate of mass growth of char dictates the rate of movement of the mesh. On defining the mathematical equation governing the physical phenomena, the next step is to define the initial and boundary conditions. Figure 5.11 illustrates that the mathematical model developed by the authors could accurately predict the dynamic heat release rate including information on peak heat release rate. While the developed approach shows a way to model heat release rate, a much-needed parameter for safety aspects, it is important to understand that while applying the same set of equations for another system, one needs to validate the assumptions and mathematical representations as applicable toward the specific system.

## 5.7 DISPERSION MODELING

Similar to the application of CFD, finite element analysis has been used in modeling dispersion of hazardous gases (Kanayama et al., 2008, 2007; Hallgarth et al., 2009). The mathematical model formulation of a dispersion study is governed by the momentum equation, continuity equation, and mass conservation equation. In addition, boundary and initial conditions need to be defined for the given model, similar to any general finite element analysis problem. The work by Kanayama et al. (2007, 2008) addressed dispersion phenomena of hydrogen using an analogy of thermal convection problems with Boussinesq approximation. In the study, the finite element method was used for discretization of space, and the backward Euler method was used for discretization of time. The analysis yielded an estimation of hydrogen concentration as a function of space and time. The problem setup is inside a hallway with the leak occurring at a location in the floor at the left end of the hallway. Figure 5.12 demonstrates the expansion of the hydrogen cloud with various concentrations (by volume) within

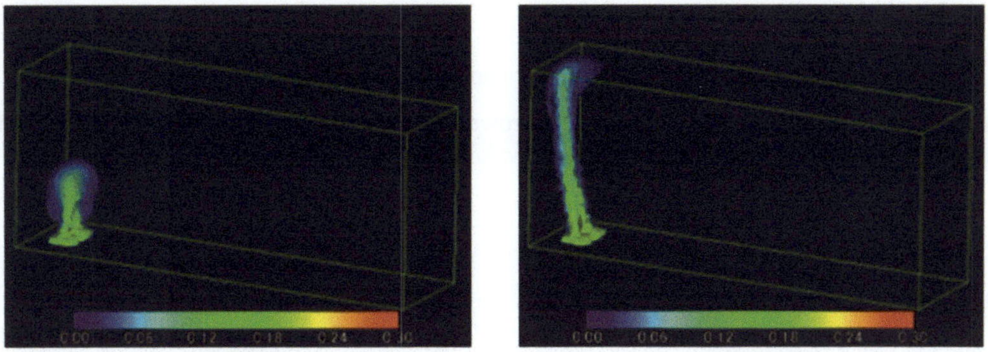

2.5[s] 5[s]

**FIGURE 5.12**

Hydrogen concentration (0–30 vol%) at 2.5 and 5 s, repectively, after release inside a hallway (Kanayama et al., 2008).

the hallway. The information can be used for understanding the dynamics of the flammable hydrogen cloud. Please note that the lower flammability limit of hydrogen is 4% by volume.

While this particular study relates to indoor dispersion, studies similar can be performed for consequence analysis in facilities.

## 5.8 CONCLUSION

Finite element analysis is powerful in process safety applications and can address a diverse set of problems, including fluid dynamics (Zienkiewicz et al., 2013b), such as dispersion modeling, and problems dealing with both fluid flow and structure, such as for studying interaction between fire and structure. Since the finite element analysis can handle structural response, studies on damage detection and mechanical failure due to thermal stress can also be studied in addition to a regular fluid flow problem. The purpose of this chapter was to provide a brief overview of finite element analysis within the context of process safety applications, as opposed to discussing the mathematical details. With the availability of modern finite element analysis–based tools such as COMSOL, ADINA, ANSYS, ABAQUS, and OpenFOAM, in general the user does not need to explicitly define the mass, momentum, and energy balance equations. General guidelines in building a model involve (Tabatabaian, 2013):

- Defining the problem, physics, and material involved
- Identifying (from a set of available) governing equations in alignment with the problem objective and a clear understanding of the scope of the solution
- Building the geometry of the problem set-up as a one-, two-, or three-dimensional system
- Assigning material and material properties in various parts of the geometry
- Assigning and governing equations as applicable in different parts of the geometry
- Creating a mesh or generally variable resolution as required in different parts of the geometry

- Running the model
- Visualize and analyze results
- Validate results based on known information and critical analysis

Although the above set of guidelines are borrowed from guidelines on using COMSOL as a finite element analysis tool, these steps are generally valid across all tools serving similar objectives.

## REFERENCES

Brebbia, C.A., 2013. Finite Element Systems. Springer Science & Business Media.

Cook, R.D., Malkus, D.S., Plesha, M.E., Witt, R.J., 2001. Concepts and Applications of Finite Element Analysis. Wiley, New York, NY.

Davidy, A., 2012. Thermo-Mechanical Analysis of Composite Material Exposed to Fire.

Eurocode 1: Actions on Structures, 2002.

Eurocode 2-Design of Concrete Structures: Part 1-1-General Rules and Rules for Buildings, 2004. BSI, London.

Ferrero, F., Beckmann-Kluge, M., Holtappels, K., 2009. COMSOL multiphysics as a tool to increase safety in the handling of acetylene cylinders involved in fires. COMSOL CONFERENCE Milan, October 14–16.

Goyal, P., Verma, V., Singh, R.K., Ghosh, A.K., 2009. Thermal Analysis of a Spent Fuel Transportation Cask.

Hallgarth, A., Zayer, A., Gatward, A., Davies, J., 2009. Hydrogen Vehicle Leak Modelling in Indoor Ventilated Environments.

Hutton, 2005. Fundamentals of Finite Element Analysis. Tata McGraw-Hill Education.

Kanayama, H., Tsukikawa, H., Ndong-Mefane, S.B., Sakuragi, O., 2008. Finite element simulation of hydrogen dispersion by the analogy of the boussinesq approximation. Journal of Computational Science and Technology 2 (4), 643–654. http://dx.doi.org/10.1299/jcst.2.643.

Kanayama, H., Maeda, K., Mino, M., Matsuura, K., 2007. Finite Element Simulation of Hydrogen Dispersion. International Society of Computational Fluid Dynamics.

Kucera, P., 2007. Thermal-Mechanical Analysis of Concrete Structure Exposed to High Temperature.

Moaveni, S., 2015. Finite Element Analysis: Theory and Application with ANSYS, Global ed. Pearson Higher Ed.

Reed, M., Peterson, E., 2012. Enhancing fire risk evaluations of plant structures using finite element analysis. Process Safety Progress 31 (4), 411–412. http://dx.doi.org/10.1002/prs.11519.

Statler Jr., D.L., Gupta, R.K., 2008. A Finite Element Analysis on the Modeling of Heat Release Rate, as Assessed by a Cone Calorimeter, of Char Forming Polycarbonate.

Tabatabaian, M., 2013. COMSOL for Engineers, Mercury Learning.

Thiagarajan, R.C., 2011. Multiscale Damage Detection in Conductive Composites, ATOA Scientific Technologies Pvt Ltd.

Zienkiewicz, O.C., Taylor, R.L., Zhu, J.Z., 2013a. The Finite Element Method: Its Basis and Fundamentals. Butterworth-Heinemann.

Zienkiewicz, O.C., Taylor, R.L., Nithiarasu, P., 2013b. The Finite Element Method for Fluid Dynamics. Butterworth-Heinemann.

CHAPTER

# DYNAMIC SIMULATION, CHAOS THEORY, AND STATISTICAL ANALYSIS IN PROCESS SAFETY

# 6

## 6.1 INTRODUCTION

> A disproportionate percentage of process safety incidents have occurred during transient operations.
>
> **Ostrowski and Keim (2010)**

As summarized in this quote, historical data on process industry incidents illustrate a nonintuitive trend. While a refinery or chemical facility spends less than 10% of its time on transient operations, more than 50% of process safety incidents have occurred during such activities (Ostrowski and Keim, 2010). Major plant accidents are more than five times likely to occur during abnormal operations (Patle et al., 2014). In that regard, understanding dynamic behavior of processes through studying its fundamental principles or analyzing process data is of significant importance. Process modeling and simulation studies capturing underlying physics of a process, albeit in a lumped fashion to various degrees along with process plant data analysis, are the commonly used approaches for addressing these concerns. These topics have been and are being widely studied as independent fields of research; here, we only briefly discuss the usefulness and applicability of these approaches in the context of process safety. For details, current status, and concerns regarding these methods itself, one is directed to available literature on these subject matters. Barring the implementation of chaos theory, as of today, both dynamic simulation-based and statistical analysis-based approaches have become increasingly common within the process industry. As an example, employment of process simulation techniques to provide flare network solutions is common in process industries to ensure safety of personnel and assets. Ideally, process modeling of a plant should capture both steady state phenomena, such the plant running under the normal process conditions, and transient state phenomena, such as the system going from one process state to another process state. Scenarios of such phenomena can arise from process upsets, startup, shutdown, emergency shutdown, or a combination of these. Hence, the modeling level should be at a scale where details of a system can be lumped to address, understand, and gain insight into the process at the time and length scale we are more familiar with, in our day-to-day process monitoring activity. While we have demonstrated the usefulness and power of modeling techniques at smaller scales in previous chapters, higher level modeling approaches, such as the ones discussed here, can provide useful insight related to the overall process and contribute to safer operations of a process facility by capturing phenomena at a higher time and length scale. Given that, a majority of the time a facility's operation revolves around steady state, steady state simulation is widely used in various process industries and is useful in this context. However, as pointed out in the beginning of this chapter, the appropriateness of such approach varies on a case-by-case basis while looking into transient operations. On the contrary, application of high-fidelity and full-scale dynamic models can be used to address transient operation-related concerns and has the ability to capture asynchronous events as the system moves from one state to other, planned or otherwise. For example, "what-if" questions related to HAZOP studies can be addressed through identifying problems during transient scenarios such as flaring, thermal oxidizers, scrubbing systems, process unit upset, startup, and shut down. In addition to experience, the best way to answer these "what-if" questions are through simulation studies of a validated model. In addition, one can also look into the role of advanced control systems in the context of process safety at a macro-scale level, but that is an entire discipline by itself, with its primary focus on advanced algorithms for pairing up variables and devising advanced level control loops for smarter control of a process. But even then, having a dynamic model of a facility can benefit a plant in lots of ways. A simple illustrative diagram (Jonas, 2004) depicts this (Figure 6.1).

The other two approaches briefly discussed here are on application of chaos theory and multivariate statistical analysis in the context of process safety. Use of chaos theory in preventing runaway reactions and utilization of statistical analysis such as principal component analysis (PCA) for early event detection are also performed at the macro-scale level, albeit are less commonly utilized in an industrial modeling and simulation environment than employing process simulation. However, that does not reduce the promising potential of these approaches in getting better information and insight from process data as explored later in the chapter.

## 6.2 DYNAMIC SIMULATION

Steady state simulations does not store information about process transient phases and is primarily focused on finding a steady and converged characteristics of various process variables. Put differently, a steady state process represents equilibrium state of the process for a given set of process conditions. Depending on the process specifics, the effect of perturbation in a steady state system might result in a new steady state with some process variables settling to a different value. What, however, these simulations fail to report are insights into the process-related specifics as it moves from one steady state to another. Since safety-related incidents are benefited significantly by accurate understandings of process transient behavior, dynamic simulation is a very powerful tool in this aspect. In dynamic

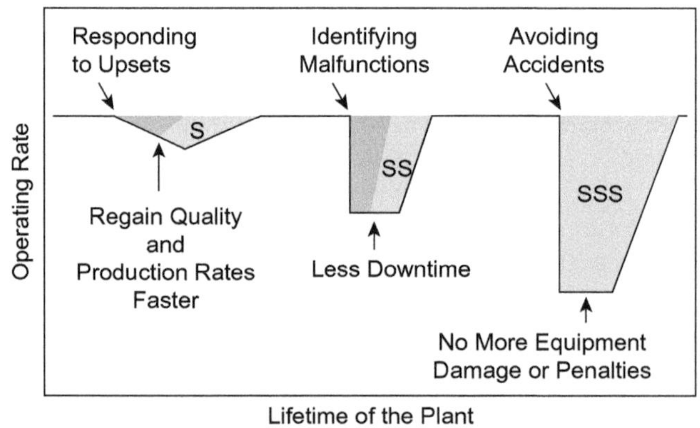


**FIGURE 6.1**

Benefits of dynamic simulation (Jonas, 2004).

simulation, the representative algebraic mass and energy balance equation set is represented by differential equation set so as to address time-dependent nature of the variables. The resulting set of equations, also referred as differential algebraic equations (DAE), increases the complexity of the problem. The combined equation set should be built so that the index of the problem does not exceed 1. Unless one is developing these set of equations from scratch for a custom model, this activity typically is not a problem for defining a problem with process simulators. Regardless, nowadays, solving dynamic simulation with increased complexity and computational requirement is widely possible and is quickly becoming a regular practice in industrial setups. Development of dynamic model of a process itself requires some expertise, and is followed by validation of the model from process plant data. It is important to ensure that the data used to build the empirical parts of the model are not used for validation purposes. The validation should be done to compare the model output to real process data under normal and abnormal operating conditions. Dynamic simulation is a powerful tool that represents a high-fidelity model of a facility and can be very useful for operator training, development of controller configuration; flare network studies, and validation studies such as those of equipment design and controller configuration. A dynamic model exhibits cross-coupling between many variables and multiple feedback loops within the equations describing the physical system. Therefore, the development of the model and the analysis thereof become very complex and challenging. Typically, the model needs to be developed incrementally and validated at each stage with plant data. This ensures identifying problems with the model assumptions, model inputs, and other concerns at the very initial stages prior to scaling it up to a bigger model.

Various benefits of performing dynamic simulation are well illustrated in Figure 6.2. As one can observe, all of the aspects shown here contribute directly or indirectly toward enhancing process safety in a facility.

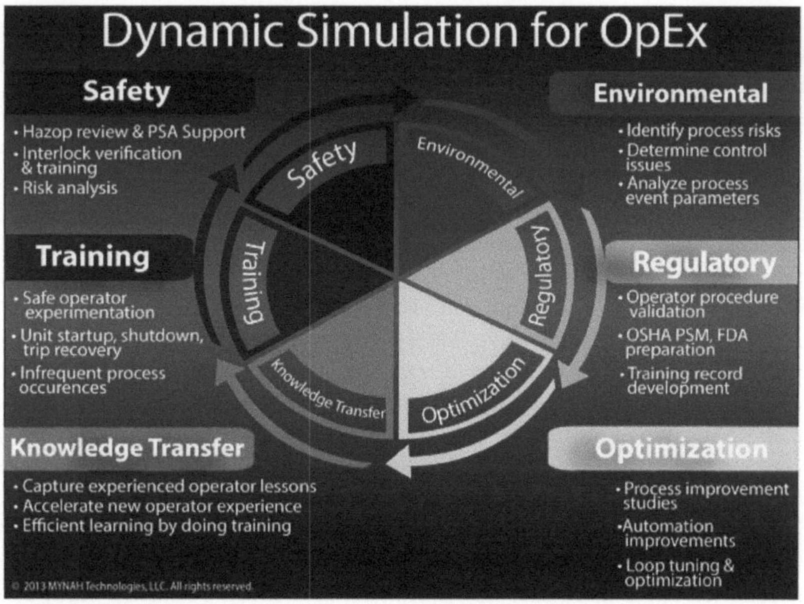


**FIGURE 6.2**

The various advantages of performing process dynamic simulation (Berutti, 2015).

### 6.2.1 HAZOP STUDY

Process hazard analysis is an essential part of any process facility. Elements of process safety management as identified and enforced by various regulatory bodies, require safety studies in process plants at regular intervals. In the United States, for example, as per Occupational Health and Safety Administration (OSHA) regulations (CFR 1910.119), facilities are required to perform process hazard analysis study at regular intervals, that is, every 5 years (as per OSHA norms in the United States) or when a significant change takes place in an existing process, whichever is earlier. There are two crucial components in process safety engineering. One is to quantitatively represent and study system behavior as a whole from process upsets and perturbation from the physics point of view. The second is to understand qualitatively how the interaction between various plant components contributes to the overall goal of the system (Wu et al., 2014). The approach in performing such a safety analysis is to examine inadequacies in design, training, procedures, and characteristics of the process itself, primarily qualitatively, to enlist potential safety concerns related to the process as resulted from deviation from normal operating conditions. Among the various methodologies in process hazard analysis, Hazard and operability study (HAZOP) is a well-recognized approach across process industries in successful identification of hazardous scenarios at different facilities. Effective implementation of regular HAZOP studies and acting on its outcome have helped in risk mitigation of facilities in a timely and efficient manner. It is a multidisciplinary team–based brainstorming activity that systematically reviews and identifies potential safety hazards in a process facility. Often, HAZOP is the preferred method of safety analysis in a facility (Vaidhyanathan, 1996) and has been adopted throughout the process industries. Regardless of the view on whether or to what extent one should allow HAZOP studies to identify potential hazards in a facility or shift toward making a culture change and emphasize inherently safer approaches early in the project lifecycle is a different discussion and is not stressed here. Even then (i.e., with facilities built on the principles of inherently safer design), PHA studies are a necessity. We attempt to illustrate how one can get better insight in HAZOP studies by implementing advanced computational tools will remain relevant. In that regard, Ostrowski and Keim (2010) developed transient operation HAZOP (TOH) approach to focus on operational task and procedural control.

In both cases (i.e., HAZOP and TOH), however, a multidisciplinary team of experts using guide words, reference lists, and a disciplined approach carries out a study. A series of guide words such as "more," "less," and so on are used to prioritize the risk associated with a particular scenario. The end goal is to identify a set of actions to mitigate the risk identified by a set of experts and experienced personnel of the plant with diverse backgrounds. The problem with this approach is that it relies hugely on the expertise, experience, and availability of process staff. As a consequence, as senior plant personnel retire, so go the years of accumulated knowledge with them (Rasel and Richmond, 2014). In the best interest of the plant, it is imperative that the knowledge is transferred to younger personnel in an efficient and timely manner. This matter of transferring knowledge related to critical plant safety issues can be significantly aided by the implementation of computer-aided simulation studies (Chiappetta et al., 2006; Wu et al., 2014). The idea is to incorporate accumulated process knowledge into developed process models. Several researchers have suggested the use of steady state analysis and dynamic simulation for HAZOP study in the past (Švandová et al., 2005; Eizenberg et al., 2006). A qualitative model-based approach in HAZOP is required to fully understand and appreciate the complexity of the process. Various qualitative methods have been implemented in performing HAZOP studies and are reported in the literature. However, a detailed specific HAZOP study can significantly benefit from integrating the typically qualitative nature of HAZOP analysis with the quantitative nature of mathematical modeling, capturing physics of the system.

The integration of the qualitative understanding with dynamic simulation based on mathematical modeling of the system has the potential to build a comprehensive model, where one part acts as the supplement for the other.

Development of a steady state process model should be regarded as the first step in developing quantitative representation of the plant. The developed steady state model, which mimics the plant behavior under normal operating conditions within acceptable tolerance, is then extended to build a dynamic model of the process under the same set of process parameters. For development of an adequate model to be used during HAZOP study, it must be kept in mind that the model should be capable of representing process deviations as identified in HAZOP, in quantifiable terms. It is equally important to ensure that the model assumptions used to develop the model is still valid while extending it to specific HAZOP scenarios. The developed model can be regarded as a base model for the HAZOP study. The advanced model can then be used in a number of ways for extracting insight during a HAZOP study:

- At an early design phase, with no or minimal supporting data from real plant, a dynamic model is a good source to refer in answering various hazard-related questions and identify potential hazardous scenarios during operations of transient nature.
- For a plant in operation, the developed model can attempt to mimic historical scenarios specific to it as identified during a HAZOP study by plant personnel. The developed model can then be modified to reproduce such phenomena for a given plant upset scenario. Afterward, potential abnormal input conditions of various intensities, such as loss of power or cooling water, can be fed to the model to predict system behavior at different parts of plants. This can help develop early warning detection systems for potential upset phenomenon.
- What-If questions, in several cases, can be precisely answered through dynamic process simulation. One of the advantages of carrying a What-If scenario in a process dynamic model is that the model takes into account all process interactions while assessing the effect of the particular failure scenario. This is one way to ensure that inexperience does not lead to oversight of a particular process interaction anywhere within the plant that might result in a safety incident. Thus, the limitation posed by traditional HAZOP studies through its requirement of experience and expertise can be diluted if not eliminated through quantitative approach.
- Quantitative dynamic simulation approach has tremendous potential in verifying and validating unacceptable risks as identified during qualitative HAZOP analysis of a process facility. As discussed, various extreme upset conditions can be quantified in process model development to observe system transient behavior and validate risk claims based on acquired knowledge of plant personnel.

Table 6.1 illustrates proposed translation of some of the typical HAZOP keywords to equivalent quantitative inputs for dynamic simulation as illustrated previously (Wu et al., 2014).

**Table 6.1 Quantitative Representation of Qualitative HAZOP Keywords**

| Parts/Components | Qualitative Failure Mode | Quantitative Representation |
|---|---|---|
| Valve | Open failure/closed failure | Digital stem position |
| Heat exchanger | Fouling, plugging | Digital plugging factor |
| Separator | Open failure/closed failure | Process parameter values |

Given that the chemistry of a reaction is well understood, scenarios such as runaway reactions, unstable steady states, and multiple steady states can be better assessed for hazard identification through use of dynamic simulation.

In conclusion, it must be noted that dynamic simulation should be used as a tool that complements the traditional qualitative HAZOP already exercised in a facility. While a dynamic simulation tool will not replace traditional HAZOP study, having such a tool will allow one to use a deeper, easier, and comprehensive HAZOP study. It must also be taken into account that the engineer(s) working on development of such model should be careful about the time required to build the same without sacrificing much detail. The magnitude of possible deviations from process conditions and quantitative assessment of its corresponding consequences can be better represented through implementation of a modeling framework as mentioned here. During such exercise, it is also likely that new consequences are identified through the developed model for the same input failure scenario.

### 6.2.2 OPERATOR TRAINING

Inadequate training has often been identified as one of the significant contributors to process safety incidents (Broadribb, 2012; Patle et al., 2014). Based on analysis of 170 hydrocarbon process industrial largest property damage losses during 1970–2000, as depicted in Figure 6.3, 28% (percentage shown within each pie) are owing to operational errors.

One promising approach is to use the operator training simulator, commonly known as OTS. OTS is an efficient and cost-effective alternative to train operators without actually endangering the plant or personnel (Patle et al., 2014). Often these training simulators are also coined as dynamic operator training simulator (DOTS) (Yang, Yang, and He, 2000) owing to the dynamic nature of the built in process models. Operators can be trained in getting insight on potential process behavior as and when it is deviated from normal process conditions. It helps them to practice, rarely used but critical, emergency operating procedures for various scenarios in a facility. The benefit can be extended in identifying various transient conditions that might act as a precursor to a safety incident. Process disturbance caused by unusual scenarios such as equipment failure (e.g., pump malfunction) can easily

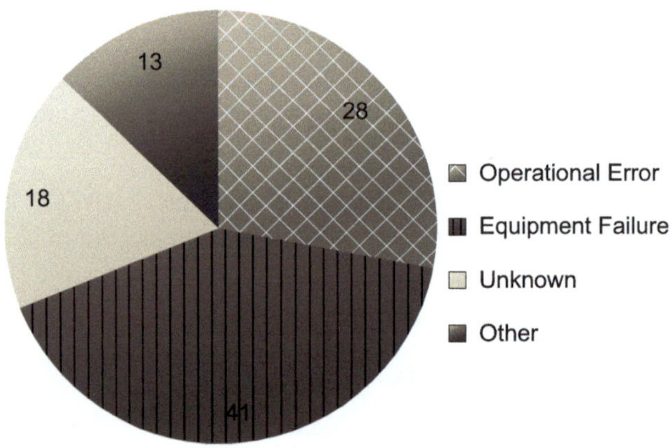


**FIGURE 6.3**

Cause and average percentage dollar loss (Yang et al., 2000).

be incorporated in within the scope of this training. The approach is not new (Jonas, 2004; Westerweele and Laurens, 2008), and extensive use of dynamic simulators has become more the norm than an exception in industry for the purpose of operator training.

A HAZOP study, as demonstrated in the earlier section, intends to identify potential hazardous scenarios, its likely effects, and development of mitigation strategies. For that, it requires a multidisciplinary team with domain knowledge and experience to appropriately identify the cause and map it with potential consequence. On the other hand, the purpose of an operator training program is to train operators, to ensure efficient running of a plant in addition to steady condition, under startup, shutdown, and emergency conditions. Put differently, several of the identified causes of failures in a HAZOP study can be better handled with a trained set of operators. Needless to say, this becomes significantly more important for operators relatively new to the equipment or process. Within the context of dynamic simulation, operator training can help in the following areas:

- Reduce incidents through operating a facility safely
- Run the plant closer to optimal conditions to minimize cost and maximize yield subject to constraints such as quality
- Validate controller configuration and plant improvements

Simulation is a proven tool to educate operators on how to respond in a given situation (Jonas, 2004). For example, in a simulator one can introduce a disturbance as a servo or regulatory problem and observe how the system behaves. Similar to step change tests in plant for determination of control parameters, a model can predict possible dynamic outcomes for a given change in parameters. As industry moves toward more sophisticated equipment and control structure with fewer operators (Emigholz, 2003), capturing realistic transient plant operation scenario through dynamic simulators for operator training has a lot of potential in improving plant safety.

For an OTS system to be most effective the plant model should be comprehensive, that is, being able to capture all four modes, startup, normal operation, shutdown, and emergency shutdown of a plant, as depicted in Figure 6.4.

In their review (Patle et al., 2014), the authors have demonstrated several examples of dynamic operator training modules as implemented across chemical industries. The article illustrated that these systems have been widely used in both batch and continuous processes.

While relatively widely implemented, challenges still exist in development of training simulator systems. As an example, there is no well-defined set of methodologies in developing a dynamic operator training system based on steady state design information. Given that being addressed, industry needs to define standard pathways for using the insights gained from these exercises toward improving process safety within a plant. In addition, development of a high-fidelity dynamic model itself requires detailed information on equipment sizing parameters, geometry, material of construction, controller tuning information, signal noise, valve hysteresis, sampling time, and shutdown categorization. Given the emphasis on intellectual property protection, it is often difficult for a third party to develop high-fidelity reactor models as the details of reaction kinetics and reactor design are difficult to get. For example, a facility may not be willing to reveal its catalyst properties for a particular set of reactions. In other scenarios, it is possible that there exists a lack of understanding about the underlying fundamental physical phenomenon that is governing a process. A black box empirical or semiempirical model is the only alternative path in developing a detailed model of the process in such scenarios.

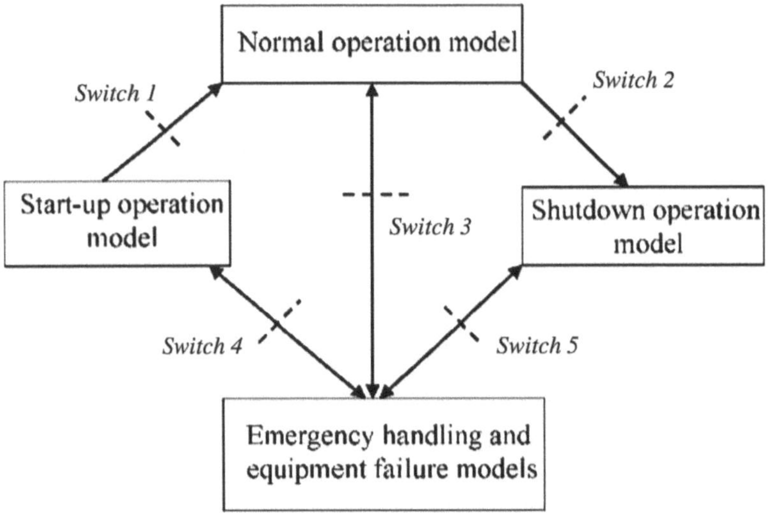


**FIGURE 6.4**

Framework of a comprehensive model suitable for OTS (Yang et al., 2000).

### 6.2.3 DEPRESSURIZATION AND BLOWDOWN STUDIES

Flare and relief systems are inherently dynamic processes. While relief systems are designed to handle planned or sudden outage in a plant, it itself can be a source of incident. Incidents in emergency relief systems are likely. It is due to the fact that for a process downstream, a process upset has led to overpressurization. For example, in an analysis, based on incidents in Taiwan, heat of the reaction has been found to be the most frequent source of incidents in emergency relief systems (Ho et al., 1998). The correlation between heat of reaction and emergency relief systems can be connected through the occurrence of runaway reactions. Runaway reactions typically occur as a direct consequence of uncontrollable heat from exothermic reactions. This in turn can lead to overpressure and hence the need for emergency relief system to handle the depressurization of the system in an optimal fashion.

Owing to the dearth of advanced simulators and lack of computational power, the initial approach by the process industry to model flare-related phenomena has been based on steady state simulations. Performance of relief devices are based on a number of factors such as the dependency of valve discharge on back pressure, rupture disc burst condition affecting flare header peak pressure, radiation-imposed limits as a function of peak relief rate, sizing of flare equipment based on relief load characteristics, and designed stages for relief. As a consequence, temporary events such back flow, overpressure, and local hot and cold spot formation (sometimes even outside design temperature) can potentially take place during a flaring event. To capture these effects, dynamic simulation can play a significant role, if not the primary role. Furthermore, flare network system in a facility is a major protection layer to protect plant from overpressure and ensure safe operations. Thus, it is critical to have a high-fidelity dynamic model for flare and relief systems.

In this context of implementing advanced modeling methodologies for flare network systems, several aspects come into play. API-521 (Zamejc, 2006) provides guidelines on designing of a pressure safety valve or pressure relief valve. However, currently published standards do not provide specific guidelines on relief system for every possible situation. To fill in the gaps, corporate standards, such as AICHE publications from Center for Chemical Process Safety (CCPS) and Design Institute for Emergency Relief Systems (DIERS), have been provided. While it is not covered here, it is important to mention that designing emergency relief system for runaway reactions is not trivial and a consortium of 29 companies developed advanced approaches to design pressure relief systems that resulted in the formation of a corporate standard (DIERS) (Fisher et al., 2010).

Application of dynamic simulation has quickly become a key element in process safety analysis for accessing flare network systems performance for a given outage scenario (Speranza and Terenzi, 2011; Mahgerefteh and Wong, 1999). The combined effects of the releases happening at various time intervals and the dynamic pressurization of the flare system translate into reduction of the peak loads and backpressures on system components. However, care should be taken so that the plant engineer ensures that the fundamental assumptions affecting the timing of each system or release are reasonable and defensible, thereby ensuring that the system is properly modeled. Participation of experienced personnel and historical data often become the key here to get the fundamental assumptions right. In addition to providing greater insight of the phenomenon, such an approach can also help reduce capital cost (Matzopoulos et al., 2010).

Owing to rapid changes in temperature during fast depressurization, the system does not have enough time to reach equilibrium. Accordingly, the model developed for this purpose should be rate based as oppose to equilibrium based. It has been shown that doing otherwise has the potential to reflect a nonconservative estimate of the wall temperature of a process vessel and thereby underestimate the risk of material failure due to sudden cooling as a result of the Joule-Thomson effect during rapid depressurization (Matzopoulos et al., 2010). To reiterate, the benefits that can be realized as a direct impact of implementing dynamic simulation are listed next:

- Steady state analysis cannot capture the fact that not all relief devices in a flare network release material simultaneously and hence overestimates the burden on a flare header. Thus, accurate assessment of transient relief load characteristics in a flare network (Chen et al., 1992; Xu et al., 2009; Singh et al., 2006) system of a facility can be done through dynamic simulation. This has a direct effect on capital expenditures or CAPEX of a facility. Through application of dynamic simulation to flare load calculation, it is possible to reduce flare load dramatically because of simultaneous consideration of relieving time, abnormal mass and heat balance, and effect of log mean temperature difference. This in turn results into minimizing the chances of overdesigning a flare system.
- Dynamic simulation can capture various transient phenomenon such as back flow, overpressure, and local hot and cold spot formation during a transient phase as the entire system moves from one steady state (normal operating conditions) to another steady state as a result of, say, an emergency failure scenario, such as loss of cooling water or power or typical equipment failure.
- Depressurization can lead to rapid cooling and the possibility that in parts of the flare the network may experience temperatures outside the design temperature cannot be ruled out. The sudden change in temperature has a direct effect in the phase equilibrium of the system. This might result in formation of droplet phase within the vessel. Local temperature gradients are likely to introduce thermal stresses that can make piping segments brittle.
- In a realistic plant scenario, flare systems rarely present the opportunity to tune controllers. Tuning flare system controllers can be benefitted much by a dynamic model of a flare network. Controller parameters can be tuned for significant upsets in an efficient and accurate manner through a well-validated dynamic model.

**Table 6.2 Available Process Simulators for Dynamic Flare Modeling**

| Software/Code | Developer |
|---|---|
| Aspen Flare System Analyzer | ASPEN Technology |
| ChemCAD | ChemStations |
| gFlare | Process Systems Enterprise |
| Flaretot | KBC |

- Of utmost importance is also capturing the possibility of two-phase flow release and its corresponding influence on the volumetric discharge through a flare header. Dynamic process simulation packages with rigorous thermodynamic property packages are very efficient in such scenarios.

Development of a dynamic model representing flare systems can be done using any standard commercially available process modeling and simulation package. Table 6.2 lists some of the available process simulators that can be used for this purpose.

The inherent dynamic nature of the process is captured through a set of differential algebraic equations that are inbuilt in any process modeling packages. On gathering data on flare network geometry, dimensions, materials, and process conditions, one needs to represent the same in the form of a model. Similar to mathematical modeling of any other system, in several situations the modeler may need to approximate the flare network representation for modeling purposes without losing important information relevant to the problem objective. For example, he or she may need to represent a multistage flash distillation unit into a simple two-phase flash drum. On development of the model, validation of the model is essential, through field-testing, historical evidence, and operator experience. Performing field-testing for validating dynamic model of flare network system can be challenging, especially for testing large flow rate scenarios. However, in addition to assess the performance of flare network by performing small scale upsets for validation of the model, one can work with experienced and knowledgeable operators to validate qualitatively the model on a case-by-case basis. Reconstructing plant historical flare incident can also be done to validate the model, at least in part, for which data are available. Once the steady state data of the plant compare well with the prediction of the dynamic model of the flare network, one can move into validating transient outcomes in a similar manner. The development of the model is iterative and continuous in nature, and model improvements or changes need to be implemented to mimic plant performance and reconcile data.

## 6.3 CHAOS THEORY AND STATISTICAL ANALYSIS

### 6.3.1 RUNAWAY REACTION PREVENTION USING CHAOS THEORY TECHNIQUES

> It is now well established that seemingly innocuous dynamical systems, dissipative or not, can produce complicated phase trajectories and, eventually chaos.
>
> **Doherty and Ottino (1988)**

The potential of a severe accident owing to a runaway reaction is well perceived. If and when the capability of heat removal becomes less than the rate of heat generation in an exothermic reaction, the imbalance in heat will act as a positive feedback mechanism to the system and act as an automatic catalyst to increase the rate of reaction. This potentially can initiate possible undesired reactions, those favorable at high temperatures, such as decomposition reactions leading to generation of overpressure and eventually ending in a disaster. While heat-removing mechanisms, such as conduction and convection, are linear in nature with respect to the temperature, the heat-generation mechanism in a reaction is typically exponential in nature. As a consequence, at higher temperatures, chances of imbalance in heat removing mechanism become greater. It must also be noted that despite having an efficient and sufficient heat removal system in place, while performing hazard analysis of a facility, scenarios such as that of cooling water failure can emerge as a potential incident, which in turn can cause a runaway reaction event. Broadly speaking, there are two different measures that can be taken to mitigate risks arising from possibility of a runaway reaction in a facility: protection and prevention. In the protection approach, the focus is on minimizing the effect of consequence from a runaway reaction. Optimum location of buildings to stay away from radiation zone, escape routes etc. falls into this category. The prevention approach, on the contrary, is to identify precursors to such event in a timely and efficient manner in order to minimize the chances of occurrence of such an event. Clearly the prevention measures are more desirable to a facility. However, both measures are equally important for risk mitigation purposes. So, in addition to having controllability of high degree around a process with a potential for runaway reaction, the benefits of having an accurate early detection system for the same cannot be discounted. Controlling a reactor between fixed process variable limits has its own merits, but since the limits are typically based on laboratory scale or pilot-scale experiments, as the system is scaled up, the limits might translate into a new range of values, unknown to the plant operators or engineers. In addition, uncontrolled and unpredictable disturbances in a real plant may pose difficulties in mimicking the system response at the smaller-scale experiments. Early warning devices are thus indispensible regardless of the degree of detailed engineering, and consideration of reaction mechanism has been incorporated for a safer design of a reactor system. Building a successful early warning detection system for a runaway reaction event typically requires the following:

- An efficient method of extracting real-time data from the relevant process. With emphasis on real-time monitoring in process facilities these days, this is a widespread practice. Typically, using OPC client/server technology real time and historical data of the plant can be extracted for analysis purposes.
- Development of well-defined criteria to accurately distinguish between alarming and nonalarming conditions. This step is perhaps one of the most vital steps in the process. The exercise can be further broken into:
  - Monitoring and collecting data—Collect process data and import it through a suitable data analysis program.
  - Cleaning data—Structured data cleaning for noise reduction.
  - Analyze data—Implementation of appropriate data analysis and interpretation of the same.
  - Detection—Develop well-defined criteria for early runaway reaction event detection. It is needless to mention that the process of analyzing real-time data has to be faster than the real process in order to make the early warning detection procedure effective. The detection system has to be quick and efficient enough to provide sufficient time for plant operations to act on the situation along with, minimization if not elimination, of the number of false alarms.

- Diagnosis and evaluation—Procedure for triggering the alarm or warning along with validation of its effectiveness.

While rigorously following the above-mentioned steps, ensures development of a robust early warning detection system, protocol for subsequent counter measures should also be in place, for mitigation and safety purposes. In this chapter, however, the focus is on development of the early warning detection system from data analysis—that is, starting from historical data and offline analysis of the same, using the concepts of parametric sensitivity analysis, development of an online tool to identify runaway reaction conditions in advance.

The analysis of parametric sensitivity refers to the fact that the system is capable of undergoing substantial changes for minuscule changes in input variable values, typically arising from disturbances. This very concept, being at the center of the chaos theory, enables one to borrow ideas and insights from chaos theory and implement in understanding system behavior with runaway reactions. As identified above, the critical step in the whole process is to develop well-defined criteria for differentiating between a normal and an abnormal condition. In some sense, it is, as identified (Strozzi et al., 1999), the bottleneck step in the whole exercise. Parametric sensitivity studies in chemical reactors, through performing offline analysis of mass and energy balance, can be used for reconstruction of states. Such a predictive model can then help in identification of possible hazardous situations in advance of the actual potential event. A large number of state variables, typical of a chemical process, such as temperature, pressure, and concentration of individual species, make it difficult for development of a real-time early warning hazard identification system based on all of these variables. There are two different kinds of inputs that may be analyzed for development of real-time early warning systems, namely measurable signals and nonmeasurable state variables or process parameters. One of the requirements is to measure these variables in real time within a reasonable amount of time to allow information to be used for the detection procedure. For instantly measureable signals, one common and straightforward approach is real-time measurement of identified set of variables and checking their values against predefined acceptable limits. As the nature of the approach suggests, comprehensive knowledge of the process is essential in this case. The checks based on predefined limits are also applicable for time derivatives of process variables. As an example, one can implement safe limits of an operation such that:

$$\left(\frac{dT}{dt}\right)_{min} < \left(\frac{dT}{dt}\right)_{process} < \left(\frac{dT}{dt}\right)_{max} \quad \text{(Through monitoring rate of change of temperature)}$$

$$\left(\frac{d^2T}{dt^2}\right)_{min} < \left(\frac{d^2T}{dt^2}\right)_{process} < \left(\frac{d^2T}{dt^2}\right)_{max} \quad \text{(Through monitoring change in rate of change of temperature)}$$

Instead of temperature, researchers have also looked into monitoring power generated by chemical reaction in a similar fashion, that is, monitoring change in power generation or change in the rate of change of power generation within predefined limits. Depending on the process, other process variables, such as pH, density, and viscosity, can also be selected as variables based on which a runaway reaction can be identified.

Another approach suggested in the literature is to detect runaway reaction criteria when the divergence of a batch or semibatch reactor becomes positive on the segment of a reaction path (Strozzi, 1997).

In other words, the continuous real-time evaluation of the quantity:

$$\frac{\partial(dT/dt)}{\partial T}+\frac{\partial(dz/dt)}{\partial z};\ \text{(where T and z are temperature and conversion, respectively)}$$

will help one in identifying conditions leading to a runaway reaction. If we examine the terms, we observe that the positive value of the individual terms implies faster rate of increase in temperature with increasing temperature or faster rate of conversion as the reaction progresses. Put differently, both terms in an individual sense are a measure for the extent of positive feedback the system will get from its inherent characteristics. Accordingly, the sum of it will dictate the overall nature of the feedback mechanism and hence the direction of reaction (i.e., runaway or not). For real-time estimation of the divergence from plant data, it requires that the data representing the phase space be instantly reconstructed through nonlinear time-series data analysis using a delay-embedding theorem (Takens, 1981) as illustrated above (Strozzi, 1997).

In another work, the authors (Strozzi et al., 1999) have implemented the method of characterizing system behavior through quantifying Lyapunov exponents of a batch reactor. In brief, the work used these exponents to quantify dynamic characteristics of the reactor system and monitored closely its runaway mechanism. This is done through studying the rate of divergence quantified by using local Lyapunov exponents. The idea, borrowed from studying chaotic systems, works on the fact that, for a runaway reaction, if the process parameters are near a runaway boundary, a slight perturbation will result in a runaway reaction, as will be reflected in the divergence of the Lyapunov exponents defining the phase space.

What is a Lyapunov exponent? In a standard scenario of a system, two closely originating trajectories (i.e., two closely spaced initial value problems) are expected to converge at the same solution at $t \rightarrow \infty$. Given sufficient time, in other words, for our perspective, a chemical process originating with similar set of initial conditions (say temperature, pressure, composition, flow rate) should attain similar final steady state values with minimal difference among the various scenarios. This also means for small disturbances in a plant, the outcome is going to be similar and is not expected to change plant performance. Lyapunov exponent, an important parameter in assessing predictability of a dynamical system, quantifies the rate of separation of infinitesimally close trajectories. The steps involved consist of the following tasks:

1. Collection of process historical data required for mass and energy balance of the system for phase-space reconstruction and performing parametric sensitivity analysis of the system. The parametric sensitivity analysis study enables one to determine the parameters to which the process is most sensitive. In other words, small changes in those parameters introduce a large shift in the system characteristics. If one follows the evolution of a sphere defining initial conditions of an *m*-parameter system, the *i*th Lyapunov exponent is defined as (Wolf et al., 1985):

$$\lambda_i=\lim_{t\rightarrow\infty}\frac{1}{t}\log_2\frac{q_i(t)}{q_i(0)};$$

A positive value of the exponent implies the divergent nature of the *i*th Lyapunov exponent. This is very important to note. It indicates that the value of $q_i(t)$, the principal axis of the *i*th Lyapunov exponent in the unit sphere, has increased over time, indicating a chaos in the system. From the equation, it is straightforward to deduce the following relation between the volume of a sphere separated by a time $t$:

$$V(t)=2^{(\lambda_1(t)+\lambda_2(t)+\lambda_3(t)+\cdots+\lambda_m(t))\cdot t}V(0)$$

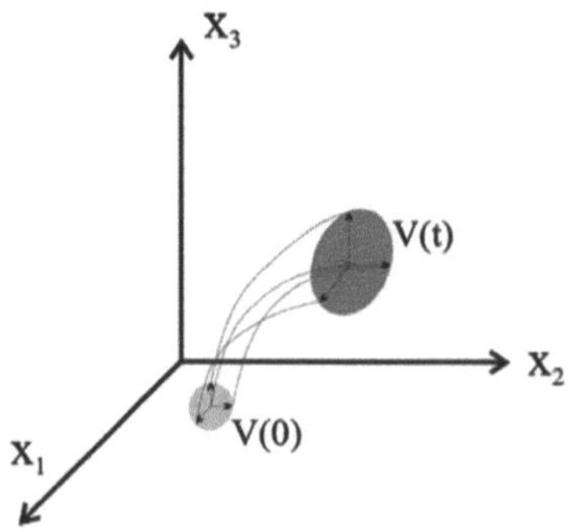


**FIGURE 6.5**

Three-dimensional representation of evolution of infinitesimal volume *V* over time (Zaldívar et al., 2003).

It can be thus realized that, more than the individual values of the Lyapunov exponents; the chaotic nature of the system is a function of the sum of these exponents. This equation can also be represented as (Strozzi et al., 1999; Zaldívar et al., 2003):

$$V(t) = e^{\int_0^t div\, F[x(\tau)]d\tau} V(0)$$

where in a *d* dimensional dynamical system:

$$\text{div}\,\{F[x(t)]\} = \frac{\delta F_1[x(t)]}{\delta x_1} + \frac{\delta F_2[x(t)]}{\delta x_2} + \cdots + \frac{\delta F_d[x(t)]}{\delta x_d}$$

As observed here, the sum of the local Lyapunov exponents is proportional to the temporal mean of the divergence of the field $F[x(t)]$. These equations thus represent the state space evolution of infinitesimal volume $V$ as a function of time similar to as illustrated in Figure 6.5:

2. On having the offline analysis data, the next step is to estimate the runaway reaction criterion. Based on the criterion that the divergence of the reactor becomes positive on a segment of the reaction path, one needs to check for the condition:

$$\text{div}\,\{F[x(t)]\} > 0$$

The advantage of this particular approach is that this does not require intricate understanding of the inherent process and can be developed based on historical data. Overall, for chemical systems with the potential to display chaotic behavior, such as massive changes in particular domain of parameter space, we can borrow established theories from chaos to analyze and better our predictability to efficiently prevent future undesired outcome.

## 6.3.2 MULTIVARIATE STATISTICS–BASED PROCESS MONITORING

The goal is to turn data into information, and information into insight.

**Fiorina (2004)**

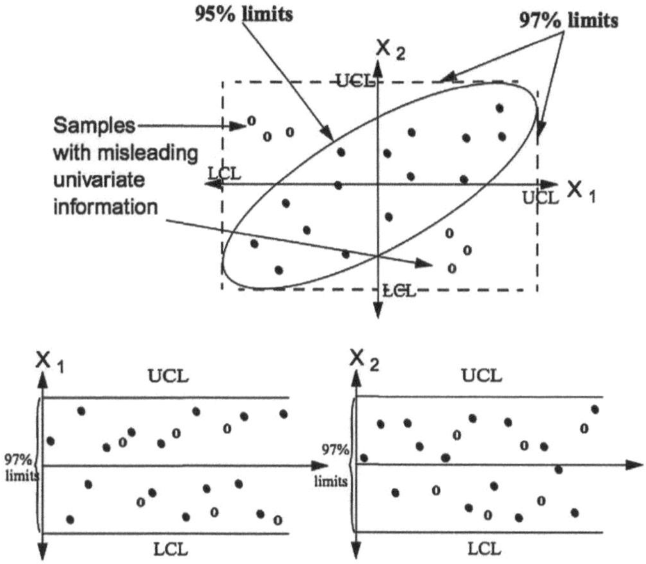


**FIGURE 6.6**

Multivariate statistical analyses versus univariate statistical analyses (Chen et al., 2001).

And for this "goal" to be achieved, collection, organization, and interpretation of data according to well-defined and validated approaches need to be performed. The field of statistical analysis encompasses all of these steps.

Continuous process monitoring is of primary importance to meet product quality and yield objectives in a production facility. Modern process computers and online data surveillance system can log massive amounts of real-time data. Given the extensive number of variables and the multivariate nature of the data set, the requirement of advanced analysis has become essential. There are several statistical approaches that help in data analysis, and several textbooks on the subject are available.

Implementation of a univariate approach in an environment involving multiple variables, typical for a chemical process, represent the process in a way where it is hard to get insight about it without the knowledge of fundamentals of the physical processes involved. Traditional univariate control charts such as Shewhart, CUSUM, and EWMA charts are based on individual variables as opposed to using information from two or more variables as a single-composite measure of the process. In several instances, use of univariate statistical process control has resulted in frequently adjusted or shutdown of processes in a Type I statistical error scenario (Chen et al., 2001). Type I statistical error occurs with the incorrect rejection of a null hypothesis. In other words, a Type I error is detecting an effect when it is not present. The multivariate correlation structure is not taken into account in this approach. Simultaneous scrutiny of several variables for monitoring and diagnostic purposes is thus of significant importance.

Figure 6.6 (borrowed from Chen et al., 2001) illustrates the difficulty in using independent univariate control charts. We observe that the individual control charts displaying lower control limit (LCL) and upper control limit (UCL) of the two correlated variables $x_1$ and $x_2$ do not show anything of concern, and the process is showing normal variation. However, in a multivariate representation, we can see that there are indeed observations, which fall outside the joint confidence region.

Multivariate statistical analysis-based approach thus rearranges the information in a way so as to illustrate the relation between the process and all the variables in the same plot. Defining acceptable parameter region under normal process variation, such as the eclipse shown in Figure 6.6, where variations of variables are due to normal process condition, is the first step. The idea is then to detect new changes or subtle changes in the covariance structure of the process during real-time monitoring (Chen et al., 2001). Both continuous and batch chemical process data can be analyzed through such approaches (Martin et al., 1996). The philosophy behind multivariate techniques is to reduce the dimensionality of the problem by forming a new set of variables (Martin et al., 1996). Accordingly, the building block of any multivariate analysis is the variate, constructed from a linear combination of process variables with empirically determined weights. This can be represented as:

Variate $(x_1, x_2,\ldots, x_n)$ value $= w_1x_1 + w_2x_2 + \cdots + w_nx_n$

where $x_1, x_2,\ldots, x_n$ are the '$n$' variables and $w_1, w_2,\ldots, w_n$ are the empirical weight determined by the multivariate technique.

Multivariate data analysis is an expanding set of techniques and the established methods can be classified as follows (Hair, 2007):

- Principal components and common factors analysis
- Multiple regression and multiple correlation
- Multiple discriminant analysis and logistic regression
- Canonical correlation analysis
- Multivariate analysis of variance and covariance
- Conjoint analysis
- Cluster analysis
- Perpetual mapping or multidimensional scaling
- Correspondence analysis
- Structural equation modeling and confirmatory factor analysis

Introducing and defining each technique are beyond the scope of this book; instead, the focus here is on illustrating possible implementation of multivariate statistical techniques as illustrated in several places for better process monitoring and enhancement of process safety aspects of a facility. Here, we have looked into implementation of PCA as a multivariate statistical tool for process facilities.

An essential assumption in the PCA approach is that an underlying structure does exist in the set of selected variables. In PCA, the components are viewed as weighted combination of variables. The idea behind performing PCA lies in the fact that the approach, through analysis of time-series data of all the variables, attempts to represent the same information or data in a way so as to capture the dominant trends and patterns in data through newly constructed components. The information contained within a large number of variables is condensed into a smaller set of variates with minimal loss of information. While the number of components or variates is equal to number of variables, by only considering the variates accounting for the largest variation in the process data, the variate set is effectively reduced with insignificant loss of information. Components analysis is most appropriate when data reduction is of paramount importance. In a process facility, owing to real-time monitoring, frequent records of thousands of process variables, many of which can significantly affect process safety, are recorded. Hence, application of PCA is very appropriate in a chemical facility for analyzing real-time process data.

In a PCA, the majority percentage of the variation in the data is explained by the first component, followed by the second component, and so on. This way, the first few components itself can explain a significant portion of the variation in process data. PCA provides means to understand relations not only between the observations and variables but also among the variables itself in an efficient manner

(Wikström et al., 1998a). By summarizing the variation in a multivariate process through few principal components, a process can be effectively monitored through extended version of ordinary control charts to display multivariate parameters, such as illustrated in Figure 6.6. Significant changes in a process can thus be instantaneously detected through monitoring these control charts (Wikström et al., 1998a,b). Monitoring chemical batch processes for ensuring product quality and operational safety has been shown to be significantly aided with the help from implementation of PCA studies (Yao and Gao, 2009). In another study (Wikström et al., 1998a,b), the authors demonstrated application of PCA for monitoring product quality and establish early warning detection system for the same. The study is illustrated in the Exercises Chapter. In a similar fashion, the approach can be implemented for monitoring process with the objective of developing early warning detection system for safety incidents.

Given the availability of commercial software for performing PCA study, the challenge is to understand and interpret the results from a PCA analysis. Scores and loadings are the two main parameters in a PCA study. Mathematically orthogonal in nature, scores represent optimal summary of process variation between observations. The contribution of the variables to the scores is explained through the loadings. The absolute value of the loading represents the magnitude of the contribution along with its positive or negative correlation with the scores. A process data set when analyzed with tools for PCA analysis will calculate scores and loadings for the given information. From the generated information, one has to distinguish and discard the outliers to avoid misinterpretation of the message conveyed by the data set.

For powerful analytical and predictive capabilities through multivariate data analysis including its interpretation, one needs to keep in mind the following concerns:

- Recognition of the fact that sample size affects all results. For small sample sizes, multivariate statistical analysis may lead into results with little statistical power for identifying significant results or overfitting of data. On the other extreme, the statistical tests may seem to be oversensitive. As a rule of thumb, whenever sample sizes exceed 400, it is necessary to examine the significant results to ensure practical significance (Hair, 2007).
- Validation of data for outliers, applicability, violations of assumptions, and missing data. Data processing and cleaning represent a very important part of the whole exercise. As put by Mlodinow (2009), "*For example, most people consider that the greatest evidence of an event one can obtain is to see it with their own eyes, and in a court of law little is held in more esteem than eyewitness testimony. Yet if you asked to display for a court a video of the same quality as the unprocessed data catptured on the retina of a human eye, the judge might wonder what you were trying to put over. For one thing, the view will have a blind spot where the optic nerve attaches to the retina. Moreover, the only part of our field of vision with good resolution is a narrow area of about 1 degree of visual angle around the retina's center, an area the width of our thumb as it looks when held at arm's length. Outside that region, resolution drops off sharply. To compensate, we constantly move our eyes to bring the sharper region to bear on different portions of the scene we wish to observe. And so the pattern of raw data sent to the brain is a shaky, badly pixilated picture with a hole in it. Fortunately the brain processes the data, combining input from both eyes, filling in gaps on the assumption that the visual properties of neighboring locations are similar and interpolating. The result - at least until age, injury, disease, or an excess of mai tais takes its toll–is a happy human being suffering from the compelling illusion that his or her vision is sharp*

**Table 6.3 Available Statistical Software for Process Data Analysis**

| Software/Code | Developer |
|---|---|
| R-Studio | Open source |
| Unscrambler | CAMO |
| STATISTICA | Dell |
| Matlab | MathWorks |

*and clear.*" An analysis arising from erroneous data will result into an analysis of similar quality and will have no significance as a consequence of that. One can also extend as a part of data quality check into misclassification of observations by looking into errors in prediction.

- Implementation of appropriate approaches toward statistical model development. Consider including variables in analysis, which are irrelevant, based on the physics of the system. While this may lead to better fitting of data, it also makes it difficult to infer a generic statement for a given population. These variables also have the potential to mask the actual causal relationship between relevant variables and process data.
- Validation of results or model validation on built of successful model is of significant importance. One can gather data outside the sample space used to develop the existing model, to validate the model's applicability. The model's predictability should be aligned with the physical understanding of the process.

An example, borrowed from Messerli (2012), illustrates a reasonably positive correlation between chocolate consumption and Nobel Laureates in a country. However, it is left to the modeler to understand and establish the role of consumption of chocolate in the underlying mechanism for the observed association. In other words, it is essential to keep in mind the underlying physics behind the mechanism studied.

The advantage of such an analysis is better explained through an example borrowed from Wikström et al. (1998b). Using MATLAB and its statistical tool box, the case study is provided in the Exercises Chapter.

## 6.4 CONCLUSION

Current practices in process safety engineering mostly hovers around implementing a macro-scale level model, largely developed through empirical relations. Experience plays a significant role in filling the gaps arising from such an approach, with instances where little or no understanding of the underlying process exists in a facility. In earlier chapters, the use of modeling methods at various scales was introduced to get better insights into understanding of processes through detailed level modeling. In this chapter, barring dynamic modeling and simulation, we have introduced data analysis-based methodologies within the context of process safety. The methodologies discussed here should be treated as brief introductions and are by no means presented comprehensively. Each of the topics such as dynamic simulation, chaos theory, and multivariate statistical analysis has classic books on them. The primary objective in this chapter was to provide an overview of the available tools to the readers that can be successfully applied to address challenging process safety concerns. Based on the nature of problem, the reader will have to dig deeper into these subjects as appropriate.

The list of available software tools, by no means comprehensive, for using the discussed approaches is given in Tables 6.2 and 6.3.

## REFERENCES


Berutti, M., 2015. Life Cycle Dynamic Simulation with Mimic. Mynah.com. Accessed March 16. https://www.mynah.com/knowledge-base/life-cycle-dynamic-simulation-mimic.

Broadribb, M.P., 2012. It's people, stupid! human factors in incident investigation. Process Safety Progress 31 (2).

Chen, F.F.K., Jentz, R.A., Williams, D.G., 1992. Flare System Design: A case for Dynamic Simulation. Offshore Technology Conference, May 4, 1992, Houston, TX. ISBN 978-1-61399-088-9.

Chen, K.H., Boning, D.S., Welsch, R.E., 2001. Multivariate statistical process control and signature analysis using eigenfactor detection methods. In: The 33rd Symposium on the Interface of Computer Science and Statistics, Costa Mesa, CA.

Chiappetta, G., Clarizia, G., Drioli, E., 2006. Analysis of safety aspects in a membrane reactor. Desalination 193 (1–3), 267–279. http://dx.doi.org/10.1016/j.desal.2005.06.064.

Doherty, M.F., Ottino, J.M., 1988. Chaos in deterministic systems: strange attractors, turbulence, and applications in chemical engineering. Chemical Engineering Science 43 (2), 139–183.

Eizenberg, S., Shacham, M., Brauner, N., 2006. Combining HAZOP with dynamic process model development for safety analysis. Computer Aided Chemical 21, 389–394.

Emigholz, K.F., 2003. Improving the operator's capabilities during abnormal operations; observations from the control house. Process Safety Progress 15 (3), 154–158. http://dx.doi.org/10.1002/prs.680150308.

Fiorina, C., 2004. Information: The Currency of the Digital Age. Oracle OpenWorld, San Francisco. http://www.hp.com/hpinfo/execteam/speeches/fiorina/04openworld.html.

Fisher, H.G., Forrest, H.S., Grossel, S.S., Huff, J.E., Muller, A.R., 2010. Emergency Relief System Design Using DIERS Technology: The Design Institute for Emergency Relief Systems (DIERS) Project Manual.

Hair, 2007. Multivariate Data Analysis. Pearson Education India.

Ho, Ta-C., Duh, Y.-S., Chen, J.R., 1998. Case studies of incidents in runaway reactions and emergency relief. Process Safety Progress 17 (4), 259–262. http://dx.doi.org/10.1002/prs.680170406.

Jonas, R.K., 2004. Design and Training Improvements through the Use of Dynamic Simulation. Light Metals, Warrendale.

Mahgerefteh, H., Wong, S.M.A., 1999. A numerical blowdown simulation incorporating cubic equations of state. Computers & Chemical Engineering 23 (9), 1309–1317.

Martin, E.B., Morris, A.J., Zhang, J., 1996. Process performance monitoring using multivariate statistical process control. IEE Proceedings–Control Theory and Applications 143, 132–144. http://dx.doi.org/10.1049/ip-cta:19960321.

Matzopoulos, M., Urban, J., Marriott, J., Marshall, B., 2010. Consider new analysis for flares. Hydrocarbon Processing 89 (11), 41–47.

Messerli, F.H., 2012. Chocolate consumption, cognitive function, and Nobel laureates. New England Journal of Medicine 367 (16), 1562–1564. http://dx.doi.org/10.1056/NEJMon1211064.

Mlodinow, L., 2009. The Drunkard's Walk. Penguin, UK.

Ostrowski, S.W., Keim, K.K., 2010. Tame Your Transient Operations. Chemicalprocessing.com. http://dx.doi.org/10.1007/978-3-642-12116-6_1.

Patle, D.S., Ahmad, Z., Rangaiah, G.P., 2014. Operator training simulators in the chemical industry: review, issues, and future directions. Reviews in Chemical Engineering 30 (2), 199–216. http://dx.doi.org/10.1515/revce-2013-0027.

Rasel, M.A.K., Richmond, P.C., 2014. Improve safety and reliability with dynamic simulation. Process Safety Progress 33 (4), 333–338. http://dx.doi.org/10.1002/prs.11667.

Singh, A., Li, K., Lou, H.H., Hopper, J.R., Golwala, H., Ghumare, S., Kelly, T., 2006. Flare minimisation via dynamic simulation. International Journal of Environment and Pollution 29 (1–3), 19–29. http://dx.doi.org/10.1504/ijep.2007.012794.

Speranza, A., Terenzi, A., 2011. Blowdown of Hydrocarbons Pressure Vessel with Partial Phase Separation. http://dx.doi.org/10.1142/9789812701817_0046. Audio, Transactions of the IRE Professional Group on January.

Strozzi, F., Zaldívar, J.M., Kronberg, A.E., Westerterp, K.R., 1999. On-line runaway detection in batch reactors using chaos theory techniques. Aiche Journal 45 (11), 2429–2443. American Institute of Chemical Engineers.

Strozzi, F., 1997. Runaway Prevention in Chemical Reactors Using Chaos Theory Techniques.

Švandová, Z., Jelemenský, L., Markoš, J., 2005. Steady states analysis and dynamic simulation as a complement in the HAZOP study of chemical reactors. Process Safety and Environmental Protection 83 (5), 463–471.

Takens, F., 1981. Detecting strange attractors in turbulence. Dynamical Systems and Turbulence 898, 366. http://dx.doi.org/10.1007/BFb0091924.

Vaidhyanathan, R., 1996. Experience with an expert system for automated HAZOP analysis. Computers & Chemical Engineering 20 (2), S1589–S1594.

Westerweele, M.R., Laurens, J., 2008. Mobatec Modeller—A flexible and transparent tool for building dynamic process models. Computer Aided Chemical Engineering 25, 1045–1050.

Wikström, C., Albano, C., Eriksson, L., Fridén, H., 1998a. Multivariate process and quality monitoring applied to an electrolysis process: part II. Multivariate time-series analysis of lagged latent variables. Chemometrics and Intelligent Laboratory Systems 45 (1–2), 233–240.

Wikström, C., Albano, C., Eriksson, L., Fridén, H., 1998b. Multivariate process and quality monitoring applied to an electrolysis process: part I. Process supervision with multivariate control charts. Chemometrics and Intelligent Laboratory Systems 42 (1–2), 221–231.

Wolf, A., Swift, J.B., Swinney, H.L., Vastano, J.A., 1985. Determining Lyapunov exponents from a time series. Physica D: Nonlinear Phenomena 16 (3), 285–317.

Wu, J., Zhang, L., Hu, J., Lind, M., Zhang, X., Bay Jørgensen, S., Sin, G., Jensen, N., August 27, 2014. An integrated qualitative and quantitative modeling framework for computer-assisted HAZOP studies. Aiche Journal 60 (12). http://dx.doi.org/10.1002/aic.14593.

Xu, Q., Yang, X., Liu, C., Li, K., Lou, H.H., Gossage, J.L., 2009. Chemical Plant Flare Minimization via Plantwide Dynamic Simulation. American Chemical Society.

Yang, S.H., Yang, L., He, C.H., 2000. Improve safety of industrial processes using dynamic operator training simulators. Process Safety and Environmental Protection 79 (6), 329–338. http://dx.doi.org/10.1205/095758201753373096.

Yao, Y., Gao, F., 2009. Multivariate statistical monitoring of multiphase two-dimensional dynamic batch processes. Journal of Process Control 19 (10), 1716–1724. http://dx.doi.org/10.1016/j.jprocont.2009.07.003.

Zaldívar, J.M., Cano, J., Alós, M.A., Sempere, J., Nomen, R., Lister, D., Maschio, G., et al., 2003. A general criterion to define runaway limits in chemical reactors. Journal of Loss Prevention in the Process Industries 16 (3), 187–200. http://dx.doi.org/10.1016/S0950-4230(03)00003-2.

Zamejc, E., 2006. API standard 521 new alternative method to evaluate fire relief for pressure relief device sizing and depressuring system design. Journal of Loss Prevention in the Process Industries 27 (2014), 21–31.

CHAPTER

# EQUIPMENT FAILURE

# 7

## 7.1 INTRODUCTION

In previous chapters, the primary area of interest was consequence modeling at multiscale levels such as assessment of thermal hazards from fire, dispersion modeling of hazardous gases, consequence modeling of explosion, and estimation of safety parameters relevant to flammable and toxic chemicals. In safety analysis, consequence modeling is necessary but not a sufficient step. Consequence modeling addresses the question: *How severe can a process safety incident be?* It does not address: *How often could it happen?* The likelihood of a process safety event is thus very important for performing a quantitative risk assessment (QRA) study of a given scenario. Some of the information that helps estimate likelihood of an event are ignition probability (early for fire and late for explosion), weather data (wind speed, direction, temperature, etc.), and equipment failure rate. Among these, the failure of equipment such as that of a pump, a compressor, a pressure vessel, or a specific-diameter pipeline typically initiates a process safety event. In the following, various approaches applicable to the estimation of equipment failure rate are explored. In addition, modeling methodologies that can help understand the underlying mechanism behind equipment failure is demonstrated.

## 7.2 FAILURE RATES FOR VARIOUS TYPES OF EQUIPMENT

As shown in Figure 7.1, equipment in general exhibits three phases during their life cycle. The first phase, known as the "burning in" or "infant mortality" phase, exhibits a decrease in the failure rate. The second phase, also known as the "useful-life" or "chance-failure" phase, exhibits a constant failure rate. The third and final stage, known as the "wear-out" phase, exhibits an increase in failure rates (Glaser, 1980).

The curve shown in Figure 7.1 is known as bathtub curve because of its cross-sectional shape. Although this curve can be useful in some cases, it loses its validity when, for example, the use of an equipment has decreased before or during the final stage. In reliability studies, the failure rate function is of great importance. This function is characterized by the probability distribution of the time to failure of the equipment, as will be described in the following sections. Most equipment will have an increasing failure rate because of deterioration with use.

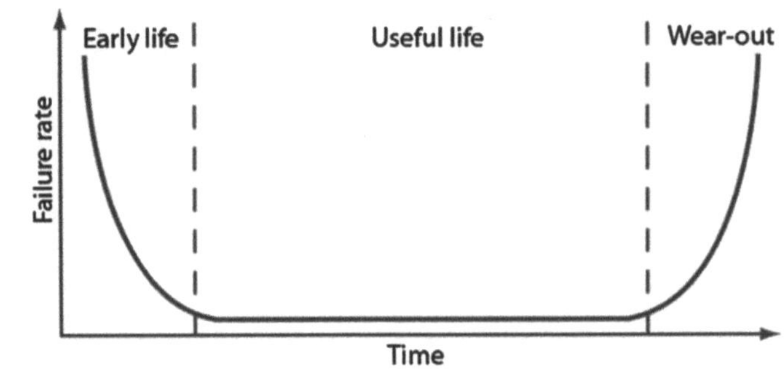


**FIGURE 7.1**

Bathtub curve.

### 7.2.1 EXAMINING EXTRAPOLATION OF CURRENT DATA OUTSIDE THEIR APPLICABLE DOMAIN OF CERTAIN TEMPERATURE AND PRESSURE FAILURE RATE OF EQUIPMENT

During the past 25 years, probabilistic safety assessment has been gradually accepted as a well-defined scientific discipline. The main reason for this acceptance has been the recognition that a consistent framework in which to analyze the safety of large industrial complexes is needed. Nuclear power plants are included within these complexes. It has also been recognized that conventional reliability theory, as developed for electronic components and manufacturing in general, cannot be applied in a straightforward way. This is because of the rarity of the events of interest and the consequent absence of good statistical data. Such lack of data has encouraged the elicitation of opinion from experts in the field: "expert judgment" techniques (Thorne and Williams, 1992). This in turn has raised the question of the reliability of such judgments, which must contain some element of subjectivity. A typical example of this situation occurs when, after a series of operations, an engineering device has not failed. One then asks what can be said about its failure rate. It is clear that unless some external factors are introduced, such as the failure rates of similar devices, one can do no more than make weak statements constraining the range of possible failure rates consistent with the observation of no failures (Apostolakis, 1981). In order to understand these kinds of problems and to give a useful assessment of the uncertainties, it is necessary to construct probabilistic models that represent the physical phenomena and then to quantify the uncertainties in terms of these models. The probabilistic models themselves do not require any interpretation (as long as they are physically plausible) since they are based on an axiomatic theory; namely, the theory of probability. The actual quantitative measure of the uncertainties depends, however, on the particular way that the assessor of the problem interprets probability concepts. Evidence and methods of analysis that may be admissible in one "school" may not be acceptable in another. There is considerable subjectivity involved.

It is worth noting that there exist two principal schools of thought regarding probability: (1) the subjectivist (or Bayesian) and (2) the frequentist (or Classicist). The subjectivist interprets probability as a degree of belief and imposes certain requirements of coherence that an assessor must associate with his probabilities (De Finetti, 1974). The frequentist school, on the other hand, defines probabilities as limits of long-run relative frequencies (v Mises, 1957). From the point of view of rare events, it is difficult to see how the frequentist view can survive since, by definition, there are no long-run data available. Indeed, in some instances, the event under investigation may never have happened and it might be easier to predict when the sun will explode than to predict when a rare accident event of a specific type will occur in some industrial process. At least, in the case of the sun, one can obtain astronomical data from other similar stars. There is no way that one can observe nuclear reactor cores melting to order. It seems, therefore, that any real progress in predicting the occurrence of rare events must be based on a large degree of judgment and a very good knowledge of the nature of the underlying statistical processes.

The introduction of judgment, and by this we mean "expert judgment," can lead to very difficult problems of interpretation, reconciliation, and aggregation of results obtained from several experts. Acceptance of probability theory will guarantee that each of the experts is self-consistent, but it cannot force consensus between analysts, or ensure coherence of their combined views. It is argued by some people that, given the same information, scientists should agree on their inferences; however, to accept this tenet as a principle, it would be necessary to define criteria that would determine under what conditions several individuals actually have the same information. This is clearly not possible and one has to accept that experts will disagree even though they essentially possess the same initial data relating to the issue under consideration. Because of this, further statistical techniques are needed to assess the worth of an expert and considerable effort has been expended in developing such techniques (Apostolakis et al., 1988).

#### 7.2.1.1 Mean time to failure

In reliability theory, there are several mean times to failure distributions that could be used. These distributions have properties that make them more useful and applicable in certain cases. In this section, we will look at some of these distributions (Table 7.1) and when they should be used.

#### 7.2.1.2 Exponential distribution

This is one of the most widely used mean times to failure distributions because of its simplicity and ease of use. The exponential distribution is characterized by the use of a constant failure rate. This is often the case with electronic and mechanical components (Finkelstein, 2008). The exponential distribution is a good choice to represent nonredundant components consisting of many interacting parts. The exponential probability density function (PDF) is defined as:

$$f(t) = \lambda \exp(-\lambda t) \tag{7.1}$$

where the instantaneous failure rate, expressed as *h(t)* throughout this chapter, is equal to $\lambda$. The mean time to failure rate, or MTTF, is given by $1/\lambda$.

Another important characteristic of the exponential distribution is that it has "no memory." This means that a component that has worked up to a certain time $t$ has no memory of its past. Therefore, the probability that an item will fail between $t$ and $t + \Delta t$ is represented by

$$P(t \leq T \leq t + \Delta t | T > t) = \frac{\exp(-\lambda t) - \exp[-\lambda(t + \Delta t)]}{\exp(-\lambda t)} = 1 - \exp(-\lambda \Delta t) \tag{7.2}$$

#### 7.2.1.3 Gamma distribution

The gamma distribution is a more generalized form of the exponential distribution. The gamma distribution is useful for instruments or equipment that needs to be calibrated. It is also useful for modeling a system with and standby components (Modarres et al., 2009; Finkelstein, 2008).

The PDF of the gamma distribution is given by:

$$\frac{1}{\beta\alpha\Gamma(\alpha)} t^{\alpha-1} \exp\left[-\frac{t}{\beta}\right] \tag{7.3}$$

**Table 7.1 Summary of the Lifetime Distributions**

| Distribution Characteristic | Exponential Distribution | Weibull Distribution | Normal Distribution | Lognormal Distribution |
|---|---|---|---|---|
| PDF, $f(t)$ | $\lambda \exp(-\lambda t)$ | $\frac{\beta(t)^{\beta-1}}{\alpha^\beta}\exp\left[-\left(\frac{t}{\alpha}\right)^\beta\right]$ | $\frac{1}{\sigma\sqrt{2\pi}}\exp\left[-\frac{1}{2}\left(\frac{t-\mu}{\sigma}\right)^2\right]$ | $\frac{1}{\sigma_t t\sqrt{2\pi}}\exp\left[-\frac{1}{2\sigma_t^2}(\ln t-\mu_t)^2\right]$ |
| CDF, $F(t)$ | $1-\exp(-\lambda t)$ | $1-\exp\left[-\left(\frac{t}{\alpha}\right)^\beta\right]$ | $\frac{1}{\sigma\sqrt{2\pi}}\int_0^t \exp\left[-\left(\frac{\theta-\mu}{2\sigma^2}\right)^2\right]d\theta$ | $\frac{1}{\sigma_t\sqrt{2\pi}}\int_0^t \frac{1}{\theta}\exp\left[-\frac{1}{2}\frac{(\ln\theta-\mu_t)^2}{2\sigma_t^2}\right]d\theta$ |
| Instantaneous failure rate, $h(t)$ | $\lambda$ | $\frac{\beta}{\alpha}\left[\left(\frac{t}{\alpha}\right)^{\beta-1}\right]$ | $\frac{f(t)}{1-F(t)}$ | $\frac{f(t)}{1-F(t)}$ |
| MTTF | $\frac{1}{\lambda}$ | $\alpha\Gamma\left(\frac{1+\beta}{\beta}\right)$ | $\mu$ | $\exp\left(\mu_t+\frac{1}{2}\sigma_t^2\right)$ |

where *h(t)* is given by:

$$\frac{t^{\alpha-1}\exp\left(-\frac{t}{\beta}\right)}{\beta^{\alpha}\left[\Gamma(\alpha)-\int_0^t y^{\alpha-1}\exp\left[-\frac{y}{\beta}\right]dy\right]} \tag{7.4}$$

And the MTTF is equal to $\alpha\beta$. The cumulative distribution function (CDF) of the gamma distribution is given by:

$$\frac{1}{\beta\alpha\Gamma(\alpha)}\int_0^t y^{\alpha-1}\exp\left[-\frac{y}{\beta}\right]dy \tag{7.5}$$

### *7.2.1.4 Weibull distribution*

The Weibull distribution is also considered to be one of the most used failure distributions. This is because of its flexibility for describing hazard rates. The Weibull distribution allows for the use of increasing and decreasing failure rates. This means that can be used to correctly model the "burn-in" and "wear-out" period of the component life cycle. The Weibull distribution is especially useful for the following applications or components (Modarres et al., 2009):

- Capacitors, relays, electron tubes, transistors, ball bearings, photoconductive cells, and motors
- Corrosion resistance

The PDF of the Weibull distribution is given by:

$$f(t)=\frac{\beta(t)^{\beta-1}}{\alpha^{\beta}}\exp\left[-\left(\frac{t}{\alpha}\right)^{\beta}\right],\quad \alpha,\beta>0,\ t>0 \tag{7.6}$$

where *h(t)* is given by:

$$\frac{\beta}{\alpha}\left[\left(\frac{t}{\alpha}\right)^{\beta-1}\right] \tag{7.7}$$

and the MTTF is calculated as follows:

$$\alpha\Gamma\left(\frac{1+\beta}{\beta}\right) \tag{7.8}$$

The CDF of the Weibull is given by:

$$1-\exp\left[-\left(\frac{t}{\alpha}\right)^{\beta}\right] \tag{7.9}$$

#### 7.2.1.5 Normal distribution

Even though the normal distribution is probably the most widely used distribution in statistics, it is a very limited life distribution in reliability studies because of its impracticality. This impracticality stems from the fact that a normally distributed random variable could have any value in the $(-\infty, +\infty)$ range (Yang, 2007). The normal distribution is very useful in modeling the wear-out region for equipment. The normal T distribution is also very useful for strength—stress modeling (Modarres et al., 2009).

The normal PDF of the normal distribution is given by:

$$f(t) = \frac{1}{\sigma\sqrt{2\pi}}\exp\left[-\frac{1}{2}\left(\frac{t-\mu}{\sigma}\right)^2\right] \tag{7.10}$$

and the CDF is given by:

$$F(t) = \frac{1}{\sigma\sqrt{2\pi}}\int_0^t \exp\left[-\left(\frac{\theta-\mu}{2\sigma^2}\right)^2\right]d\theta \tag{7.11}$$

where *h(t)* is given by:

$$\frac{f(t)}{1-F(t)} \tag{7.12}$$

and $\mu$ is the MTTF, and $\sigma$ is the standard deviation of failure time.

#### 7.2.1.6 Lognormal distribution

The lognormal is a modification to the normal distribution. The lognormal distribution truncates the random variable range, allowing for only the use of positive values. This makes the lognormal more suitable for use in reliability studies. The lognormal is useful in modeling maintenance activities associated could lead to mechanical fatigue (Modarres et al., 2009). It is also useful in modeling breakage processes.

The normal PDF of the lognormal distribution is given by:

$$f(t) = \frac{1}{\sigma_t t\sqrt{2\pi}}\exp\left[-\frac{1}{2\sigma_t^2}(\ln t - \mu_t)^2\right] \tag{7.13}$$

and CDF is given by:

$$F(t) = \frac{1}{\sigma_t\sqrt{2\pi}}\int_0^t \frac{1}{\theta}\exp\left[-\frac{1}{2}\frac{(\ln\theta - \mu_t)^2}{2\sigma_t^2}\right]d\theta \tag{7.14}$$

where *h(t)* is calculated by the same formula than in the normal distribution, and

$$\mu_t = E(\ln t) \text{ and } \sigma_t^2 = Var\,(\ln t).$$

The MTTF is given by:

$$\exp\left(\mu_t + \frac{1}{2}\sigma_t^2\right) \tag{7.15}$$

The evaluation of the reliability is, in most cases, a prediction. Prediction is both an analytical and systematic process based on the knowledge and combination of the failure rate values for the individual components and plant items that comprise the system. In many cases, field data are few, even for individual components, and so a prediction exercise is necessary. For the prediction to have credence, some form of comparison is necessary between the predicted data and those derived from observations on an operating system – where this is possible. Such comparisons are useful even if they do not correspond precisely to the component of interest since they may give a trend. An extensive survey was carried out by Snaith (1981) in which the predicted and observed failure rates on 130 different engineering systems were studied. In all, a total of 140 items of data were assessed. The predicted values include both direct assessment by experts and the results of analysis. It was noted that 63% of all predicted values are within a factor of 2 of the observed values, and 93% are within a factor of 4. It was also noted that the ratio (observed/predicted) tends to the log normal distribution. A further result was that for equipment having low actual failure rates, the failure rate predictions tended to be high and vice versa: the error is in the conservative direction for low failure rates.

There is also evidence to suggest that an individual's probability distribution tends to be too tight and the assessment of extreme fractiles is particularly prone to error (Mosleh et al., 1987); this is of concern as the use of 5% and 95% limits is common.

#### 7.2.1.7 Failure rates without failure

The prediction or estimation of a failure in which failure is missing can be very complicated. There exists different methods to deal with this problem; some of them are described next (Freeman, 2011):

- **To assume the occurrence of a failure**: Assuming that one failure has happened in $n$ years, the failure rate is given by:

  $$\lambda = 1/n \tag{7.16}$$

  In this case, the failure rate is a discrete variable.
- **Bayesian approach**: Lipow and Welker (1974) introduced some interesting new ideas on the prediction of failure rates when no failures have been recorded. They proposed a Bayesian approximation of the "best guess" for the failure rate, which is given by:

  $$w(\lambda|c=f) = \frac{P(\lambda|c=f)w(\lambda)}{P(c=f)} \tag{7.17}$$

where $\lambda$ is the failure rate, $c$ is the number of observed failures during a test of duration $T$, $f$ is the observed failures in a particular test. $w(\lambda)$ is the assumed density function, $P(\lambda \mid c=f)$ is the conditional probability of having $f$ failures given a failure rate $\lambda$, $P(c=f)$ is the unconditional probability of having $f$ failures on the assumed $w(\lambda)$, and $w(\lambda|c=f)$ is the posterior distribution for $\lambda$ conditioned on the observation of $f$ failures. Using the exponential distribution, the authors calculated a guess for $\lambda$ and used that guess as prior distribution for a new guess of $\lambda$. Therefore, Welker and Lipow calculated $\lambda$ as:

$$\lambda = k/(n) \tag{7.18}$$

where $k$ is a constant estimated by the repetitive Bayesian solution. The authors found that the values of $k$ were between 0.24 and 0.51 and proposed to estimate $\lambda$ by:

$$\lambda = 1/(3n) \tag{7.19}$$

- **Confidence limit on the failure rate**: A confidence bound of C% for $\lambda$ is given by the chi-squared distribution $\chi^2$ with $2 + n$ degrees of freedom (Kececioglu, 2002).
  The one-sided confidence limit when failures have not been observed is given by (Green and Bourne, 1972):

$$\lambda = X^2_{\alpha;2}/(2n) \tag{7.20}$$

where $X^2_{\alpha;2}$ is the value of the chi-squared distribution with two degrees of freedom and a confidence level (CL), given by:

$$\text{CL} = 1 - \alpha \tag{7.21}$$

As $\alpha$ (parameter related with the area of the tail of the chi-squared value) is always between 0 and 1; therefore:

$$1 > \text{CL} > 0 \text{ or } 0 < \text{CL} < 1 \tag{7.22}$$

For example, if the confidence level is set to 95%, $X^2_{\alpha;2} = \chi^2 = 5.991$, and the estimator for the failure rate can be calculated as:

$$\lambda = \frac{X^2_{\alpha;2}}{2n} = \frac{5.991}{2n} \tag{7.23}$$

There are other methods that have been proposed to estimate failure rate in the absence of failure rate. These methods are well explained elsewhere (Bailey, 1997; Freeman, 2011).

### 7.2.2 BAYESIAN LOGIC

There are several different types of information that can be used in the process industries to estimate failure rates. Plant-specific data can be gathered through the use of documentation systems, particularly through work and maintenance requests. Generic data can also be found through incident databases and tabulated data. Furthermore, it is possible to incorporate expert judgment into the analysis, basing the information on experience and sound knowledge.

On the surface, each of these solutions looks good. Plant-specific data must be useful because these are information that are unique and useful for the plant. Generic data must be useful because these give insight where data on plant-specific functions may not be available. Expert judgment must be useful because it comes from an expert. All of these statements are half-truths.

Plant-specific data suffer from several downfalls. First, there may not be as much information as is needed, and if the information exists, it may not be the right information. It is possible that the information gathered, say, through work orders is not complete, consistent, or correct. Further, even if the information is correct, consistent, and complete, there may not be enough failures to give an accurate picture of reliability. Take, for example, the failure of a tank containing a corrosive material. It is possible that in many years, this failure will never occur, especially with good maintenance procedures. However, just because there has never been a failure does not mean that there can never be one in the future.

Generic data can be used for many applications, especially where failures are infrequent or hard to detect. However, generic data do not necessarily accurately reflect the process parameters and therefore may not be as representative of the actual process as is expected. This can lead to false confidence in the case of an overestimation of reliability or added expense in the case of an underestimation of reliability.

Expert judgment is useful as a tool in conjunction with other techniques, but by itself it may be less than adequate. It can be very subjective, and it is likely that if several experts were asked their quantitative opinion of the reliability of a component, they would all have a different answer.

Since no approach is infallible (and it could even be argued that each is deficient when used alone), it is important to use a technique that allows for multiple approaches to be used but, even more importantly, allows for the conclusions that have been gathered to be updated on the collection of more data. This technique is known as the Bayes theorem, a method by which a prior probability or probability distribution is updated by observance, such as through testing or data collected through process operation.

#### *7.2.2.1 Introduction to the Bayes theorem*

The idea behind the Bayes theorem – the concept of using probabilistic theory to update beliefs and judgments – was first suggested by Thomas Bayes in the eighteenth century and was extended independently by Pierre-Simon Laplace in the early nineteenth century. It has since been regarded as one of the fundamental and most important theorems in all of probability theory. The Bayes theorem is derived from the law of total probability and a few probability identities. A simple derivation is presented as follows:

Two events are considered to be "statistically independent" if the occurrence of one event does not depend on the occurrence of the other event. For example, if one rolls two dice on a table, the value shown by each dice is independent of the other. The dice may collide and influence the path of one another, but the value shown on the upward face of one dice has absolutely no impact on the other. The values are independent. If two events, A and B, are independent then the probability of A given that event B has already occurred is simply equal to the probability of A (since B has no influence on A). This can be notated as $P(A|B) = P(A)$. The probability of both A and B occurring would simply be the product of the two events happening independently, or $P(A \cap B) = P(A)P(B)$.

Two events may also be dependent on each other, in which case the previous expression would not hold. $P(A|B)$ would be its own value, and the probability of both A and B occurring would be $P(A \cap B) = P(A|B)P(B)$, which is equivalent to $P(A \cap B) = P(B|A)P(A)$. This can be thought of as asserting that the probability of both A and B occurring is equal to the product of the probability that A occurs independently and the probability that B occurs dependently on A once A has occurred (or the other way around).

Equating these terms, P(A|B)P(B) = P(B|A)P(A). This can then be rearranged to P(A|B) = P(B|A) P(A)/P(B) or P(B|A) = P(A|B)P(B)/P(A). Because the total probability of A and B are dependent, their total probability of occurrence is a function of all of the possible scenarios of the other. Therefore, in the case of the expression for P(A|B), the denominator must be expanded to capture the dependency, giving the expression:

$$\mathrm{P(A|B)} = \frac{\mathrm{P(B|A)P(A)}}{\sum_{i=1}^{n} \mathrm{P(B|A}_i\mathrm{)P(A}_i\mathrm{)}} \tag{7.24}$$

where $A_i$ is the full set of events A that can occur that are dependent on B. A similar derivation can be used to find $P(B_i|A)$. If B depends on mutually exclusive and collectively exhaustive events $A_i$ (meaning that set A is completely defined by all events $A_i$ and no two events $A_i$ and $A_j$ can occur concurrently), then Bayes equation can be put into its most useful form:

$$\mathrm{P(A}_i\mathrm{|B)} = \frac{\mathrm{P(B|A}_i\mathrm{)P(A}_i\mathrm{)}}{\sum_{i=1}^{n} \mathrm{P(B|A}_i\mathrm{)P(A}_i\mathrm{)}} \tag{7.25}$$

which describes the probability of a single event $A_i$ as it depends on event B. Several different components of this equation have special meaning. $P(A_i)$ is the "prior probability." This is the belief of how the system dependencies are defined prior to more information being integrated into the analysis. The remaining factor on the right-hand side is the "likelihood function," the scaling factor for the new probability value. The left-hand side term $P(A_i|B)$ is the "posterior probability," which is the new belief of system dependencies based on the observation of an event. Once an event occurs, perhaps a test being done, the $P(A_i|B)$ value can be used as the new prior probability value for $P(A_i)$.

#### 7.2.2.2 An illustrative example of Bayes theorem

Consider a (rather dull) dice game where a single die is rolled and rolling a 6 wins, while rolling anything else loses. Naturally there would be some temptation to cheat at this game, such as by the use of loaded dice. By using prior data (perhaps generic data, or perhaps expert judgment) as to the properties of a loaded dice and observations as to the nature of the result of the dice rolls, Bayes theorem can be used to determine whether a die is likely to be loaded.

Consider that a fair die will land on each of its sides with roughly equal probability, 1/6 (16.7%). Therefore, landing on 6 would have a 1/6 probability and landing on anything but 6 would have a 5/6 probability. Perhaps one would estimate that the probability of a loaded die landing on 6 is 9/10 (90%) and the probability of it landing on anything else would be 1/10. Further, say that you have no certainty of whether the dice is loaded, it is 50/50 that it could be either. By throwing a die and observing the outcome (perhaps termed as a "test"), your confidence of the dice being loaded or not (which can perhaps be termed as the "reliability" of the dice) can be modified and improved. The prior information is shown in Table 7.2.

Now the die is thrown. Consider that the result is a 6. Using Bayes equation, the updated probability that the die is loaded can be found using the equation:

$$\mathrm{P(L|S)} = \frac{\mathrm{P(S|L)P(L)}}{\mathrm{P(S|L)P(L) + P(S|F)P(F)}} = \frac{0.9 \times 0.5}{0.9 \times 0.5 + 0.167 \times 0.5} = 0.843 \tag{7.26}$$

**Table 7.2 Initial Values for the Die Roll Example**

| Components | Defined Probabilities | Value |
|---|---|---|
| "Reliability" | P(F) = probability of fair die | 0.5 |
| | P(L) = probability of loaded die | 0.5 |
| "Test" | P(S\|F) = probability of 6 given a fair die | 0.167 |
| | P(S′\|F) = probability of not 6 given a fair die | 0.833 |
| | P(S\|L) = probability of 6 given a loaded die | 0.9 |
| | P(S′\|L) = probability of not 6 given a loaded die | 0.1 |

And the updated probability that the die is fair is simply the complement of 0.843 by the law of total probability, which is 0.157. This initial test makes the die seem as though it is probably loaded. But more testing can be done by using the posterior distribution as the new prior and throwing the die again. Say the result of the second throw is a 2. What are the chances the die is loaded after two data points?

$$\mathrm{P(L|S')} = \frac{\mathrm{P(S'|L)P(L)}}{\mathrm{P(S'|L)P(L) + P(S'|F)P(F)}} = \frac{0.1 \times 0.843}{0.1 \times 0.843 + 0.833 \times 0.157} = 0.392 \tag{7.27}$$

And the updated probability of a fair die is the complement at 0.608. By using the prior probability as our new guess and scaling it by the likelihood that the roll would not be a 6, we have come up with a much more optimistic guess of whether the die is loaded.

This example may seem trivial, but by relating a loaded die to reliability and rolling to testing, it can be seen how this approach may be useful in the case of a process plant. Say, instead of a loaded die, a plant has a tank that has a chance of a defect due to corrosion, and instead of rolling a die to test, several small samples are taken at various intervals to perform corrosion tests. Continuous testing and updating through the Bayes theorem can identify trends in important fields such as reliability as simply as it can predict whether dice are loaded.

#### *7.2.2.3 Bayes theorem and application to equipment failure*

Bayes theorem can also be extended to a continuous form where probability distributions are used:

$$f''(\theta|x) = \frac{\mathrm{P}(x|\theta)f'(\theta)}{\int_{-\infty}^{\infty} \mathrm{P}(x|\theta)f'(\theta)d\theta} \tag{7.28}$$

where $f'$ represents the prior probability *distribution*, $f''$ represents the posterior probability *distribution*, and $\mathrm{P}(x|\theta)$ is the probability of $x$ as a function of $\theta$. The idea is the same: update prior beliefs about the performance of the system with observed information, but in this case probability distributions are used and updated.

Typically, the selection of the prior distribution is somewhat subjective, so a selection of a conjugate prior from the same family of distributions as the posterior can make the choice more objective for easier computation of the posterior parameters. An example of conjugate distributions are a beta prior and binomial likelihood (leading to a beta posterior), and a gamma prior and Poisson likelihood (leading to a gamma posterior). According to the literature (Shafaghi, 2008), the gamma distribution and Poisson likelihood are best suited to finding a failure rate of components, while the binomial likelihood and beta distribution are best suited for the calculation of rate of failure on demand. Explanation of the Poisson/gamma process follows.

The discrete Poisson distribution assumes that events of interest are randomly dispersed in time or space with a constant intensity of $\lambda$ (Modarres et al., 2009). The distribution can be expressed as:

$$P(x) = \frac{\rho^x \exp(-\rho)}{x!} \tag{7.29}$$

where $x$ is a discrete variable that describes the number of events observed at a certain time interval, and $\rho = \lambda t$, where $\lambda$ is as described earlier and $t$ is the time interval of interest. The Poisson distribution can be used as the likelihood function gathered from observed data in the continuous Bayes formulation.

The gamma distribution is the continuous analog of the Poisson, and models time for $\alpha$ events to occur randomly with an observed mean time of $\beta$ between events. This is used as the prior probability in the continuous Bayes formulation. This distribution can be expressed as:

$$f(t) = \frac{1}{\beta^\alpha \Gamma(\alpha)} t^{\alpha-1} \exp\left(-\frac{t}{\beta}\right) \tag{7.30}$$

$\Gamma(\alpha)$ can be simplified to $(\alpha - 1)!$ for all discrete values of $\alpha$. As a simple example of the use of this continuous formulation, consider a set of tanks that are believed to crack due to corrosion in a gamma-distributed manner. It is believed that three events will occur randomly with a mean time or 5 years in between each event ($\alpha = 3$, $\beta = 5$). The prior probability function can be stated as:

$$f(t) = \frac{1}{5^3(3-1)!} t^{3-1} \exp\left(-\frac{t}{5}\right) = \frac{1}{250} t^2 \exp\left(-\frac{t}{5}\right) \tag{7.31}$$

Suppose that the tanks are observed for 1 year and an event occurs. The observed $\lambda$ value would now be one per year. The Poisson distribution of this observation would be:

$$P(x) = \frac{(1 \times t)^1 \exp(-1 \times t)}{1!} = t\exp(-t) \tag{7.32}$$

Using the above formula to calculate the posterior distribution, we find:

$$f''(\theta|x) = \frac{P(x|\theta)f'(\theta)}{\int_{-\infty}^{\infty} P(x|\theta)f'(\theta)d\theta} = \frac{t\exp(-t)\cdot\frac{1}{250}t^2\exp\left(-\frac{t}{5}\right)}{\int_0^{\infty} t\exp(-t)\cdot\frac{1}{250}t^2\exp\left(-\frac{t}{5}\right)dt} \tag{7.33}$$

this can be simplified to:

$$f''(\theta|x) = \frac{t^3\exp\left(-\frac{6t}{5}\right)}{\int_0^{\infty} t^3\exp\left(-\frac{6t}{5}\right)dt} \approx \frac{1}{2.89} t^3 \exp\left(-\frac{6t}{5}\right) \tag{7.34}$$

which fits the gamma distribution with parameters $\alpha = 4$ and $\beta = 5/6$. The beta parameter has gone from a prediction of a mean time between events of 5 years and 10 months. The updated distribution compared with the prior distribution is as follows (Figure 7.2).

As can be seen, the probability of a failure has gone from relatively uncertain to having a significant spike early in operation as a result of the early failure. The Bayes model has predicted that the failures

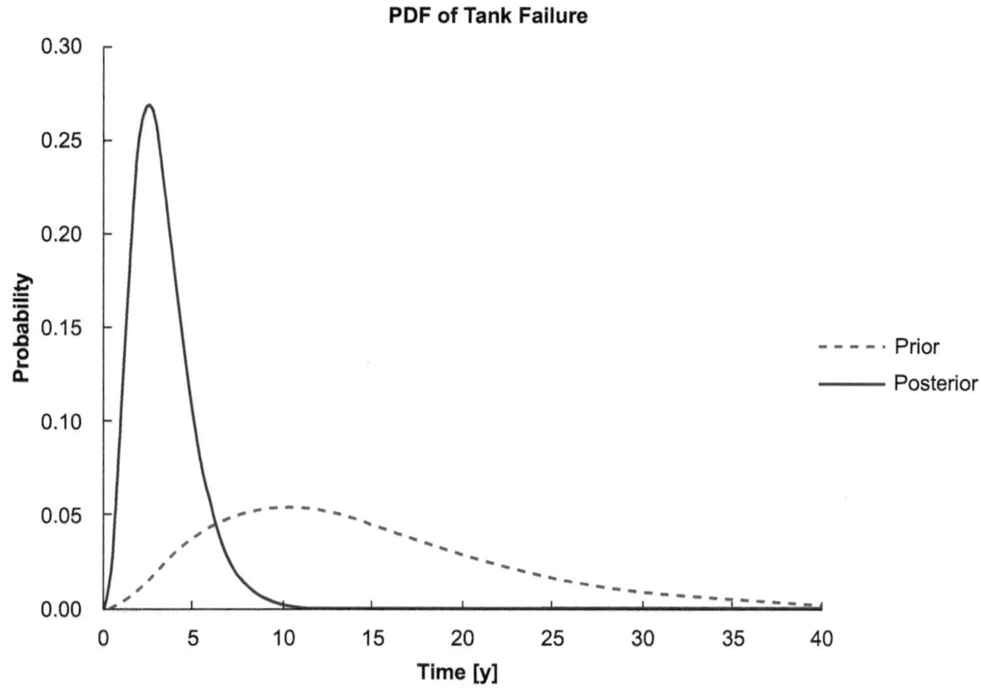


**FIGURE 7.2**

Probability distribution function of tank failure.

are expected to occur much earlier than originally expected. With further monitoring and continuous testing, it is possible that the model can once again be updated with more information as the time goes on to provide an even more accurate distribution.

#### *7.2.2.4 Sources of generic data*

Generic data may be obtained from a large number of sources, including CCPS Guidelines for Process Equipment Reliability Data (CCPS, 1988), IEEE Standard 500 – Guide to the Collection and Presentation of Electrical, Electronic, Sensing Component, and Mechanical Equipment Reliability Data for Nuclear-Power Generation Stations (IEEE, 1984), and the Offshore Reliability Data Handbook (OREDA, 2002). Further sources may be found in Lees' Loss Prevention in the Process Industries (Mannan, 2013).

### 7.2.3 RISK ASSESSMENT OF LNG TERMINALS USING THE BAYESIAN–LOPA METHODOLOGY

LNG refers to natural gas converted into its liquid state by super cooling to 260 °F (162.2 °C). LNG consists of 85–98% methane and the remaining is made up of smaller concentrations of heavier hydrocarbon components. LNG provides a cost-effective containment as well as transportation across great distances onshore and offshore at atmospheric pressure. Moreover, LNG is environmentally friendly because of its clean burning.

Because of these properties, the LNG demand has been growing to diversify the energy portfolios and fulfill the energy demand for LNG fuel for many applications, including heating, cooking, and power generation. Following the increasing demand for LNG, there are at least 100 currently active LNG facilities across the United States, including importation terminals, peak-shaving facilities, or base load plants. In addition, there are also a number of proposed projects for LNG terminals in North America. This analysis will focus on LNG terminals.

Even though LNG has several advantages, it may cause fire or explosion when it is discharged at undesired conditions. Especially in LNG terminals, large consequences may occur due to the huge amounts of LNG stored. Although the LNG industry has had an excellent safety record over the past 40 years, the risk related to LNG terminals may be increasing with the growing LNG industry. Consequently, risk-informed decisions found on sound science are critical to controlling risks related to LNG terminals.

In order to control risk, it is necessary to quantify the risk by applying a risk assessment methodology. LOPA provides a straightforward and systematic approach to obtain quantified risk results with less effort and time than other methods, especially QRA.

For the application of LOPA methodology, failure data of equipment and facilities are required to quantify the risk. However, the LNG industry has a relatively short operational history compared to other industries, such as the chemical industry, and there are only a few incident records. Therefore, available plant-specific failure data of LNG system are very sparse. The risk values estimated with these insufficient data may not show statistical stability or represent specific conditions of an LNG facility. Generic failure data from other industries such as the petrochemical and nuclear industries may be used for the LNG industry because they have sufficient and longer-term historical records. However, these data also may not provide appropriate risk results for the LNG industry because the operational conditions and environment of LNG facilities are different from those of other industries.

Therefore, the Bayesian estimation can be used to obtain more reliable risk values. Bayesian logic can produce updated failure data using the prior information of generic data from other industries and the likelihood information of LNG plant-specific data. The updated data can reflect both statistical stability from the generic data and the specific conditions from the LNG plant data. In addition, Bayesian estimation includes variability and uncertainty information to result in Bayes credibility or probability intervals, such as a 90% credible interval of updated data, even though LNG plant failure data are sparse. In other words, the weakness of failure data from the LNG industry can be reduced or overcome by the use of Bayesian estimation. Therefore, the Bayesian–LOPA methodology, which was newly developed in this paper, was used to conduct risk assessments for an LNG terminal.

#### *7.2.3.1 Methodology development*

LOPA is a simplified form of QRA that uses initiating event frequency, consequence severity, and the probability of failure on demand (PFD) of independent protection layers (IPLs) to estimate the risk of an incident scenario (Crowl, 2001). Typically, LOPA builds on the information developed during process hazard analysis (PHA) where techniques, such as HAZOP and What-If methods, are used to develop incident scenarios. The purpose of LOPA is to estimate risk values for a facility so that risk decisions can be made based on tolerable risk criteria adopted by the facility. LOPA can also be used to rank the estimated risk values of identified incident scenarios to give priority to safety measures for higher-risk scenarios and for critical equipment that contribute significantly to the risk levels.

The Bayesian–LOPA methodology is an advanced LOPA method, because it can yield more statistically reliable or concrete risk results in an LNG facility than normal LOPA methods. Figure 7.3 shows the procedure of the Bayesian–LOPA method.

As a PHA method, a hazard and operability (HAZOP) study (Crowl and Louvar, 2001) will be used because it is one of the most systematic hazard identification methods. The HAZOP method requires process information such as piping and instrumentation diagrams (P&IDs) and process flow diagrams, which include basic designs and minimum specifications adopted from industrial standards such as NFPA 59A (NFPA, 2006) and EN 1473 (Committee of European Nations, 1997) for an LNG terminal.

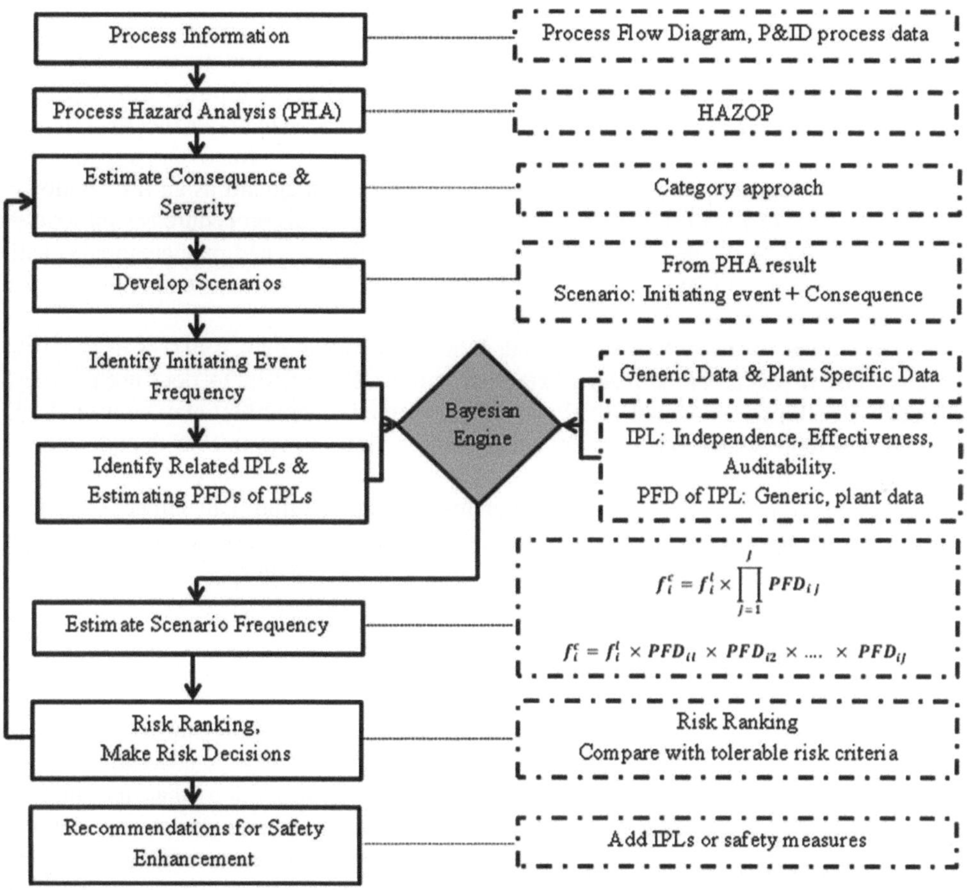


**FIGURE 7.3**

Procedure for the Bayesian–LOPA method.

Results from an HAZOP study are used to make incident scenarios, which can be pairs of causes and consequences. The incident scenarios are screened according to severity by the category method, which is a qualitative way to classify consequences using engineering expert judgment.

Causes found in the HAZOP results may be initiating events of incident scenarios. After identifying an initiating event, its occurrence frequency should be obtained for a LOPA application. Based on Bayesian estimation, the frequency of an initiating event can be obtained by using a conjugate gamma distribution for the prior information and a Poisson distribution for the likelihood function as shown in Figure 7.4. Offshore reliability data (OREDA) (OREDA, 2002) can be used in this manner as prior information, because data were produced from a gamma distribution. Figure 7.5 also shows some formulas for not only gamma and Poisson distributions, but also mean value and 90% credible interval of Bayesian-updated posterior values. If there is no prior information, the Jeffreys noninformative prior (Jeffreys, 1961) may be used. For the plant-specific likelihood information, the LNG plant failure database (Johnson and Welker, 1981), which was established from 27 LNG facilities, is used. This database provides operating hours, number of failures, and mean time between failures (MTBFs) of equipment or facilities.

After obtaining the frequency data of an initiating event, it is necessary to identify each independent protection layer (IPL), which can be found and chosen in the list of safeguards of each incident case in the HAZOP results. However, even though all IPLs can be safeguards, not all safeguards are IPLs, because IPLs must meet the three requirements: independence, effectiveness, and auditability. Thus, very careful consideration should be given to designate a safeguard as an IPL. Following IPL identification, the PFD of each IPL should be obtained. Generally, this procedure is similar to the case of initiating events, but there are a few differences in the distributions and mathematical calculations. The PFDs of IPLs can be estimated using the conjugate beta prior distribution and binomial likelihood distribution, as shown in Figure 7.4. (Table 7.3)

EIReDA (European Industry Reliability Data Bank) (Procaccia et.al., 1997) is used for prior information, because it was prepared from a beta distribution. When EIReDA does not provide failure data, the newly developed frequency−PFD conversion method (Yun, 2007) may be used. LNG plant failure data will be represented by a binomial likelihood function. However, even though some types of failure data are available, the EIReDA database does not provide the number of demands, which is needed together with the number of failures for the binomial distribution. The number of demands can be estimated, however, by correlation using the PFD estimating equation as shown in Figure 7.5, provided periodic tests of equipment are performed that reveal the failures (Crowl and Louvar, 2001; Yun, 2007).

The next step is to use a spreadsheet to calculate the frequency of an incident scenario from the equation given in the "Estimate scenario frequency" step of Figure 7.3. Microsoft EXCEL software can be used to create the spreadsheet.

The last step is to make risk decisions from comparison of an obtained scenario frequency with the tolerable risk criteria, which may be specified by companies, industries, or government. If the estimated frequency cannot meet the tolerable criteria, some recommendations that may include additional IPLs or more frequent proof tests must be given to reduce the incident frequency or mitigate the severity of consequence. As shown in Figure 7.3, these steps will be repeated for every incident scenario. The frequencies of all incident scenarios is estimated and then compared to each other to rank the risks among all scenarios with similar consequence severities. This risk ranking may be used also to develop strategies for maintenance or safety measures.

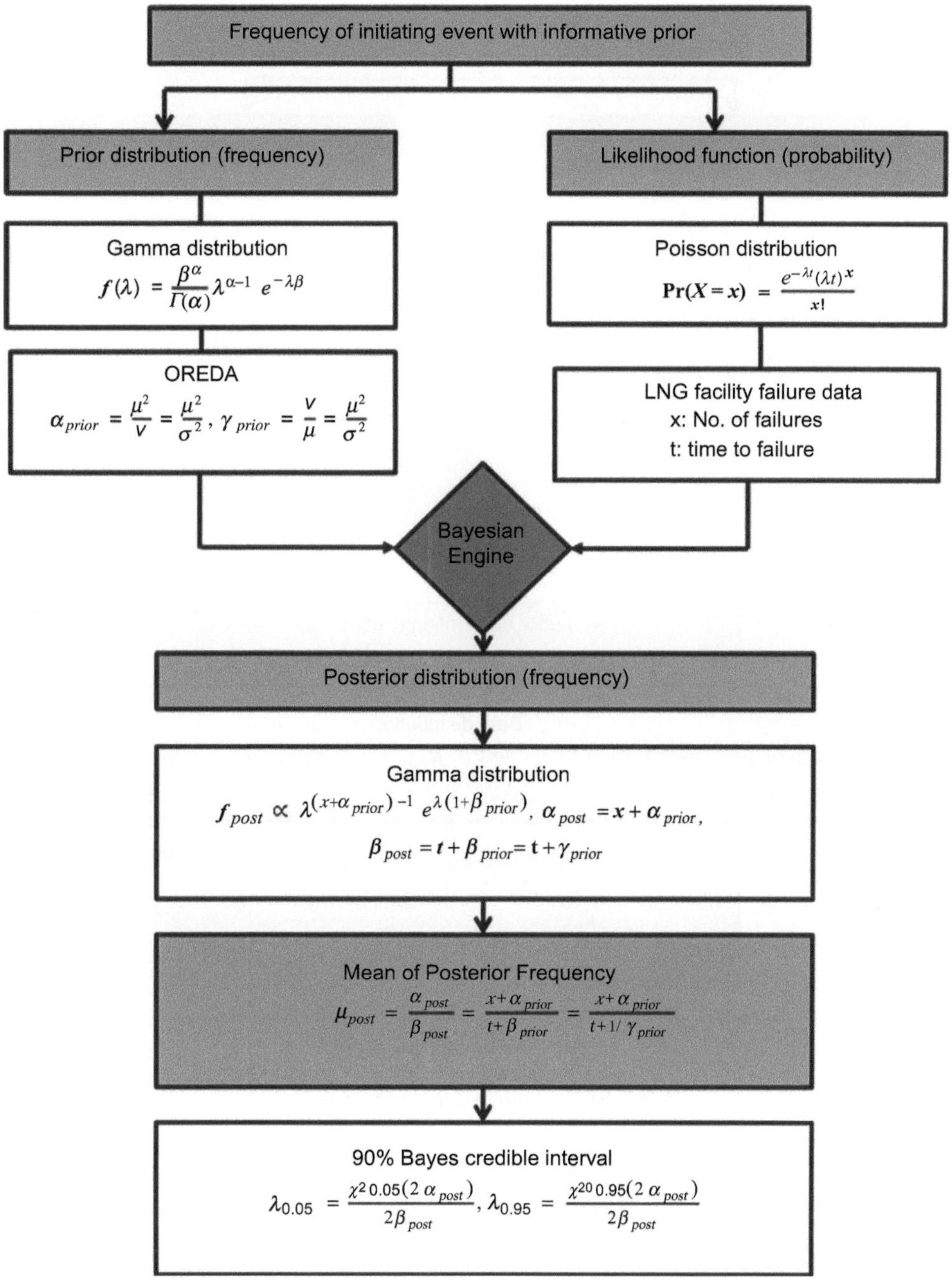


**FIGURE 7.4**

Schematic diagram of Bayesian estimation for IPLs.

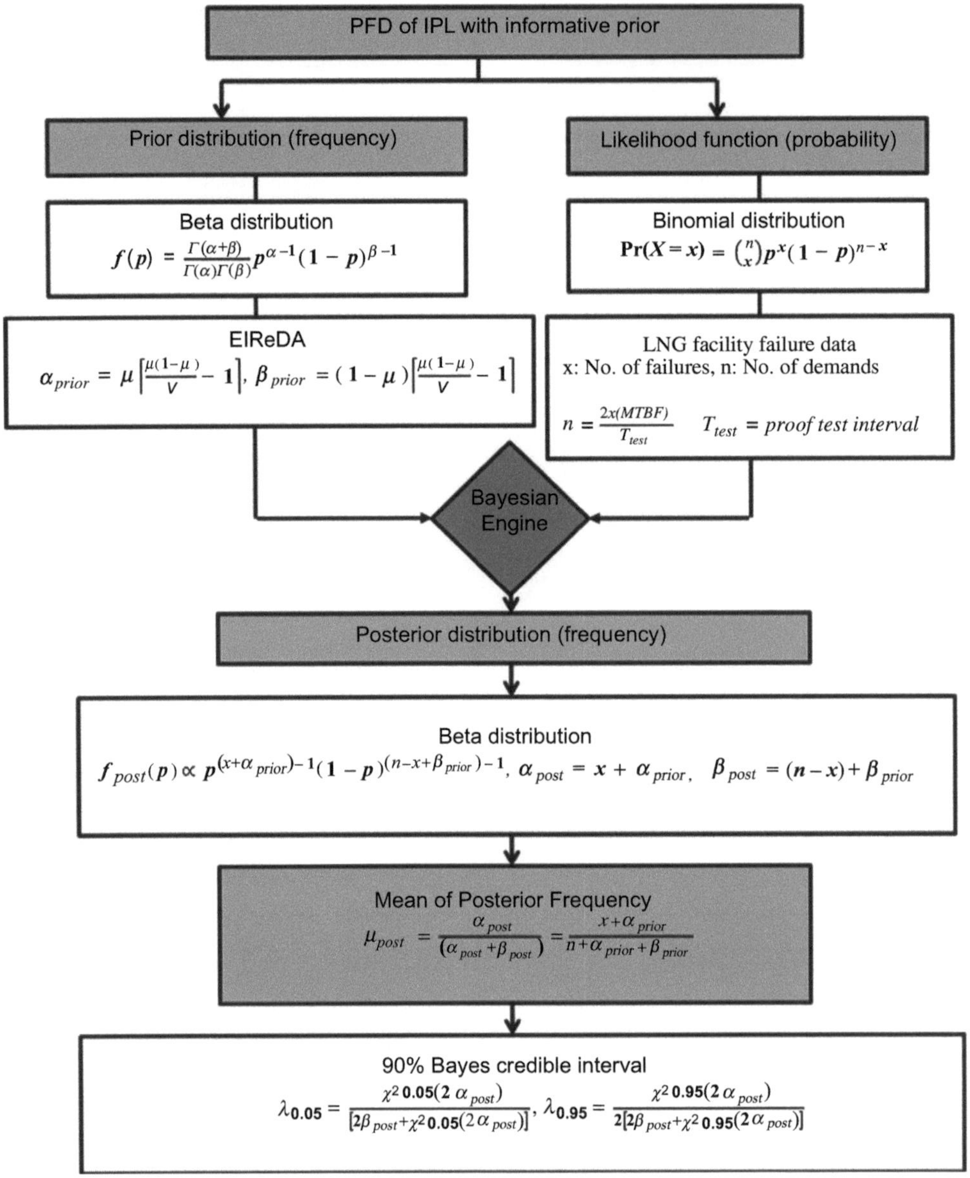


**FIGURE 7.5**

Schematic diagram of Bayesian estimation for initiating events.

**Table 7.3 LOPA Incident Scenarios in an LNG Terminal**

| No. | Scenarios |
|---|---|
| 1 | LNG leakage from loading arms during unloading. |
| 2 | Pressure increase of unloading arm due to block valve failure closure during unloading. |
| 3 | HP pump cavitation and damage due to lower pressure of recondenser resulting from block valve failed closure. Leakage and fire. |
| 4 | Higher temperature in a recondenser due to more boils off gas (BOG) input resulting from flow control valve spurious full open. Cavitation and pump damage leading to leakage. |
| 5 | Overpressure in tank due to rollover from stratification and possible damage in tank. |
| 6 | LNG level increases and leads to carry over into the annular space of LNG, because operator lines up the wrong tank. Possible overpressure in tank. |
| 7 | Lower pressure in tank due to pump-out without BOG input resulting from block valve failed closure. Possible damage tank. |

## 7.3 MULTISCALE MODELS

Modeling failure mechanisms is important for predicting the future performance of materials in a system. Failure data from past incidents alone are not sufficient by itself in predicting material failure, due to process changes and changes in the microstructure of materials with time (Taylor, 2015). The mechanical deformation and material failure of many engineering materials are inherently multiscale in nature (Curtin and Miller, 2003). Given that, multiscale materials modeling have been proved to be a powerful approach in combining knowledge from historical data, experimental results, and first principle-based calculations for predicting the risk of material failure. However, model development for material degradation or failure using all available information and underlying fundamental physics is a challenging problem. In addition to addressing material failure, development of models should also incorporate abilities to assess various risk mitigation strategies. One such example could be the addition of corrosion inhibitors for corrosion issues as illustrated in the later part of this chapter. Among several set of modeling approaches in addressing material failure problems, atomistic and quantum mechanics–based simulations stand out as the most fundamental class of modeling approaches. While the failure of material at the macroscale happens at a very different time and length scale than that can be modeled at the atomistic level, the underlying physics between the two events are connected, albeit sometimes in a nonobvious fashion. It is well established that processes governing observed macroscopic material behavior occur at various time and length scales (Curtin and Miller, 2003). At the nanoscale, the origin of failure may initiate from the nonuniform distribution of the electrons (Sih, 2005, 2008). Examples of defects at the nanoscale include vacancies, impurities, dislocations, voids, crack tips, and grain boundaries. There is an intimate coupling between the disparate length scales as the long-range defect stress fields can introduce nucleation of new defects at the atomistic scale (Curtin and Miller, 2003). The collective effect of plethora of defects interacting at the atomistic level translates into material's macroscopic response to external forces. Fidelity of the mathematical model developed for describing a failure mechanism can be improved by incorporating details from multiscale methods. Fully atomistic simulation of most collective defect processes is not feasible due to the requirement of simulating a large number of atoms.

### 7.3.1 MATERIAL FAILURE

Owing to its inherent multiscale nature, phenomena such as fracture are best studied through application of coupled multiscale models. Figure 7.6 depicts the fracture phenomenon at various characteristic length scales.

Typically, multiscale methods in modeling material failure require modeling at the quantum mechanical, atomistic, and finite element analysis level. While detailed molecular-level interactions are dealt at the quantum and atomistic simulation level, finite element analysis allows one to model defects at the continuum level. Accurate estimation of energetics lies at the core of the model development for material failure. In theory, atomistic simulation should be able to accurately estimate zero energy positions of the atoms constituting the material of interest. Estimation of potential energy of a system can be done by performing quantum mechanical calculations or by using a suitable force field for the system of interest as discussed in Chapter 3.

The response of a material to an external stress is governed by its stiffness constants. By applying the formal definition in Voigt notation, the elastic stiffness matrix can be represented as:

$$C_{ij} = \frac{1}{V_0} \frac{\partial^2 E}{\partial \varepsilon_i \partial \varepsilon_j}$$

where $C_{ij}$ is the elastic constant, $E$ is energy, and $\varepsilon_i$ is the strain in the $i$th direction. Performing Taylor series expansion of energy around minimized structure as a functional form of the strain, $\varepsilon$, energy of a system can be represented as:

$$E(\varepsilon) = E_0 + \sum_{i=1}^{6} \left( \frac{\partial E}{\partial \varepsilon_i} \right)_{E_0} \varepsilon_i + \frac{1}{2} \sum_{i,j=1}^{6} \left( \frac{\partial^2 E}{\partial \varepsilon_i \varepsilon_j} \right)_{E_0} \varepsilon_i \varepsilon_{j+\cdots}$$

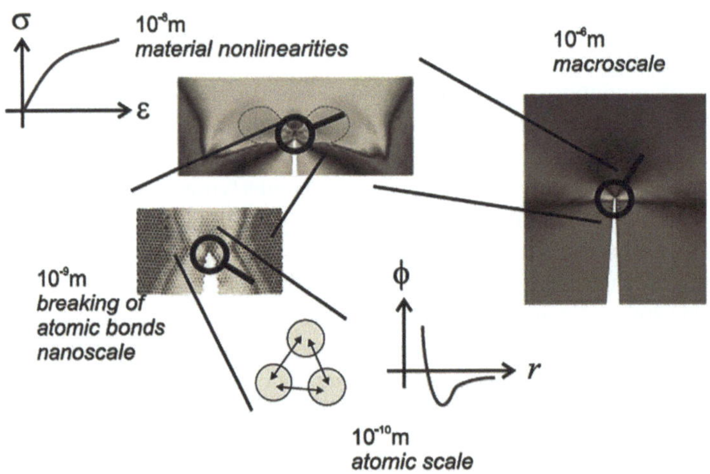


**FIGURE 7.6**

Characteristic length scale associated with material fracture.

where $E_0$ refers to the energy of zero−strain equilibrium configuration. Since the force on the atoms of a molecular structure in its energy minimized configuration is zero, neglecting the higher order terms in the Taylor series expansion, the elastic constant $C_{ij}$ can be estimated by implementing molecular-level modeling approaches. Figure 7.7 depicts a typical outcome of energy−strain curve as estimated based on energy minimization of a molecular system under a given strain either through quantum mechanics or molecular mechanics-based approach. Fitting the curve will estimate the elastic constant of interest.

In general, the steps involved are:

- Define the unit cell of the system of interest
- Choose a suitable classic force field with modified interaction potential as applicable
- For various values of strains, estimate the energy of the unit cell and express it in terms of strain
- Use the expression, $C_{ij} = \frac{1}{V_0}\frac{\partial^2 E}{\partial \varepsilon_i \partial \varepsilon_j}$, as mentioned before in the Taylor series expansion for estimation of the elastic constants

At the continuum level, the potential energy of a system can be estimated as:

$$E = \int_V W(x)dV$$

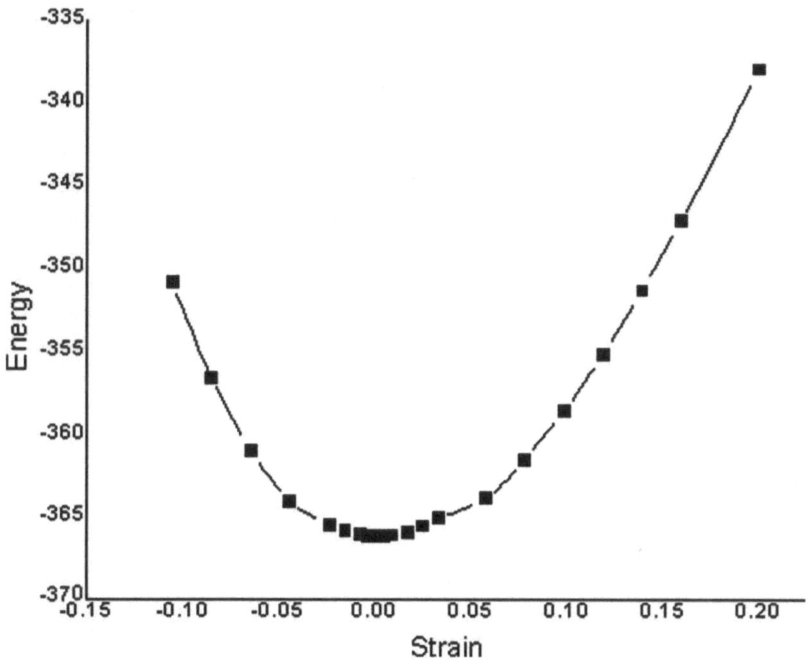


**FIGURE 7.7**

Typical energy−strain curve of a system.

where $E$ is the potential energy and can be estimated through integrating $W(x)$, the energy density at location $x$, over the total volume $V$. The energy density can be defined as:

$$W(x) = \frac{1}{2}(C_{ij} \cdot \varepsilon_{ij})$$

The value of $C_{ij}$ can be obtained from experimental or molecular modeling approaches. In a similar fashion, at the continuum level, finite element analysis is implemented for estimation of the strain field of a deformed body due to application of external forces (Curtin and Miller, 2003). The minimization of energy of a given system under external forces allows one to estimate the strain field. At the continuum level the energy equation can be rewritten in terms of the elements in the finite element analysis approach:

$$E = \sum_{i}^{n} E_i$$

$$E_i = \int_{V_i} W_i(x) dV_i$$

where $E_i$ is the energy of the $i$th element, $V_i$ is the volume of the $i$th element, and $W_i(x)$ is the strain energy density at location $x$ at the $i$th element.

At each node, displacement due to deformation is calculated. Deformation between nodes can be calculated through shape functions or interpolation functions. Accordingly, the displacement of a point within an element depends only on the displacement of the two nodes corresponding to the vertices of the element. The energy is minimized with respect to the nodal degrees of freedom. The strain energy density, $W_i(x)$, can be written as:

$$W_i(x) = W_i(F(x))$$

where $F(x)$ is the deformation gradient, the fundamental quantity of continuum deformation. For material with no defects, the deformation gradient can be assumed to be homogeneous. Homogeneous deformation gradient in a crystal can be calculated using molecular modeling approaches either through classic force field-based approaches or through quantum mechanical calculations by implementing the Cauchy–Born rule (Weiner, 2012; Ericksen, 2008). For crystals, the Cauchy–Born rule is effective in correlating the deformation energy at the continuum level to atomistic potential energy from stretched crystals at the atomistic level. The rule states that the atomistic positions within the crystal lattice follow the overall strain of the medium and that it does not allow relaxation of the lattice atoms within the unit cell. Accordingly, it must be borne in mind that the Cauchy–Born rule is only valid for small deformations. Figure 7.8, illustrates the deformation of a crystal from atomistic and continuum viewpoints (Buehler, 2008). At the atomistic level, one is interested in shifting of the individual atoms, whereas at the continuum level, it is the overall deformation resulting in a strain field within the macroscopic body.

One of the challenges in multiscale model development of material failure through atomistic simulation and finite element analysis lies in the formulation of the strain energy density functional (Curtin and Miller, 2003). The true atomistic energy is nonlocal, whereas at the continuum level, the deformation around an element is assumed homogeneous. Owing to the scale difference, at the finite

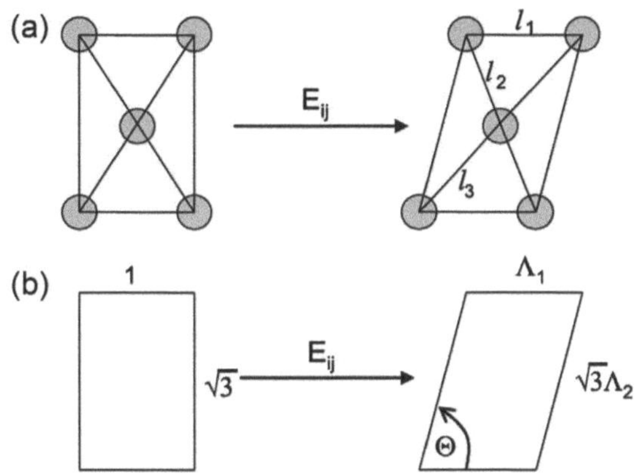


**FIGURE 7.8**

Deformation of a crystal. (a) atomistic viewpoint, (b) continuum viewpoint (Buehler, 2008).

element level it is not possible to represent the true inhomogeneous deformation representation as assessed by atomistic simulation. One approach to address this is through implementation of homogeneous deformation within an element consisting of several nodes, each represented by an atom, to estimate the approximate atomic positions in the deformed body. While the resultant atomic positions, represented by nodes, will only be approximate, the continuum strain energy density calculated in this approach corresponds to an atomistic calculation of the energy for the approximate set of atom positions. Thus the deformation at the continuum level accounts for the nonlocal and finite range-interatomic potentials with an approximation to the actual atomic positions. For achieving equilibrium, the energy estimated in this approach needs to be minimized with respect to the nodal degrees of freedom. In order to estimate the energies for displacement of the nodes, it must be taken into account that, in a finite element analysis, a change in the position of a node affects the deformation gradient of all the elements connected to it.

For an integrated coupled model, the idea is to use atomistic-level description at regions where details are important and continuum descriptions in other regions as illustrated in Figure 7.9.

Approximations are necessary while formulating a problem in this manner due to the fundamental incompatibility between atomistic and finite element analysis approach as noted before. The "pad" region in Figure 7.9 consists of pseudo atoms and overlaps physical space with the mesh of the finite element analysis approach. At the interface, there is a one-to-one mapping between atoms and nodes from mesh of the finite element approach. Various approaches (Miller et al., 1998; Shenoy et al., 1998; Rudd and Broughton, 2000; Kohlhoff et al., 1991; Shilkrot et al., 2002; Buehler, 2008; Ersland et al., 2011) exist to couple the two modeling approaches at the transition region lying between the atomistic and the continuum region. Instead of delving deeper into these topics, the goal here is to make the reader aware of various detailed and coupled modeling approaches that can be implemented in understanding and estimating characteristics of material failure. Owing to the scale of deformation, often an all-atomistic model is not feasible in today's date and hence one can consider the coupled multiscale modeling approach as an alternative that can provide significant insight using reasonable amount of resources. There are also studies involving only molecular dynamics to investigate the effect on

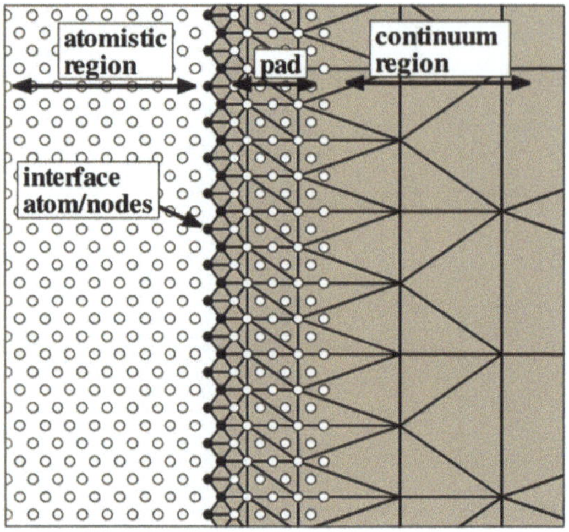


**FIGURE 7.9**

Atomistic finite element analysis-based coupled multiscale modeling approach.

structural properties of steel such as effect of grain boundary sliding on the toughness of steel (Xie et al., 2013). Grain boundaries are defects in crystal structures and lies at the interface between two crystallites. Implementing molecular dynamics, the work could assess the crack propagation dynamics with time as depicted in Figure 7.10.

A similar problem has also been addressed by coupling atomistic and finite element analysis (Park et al., 2005) in a similar way discussed before.

The formation and breaking of covalent bonds at and in the vicinity of the crack tip can be addressed by a quantum mechanical approach or tight binding formulation. The region somewhat further away from the crack tip, may experience large strain but is not expected to result in formation or breaking of bonds. Hence, these regions can be treated through interatomic potentials in a classic force field. Even further, the strain fields tend to become homogeneous and the strain gradients are small. Hence, a continuum finite element analysis is appropriate for the far-field region. With three different modeling approaches, two interfaces show up. The interface between finite element and molecular dynamics has been discussed before. The FE mesh dimensions are brought down to atomistic dimensions prior to switching over to the atomistic simulation domain. The second interface lies between the atomistic and tight binding/quantum mechanical calculations. Given that this interface is in the vicinity of the crack, it is more dynamic in nature and complex to model. However, modeling methods have been developed to tackle such challenges on a system-by-system basis and is nontrivial in nature (Buehler, 2008). Using ReaxFF force field is another possible approach where one can address an atomistic-level phenomenon of both reactive and nonreactive atoms.

The fracture response of materials is inherently linked to process safety incidents. For example, fracture response of pipelines subject to large plastic deformation is of interest when looking into designing long-distance offshore pipelines (Østby et al., 2004; Jayadevan et al., 2003). This chapter introduces the readers to the existing model development approaches incorporating detailed atomistic

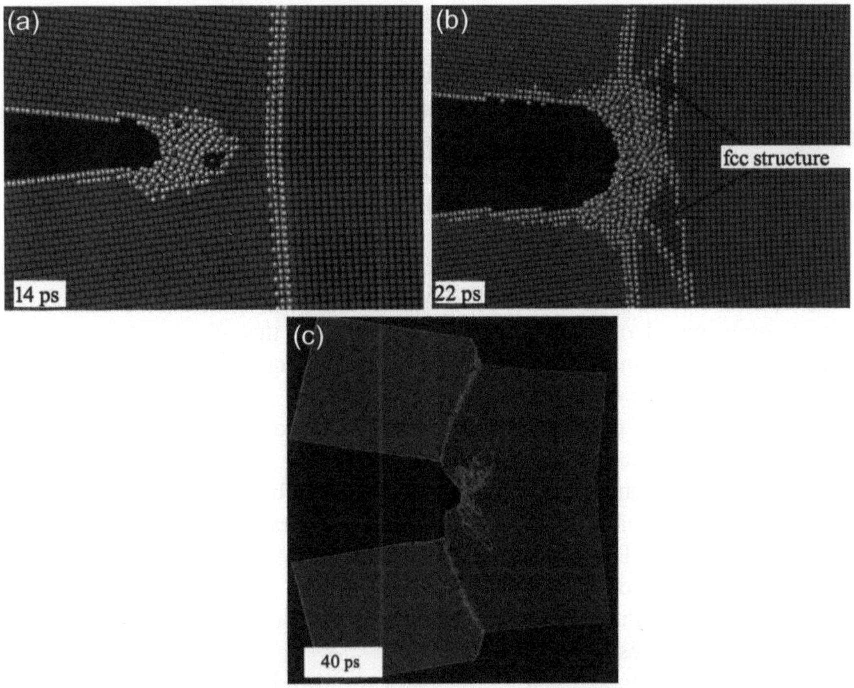


**FIGURE 7.10**

(a–c) Crack propagation as estimated by implementing MD studies on ultrafine grain structure steel (Xie et al., 2013).

studies that can be used in simulating these processes (Buehler, 2008). The details discussed here are merely an outer layer of the topic but should provide an idea about the immense potential of multiscale modeling methods in assessing fracture and deformation characteristics of material under various process conditions. Further readings, such as the book by Buehler (2008), are suggested.

### 7.3.2 CORROSION

Among various failure mechanisms, equipment or pipeline failure from corrosion resulting in process safety incident is a very likely phenomenon (Fire, 2012). Accordingly, corrosion monitoring and corrosion control are of substantial importance in the process industries. Corrosion control can involve any or all of the following major approaches (Center for Chemical Process Safety [CCPS], 2012):

- Change in materials
- Change in the process environment
  - Change in process conditions
  - Addition of corrosion inhibitors
- Use of barrier linings or coatings
- Application of electrochemical techniques

In all of these changes, one needs to understand the sensitivity of the corrosive action with respect to the change(s) introduced.

Addition of corrosion inhibitors to the process fluid requires understanding of the underlying mechanism of its action. Their classification according to the mechanism of their action is provided in Table 7.4 (Center for Chemical Process Safety [CCPS], 2012).

Quantum mechanical approaches have been proved to be a powerful tool for studying the structure and behavior of corrosion inhibitors (Gece, 2008). The idea is to assess the efficiency of corrosion inhibitors, through implementing molecular modeling studies on those chemicals with a desired set of properties. In the case of organic inhibitors, the inhibition mechanism is dictated by physicochemical and electronic properties of the inhibitor. The combined effect of the facts that a molecular system can be characterized solely based on its molecular structure and that chemical reactivity of a set of molecules will account for any inhibitor mechanism of interest enables one to use molecular modeling approaches as powerful tools in corrosion inhibitor studies.

Upon performing quantum chemical calculations on corrosion inhibitors, the approach is to discover a strong correlation between the inhibition mechanism and a quantum chemical parameter. Once that is determined, one can assess the effect of inhibitors by monitoring the quantum chemical parameter. Quantum chemical parameters can be atomic charges, atomic energies, molecular orbital energies, and dipole moment. The usual steps in estimating these parameters start with geometry optimization including calculation of atomic charges, similar to any quantum mechanical calculation as illustrated in Chapter 3. Further studies are only performed with the already optimized molecular structure of the system of interest.

Semiempirical calculations to assess the effectiveness of imidazole derivatives as acidic corrosion inhibitors for zinc and iron have been performed using AM1, PM3, MNDO, and MINDO/3 methods by Bereket et al. (2000, 2001). Several quantum chemical parameters such as total energy, atomic charges, ionization potential, dipole moment, HOMO (highest occupied molecular orbital), and LUMO (lowest unoccupied molecular orbital) energies were calculated.

In a similar manner, El Issami et al. (2007) showed that less negative HOMO energy along with a smaller energy gap between HOMO and LUMO indicated greater inhibitor efficiency. Inhibition efficiency can be estimated in several ways including using change in current densities and polarization resistance. For example, for a 100% efficient inhibition, the current density measured should be negligible or close to zero. Figure 7.11 illustrates the correlation found in the study between inhibition

**Table 7.4 Corrosion Inhibitors (Center for Chemical Process Safety [CCPS], 2012)**

| Inhibitor | Method of Action | Example |
|---|---|---|
| Absorption type | Suppresses metal dissolution and reduction reactions | Organic amines |
| Hydrogen evolution poison | Retards hydrogen evolution but can lead to hydrogen embrittlement | Arsenic ion |
| Oxygen scavenger | Removes oxygen from aqueous solution including water | Sodium sulfite, hydrazine |
| Oxidizer | Inhibits corrosion of metals and alloys that demonstrate active–passive transition | Chromate, nitrate, and ferric salts |
| Vapor phase | Blanket gas for machinery or tightly enclosed atmospheric equipment | Nitrogen blanket |

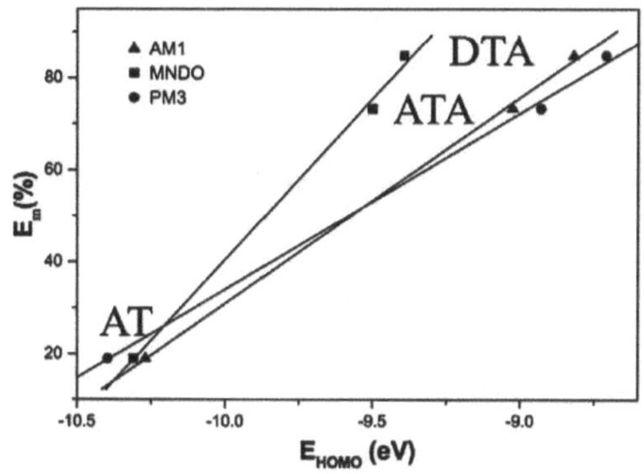


**FIGURE 7.11**

Correlation between HOMO ($E_{HOMO}$) energy and inhibition efficiency ($E_M$) for three (AT, DTA, and ATA) scenarios.

efficiency, determined through experiments, and HOMO energies for three different systems with the corresponding inhibitors.

In addition to using quantum mechanical methods, molecular dynamics approaches have also been used to investigate corrosion phenomena (Susmikanti et al., 2012). The study looked into the effect of increasing molybdenum (Mo) content in steel with the objective of increasing corrosion resistance. Since a molecular dynamics approach was used, the effect of temperature was also investigated. A visual approach showed increase of corrosion with increasing temperature and increase of corrosion resistance with increasing Mo percentage. Figure 7.12 illustrates the increase of corrosion rate for

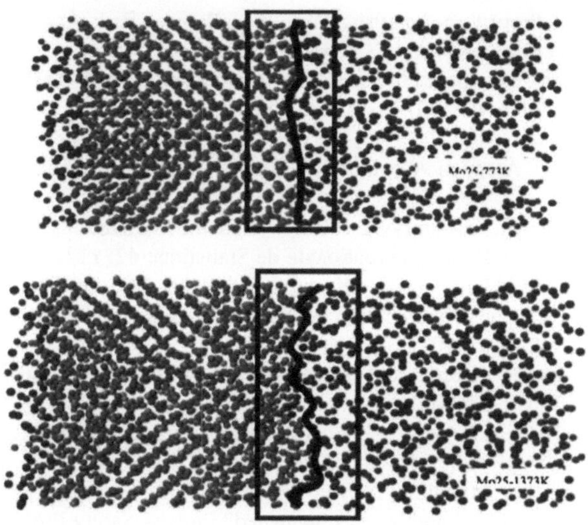

**FIGURE 7.12**

Corrosion characteristics for steel with Mo 2–3% content at 773 K (top) and 1373 K (bottom).

change in temperature as can be observed by inspecting the surface (depicted by the dark line inside the rectangular box at the interface) between crystal steel-containing Mo (left) and coolant material Pb-Bi (right).

Computational chemistry can be used as a powerful tool for corrosion inhibitor studies. Both quantum mechanical calculations and molecular dynamics study can be implemented in assessing inhibitor efficiency, corrosion resistance, and potential to corrosion studies. While experimental studies are needed for validation, computational-chemistry approaches can narrow down the number of experimental studies required in a significant way.

## REFERENCES


Apostolakis, G., 1981. Bayesian Methods in Risk Assessment. Advances in Nuclear Science and Technology. Springer, 417–465.

Apostolakis, G., Bier, V., Mosleh, A., 1988. A critique of recent models for human error rate assessment. Reliability Engineering & System Safety 22 (1), 201–217.

Bailey, R.T., 1997. Estimation from zero-failure data. Risk Analysis 17 (3), 377–380.

Bereket, G., Öğretir, C., Yurt, A., 2000. Quantum mechanical calculations on some 4-methyl-5-substituted imidazole derivatives as acidic corrosion inhibitor for zinc. Journal of Molecular Structure: THEOCHEM 571 (1), 139–145. http://dx.doi.org/10.1016/S0166-1280(01)00552-8.

Bereket, G., Hür, E., Öğretir, C., 2001. Quantum chemical studies on some imidazole derivatives as corrosion inhibitors for iron in acidic medium. Journal of Molecular Structure: THEOCHEM 578 (1), 79–88. http://dx.doi.org/10.1016/S0166-1280(01)00684-4.

Buehler, M.J., 2008. Atomistic Modeling of Materials Failure. Springer Science & Business Media.

CCPS, 1988. CCPS Guidelines for Process Equipment Reliability Data. AIChE.

Committee of European Nations, 1997. Installation and equipment for liquefied natural gas—design of onshore installations, EN-1473. Paramus, NJ.

Crowl, D.A., 2001. Layer of Protection Analysis: Simplified Process Risk Assessment. Center for Chemical Process Safety of the American Institute of Chemical Engineers.

Crowl, D.A., Louvar, J.F., 2001. Chemical Process Safety: Fundamentals with Applications. Pearson Education.

Center for Chemical Process Safety (CCPS), 2012. Guidelines for Engineering Design for Process Safety. John Wiley & Sons.

Curtin, W.A., Miller, R.E., 2003. Atomistic/Continuum Coupling in Computational Materials Science. Institute of Physics Publishing.

De Finetti, B., 1974. Bayesianism: its unifying role for both the foundations and applications of statistics. International Statistical Review/Revue Internationale de Statistique 42, 117–130.

El Issami, S., Bazzi, L., Mihit, M., Hammouti, B., Kertit, S., Ait Addi, E., Salghi, R., 2007. Triazolic compounds as corrosion inhibitors for copper in hydrochloric acid. Pigment & Resin Technology 36 (3), 161–168. http://dx.doi.org/10.1108/03699420710749027.

Ericksen, J.L., 2008. On the Cauchy–Born rule. Mathematics and Mechanics of Solids 13 (3–4), 199–220. http://dx.doi.org/10.1177/1081286507086898.

Ersland, C.H., Thaulow, C., Vatne, I.R., Østby, E., 2011. Atomistic modeling of micromechanisms and *T*-stress effects in fracture of iron. Engineering Fracture Mechanics 79, 180–190.

Finkelstein, M., 2008. Failure Rate Modelling for Reliability and Risk. Springer.

Freeman, R.R., 2011. What to do when nothing has happened? Process Safety Progress 30 (3), 204–211.

Fire, Chevron Richmond Refinery, 2012. Interim Investigation Report.

Glaser, R.E., 1980. Bathtub and related failure rate characterizations. Journal of the American Statistical Association 75 (371), 667–672.

Green, A.E., Bourne, A.J., 1972. Reliability Technology. Wiley-Interscience, New York.

Gece, G., 2008. The use of quantum chemical methods in corrosion inhibitor studies. Corrosion Science 50 (11), 2981–2992.

IEEE, 1984. IEEE Guide to the Collection and Presentation of Electrical, Electronic, Sensing Component, and Mechanical Equipment Reliability Data for Nuclear-power Generating Stations. IEEE.

Jefreys, H., 1961. Theory of Probability, 3rd ed. Oxford University Press, Oxford, MR0187257.

Johnson, D., Welker, J., 1981. Development of an Improved LNG Plant Failure Rate Data Base. Final report March 1980–June 1981, Applied Technology Corp. Norman, OK (USA).

Jayadevan, K.R., Østby, E., Thaulow, C., 2003. Fracture response of pipelines subjected to large plastic deformation under tension. International Journal of Pressure Vessels and Piping 81 (9), 771–783. http://dx.doi.org/10.1016/j.ijpvp.2004.04.005.

Kececioglu, D., 2002. Reliability and Life Testing Handbook. DEStech Publications, Inc.

Kohlhoff, S., Gumbsch, P., Fischmeister, H.F., 1991. Crack propagation in BCC crystals studied with a combined finite-element and atomistic model. Philosophical Magazine A 64 (4), 851–878. http://dx.doi.org/10.1080/01418619108213953.

Lipow, E., Welker, M., 1974. Estimating the exponential failure rate from data with no failure events. In: 1974 Annual Reliability and Maintainability Symposium: Proceedings, Los Angeles, California, January 29, 30, 31, 1974, Institute of Electrical and Electronics Engineers.

Mannan, M.S., 2013. Lees' Loss Prevention in the Process Industries.

Modarres, M., Kaminskiz, M., Krivstov, V., 2009. Reliability Engineering and Risk Analysis: A Practical Guide. CRC Press.

Mosleh, A., Bier, V., Apostolakis, G., 1987. Methods for the Elicitation and Use of Expert Opinion in Risk Assessment: Phase 1, A Critical Evaluation and Directions for Future Research. Pickard, Lowe and Garrick, Inc., Newport Beach, CA (USA). Nuclear Regulatory Commission, Washington, DC (USA). Div. of Reactor Accident Analysis.

Miller, R., Tadmor, E.B., Phillips, R., 1998. Quasicontinuum simulation of fracture at the atomic scale. Modelling and Simulation in Materials Science and Engineering 6 (5), 607–638.

NFPA, 2006. NFPA 59A: Standard for the Production, Storage and Handling of Liquefied Natural Gas (LNG).

OREDA, 2002. OREDA: Offshore Reliability Data Handbook. OREDA SINTEF Industrial Management.

Østby, E., Jayadevan, K.R., Thaulow, C., 2004. Fracture response of pipelines subject to large plastic deformation under bending. International Journal of Pressure Vessels and Piping 82 (3), 201–215. http://dx.doi.org/10.1016/j.ijpvp.2004.08.012.

Park, H.S., Karpov, E.G., Liu, W.K., Klein, P.A., 2005. The bridging scale for two-dimensional atomistic/continumn coupling. Philosophical Magazine 85 (1), 79–113. http://dx.doi.org/10.1080/14786430412331300163.

Procaccia, H., Aufort, P., Arsenis, S., 1997. The European industry reliability data bank EIReDA. Electricite de France (EDF), 92-Clamart (France). Direction des Etudes et Recherches 1997.

Rudd, R.E., Broughton, J.Q., 2000. Concurrent coupling of length scales in solid state systems. Physica Status Solidi (B) 217 (1), 251–291. http://dx.doi.org/10.1002/(SICI)1521-3951(200001)217:1<251::AID-PSSB251>3.3.CO;2–1.

Shafaghi, A., 2008. Equipment failure rate updating—Bayesian estimation. Journal of Hazardous Materials 159, 87–91.

Snaith, E., 1981. The Correlation between the Predicted and the Observed Reliabilites of Components, Equipment and Systems. UKAEA National Centre of Systems Reliability.

Shenoy, V.B., Miller, R., Tadmor, E.B., Phillips, R., Ortiz, M., 1998. Quasicontinuum models of interfacial structure and deformation. Physical Review Letters 80 (4), 742.

Shilkrot, L.E., Miller, R.E., Curtin, W.A., 2002. Coupled atomistic and discrete dislocation plasticity. Physical Review Letters 89 (2), 025501. http://dx.doi.org/10.1103/PhysRevLett.89.025501.

Sih, G.C., 2005. Signatures of rapid movement of electrons in valence band Region: interdependence of position, time and temperature. Theoretical and Applied Fracture Mechanics 45 (1), 1–12. http://dx.doi.org/10.1016/j.tafmec.2005.11.006.

Sih, G.C., 2008. Multiscale Fatigue Crack Initiation and Propagation of Engineering Materials: Structural Integrity and Microstructural Worthiness. Springer Science & Business Media.

Susmikanti, M., Andiwijayakusuma, D., Ghofir, Maulana, A., 2012. Molecular dynamics simulation to studying the effect of molybdenum in stainless steel on the corrosion resistance by lead-bismuth. AIP Conference Proceedings 1448, 185–192. http://dx.doi.org/10.1063/1.4725454.

Thorne, M.C., Williams, M.M.R., 1992. A review of expert judgment techniques with reference to nuclear safety. Progress in Nuclear Energy 27 (2), 83–254.

Taylor, C.D., 2015. Integrated Multiscale Modelling of Materials. https://www.dnvgl.com/technology-innovation/latest-publications.html.

v Mises, R., 1957. Probability, Statistics, and Truth. DoverPublications.com.

Weiner, J.H., 2012. Statistical Mechanics of Elasticity. Courier Corporation.

Xie, H.-X., Liu, B., Yin, F.-X., Yu, T., 2013. Effect of grain boundary sliding on the toughness of ultrafine grain structure steel: a molecular dynamics simulation study. Chinese Physics B 22 (1), 010204. http://dx.doi.org/10.1088/1674-1056/22/1/010204.

Yang, G., 2007. Life Cycle Reliability Engineering. John Wiley & Sons.

Yun, G.W., 2007. Bayesian-LOPA Methodology for Risk Assessment of an LNG Importation Terminal. Texas A&M University.

CHAPTER

# INHERENTLY SAFER DESIGN

8

The application of a series of principles to inherently mitigate the potential hazards during process development is popularly known as inherently safer design (ISD). ISD principles can be categorized as the following four: minimization (intensification), substitution, moderation (attenuation), and simplification. Based on these principles, protection systems are adopted in designing inherently safer process. However, protection measures that are needed to be adopted for ISD are considered as either active or passive protection systems. A passive protection measure is usually considered as an ISD rather than an active protection measure. Optimization of facility layout and siting, accurate quantitative risk analysis (QRA), proper process control, material design, and reactor design are some of the advanced features of ISD. This chapter presents a general briefing on the state-of-the-art tools to perform ISD. Section 8.1 presents the application of Monte Carlo methods in facility siting, transportation risk assessment, and reliability analysis of large complex systems. Section 8.2 discusses the application of computational fluid dynamics (CFD) in QRA. Sections 8.3 and 8.4 present multiscale modeling approaches in process control and in material design. Finally, Section 8.5 presents advances in mesoscale reactor design with a focus on ISD.

## 8.1 APPLICATION OF MONTE CARLO METHODS IN ISD

Random and pseudorandom numbers are used in the computational algorithms of Monte Carlo methods (Metropolis and Ulam, 1949; Amar, 2006) to solve numerous kinds of problems such as optimizing the plant layout, transportation risk assessment, and reliability analysis of large complex systems. It has been used extensively in the fields of material science, physics, and quantum biology (Kahn, 1957), but application to process safety is relatively new. Use of Monte Carlo methods is advantageous in solving problems that pose a very large number of degrees of freedom. For example, performing numerical integration of a function in $N$-dimensional domain, for example, $N = 100$ would require the estimation of 10,100 points if only 10 grid points is considered in each direction, which is extremely large to be computed. Stochastic optimization of complex systems is another important application of Monte Carlo simulations. For example, a traveling salesman problem (TSP) where the number of cities needed to be traveled by a salesperson is greater than 20 is not trivial to solve using traditional methods but can be solved with a reasonable degree of accuracy via Monte Carlo methods. Being very useful in solving difficult problems, a large number of Monte Carlo methods have been developed over the years to handle difficult optimization problems. Well-known optimization algorithms such as stochastic tunneling, simulated annealing, genetic algorithm, and parallel tempering (Sobol, 1974) are some of the few existing Monte Carlo methods. Process safety being a new arena in science and engineering, the applications of Monte Carlo in process safety is also relatively new. In this section, a brief overview of the Monte Carlo applications in process safety is discussed.

**Multiscale Modeling for Process Safety Applications.**

## 8.1.1 APPLICATION OF MONTE CARLO METHODS IN FACILITY SITING STUDY

Ideally, the space management and equipment layout design should be a balance between potential risks and estimated costs (AICHE/CCPS, 2003), but in practice optimal economic design prevails in decision making without considering the potential risk associated with the design (Jayakumar and Reklaitis, 1994, 1996; Gerorgiadis and Macchietto, 1997). As a result, many industrial disasters have occurred, such as Flixborough (1974) caused 28 deaths and Pasadena, Texas (1989) caused 24 deaths (Venart, 2007). One of the major causes of these incidents was due to insufficient space allocation between occupied buildings such as control rooms, and nearby process equipment or facilities. Proper plant layout based on identified hazard and risk assessment in the early stage of facility siting or plant layout can significantly reduce losses during an industrial disaster. Unfortunately, integrated risk assessment and layout configuration optimization research have not sufficiently flourished because of the fact that the formulation of the safety objective function requires a large number of variables which are stochastic in nature. As an example, to calculate the risk function due to the toxic release, there might be thousands of potential release locations depending on the size of the plant, the effect and dispersion of released substance would be influenced by the environmental conditions such as wind speed, wind direction, ambient temperature, humidity, and atmospheric stability. With the advent of high-speed computing technology, Monte Carlo methods can be used to simulate these stochastic events in designing a safer plant layout. Thus an earlier rule-based plant layout (Moore, 1962; Mecklenburgh, 1973) is substituted with highly efficient optimization techniques where process safety consideration is of major importance.

This example study examines a hypothetical toxic release taking place in Corpus Christi, TX, published by Jung et al. (2010). In this example, Monte Carlo simulation is used to simulate weather conditions using a large bank of data. The data used were obtained from reliable sources such as the National Weather Service. The random variables considered are wind speed, wind velocity, and atmospheric stability. Monte Carlo simulation gives a great advantage over the conventional simulation techniques, especially in cases where direct simulation of every possible combination or scenario is not feasible. Apart from that, simulation of a few selected events may not capture the hazards of the real scenario. On the other hand, stochastic methods can be used to gain insights into as many different scenarios as needed and can be weighted by using probability theories resulting in a greater chance to capture even low-probability events, without exhausting all scenarios.

### *8.1.1.1 Problem description*

The general problem is that of facility layout—simply the arrangement of buildings and equipment to maximize efficiency, minimize space, and minimize risk. This problem generally takes experience and expert judgment from a person qualified for such a task. There are many subjective points to be taken into account, and the question of what is "best" can often vary from person to person based on their objectives, values, and experiences. In this example, the subjective facet of the judgment is replaced with numerical methods for optimization, where data are fed to a model that, based on an objective function, is used to optimize the layout. The objective function is the mathematical construct that defines what is maximized or minimized to constitute an optimal solution. Constraints are also necessary to ensure the practical feasibility of any such layout. Components such as road width, separation distances for emergency access, nonoverlap of buildings and equipment, and property boundaries must also be taken into account.

The specific problem is the layout of buildings with respect to minimizing risks posed by a toxic release. In a toxic release such as this one, humans are primarily at risk; and it is assumed that there is no risk to building integrity or equipment function. In this example, a rectangular plot of land of maximum dimension 250 × 500 m is to be laid out with three new facilities. In addition to these new facilities, two existing facilities are already present. One of these facilities contains chlorine gas that may be released, thus posing a hazard to the other facilities. There is a certain cost that is associated with piping between the facilities ($198.8 per meter of piping), a cost for the purchase of land ($6 per square meter), and a cost associated with a hypothetical fatality ($10 million per fatality). The goal of this problem is to identify where the midpoints of the buildings should be located, the amount of land necessary to lay out the facility, and the final expected cost of the chosen layout.

Though some may dislike the idea of putting a dollar value on a fatality, it is necessary to account for cost in the objective function. However, money is just a very simple measure of the concept of "utility," where the relative desirability or undesirability of outcomes is prioritized in proportion to one another. If desired, an alternate formulation of a utility function can be used to alleviate the stigma of a monetary value being used for fatality, as long as the piping and land costs are also converted to the alternate utility scale as well. The main idea is that there has to be some way to weigh outcomes against one another for the objective function to work.

### 8.1.1.2 Mathematical formulation

This text will not include all of the details of the formulation but will provide enough background to understand the application of the stochastic modeling. The reader may refer to the citations for more information. Jung et al. (2010) defines the following sets and parameters for the model (among others) (Table 8.1):

Key constraints are as follows:

1. Land constraints:

$$\frac{Lx_s}{2} + st \leq x_s \leq Lx - \left(\frac{Lx_s}{2} + st\right)$$

$$\frac{Ly_s}{2} + st \leq y_s \leq Ly - \left(\frac{Ly_s}{2} + st\right)$$

**Table 8.1 Parameters and Their Meaning**

| Parameter | Meaning |
|---|---|
| $i \in I$ | Set of existing facilities |
| $s \in S$ | Set of new facilities to be sited |
| $ri \in I$ | Subset of existing facilities that may release a toxic material |
| $Lx_s$ and $Ly_s$ | Length and depth of new facility |
| $Lx_i$ and $Ly_i$ | Length and depth of existing facility |
| $x_i$ and $y_i$ | Midpoint of each existing facility |
| $x_s$ and $y_s$ | Midpoint of each new facility to be placed |
| $Lx$ and $Ly$ | Maximum available land in each direction |
| $st$ | Separation distance (for streets) |

These constraints force the midpoints of all newly sited facilities ($x_s$ and $y_s$) to be placed such that the boundaries of the new facilities will not penetrate the boundaries of the land in the $x$ or $y$ direction.

**2.** Nonoverlapping constraints:

$$\begin{bmatrix} \text{"}L\text{"} \\ x_s \le x_k - D_{sk}^{minx} \end{bmatrix} \vee \begin{bmatrix} \text{"}R\text{"} \\ x_s \ge x_k + D_{sk}^{minx} \end{bmatrix} \vee \begin{bmatrix} \text{"N", "S"} \\ x_s \ge x_k - D_{sk}^{minx} \\ x_s \le x_k + D_{sk}^{minx} \\ \begin{bmatrix} \text{"}N\text{"} \\ y_s \ge y_k + D_{sk}^{miny} \end{bmatrix} \vee \begin{bmatrix} \text{"}S\text{"} \\ y_s \le y_k - D_{sk}^{miny} \end{bmatrix} \end{bmatrix}$$

where

$$D_{sk}^{minx} = \frac{Ls_s + Lx_k}{2} + st$$

$$D_{sk}^{miny} = \frac{Ly_s + Ly_k}{2} + st$$

This set of constraints is called "disjunctive." It is a set of four separate constraints that are illustrated in Figure 8.1.

The idea is that a new facility ($s$), in order not to overlap with an existing facility ($k$), must be either far enough left (labeled $L$ in the formulation) OR right ($R$) OR north ($N$) OR south ($S$) of the existing facility. More than one of these can be true, but at the very least one must be true. The disjunctive formulation makes it so that if one constraint is satisfied, the rest are relaxed and no longer need to be satisfied. Through clever use of the Big-M formulation, this can be put into an mixed integer nonlinear programming (MINLP) formulation, as is described in the reference.

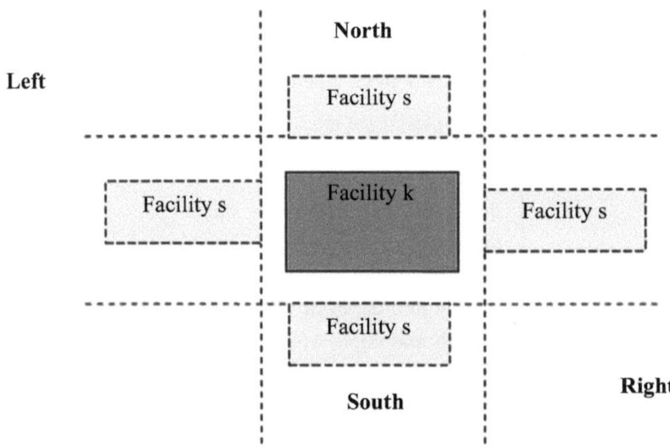


**FIGURE 8.1**

Illustration of disjunctive nonoverlapping constraint.

*Vazquez-Roman and Mannan (2010).*

#### 8.1.1.3 Risk-related equations

The response versus dosage is calculated by a probit function. A probit function is of the form

$$\mathrm{Pr} = k_0 + k_1 \ln V$$

with properly selected values for the $k_0$ and $k_1$ parameters, which are readily available. Pr represents the probit value and $V$ is the dosage at a given distance from release. The probit value can then be related to the probability of death, which can then be used as a weighting factor in the objective function. The objective function depends on the probability which in turn depends on the probit value. Again, the probit value depends on the dosage at a given point which is determined by Monte Carlo sampling, which is the beginning point for our study of Monte Carlo methods in this example.

#### 8.1.1.4 Objective function

The objective function of this formulation is as follows

$$\min C = c_{land} A_x A_y + \sum_{(i,j)} C_p d_{ij} + C_{pp} t_l \sum_{s} \sum_{ri(i,r)} f_{i,r} P_{i,r,s}$$

where the first term is the cost per unit area of land multiplied by the actual boundary size of the land, the second term is the cost of piping per unit length multiplied by the distance of all connected piping between facilities, and the final term is the cost of a fatality multiplied by the expected life of the plant multiplied by the sum of product of the frequency of release and probability of death due to that release.

The objective is to minimize this balance of costs to an optimal solution. It can be seen that while spacing the facilities closer together would cause the piping and land costs to go down, it would also increase the probability that a fatality would occur in a release scenario. Likewise, greater spacing will decrease the probability of a fatality, but will increase land and piping costs. This balance is greatly affected by the multiplying cost factors, and thus if changed could give drastically different results.

#### 8.1.1.5 Monte Carlo simulation of weather conditions

Monte Carlo sampling is used to create a series of scenarios based on probability distributions of particular events. For example, one of the random variables that is examined is wind direction. Wind direction can vary depending on many factors including time of year and annual weather patterns. It is impossible to know beforehand exactly what the wind direction will be for a given day with perfect certainty. However, it is possible to have an idea of the probability that the weather may act a certain way on a given day through accumulated data. It is in this way that Monte Carlo sampling is useful: based on that accumulated data, scenarios are weighted by their probability of occurrence, selected randomly, and then simulated. The benefit is that the most likely scenarios are chosen most frequently, but less likely scenarios are also taken into account, albeit much less frequently.

One extreme alternative would be to totally enumerate all possible scenarios. This would be infeasible, as there are an infinite number of scenarios to be considered. On the other hand, the opposite extreme would be to choose one scenario based upon likelihood and optimize the layout based upon that information. This would be inadequate, as many probable (and improbable) scenarios would be left out. It is clear that Monte Carlo simulation can be a powerful tool in data analysis.

Jung et al. (2010) and Vázquez-Román et al. (2010) proposed new approaches to optimize the plant layout where the toxic release is considered inside the installed facilities. Meteorological conditions provide direct influence under release scenarios. Scenna and Cruz (2005) also demonstrated the advantages of using Monte Carlo simulations in predicting the weather uncertainty.

Use of weather data in consequence analysis is not trivial. This is because there is no clear expected mean or worst-case value that can immediately be used. Weather services such as the National Climatic Data Center provide reports of properties at the earth's surface between the ground level and 10-m height on an hourly basis. The reports contain information about temperature, dew point, wind direction, wind speed, cloud layers, ceiling height, visibility, precipitation amount. Finding the useful information from the last 20–30 years' data is not necessarily very straightforward. In their work, they formulated a probability distribution function for each of the variables using the provided data. Then a Monte Carlo simulation was used to randomly create thousands of samples within the distribution to calculate expected values that were used for further calculation of probit functions and probability of death based on receptors' locations.

The wind rose was generated from 86,096 records from Solar and Meteorological Surface Observation Network recorded from 1981 to 1990. The direction of the wind was qualified into 36 sectors of 10 degrees to create a probability distribution of the possible wind directions. The resulting wind rose is shown in Figure 8.2. Wind speed data were analyzed and fit into a Weibull probability distribution. The resulting graphical distribution is shown in Figure 8.2. Finally, the air stability is determined from typical parameters collected at NWS stations using the method given by Turner (1964). This probability distribution is shown in Figure 8.2. The Monte Carlo simulation must be as large as practical in order to ensure good results. Too few samples and the generated data are not likely to be significant or useful. Too many samples and the generated data will take too much time to accumulate. Using the sampled weather data, a dispersion model is used to estimate toxic gas concentration at each point in the simulation volume. This, when used in conjunction with the probit function, can give an estimate of probability of fatality. This is then used in the objective function to balance risk of death with financial considerations.

#### *8.1.1.6 Results*

Results may be dependent on the number of Monte Carlo data points taken. This will be true to a higher degree if the number of data points considered is relatively few, and the variation will level out as the number of points taken is increased. Thus, it is vital to ensure sufficient data points for an accurate reflection of true solution.

### 8.1.2 APPLICATION OF MONTE CARLO METHODS IN TRANSPORTATION SAFETY

The TSP often applied in operations research and computational physics is basically route optimization, which can be directly applied during transportation of hazardous materials. The shortest tour during the transportation of hazardous material is inherently safer, provided that the other variables remain the same. Such variables could be the road conditions, the population size in nearby roads, weather conditions of that area, the traffic congestions. In safer transportation of hazardous materials, population size is an important factor because of the risk associated with the density of the population. The problem becomes even more difficult if other variables such as time dependency, road conditions, and weather conditions are taken into account. To overcome these added difficulty levels, the Monte Carlo method is applied to choose the moves from one city to another in a random manner (Bentner et al., 2001).

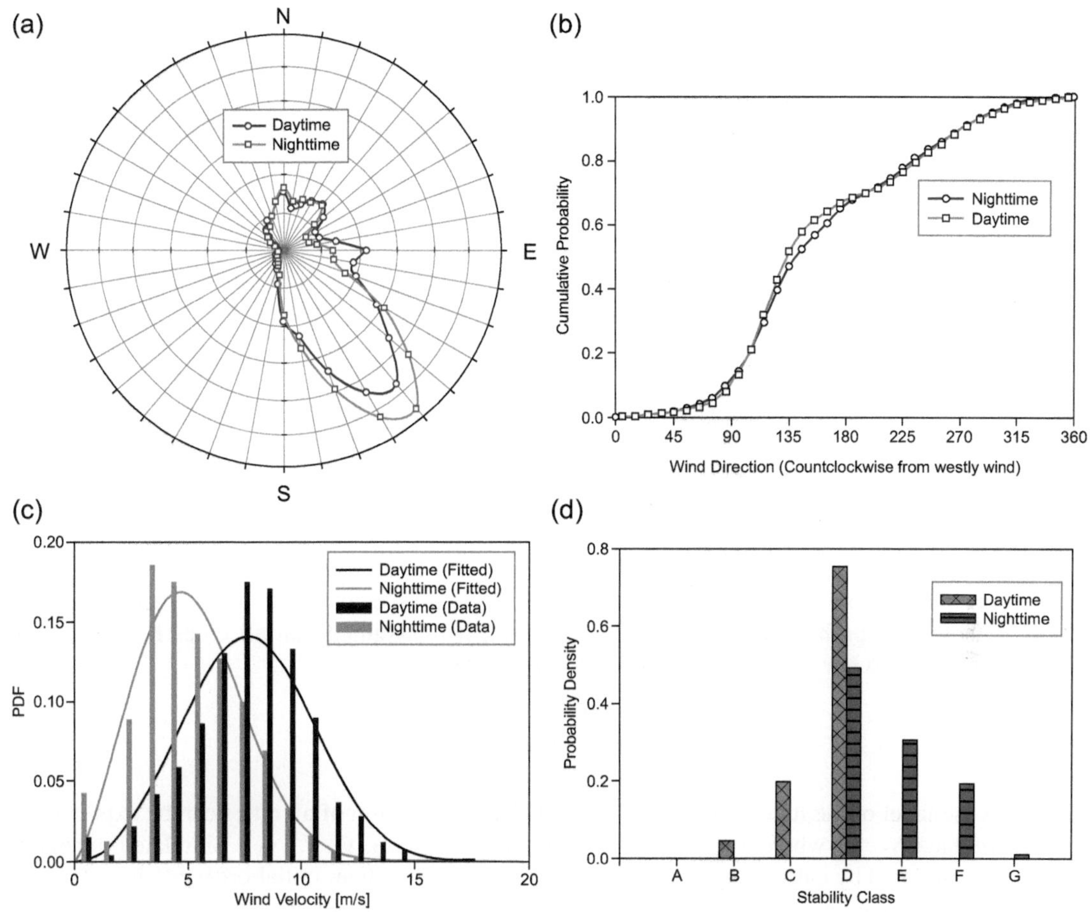


**FIGURE 8.2**

A typical (a) wind direction, (b) wind direction distribution, (c) wind velocity distribution, and (d) probability distribution of air stability.

*Vázquez-Román et al. (2010).*

For a given number of cities or places or destinations say $N$, the shortest closed route that a traveling salesman wants to travel is simply known as the TSP (Reinelt, 1994). Thus, the objective is to minimize the cost of transportation or the traveling time. In process safety application, the objective function can be to minimize the total risk associated with the delivery of hazardous goods or products to all customers. In order to formulate the cost function, the distances $D\ (i,j)$ between the cities are known a priori. Therefore, the cost function is given by the length of the route, that is, by the sum over the lengths of the edges and it is used with the cost associated with each mile. If one considers the traveling is at constant velocity, then minimizing distance or minimizing time both are exactly the same problem.

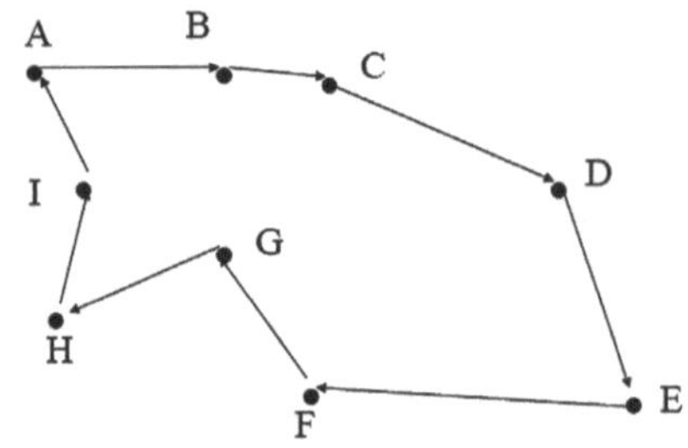


**FIGURE 8.3**

Traveling salesman.

*Caldwell (2013).*

Figure 8.3 shows an eight-point map of a closed tour. For $N$ cities, there are $(n-1)!/2$ possible combinations of city tours. Each possible tour is represented by an element set of all cyclic permutations of A, B, C ...H and denoted by $P_N$. Let us assume that all other variables except the distances remain the same between all travel routes. The optimization brute forces become impossible as the number of cities grows faster than any power of $N$. Different types of Monte Carlo methods can be applied to solve the problem. In this example problem, it is assumed that all cities have the same vertical elevation. So our objective function is to minimize,

$$l = \sum_{i=1}^{N} d_{L(i),L(i+1)}$$

where $L$ is the label of the $i$th city and $L(N+1) = L(1)$. The position of all cities can be expressed as $(x,y)$ coordinates vector where $x = [x_1, x_2, \ldots x_N]$ and $y = [y_1, y_2, y_3, \ldots y_N]$. Simulated annealing algorithm is initiated by calculating the distance matrix, $d$. and with an initial order of cities $L(i) = i$.

The new permutation is created by randomly selecting the city and swapping its order with the adjacent one. Due to a large number of swapping any city can end up in any positions of $L$. If the new order of cities gives a reduced total distance, then the new order is accepted; otherwise, it is accepted with a probability of $\exp\left(-\frac{\Delta d}{T}\right)$. For the problem shown in Figure 8.4(a), one of the optimal solutions is shown in Figure 8.5. Figure 8.4(b) refers to the initial labels of the cities and initial routes, while after the optimization, the routes become smooth as shown in Figure 8.5.

### 8.1.3 APPLICATION OF MONTE CARLO METHODS IN RELIABILITY ENGINEERING

Safety engineers often have to deal with determining reliability or availability of a complex system, which can be daunting. By applying Monte Carlo simulation, this can be done easily. This section discusses Monte Carlo applications in assessing reliability or availability of complex systems. Reliability of an item or a piece of equipment or a system of equipment can be defined as the ability of that piece to perform the required task under operating conditions for a given period of time (Lewis and Zhuguo, 1986). For example, if a system has $n$ equipment operating without replacement in the system's operating lifetime $t$, then after the lifetime period $t$, the number of equipment that has

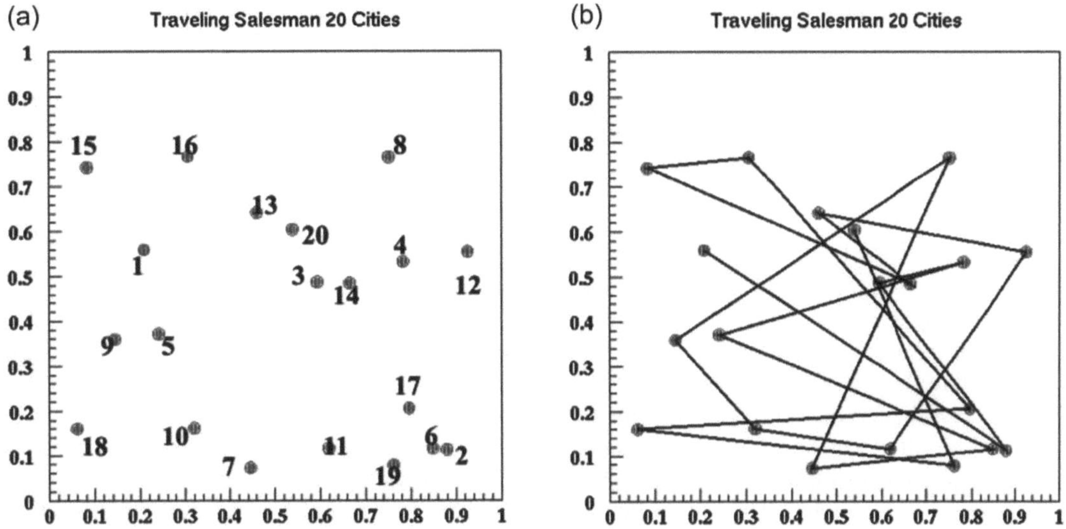


**FIGURE 8.4**

(a) Coordinates of cities. (b) Initial label or routes (before applying Monte Carlo methods).

*Caldwell (2013).*

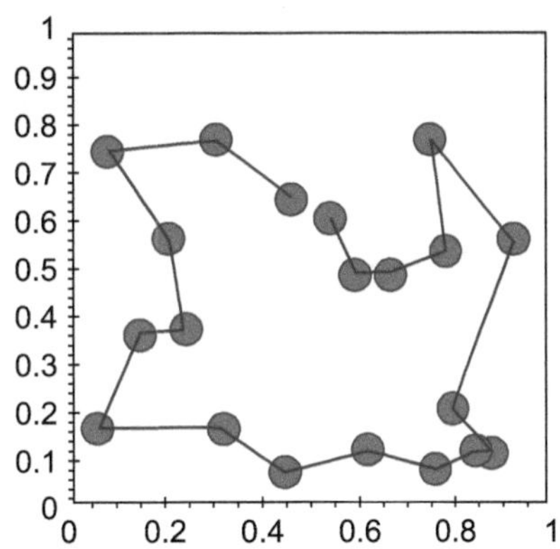

**FIGURE 8.5**

Optimum solution of a 20-city TSP.

*Caldwell (2013).*

survived is $n_s(t)$ and the number of equipment that failed is $n_f(t)$, then the reliability or the probability of the survival, $R(t)$, is:

$$R(t) = 1 - \frac{n_f(t)}{n}$$

So the probability of failure is

$$F(t) = 1 - R(t)$$

To calculate the reliability of a complex system using a stochastic Monte Carlo method, information such as configuration of the system, the failure distribution and repair characteristics of each system component, and period of interest are needed to be given (Lanore and Kalli, 1977). In this approach, the reliability of the whole system is determined for a large number of trials. At each trial, it is observed that the overall system survives or fails. With the increase of the number of trials, the ratio of the successful survivals to the total trials asymptotically approaches a constant. This asymptotic value is the required system reliability or availability. For each individual trial, the operating time is divided into a number of time increments. For the first incremental time, the failure probability of the first equipment is calculated. Then this calculated probability is compared with a generated random number from 0 to 1 using a uniform distribution. If the generated random number is less than or equal to the estimated probability of the first equipment for the first incremental time, then it is considered that the equipment fails and vice versa. Similarly, survival and failure for other components are computed. After computing the state of all equipment in the system, the overall state of the complex system is estimated. If the overall system survives, then the computation is carried out for the next incremental time. If the overall system fails, then change in system such as equipment repair for the failed equipment is applied (Muth, 1970). The calculation process is continued to cover the specified time period (Soon Heung, et al., 1985).

#### 8.1.3.1 Illustrative example

Before illustrating the reliability analysis of a complex system using Monte Carlo simulation, let us consider a single equipment system (Rubinstein, 1981). For a specified period of time, it is assumed that the failure distribution is exponential. Thus,

$$R(t) = \exp(-\lambda t)$$

where $\lambda$ is the failure rate.

Let us consider that the equipment will be operated for 6 months and the rate of failure of the equipment is 0.2 fault/year. With this analytical function, we can calculate the reliability as

$$R(t = 0.5\ year) = \exp(-\,0.2 \times 0.5) = 0.90$$

To simulate Monte Carlo, the first step is to divide the time step into small increments. We will use six incremental times. The probability of failure in each incremental time step is

$$-q = \lambda \times \mathrm{d}t = 0.2 \times 0.5 \times \frac{1}{6} = 0.0167$$

MATLAB pseudo random function *unidrnd* is used to generate random numbers $v$, which is given in Table 8.2.

**Table 8.2 Generated Pseudorandom Numbers for Monte Carlo Simulation of Reliability Analysis**

| Trial | Random Numbers | | | | | | Survival? |
|---|---|---|---|---|---|---|---|
| 1 | 0.8190 | 0.4330 | 0.5270 | 0.3730 | 0.5480 | 0.1790 | Yes |
| 2 | 0.8180 | 0.8260 | 0.4170 | 0.1990 | 0.9430 | 0.1290 | Yes |
| 3 | 0.7230 | 0.0840 | 0.6570 | 0.4900 | 0.4180 | 1.0000 | Yes |
| 4 | 0.1500 | 0.1340 | 0.6280 | 0.3400 | 0.9840 | 0.1720 | Yes |
| 5 | 0.6600 | 0.1740 | 0.2920 | 0.9520 | 0.3020 | 0.0330 | Yes |
| 6 | 0.5190 | 0.3910 | 0.4320 | 0.9210 | 0.7020 | 0.5620 | Yes |
| 7 | 0.9730 | 0.8320 | **0.0160** | 0.0530 | 0.6670 | 0.8820 | No |
| 8 | 0.6490 | 0.8040 | 0.9850 | 0.7380 | 0.5400 | 0.6700 | Yes |
| 9 | 0.8010 | 0.0610 | 0.1680 | 0.2700 | 0.6990 | 0.1910 | Yes |
| 10 | 0.4540 | 0.4000 | 0.1070 | 0.4230 | 0.6670 | 0.3690 | Yes |

The probability of failure for the first trial is $q$ (=0.0167) and compared with the generated random number $\nu$. The occurrence of survival or failure is determined based on the following criteria:

$$\nu > \text{q survival;}$$
$$\nu < \text{q failure.}$$

In the first line of Table 8.2, for the first month, the equipment survives because $\nu = 0.8190$, which is greater than $q$. It is repeated for each of the five remaining months. From the generated table, we find that for all cases $\nu > q$, which indicates that the equipment survives for the whole 6 months. Thus for the first trial, the equipment has survived. A similar analysis is carried out for the second, third and so on. The ratio of the number of survival to the total number of trials is the reliability of the equipment. In the seventh trial we find a case (as identified by bold) where nu < q (0.0167), thereby indicating a 'no' survival scenario. Hence, in this particular example, only 10 trials have been done and the equipment survives for 9. Thus, the reliability is 9/10 = 0.9, which is only accurate to one decimal place. In this case, fortunately, the Monte Carlo reliability and theoretical calculations are exactly the same, but in practice, a large number of trials must have been used to be more accurate. Figure 8.6 shows the accuracy of Monte Carlo simulations with the number of trials for this example.

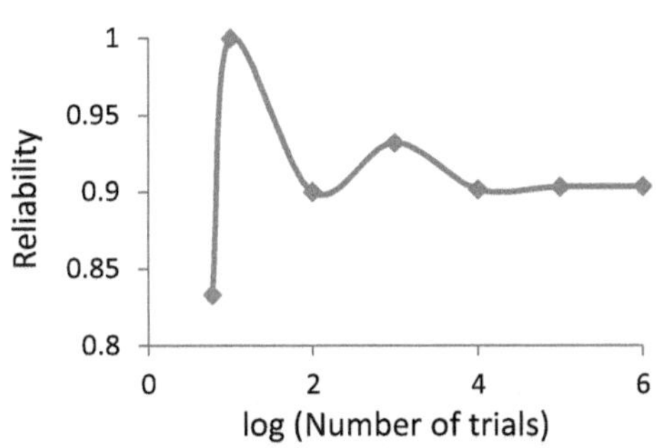


**FIGURE 8.6**

Accuracy of Monte Carlo simulation with the number of trials.

#### *8.1.3.2 Applications of the method*

This application example has been taken from Lees' Loss Prevention in the Process Industries (Mannan, 2005). The Monte Carlo method is applied to answer critical questions that are often necessary for the risk assessment of an olefin plant system shown in Figure 8.7. The elements of the system includes four cracking furnaces in parallel followed by a compressor and then by a separation train. The characteristics of the furnace are shown in Table 8.3.

Here, compressor failure rate is assumed to be 0.3 fault/year, and the failure of separation train is considered as 3 faults/year. Table 8.4 shows the repair characteristics of the equipment. It can also be assumed that the only restriction applied to a repair furnace is that only one furnace can be repaired at a time. To do the simulation, the information required are (1) the probability of achieving full availability, (2) the mean availability achieved, and (3) the range of availabilities and the corresponding probabilities, over a period of 90 days.

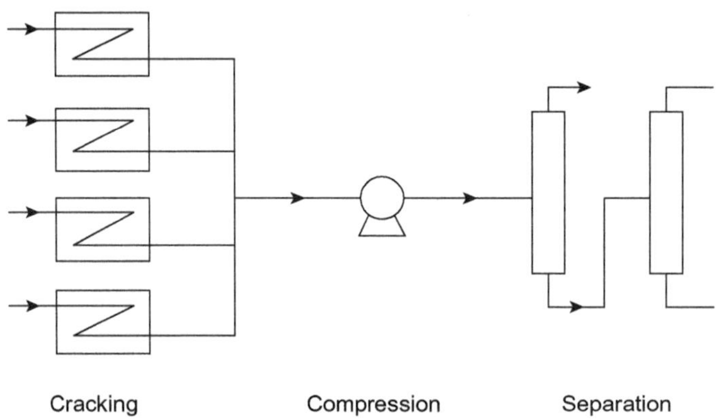


**FIGURE 8.7**

Olefin plant systems.

*Mannan (2005).*

**Table 8.3 Characteristics of Furnace Used in the Monte Carlo Simulation Example**

| Time Since Startup (days) | Furnace Failure Probability |
|---|---|
| 0–50 | 0.08 |
| 50–100 | 0.05 |
| 100–200 | 0.05 |
| 200–300 | 0.10 |
| 300–400 | 0.10 |

*Source: Mannan (2005).*

**Table 8.4 Repair Characteristics of the Equipment for the Monte Carlo Simulation Example**

| | Repair Time (days) | Proportion of Occasions |
|---|---|---|
| Furnace | 20 | 0.7 |
| | 30 | 0.2 |
| | 40 | 0.1 |
| Compressor | 10 | 0.8 |
| | 30 | 0.2 |
| Separation train | 10 | 0.7 |
| | 20 | 0.3 |

*Source: Mannan (2005).*

### 8.1.3.3 Features of the method

The Monte Carlo simulation method is necessary to solve practical reliability and availability problems where the number of constraints is large and other complex functions such as failure and repair data are involved. The histogram as shown in Figure 8.8 is the preferred form of failure data in practice. Also, the probability of failure $q$ is required to compare with the generated random number. Let us define the fraction of failure over an incremental time as $\delta F$: then

$$q = \frac{\delta F}{1 - F}$$

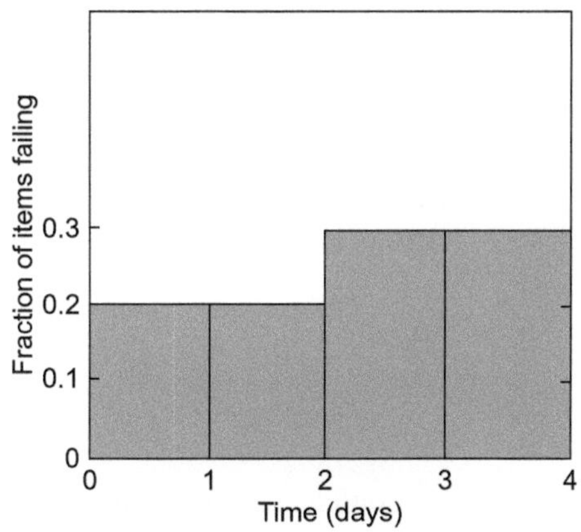


**FIGURE 8.8**

Failure histogram of the equipment.

*Mannan (2005).*

Using the above equation, failure probability for the consecutive 4 days are 0.2, 0.25, 0.5, and 1.0. Often, for convenience, simulation is performed for one set, and when the results are analyzed, simulation is performed for further sets. However, the most important issue is to make sure that the subsequent sets do not start with the same random number. To avoid this situation, the seed of random number generator is continuously changed.

#### *8.1.3.4 Number of trials*

One of the major bottlenecks of the Monte Carlo method is to estimate the number of trials required to reach the specified confidence level. Applying certain simplifying assumptions, an approximate number of trials can be known prior to simulation. This method of estimation of number of trials has been described (Shooman, 1968).

While determining the reliability using Monte Carlo simulation with a series of $n$ trials, the probability of true success is assumed to have a binomial distribution. Let us denote the probability of success with $p$ and of failure with $q$. As $p$ and $q$ approach $p > 0.5$ and $q > 5$, the binomial distribution approaches a normal distribution. If the numbers of failures in $n$ trails are $n_f$, then the probability of failure $q$ is given by $n_f/n$. Hence, fractional error $e$ can be defined as follows:

$$e = \frac{\frac{n_f}{n} - q}{q}$$

Here, with the expected number of failures $\mu$ and the expected standard deviation $\sigma$, $(1-\alpha)$ confidence interval is:

$$P(-\gamma\sigma < n_f - \mu < \gamma\sigma) = 1 - \alpha$$

where $\gamma$ is the number of standard deviations corresponding to the confidence interval. Expressed in terms of $p$ and $q$:

$$\mu = nq$$

$$\sigma = (npq)^{0.5}$$

then

$$P\left[-\gamma(npq)^{0.5} < n_f - n_q < \gamma(npq)^{0.5}\right] = 1 - \alpha$$

$$P\left[-\gamma\left(\frac{p}{nq}\right)^{0.5} < e < \gamma\left(\frac{p}{nq}\right)^{0.5}\right] = 1 - \alpha$$

Thus, at the 95% confidence level, for which $\gamma = 2$:

$$e > 200\left(\frac{p}{nq}\right)^{0.5}$$

Table 8.5 shows that to achieve a 95% confidence level with an error of $\not> 5\%$, a large number of simulations are necessary for the above-defined complex system. Detailed analysis of the number of trials required for sufficiently accurate estimation of reliability using Monte Carlo simulation is given (Shreider, 1964).

**Table 8.5 Number of Trials Required in Monte Carlo Simulation Example to Give an Error of ≯5% at a 95% Confidence Level**

| Probability of Success | No. of Trials Required |
|---|---|
| 0.9 | 14,400 |
| 0.99 | 158,400 |
| 0.9999 | 1,598,400 |

*Source: Mannan (2005).*

## 8.2 APPLICATION OF CFD IN QUANTITATIVE RISK ANALYSIS

Quantitative risk analysis (QRA) is widely used in the industry to help managers make proper decisions to improve safety or manage the remaining risk. It is a systematic approach to analyze the risk associated with an event (Spouge et al., 1999; Arendt and Lorenzo, 2000). Figure 8.9 shows the development of QRA that includes hazard identification, frequency analysis, consequence analysis, and risk evaluation (Arendt and Lorenzo, 2000). It combines event frequency and consequence data to estimate a risk value. Decision makers should know its limitations and use it with rational evaluation. QRA can also be used throughout every stage of a process or a facility.

There are four activities associated with consequence analysis (Arendt and Lorenzo, 2000), including release of material or energy (i.e., source term), dispersion phenomena, identifying effects of release (i.e., toxic, fire, or explosion), and quantification of effects. As complexities in a facility are getting higher and higher, it may be impractical to do experiments to quantify all the effects of release

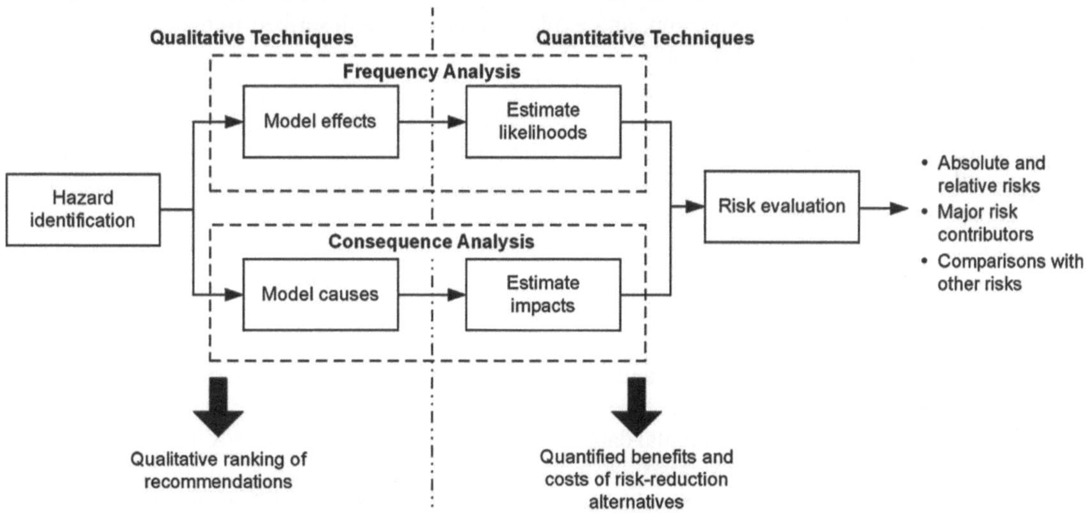


**FIGURE 8.9**

The process of risk analysis (Arendt and Lorenzo, 2000).

and dispersion phenomena. Since computing power is also getting stronger, computer software can be used to aid the consequence modeling. If more complex analysis is needed after an initial screening study, CFD can be performed for more decisive or precise data.

### 8.2.1 APPLICATION OF CFD SIMULATION IN QRA

CFD (Versteeg and Malalasekera, 1995) is the science of simulating the flow of fluids under various conditions, including heat transfer, mass transfer, and chemical reactions. It replaces partial differential equations (PDEs), which govern the fluid dynamics by using a set of numerical methods with the help of computers. CFD is widely used in many fields, such as study of new design, troubleshooting, product development, and consequence modeling in QRA. The simulation results complement experimental and theoretical data to reduce the expensive cost and time-consuming laboratory experiments.

CFD simulates the actual flow dimension instead of a laboratory scale. CFD can deal with the effect of complex geometry and provide more detailed information than integral models or simple correlations. However, the quality of CFD simulation will be influenced by physical models, numerical errors, and boundary conditions. A further check is needed before an actual application.

There are a number of commercial or free CFD software available, for example, FLUENT, COMSOL, CFX, FLACS, and PHOENICS. Many of them implement the finite volume method to solve the complex set of numerical problems which are divided into small volumes, known as a mesh, grid, or cell.

Application of CFD simulation in QRA focuses on consequence modeling, including dispersion of flammable vapor, ventilation, evaporation, thermal or overpressure effects of fire, explosion or BLEVE, health effect of dispersion of toxic vapor, or reactor design for runaway reactions. Figure 8.10 (OGP, 2010) shows possible scenarios of loss containment of hazardous materials.

Applications of CFD simulation in jet fire with or without mitigation barrier, heat load, LNG pool fire, flammable or toxic gases dispersion, and explosion have been described in detail in Chapter 4. Here, more applications are introduced.

CFD simulation can be applied to simulate the development of a fire in a mobile home and smoke migration (Schmidt et al., 2012). Figure 8.11 shows the geometry of the mobile home, location of fire, thermocouples, and slices for CFD simulation. The location of fire was at armchair, which was extinguished 180 s after ignition. But thermocouples and smoke detectors continued to record the data. Thermocouples were located at six "tree labels," each with thermocouples at seven different heights. The heat release (source term) from the armchair was polyurethane foam loss rate per second (shown in Figure 8.12) multiplied its calorific value.

The temperature distribution along slice 2 simulated by CFX and FDS is shown in Figures 8.13 and 8.14, respectively. The simulation results were based on the upward buoyant flow induced by heat source in a confined space. Figures 8.15 and 8.16 further exhibit the comparison of experimental data and modeling results by CFX and FDS, respectively, at tree E with three different heights (E2, E4, and E7). CFX results agreed with experimental data for the initial 120 s. However, CFX underestimated the temperatures at these thermocouples after 120 s. The deviation was up to 20 K. In contrast, points E4 and E7 of FDS results fitted very well with the experiment. Nonetheless, E2 (near the ceiling, high temperature) of FDS results showed strong oscillation; the increase in turbulence may be from numerical reasons for gradients in the near-wall region instead of physical reasons.

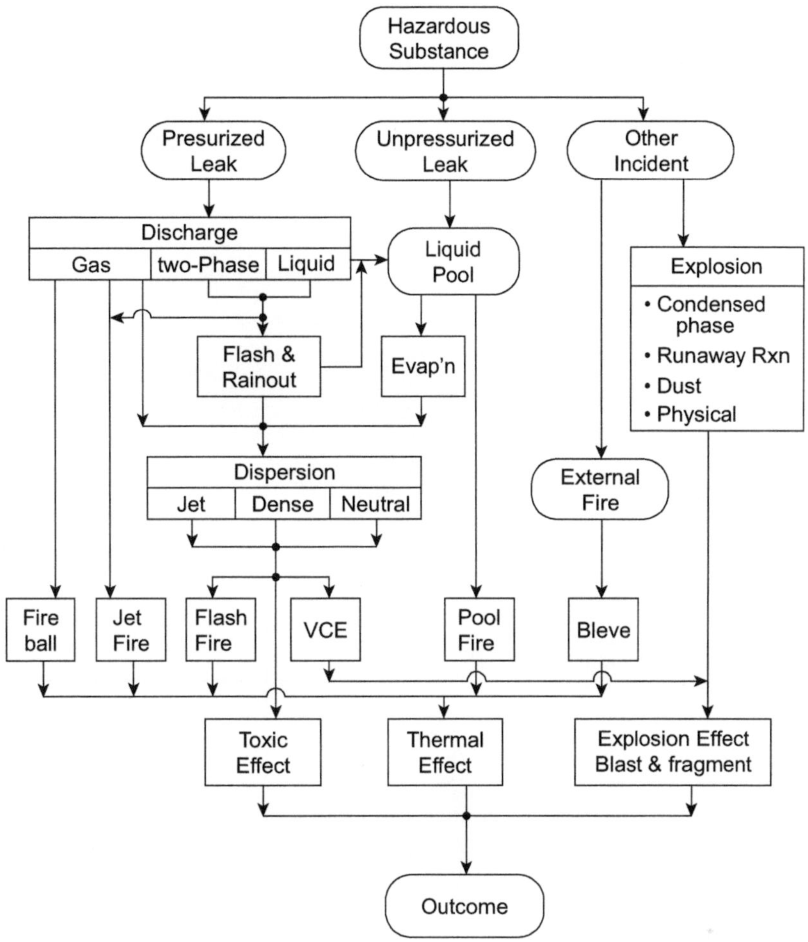


**FIGURE 8.10**

Consequence phenomena and their interrelationship (OGP, 2010).

CFD can also be applied to simulation of smoke migration, which can help to optimize the location of fire detectors and further reduce the response time and cost. It can also simulate a smoke-free layer, which is imperative for safe evacuation. But smoke is a mixture of various gases, liquids, or solid particles from combustion and is not a quantity that can be derived from governing equations, which make it difficult to analyze. However, simplification can be made to circumvent this situation. All fire products were viewed as "smoke," a quasi-component. Figure 8.17 shows the smoke concentration across slice 1. The left side was simulation by CFX, and the right side was simulation by FDS. Despite obvious differences in temperature distribution and smoke distribution, the height of smoke-free layers was similar in these results.

Another important application of CFD is to simulate the reaction and transportation phenomena (i.e., heat transfer and mass transfer). This is widely used in the reactor design (Baldyga et al., 2001;

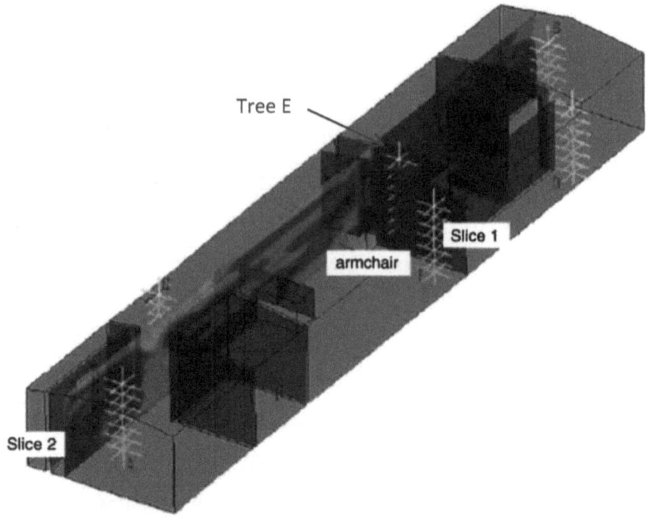


**FIGURE 8.11**

Geometry of the mobile house, location of fire, thermocouple, and slice for CFD simulation (Schmidt et al., 2012).

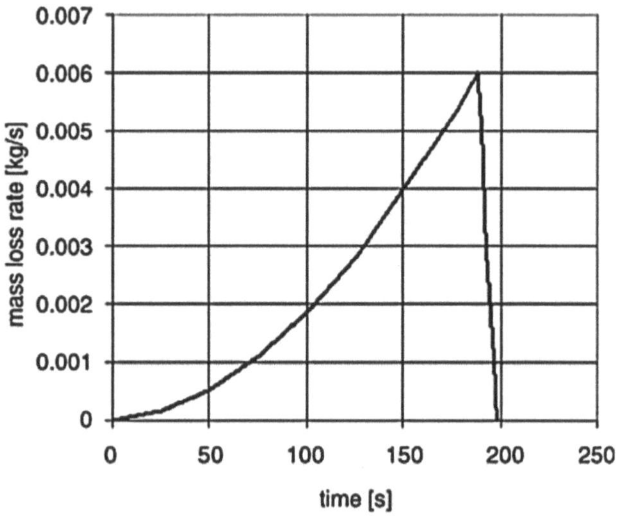


**FIGURE 8.12**

PU foam loss rate per second (Schmidt et al., 2012).

Ding et al., 2010; Patel et al., 2010; Schmidt et al., 2012). It is also critical to understand the concentration and temperature profiles for a runaway reaction in a CSTR. Usually, rapid mixing is one way to obtain a uniform concentration profile and hence a temperature profile. The design of the paddles, inlet and outlet location, residence time, and the impeller speed are important parameters for mixing, which can influence the chemical reaction rate and quality of final products (Patel et al., 2010). CFD can help to design suitable reactors or identify problems at the early stage. Figure 8.18

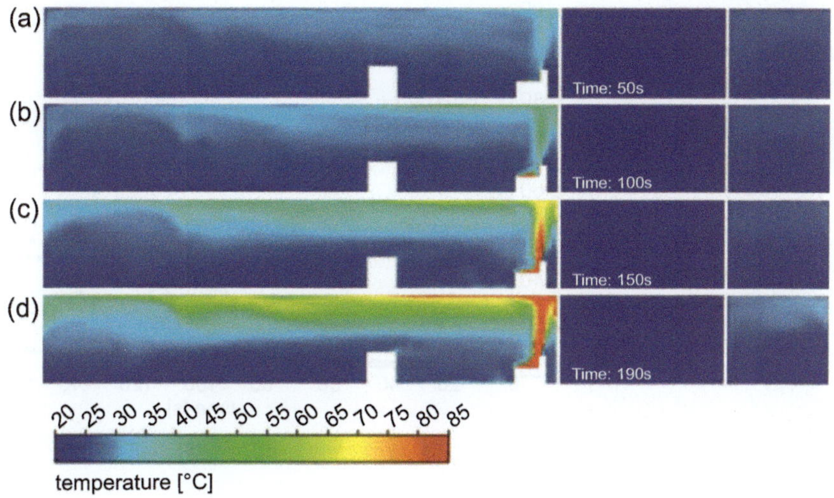


**FIGURE 8.13**

Temperature distribution along slice 2 simulated by CFX, (a) 50 s after ignition; (b) 100 s; (c) 150 s; (d) 190 s (Schmidt et al., 2012).

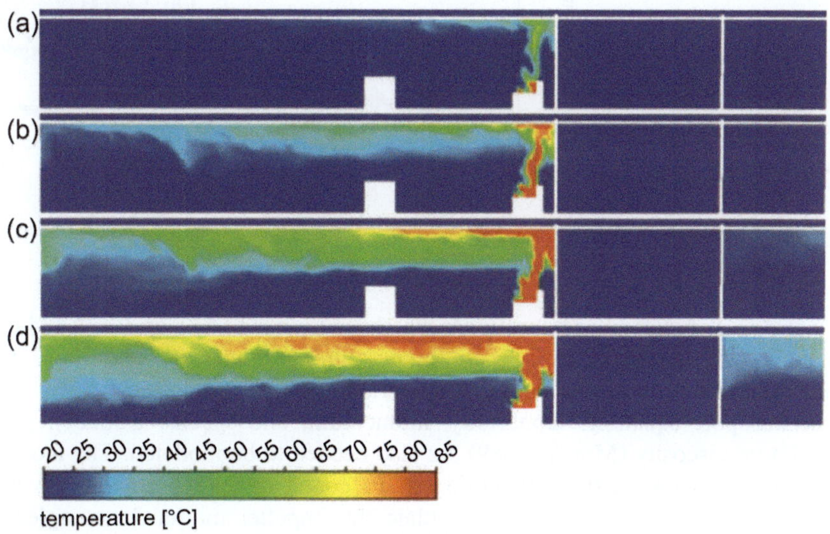


**FIGURE 8.14**

Temperature distribution along slice 2 simulated by FDS, (a) 50 s after ignition; (b) 100 s; (c) 150 s; (d) 190 s (Schmidt et al., 2012).

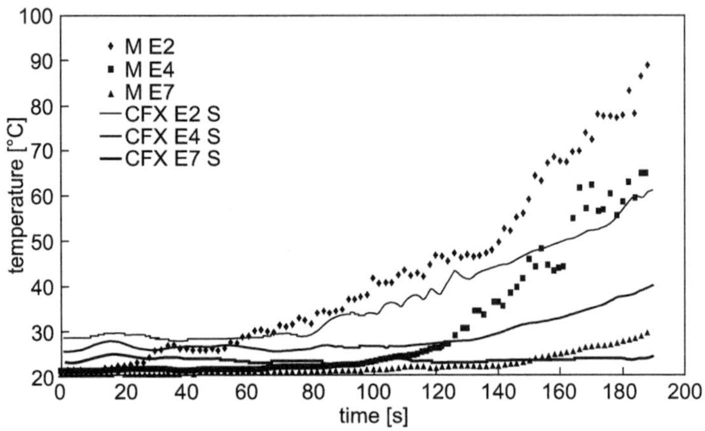


**FIGURE 8.15**

Comparison of experimental and CFX simulated results along tree E (Schmidt et al., 2012).

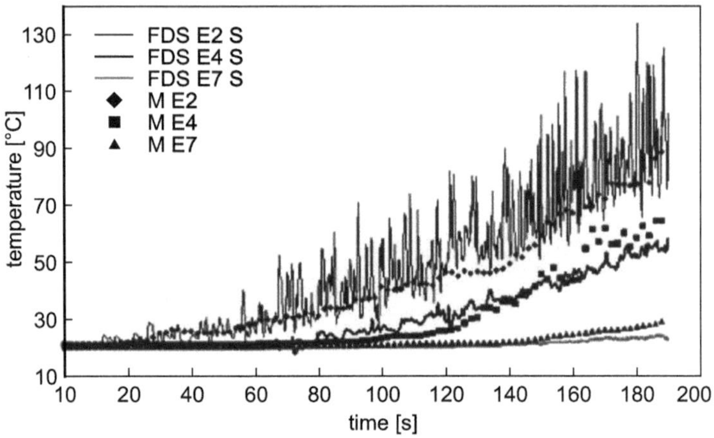


**FIGURE 8.16**

Comparison of experimental and FDS simulated results along tree E (Schmidt et al., 2012).

indicates two different reactor configurations, which contain 372,571 and 371,784 cells, respectively. Styrene is polymerized due to high temperature (isothermal conditions, 140 °C). The unreacted monomer and product mix together and leave through the outlet.

Other than transport equations (continuity, momentum, and species balance), polymerization reaction and mixture viscosity (Moritz, 1989) are also needed to be modeled. The model was solved by FLUENT 8.3 plus a user-defined function (UDF) coded in the C language to deal with the polymerization kinetics. In order to easily simulate the impeller motion in complex geometry, a multiple reference frame (MRF) (Luo et al., 1994) technique was taken to deal with the impeller motion, the rotating frame region is shown in green color in Figure 8.18. Under these conditions, the impeller can be treated as still compared to the MRF.

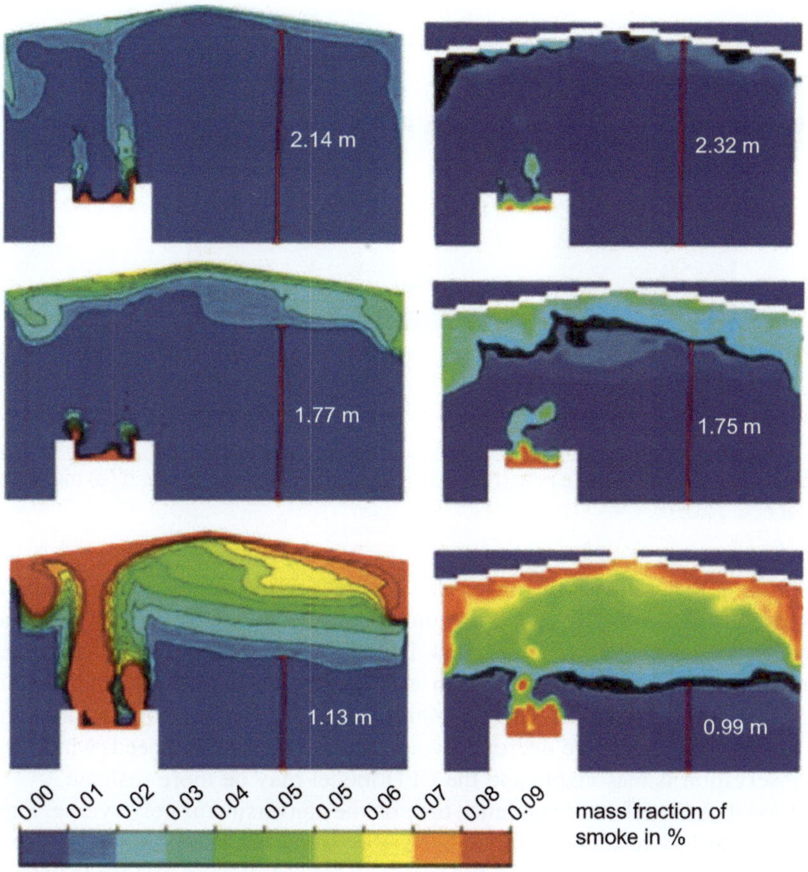


**FIGURE 8.17**

Simulation of mass fraction distribution of smoke along slice 1, left side: CFX, right side: FDS, time from top to bottom: 50, 100, and 190 s after ignition, the vertical red line (dark gray in print versions) showed the height of the smoke-free layer (Schmidt et al., 2012).

Simulation of the effect of the impeller speed (*N*) for the side inlet reactor (shown in Figure 8.18(a)) is depicted in Figure 8.19. The simulation was conducted at 144 min residence time and the rotation speed was 100, 500, and 1000 rpm, respectively. It is clear that the concentration distribution is not uniform in the reactor even at the highest mixing speed, which is different from a perfect CSTR. After styrene monomer enters the reactor, it is pushed up by the upward axial flow generated by the impeller. This is followed by a reverse flow of the styrene monomer, down the reactor due to circulation while polymerization of the same takes place. As shown in Figure 8.19, the higher the rotation speed, the better the uniform mixture distribution.

Styrene conversion would be influenced by various parameters. The parameters include impeller rotation speed, residence time, and inlet-outlet locations. The effect of the impeller speed is shown in Figure 8.20. Since no experimental CSTR data are available, batch data were used as reference

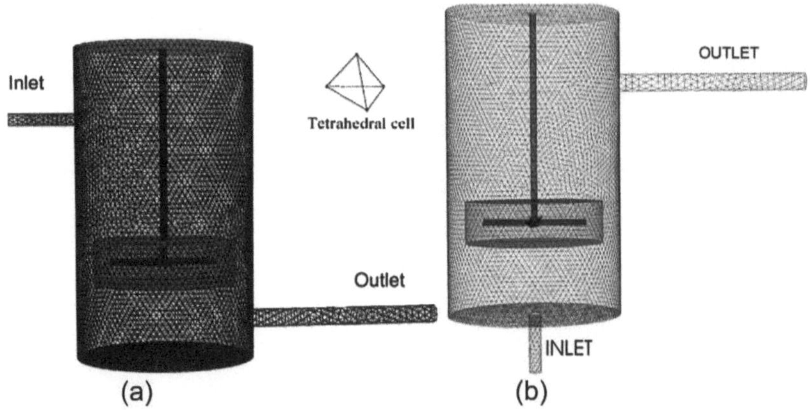


**FIGURE 8.18**

Discretization of the reactor: (a) the side inlet (near liquid level) and bottom outlet and (b) the bottom inlet and side outlet (Patel et al., 2010).

(Dhib et al., 2000). CFD simulation showed that the conversion decreased with increasing rotation speed. One possibility is that an unreacted monomer is pushed away from the styrene-rich region and escaped easily to outlet at higher impeller speed. In addition, the reaction rate deteriorated with lower monomer concentration (better mixing), which also causes the lower conversion at higher rotation speed. Compared to the CSTR model which assumes perfect mixing, the conversion estimated by CFD simulation is higher. The difference decreases with increasing impeller speed, which results in better mixing. The observation is reasonable and the CFD model may be more realistic.

Figure 8.21 exhibits the effect of residence time on the conversion of polystyrene. The difference of conversion is small at low residence time. With higher residence time, the difference is larger. The possible explanation is that the conversion and viscosity are low at short residence time. So the mixing is relatively uniform compared to longer residence time, which makes the styrene concentration and conversion similar at low residence time in these models.

The inlet-outlet location also has significant influence on the mixing effect in a continuous process (Saeed and Ein-Mozaffari, 2008). Figure 8.22 indicates the styrene concentration of the bottom inlet reactor. In contrast to Figure 8.19, the mixing is greatly improved by placing the inlet against the impeller pumping direction. Since styrene distribution is similar in either impeller speed, conversion is also the same in this case, which suggests that conversion is independent of impeller speed as shown in Figure 8.23.

CFD simulation is capable of considering the inhomogeneity in CSTR and makes better predictions. It can also provide comprehensive data that are not easy to measure in real situation. CFD can be further used to exploit other parameters including inlet-outlet location, residence time, and impeller speed. Most suitable design of the reactor may be simulated and other problems may be solved in the early stage. However, polymerization is a highly coupled transport phenomenon. Simplification of reaction mechanisms and transport equations are needed to reduce the computing time.

CFD is also used to solve the multiphase flow including phase changes, for example, two-phase boiling (Kunza et al., 2000; Končar et al., 2004; Liu et al., 2008; Ustinenko et al., 2008). Liu et al.

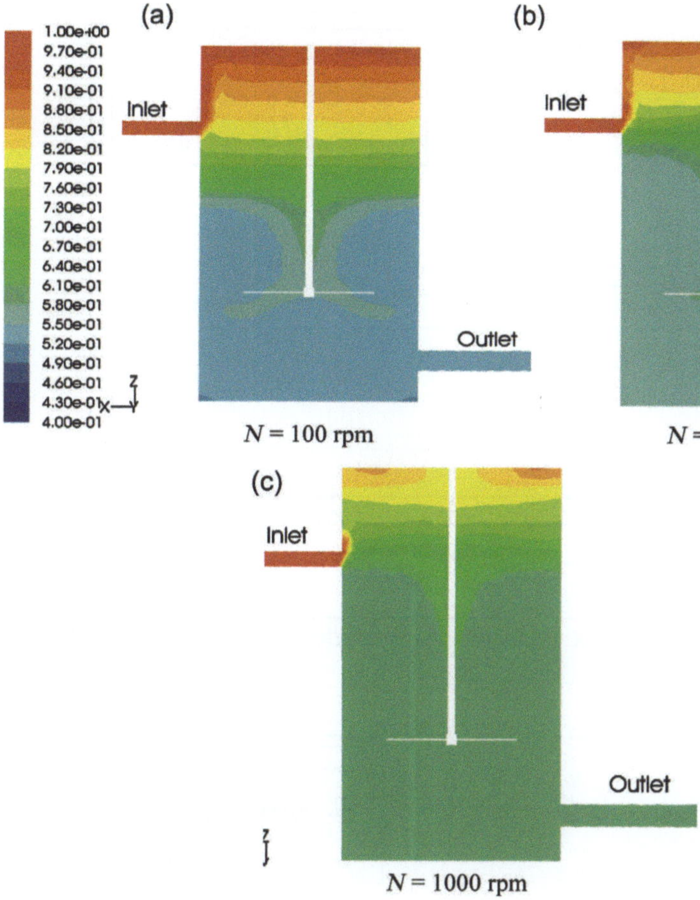


**FIGURE 8.19**

Simulated styrene mass fraction of different impeller speed ($N$) for the side inlet reactor at 144 min residence time, (a) $N = 100$ rpm, (b) $N = 500$ rpm, (c) $N = 1000$ rpm (Patel et al., 2010).

(2008) built a mathematical model using UDF to describe the phase change (i.e., boiling). The model can describe the effect of ground temperature on the boiling rate. Different boiling patterns, such as film boiling, transition boiling, and nucleate boiling, have been identified. The Leidenfrost point obtained is comparable to the experimental data. There is only one adjustable parameter, the relaxation time, in this model. The other variables in this boiling model employ some real data of the physical properties of the liquid and vapor. It can potentially be expanded to evaluate the risk of vaporization of cryogenic liquids, such as LNG, if the effect of multiple components is introduced.

In general, CFD software applies the $\kappa-\varepsilon$ model, which is one of the Reynolds-Averaged Navier–Stokes (RANS) models to solve turbulence equations, but other advanced models can also be used, such as large eddy simulation (LES) or direct numerical simulation (DNS). The LES relies on the spatial filtering applied to the governing equations. The filtering excludes eddies smaller than the

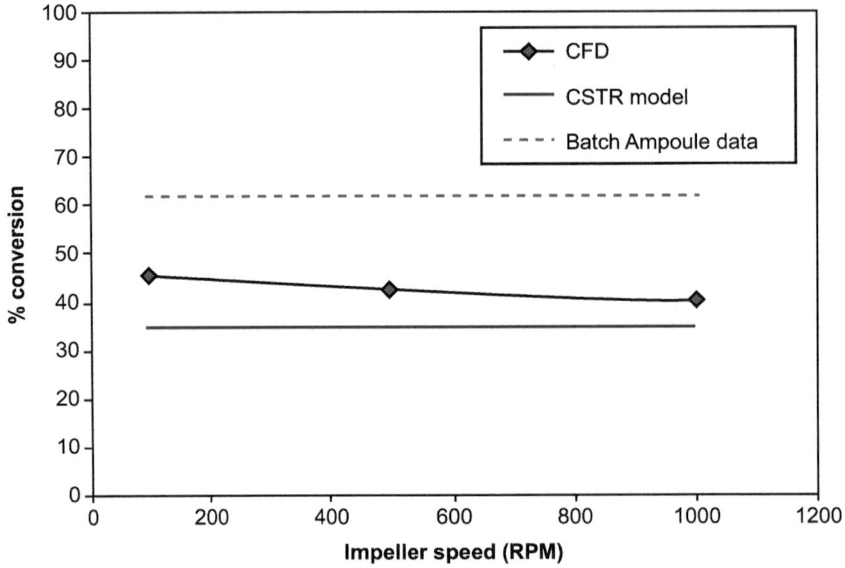


**FIGURE 8.20**

Conversion versus rotation speed for side inlet reactor at 144 min residence time (Patel et al., 2010), with batch data from Dhib et al. (2000).

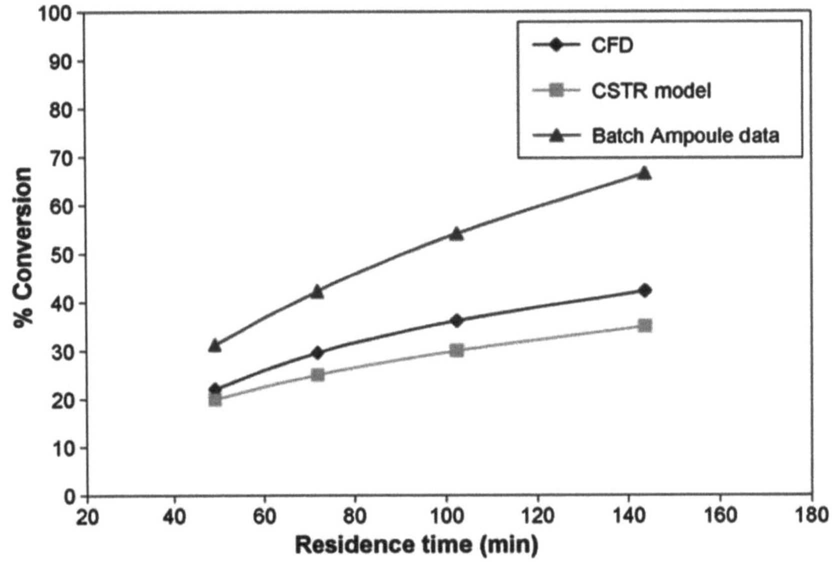


**FIGURE 8.21**

Conversion versus residence time for the side inlet reactor at 500 rpm rotation speed (Patel et al., 2010), with batch data from Dhib et al. (2000).

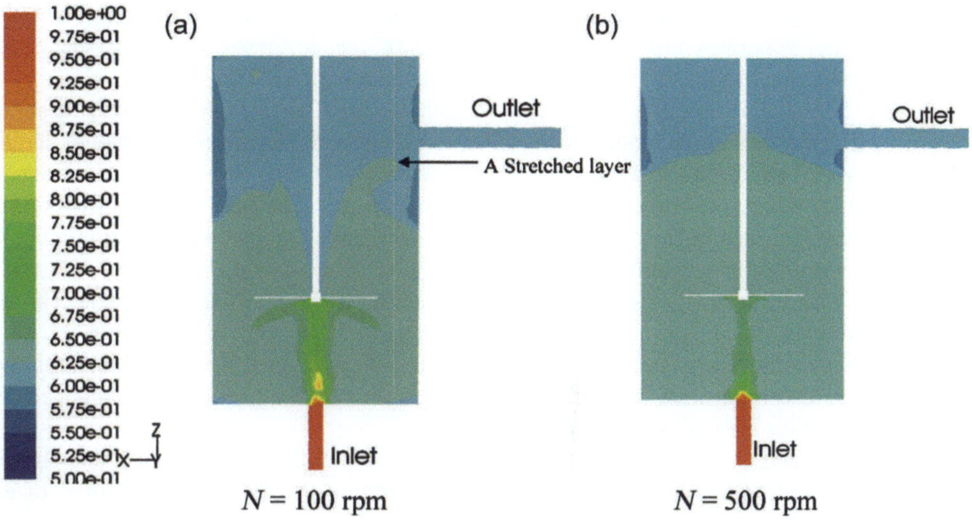


**FIGURE 8.22**

Simulated styrene mass fraction of different impeller speeds ($N$) for bottom inlet reactor at 144 min residence time, (a) $N = 100$ rpm, (b) $N = 500$ rpm (Patel et al., 2010).

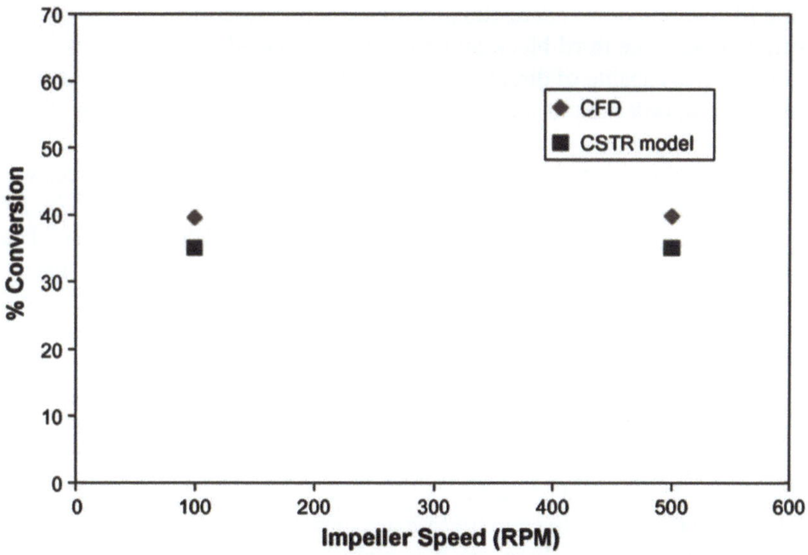


**FIGURE 8.23**

Conversion versus rotation speed for the bottom inlet reactor at 144 min residence time (Patel et al., 2010).

filter width, and the dynamics of the large eddy is governed by the resulting equations. Nevertheless, the filtering process generates unknown terms because of the nonlinear nature of the governing equations. The unknown terms can be modeled at the subgrid scale (SGS).

The closure of the SGS stress tensor can be reached based on a scale-similarity assumption: Smagorinsky-Lilly eddy viscosity model (Lilly, 1992). During the LES process, the model parameters were in situ computed based on the dynamical flow structures obtained from the smaller scales of the resolved field, which allows the eddy viscosity to react accordingly. The SGS model allows simulation of the complex physical phenomena at different scales. It increases the accuracy of simulation while not obviously increasing the computation time.

Di Sarli et al. (2009) used LES and SGS models to simulate the premixed flames accelerated by obstacles. The vortices ahead of the flame front were studied. They were created from the unsteady combination effect of the flame propagation and the flow field caused by the obstacles along the propagation route. The intensified rate of flame propagation and the pressure rise induced by this flame–vortex interaction are well explained.

In Figure 8.24(a) and (b), the experimental data and LES simulation results of flame shapes and arrival times are presented respectively. It is clear that the flame shapes and arrival times of experimental results were predicted well by the LES simulation.

Figure 8.25 further exhibits the flame-speed profile and pressure time history. In Figure 8.25(a), the black rectangles on the $x$-axis = 100, 200, and 300 mm indicate the locations of the obstacles. Since the LES model predicted the position and shape well, the flame speed was also described by the LES model qualitatively and quantitatively. The results show that flame speed was accelerated when it passed each obstacle and decelerated slightly because of flame expansion between obstacles. This flame acceleration phenomenon may result in deflagration to detonation. Figure 8.25(b) indicates that the pressure after 39 ms (bottom part of the chamber) was also predicted well by the LES model. Nevertheless, the LES simulation underestimated the peak pressure (near 37 ms), at the time that the flame had traveled through the third block and almost reconnected.

The LES simulation is capable of directly resolving the influence of the large vortices on the flame propagation. However, in order to consider the influence of the small vortices (i.e., the SGS vortices), the SGS wrinkling factor ($\Theta_\Delta$) according to the combustion model by Charlette et al. (2002) was quantified. By assuming $\Theta_\Delta = 1$ in the combustion model, the effect of SGS vortices was neglected in the flame propagation. As shown in Figures 8.24(c) and 8.25, the results obtained by LES simulation with $\Theta_\Delta = 1$ are obviously different from experimental data and LES simulation. The flame speed with $\Theta_\Delta = 1$ is much slower compared to other methods. Nonetheless, the flame shape is still similar to LES simulation and experiment. The trend of flame speed is also similar but a little bit slower in LES simulation with $\Theta_\Delta = 1$. However, the prediction of peak pressure is delayed and significantly underestimated when SGS vortices are neglected. It can be concluded that the large vortices contribute to the qualitative properties (i.e., flame structure and shape, flame acceleration and deceleration with blocks, flame speed trend). In contrast, SGS vortices dictate the quantitative part for the peak pressure and the flame propagation.

Although the CFD technique is a powerful tool to predict the consequence of loss containment and provides insight into the reactor design and fire/gas detector optimization, it requires further validation before it can be used in practice. There are two main types of methods to check the simulation: (1) verification and (2) validation. Verification is to make sure whether the simulation works correctly, while validation is to make sure whether the simulation works well to predict the phenomena in different conditions.

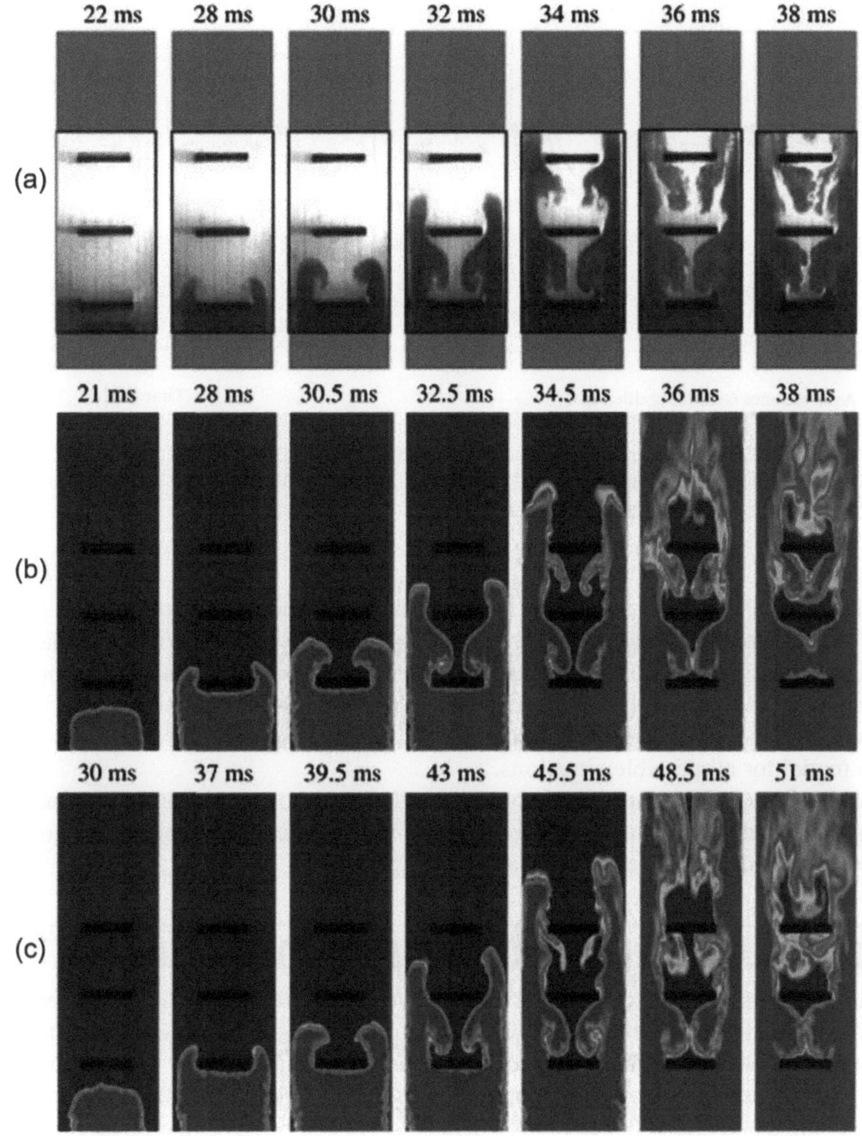


**FIGURE 8.24**

Flame shapes and arrival times: (a) experimental data (Patel et al., 2002), (b) LES simulation (Charlette et al., 2002), (c) LES simulation by assuming $\Theta_\Delta = 1$ (part c will be explained later) (Di Sarli et al., 2009).

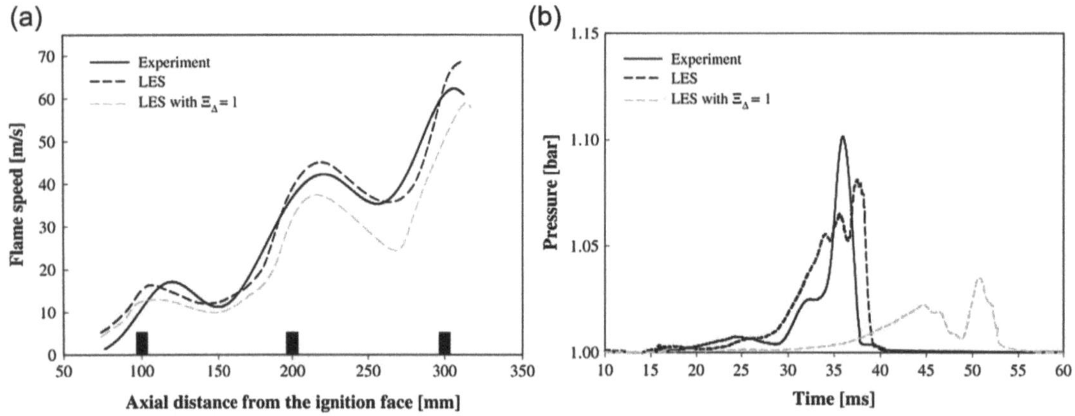


**FIGURE 8.25**

(a) Measured flame speed compared with LES simulation and (b) pressure time history at the bottom end of the combustion chamber (Di Sarli et al., 2009).

To validate the CFD model, much experimental data are needed to compare with the simulation results. However, sometimes the gaps between the simulation and experimental may come from unsuitable input conditions in CFD simulation. So verification should be taken to identify the causes and to improve the simulations. Even if the models have been validated by some experimental data at specific situations, the modeling results should still be used judiciously since it is impossible to validate the model for all possible situations.

It should also be noted that the purpose of any simulations is to complement experiments instead of substituting experiments. Simulation can reduce the time and cost to achieve the goal and obtain valuable properties that are not easy to measure in experiments.

## 8.3 MULTISCALE MODELING APPROACH IN PROCESS CONTROL

In a plant, there are many multiscale interactions between materials, devices, and systems, which can be complex. For example, for some oxidation reaction processes, the profiles of temperature, pressure, and concentration will affect the material properties, such as reaction activity and selectivity, usually in a microscale or mesoscale. The reaction activity and selectivity will also affect the heat release rate, and ultimately the temperature, concentration, and pressure conditions in a microscale or macroscale.

Many tasks in the industry, like fault diagnosis, scheduling of control strategies, and planning of operating procedures, require process models with different rates of dynamics. These process models should be consistent with the models used in the design of controllers. Control actions occur at different rates, since the sensors have a wide range of sampling time. The optimal fusing of control action and measurement collection relies on the availability of process models at different time scales (Stephanopoulous et al., 2000).

To develop the process control system, it is necessary to understand the different phenomena in a multiscale. A hierarchy of models, including the integration of the computational modeling and

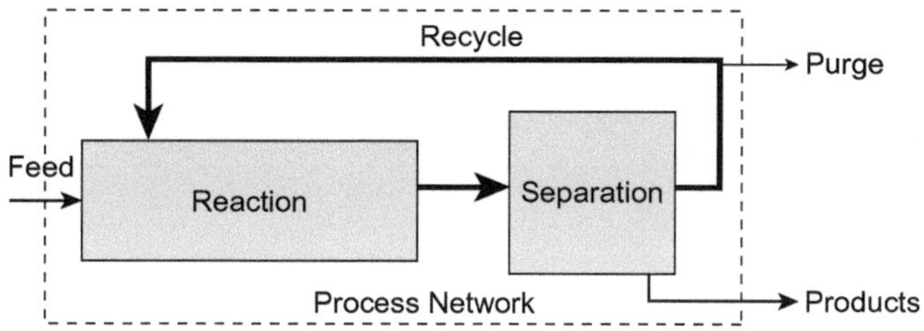


**FIGURE 8.26**

Generic reactor–separator process network with large recycle and purge (Baldea and Daoutidis, 2007).

simulation, optimization and control, and the representation of uncertainty and its propagation, is needed to describe the phenomena in a multiscale.

Process networks (Figure 8.26) consisting of reactions and separation sequences connected through material and energy recycle streams are very common in the modern process industry. Though it gives a significant economic benefit, an integrated process network poses a problem in control and analysis, as recycle streams affect feedback interactions among process units (Baldea and Daoutidis, 2007).

A process plant is a huge network of small processes joined together. This huge network has strong interactions and network dynamics, which initiate coupling between different control loops in the plant. This complex dynamics hinders the implementation of advanced control strategies like model-based control since the dynamic models of these networks become overwhelmingly huge and complex. This control loop coupling in plant-wide control is a problem of multiscale control, which has gained the notice of many leading researchers (Baldea and Daoutidis, 2007).

Most of the processes involved in the chemical process industry necessitate considering the fluid, heat, and mass transfer of the process along with the chemical reactions. They are characterized by phenomena that evolve at both macroscopic and microscopic length scales. Processes to be considered range from large-scale processes like reactive distillation in petroleum engineering to small and sophisticated processes like plasma-enhanced chemical vapor deposition in the semiconductor industry. Several proposed applications of multiscale modeling process control are available in the literature.

Researchers and engineers are thinking of many ways to mitigate the carbon dioxide emissions to the environment. One of the most innovative ways explored is carbon sequestration or carbon storage. Carbon storage in the deep ocean has a multiscale nature associated with it. Predicting its characteristics involves understanding the process that affects carbon storage, which has different time-scales. The reservoir itself has four length/time scales—the molecular, pore scale, sedimentary material/bulk phase scale, and reservoir scale. The reservoir scale model is again complicated by reaction/phase equilibrium, $CO_2$ gas hydrate formation, and multiphase flow through porous media. $CO_2$ levels in the ocean is again controlled by the global vertical circulation of ocean waters, which has a time constant of hundreds of years. Regional and local level ocean effects should also be considered (Lucia, 2010).

Balaji et al. (2010) developed a multiscale model to predict the growth of silicon particles in a fluidized bed reactor (FBR) due to chemical vapor deposition (CVD). This is used for making poly-silicon for solar cell applications. They have used three modules to describe the relationships between continuous and disperse phase and the CVD onto the silicon particles in the FBR model, which are computational fluid dynamics (CFD), reaction module, and population balance module. By coupling these models, they build a very comprehensive multiscale model for the particulate CVD (Balaji et al., 2010).

### 8.3.1 MODEL BUILDING

There are two important classes of approaches in multiscale model building. The first one is considered hierarchical, and the information synthesized at lower scales is passed on to the next level up (i.e., the higher-scale, bottom-up approach). The information flow is unidirectional. This information flow could also go from top to down. The "average" values are calculated at immediately adjacent scale and the necessary information is sequentially transferred to the upper or lower scale, respectively. A perfect example where this is used widely is the CVD, where unidirectional coupling is required (Raimondeau and Vlachos, 2002; Bi and Li, 2004). It can also be seen in the multiscale model for deactivation of immobilized glucose isomerase in packed bed reactors (Dadvar and Sahimi, 2003). However, in many complex systems, it is very difficult to model the system without the flow of information in both directions. There are strong and complex interactions between components at the same or different scales. So, the iterative or concurrent multiscale model, which couples models at various scales to allow for the interaction between scales iteratively or simultaneously, is thus required (Bi and Li, 2004). Most heterogeneous reactors fall under the second category. For example, to simulate the thermal deposition of silicon dioxide from tetraethoxysilane over patterned wafers, coupled and iterative multiscale models are developed (Merchant et al., 2000; Cale et al., 2002). Integrated multiscale process simulation has been discussed by Cale et al. (2002), which gives a very good example of the coupled solution of the latter model approach category. Compared to sequential multiscale models, multidirectional multiscale models require solving simultaneous differential equations across multiple scales, which require a high computing capability. Zhao et al. (2012) discuss about this problem in detail, where they have reviewed and bettered generic algorithms to model and solve specific frameworks of multiscale models (Figure 8.27).

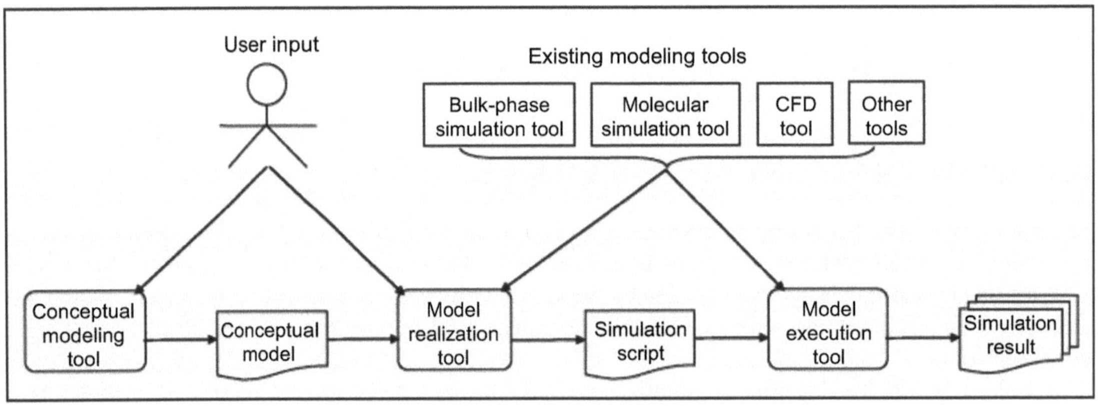


**FIGURE 8.27**

The workflow of simulating multiscale models (Zhao et al., 2012).

## 8.3.2 MODEL REDUCTION

Model reduction techniques in process control operation is one of the major issues. The computational power used in controlling even a small reactor operation is huge. So, we could imagine the trouble of handling the whole plant with detailed model description, even more if we add the multiscale phenomena to the complexity. In plant-wide control, apart from control of the whole plant, many control loops are local to certain units. The model created has to be solved many times until a solution to the plant-wide control structure is reached (Hannemann-Tamás et al., 2013).

Model reduction techniques are based on retaining the significant variable functions in the model. It should ensure that the resultant model is accurate yet computationally less expensive. The model reduction techniques can be broadly classified into two categories: (1) Conceptual approaches, which are based more on the physical interpretation of the system (Kang and Weinberg, 1992), and (2) Purely mathematical, numerical, and data-driven techniques in systems engineering. Table 8.6 provides a list of these approaches. Bogdan et al. suggest a systematic model reduction approach, where the physical (e.g., units, domains, phases) and the systemic (connectivity) levels of chemical process model are simplified first. Gradually, mathematical and numerical scheme reduction approaches are dealt with. This ensures that the basic structural feature of the model is preserved. Chin et al. discuss about local and global model reduction techniques, which are based on the reduced basis approach (Efendiev et al., 2012).

Bodgan et al. give a very good review about model reduction for plant-wide control. Model reduction has been categorized as (1) model simplification, which preserves the number of equations, but tries to

**Table 8.6 Model Reduction Options and Approaches in Chemical Engineering Applications (Efendiev et al., 2012)**

| Options | | | Approach |
|---|---|---|---|
| Level | | Attribute | |
| Physical | Structure | Resource | Ignore resource |
| | | Unit | Lumping |
| | | Compartment | |
| | | Domain | |
| | | Region | |
| | | Phase | |
| | | Coordinate | Steady state assumption: integrate/ average out; symmetry assumption |
| | Connectivity | Transfer and transport | Ignore terms; average |
| | System | Unit | Lumping/aggregation |
| | Behavior | Source and sinks | Lumping; timescale analysis; sensitivity analysis |
| Numerical | Mathematical | Differential equation | Order-reduction \| Linear; nonlinear |
| | | Algebraic equation | Linearization; approximation of functional expressions; simplification |
| | Numerical scheme | | Discretization; functional approximation |

reduce the complexity of the functional expressions in the model equations, and (2) model order reduction, which substitutes a large-scale model with fewer numbers of equations. Under model simplifications techniques, there are some important techniques to take note of. The model could be simplified by simplifying the physical properties model (i.e., simplifying the local thermodynamic models). Eliminating insignificant reactions and species from reaction networks also helps in model simplification (Dorneanu et al., 2009). Hannemann et al. discuss this in detail, where they have focused on the optimization-based kinetic model reduction, which leads to MINLP. Here, a number of reactions to be considered are reduced by making a trade-off between the model complexity and the modeling error (Hannemann-Tamás et al., 2013). A multiscale approach for model reduction of random microstructures has been described by Wen and Zabaras (2012).

### 8.3.3 KINETIC MONTE CARLO

Due to the stochastic nature of multiscale phenomena, the dimension of the PDEs developed become very large as the systems become meaningful. It is almost impossible to directly solve using single numerical methods for integration. Kinetic Monte Carlo methods provide a numerical solution to the PDEs (Kang and Weinberg, 1992). Monte Carlo simulation has been used extensively in engineering for simulating random phenomena. The kinetic Monte Carlo combines the power of Monte Carlo with the probabilities of the events. The kinetic Monte Carlo method is extremely fast and accurate (Gupta, 2005).

Kinetic Monte Carlo has been applied to numerous multiscale phenomena. Chou and Falk (2006) use accelerated kinetic Monte Carlo simulation for molecule beam epitaxial growth. It has also been used to study the molecular features of copper electrodeposition and has been coupled with a continuous method to form a multiscale approach (Figure 8.28).

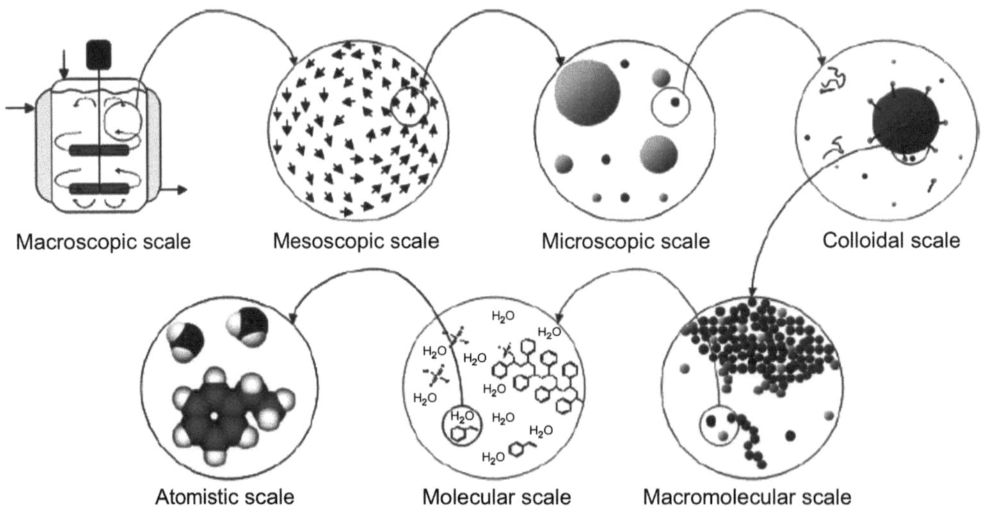


**FIGURE 8.28**

General multiscale structure of emulsion polymerization (Hernandez and Tauer, 2008).

David et al. have studied the multiscale simulation of the coarsening of the Ca-rich precipitate in Fe combining kinetic Monte Carlo, Molecular Dynamics (MD), and phase-field simulations. Iron alloyed with copper finds application in many areas like pipe materials in power plants (Molnar et al., 2012). Hugo et al. use kinetic Monte Carlo for simulating emulsion polymerization. They represent scale in emulsion polymerization with seven different characteristics scale, with different simulation techniques at each level. The kinetic Monte Carlo has been used at the molecular level, where the properties from quantum mechanics simulation like Lennard-Jones potential parameters can be used for determining transport properties like viscosity, diffusivity, and thermal conductivity (Hernandez and Tauer, 2008).

### 8.3.4 MODEL PREDICTIVE CONTROL

The first articles on model predictive control came from the industry in the 1970s (García et al., 1989). In 1978, Richalet et al. described applications of "model heuristics control," later known as model algorithmic control (MAC), and in 1979 Shell engineers proposed dynamic matrix control (DMC) and applied it to fluid catalytic cracker (Cutler and Ramaker, 1979; Prett and Gillette, 1979). The common theme of these articles was to explicitly use the dynamic model of the process and to predict the effect of the future control actions on the output and then take the best possible control actions (thus called "model predictive control"). The future control actions are determined by optimization with the objective of minimizing the predicted error. The optimization is repeated at each sampling time. Generally, a suitable future horizon is taken to make the prediction fast and performance of the controller better. After the initial papers, academicians saw this as a very important development in the control in process industries and within a decade many papers were published and are still published. Model predictive control is now well established in both academia and industry, but still is very new. But for controlling complex processes and having very high performance, model predictive controllers are generally used in multiscale process control applications.

### 8.3.5 CONTROL STRATEGIES

Multiscale process control's general steps includes first building the multiscale process model as discussed earlier. The models we prepare are generally very complex for a multiscale process. We generally use model-based controllers for multiscale processes, so the controller needs to predict the current and the future values to control the process. If the model is too complex, the controller will take time to get the prediction and take a control action, which we definitely do not want to occur. To make it work for control, we need to reduce the model size. We use various model reduction techniques as mentioned earlier to make it simple enough for the controller. Generally, kinetic Monte Carlo models are used for multiscale process control, but for some processes like control of thin film microstructures, where closed-form model are needed, we also form process models in the form of linear or nonlinear stochastic PDEs. Simple controllers like PI or PID does not show good performance for multiscale processes, since many variables are required to be controlled. Model-based controllers perform better since so much effort is given into modeling these complex processes. Model predictive control has been widely used in multiscale process control. It uses the prediction of the values from the model as well as the real output values from the process in real time to make a control step.

Multiscale process control is still a very new topic in the academic world. Many researchers are enthusiastic about this area. Practical applications in this area are rare, since it takes a lot of effort and resource to build these models and also to implement them in a controller.

## 8.4 MULTISCALE MODELING APPROACH IN MATERIAL DESIGN

Material design is critical for process industries and has extensive impact on process safety issues. For a hazardous material (e.g., a reactive chemical), the processing equipment and plants used to scale up the production of it may pose a significantly higher level of hazards than a common chemical. Even for nonhazardous materials, if it is used to deal with hazardous materials (e.g., a toxic chemical), a failure of this nonhazardous material may also lead to a loss of containment of hazardous materials if it is not well designed or maintained. In this section, a multiscale modeling strategy is discussed with regard to the prediction of corrosion and physical properties.

### 8.4.1 MULTISCALE MODELING IN CORROSION PREDICTION

Corrosion is a destructive damage factor of metal vessels and pipelines. The cause can be chemical corrosion or mechanical erosion. Effective and timely prediction of erosion or corrosion can provide good protection of vessels/pipelines. Erosion/corrosion prediction is especially relevant because of the difficulties in detection. Some types of corrosion, like pitting and crevice corrosion, occur below the pipe surface, making them very difficult to detect. Also, depending on the flow conditions, erosion/corrosion can occur very quickly. For these reasons as well as others, a reliable prediction method could help the industry prevent erosion/corrosion accidents.

Corrosion is the gradual destruction of a material due to chemical reaction. Erosion, on the other hand, is the mechanical process by which material is removed. Corrosion is one of the most important mechanical integrity issues the petrochemical industry has to deal with. While significant research has been dedicated to studying corrosion, it is still the leading cause of pipeline failure in the oil and gas industry. Not only is it the main contributor to maintenance costs, but also it accounts for about 15–20% of releases from the petrochemical industry and 80% of pipeline leaks. Enormous costs are directed toward fixing corrosion-related problems in facilities across the globe every year. Corrosion has caused some of the worst incidents in the history of the industry and is still causing more incidents every year. This shows that the problem is still not clearly understood and that the methods that are being used to control it are not sufficient.

Because the mechanisms for corrosion and erosion are different, different approaches must be used to predict their effects. Erosion is a mechanical process related to the collision of particles to remove material and is therefore highly dependent on the fluid dynamics of the system. In order to predict erosion, the fluid flow must be well understood. CFD can be used to generate flow maps, which are used to study erosion. Corrosion is a chemical process by which the pipe material is degraded into a soluble material, which can then be removed by the flowing liquid. This process is not a macroscale flow problem but a microscale reaction and transport problem. For this reason, most corrosion models pay little attention to flow patterns and instead focus on pH, $pCO_2$, temperature, liquid velocity, total pressure, as well as other important factors in reaction kinetics (Nyborg, 2002).

A second difference in erosion and corrosion is the effect on the pipe. Because erosion removes material due to contact, it only removes material that the fluid is colliding into. This means that exposed surfaces of pipe are slowly worn away from the inner pipe wall to the outer pipe wall, in a relatively uniform manner. Corrosion is reaction based, so it can occur anywhere where the reactants are present and the conditions are right. Uniform corrosion is common; however, other forms of corrosion also exist, such as pitting corrosion or crevice corrosion. Pitting corrosion occurs when corrosion begins in a very small hole, and corrosion continues moving deeper through the hole. Generally pitting corrosion follows the breakdown of a passive protective film at a localized point, but it is possible in other scenarios as well (Pitting, 2003). Pitting corrosion is especially dangerous because the material can look intact to an observer, even though a large amount of the material under the surface has corroded away. Crevice corrosion is very similar; however, it occurs on a secluded site, as opposed to on an open surface.

A third difference is in the types of fluids that cause erosion and corrosion. Erosion is mostly caused in multiphase flow scenarios, either solid liquid slurry or a liquid suspended in a gas. Erosion can occur in single-phase flow, but this effect is normally limited cavitation in the fluid. Corrosion can occur with any type of fluid as long as the reactants are present for the reaction to occur (Figure 8.29).

#### *8.4.1.1 Corrosion models*

There are many different models to predict pipeline corrosion (Nyborg, 2002). Some models are based on empirical correlations, while others are based on mechanistic models, though these are generally validated with laboratory data (Nyborg, 2002; Marsh and Teh, 2007). Generally, most of these models predict very similar results with the main differences coming from scenarios with oil wetting or the formation of protective coatings. Some models predict very strong effects of oil wetting, while others do not consider the phenomenon at all. Some models predict strong effects of iron carbonate protective layers, and others do not account for the protective layer due to the likelihood of a localized corrosion attack. New models are frequently created to handle slightly different scenarios or to put an emphasis on different corrosion phenomena. Following are some of the more prominent models.

The first model to predict the corrosion is the de Waard model (Nyborg, 2002). This model has been used in industry since it was first created in 1975. Since that time it has seen many revisions and corrections up to the present day. The original model was based only on the temperature and partial pressure of $CO_2$. Later, the effects of pH and corrosion product scale were added, followed by the effects of fluid velocity, mass transport, and steel composition.

Another commonly used model is the Cassandra model. This model was developed by BP and is an adaptation of the de Waard model (Marsh and Teh, 2007). This model is openly available, unlike many other models. The main differences between this model and de Waard model are the way they handle protective films at high temperatures, and the way they handle oil wetting effects. The Cassandra model attributes less effect to the formation of protective films at high temperatures, and does not consider oil wetting effects.

Another model that pays more attention to flow regimen is the Lipucor corrosion prediction program (Nyborg, 2002). This model is based on both laboratory data and field data. One version of this model calculates the flow regimen at a point, and another that is coupled with an external fluid flow model. Because this model relies heavily on field data, over 90 case histories, it is less conservative than laboratory models where oil wetting is not considered.

#### *8.4.1.2 Tribo-corrosion*

Tribo-corrosion is the combined effects of chemical degradation and mechanical wear. Until recently, the effects of erosion and corrosion were handled together but without a parameter related to fluid flow (Stack and Stoti, 1993). This approach limited the information about erosion. Recently, Stack et al. (2010) combined the effects of erosion and corrosion and incorporated them in a simulated flow environment using CFD. The result allows for a better understanding of erosion/corrosion on real pipes. In the recent article by Stack, many of the CFD readouts have been provided, breaking up the

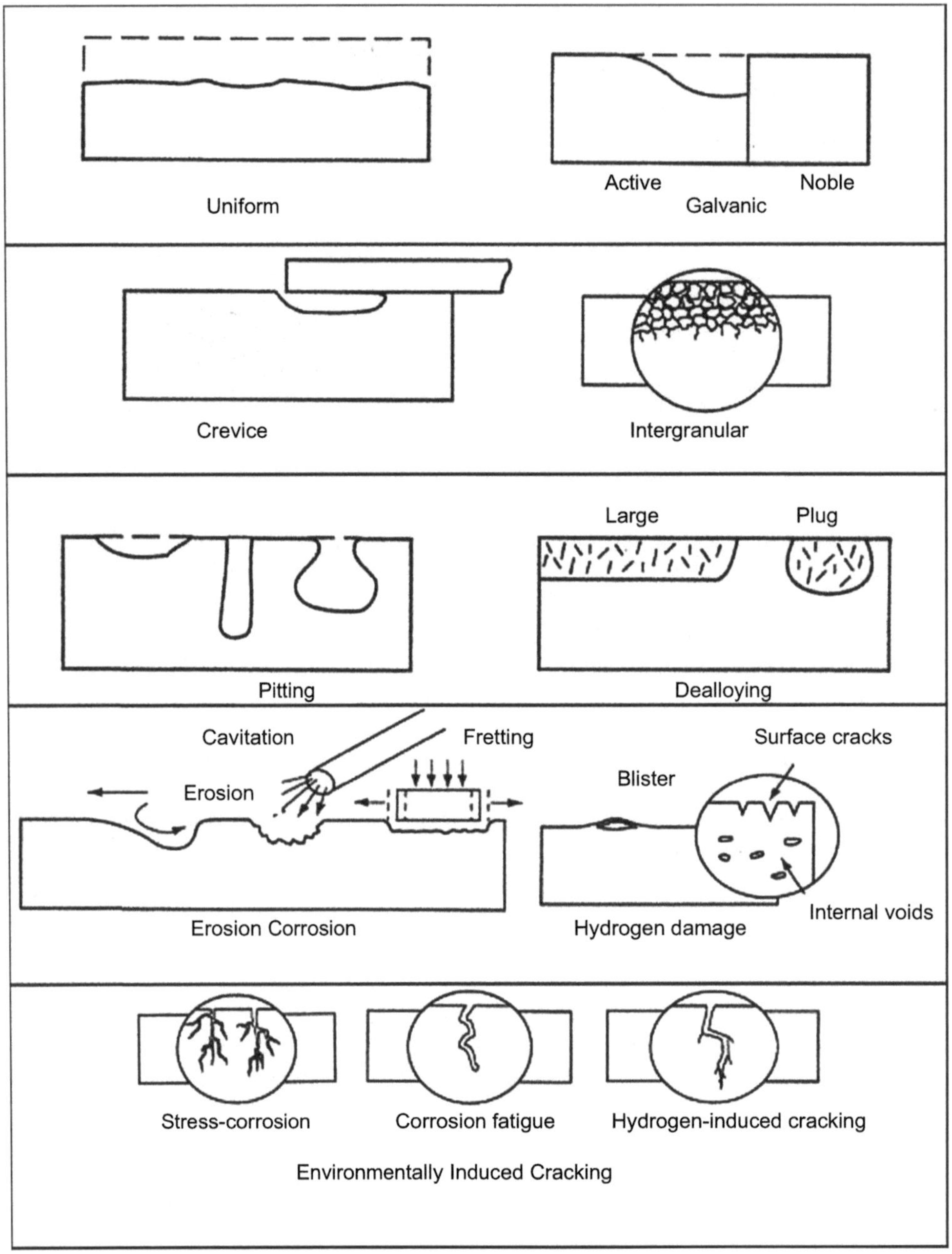


**FIGURE 8.29**

Types of corrosion.

different zones in the pipe into corrosion-dominated, erosion-dominated, combination of corrosion and erosion, and erosion free.

Early tribo-corrosion studies looked at trends in erosion/corrosion with respect to different parameters. Some of these flow parameters are velocity, pH, oxygen content, temperature, pipe geometry, and pipe material (Nicola, 2012).

Flow velocity is an important factor because it is closely related to flow turbulence, which is a determining factor in erosion. As the velocity increases, the turbulence of flow increases, causing the influence of erosion to increase. At low flow velocities, corrosion is the dominant process, but as the velocity increases, a combination of corrosion and erosion increases the rate of damage. At a certain point, the reverse of this effect takes place, and the rate of erosion/corrosion begins to diminish. At lower near-wall flow velocity, the metal concentration difference is higher, increasing the metal loss rate. However, at high turbulence, the metal concentration is fairly uniform decreasing the metal concentration difference, and decreasing the metal loss rate (Ferng and Lin, 2010).

Two important factors that affect the chemical condition of the flow are pH and oxygen content. Chemical condition is a term for the general status of the flow in relation to providing an environment for chemical reaction. These parameters mainly affect the corrosion aspect of erosion/corrosion. Above a pH of 8–9, the rate of corrosion starts gradually decreasing until it reaches a level around 9.2–9.3, where the drop off is sudden. As the amount of dissolved oxygen in the flow increases, the erosion/corrosion rate decreases. This effect is especially pronounced below 150 ppb, as shown in Figure 8.30.

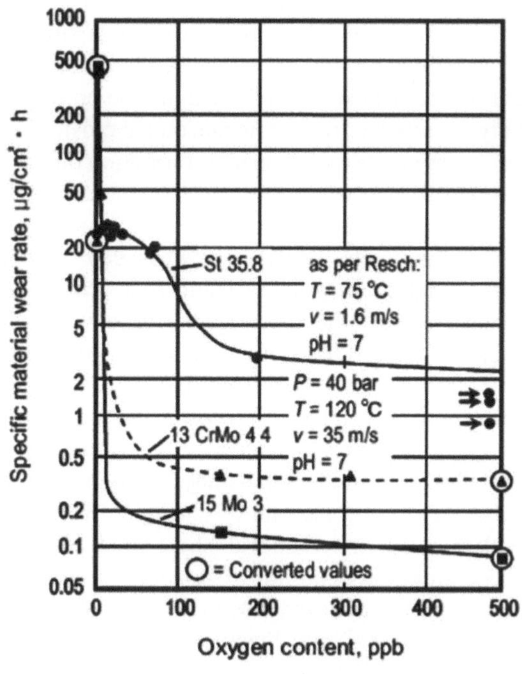


**FIGURE 8.30**

Relationship between erosion/corrosion rate and dissolved oxygen concentration for different pipe materials (ClassNK, 2008).

Temperature is another important factor in determining the erosion/corrosion rate. Temperature can have a positive or negative effect depending on the range of temperatures considered. The general shape of the temperature dependence of erosion/corrosion rate is convex with a maximum around 135 °C for single-phase flow and 180 °C for two-phase flow. These values are just guidelines, as the pipe material also plays a role in temperature effects.

The rate of erosion/corrosion is also influenced by pipe geometry and pipe material. Although these are both pipe characteristics they are related to different phenomena. Pipe geometry is important because it affects the flow pattern of the fluid, which is important in erosion rate. Turbulence can be generated near pipe fittings, behind orifices or valves, and at tees, bends, or elbows. The higher turbulence makes these areas hot spots for erosion/corrosion, much like the effects of fluid velocity. Pipe material is important on the chemistry side. Different materials form different types of protective films. The erosion/corrosion process is focused on removing this film, so the type of film and the resistance it provides are integral in determining the rate. Stainless steel, for example, forms a very strong passive film that is ideal for withstanding oxidation, making it one of the most corrosion-resistant materials. In many scenarios, different components can be added to pipe material to increase its corrosion resistance, such as copper, molybdenum, and chromium. Sometimes a protective layer can do a very good job against uniform corrosion but will develop a small hole, which allows corrosion to take place below the protective layer. This type of corrosion is called pitting corrosion and can be especially dangerous.

#### 8.4.1.3 CFD modeling of erosion/corrosion

Although trends are important in understanding the fundamentals of erosion/corrosion, in order to really move forward the complex fluid flow patterns must be understood.

Density functional theory (DFT), at a microscale, can be also used to study the corrosion process. DFT can be used to study the mechanical stress and the dynamic process of corrosion. Besides, CFD simulation can be used to study the effect of fluid flow rate on the erosion, in a macroscale, especially if there are some solid particles. It can provide some brief information on the erosion at an initial stage of process design, which can identify some bad design and prevent unexpected accidents.

The CFD methodology is used to predict fluid flow behaviors by solving partial differential equations with numerical analysis methods. This methodology can be applied to flows that are either very difficult or very tedious to solve analytically using first principles. The CFD solution is an imperfect solution that is based around the convergence of many nodes by numerical solution; however, depending on the circumstances, the solution is a very good approximation.

Nicola (2012) conducted research focusing on studying the behavior of erosion/corrosion using CFD. This would allow for determining the erosion/corrosion rate and when it would take place before it starts happening. Here, the operating conditions that led to erosion/corrosion (from the literature) are used in FLUENT, the specific CFD tool used in the study, to predict the flow hydrodynamic factors. The relationship between these factors and the rate of erosion/corrosion is studied. The results from this work show that along with the turbulence and wall shear stress, the dynamic pressure imposed by the flow on the walls also has a great effect on the erosion/corrosion rate.

Some other researchers also have done some work on the prediction of erosion by CFD simulation, such as the groups of Graham et al. (2009) and Stack et al. (2010). CFD is an important tool in the understanding and prediction of erosion. Previously, researchers used geometry scanning equipment to find erosion scars, which they used to develop erosion models (Graham et al., 2009). CFD is a useful tool, in combination with erosion models, to quantitatively predict erosion distribution. Although CFD modeling is very useful, experimental erosion data are still important for erosion prediction (Graham et al., 2009). Although much data exist, very little of it is in the form of quantitative erosion distribution measurements, which are valuable in comparison with CFD predictions.

Ferng and Lin (2010) applied CFD to a pipe elbow in order to predict wall thinning distribution. The predictions were compared to plant measurements for validation. The results are shown below in Figures 8.31 and 8.32. These figures show clearly what the outputs from CFD modeling look like and how they can be visually useful. The red areas on the pipe elbow depict the areas of most severe wall thinning, while other colors depict reduced levels of erosion/corrosion thinning. This information is very useful to the industry in that it can be used to determine the sections of pipe that are most likely to fail and inspections can be focused on these areas.

It is also evident from the diagrams where the model does a reasonable job predicting wall thinning and where it is a poor predictor. On the positive side, the model does a good job predicting the areas of

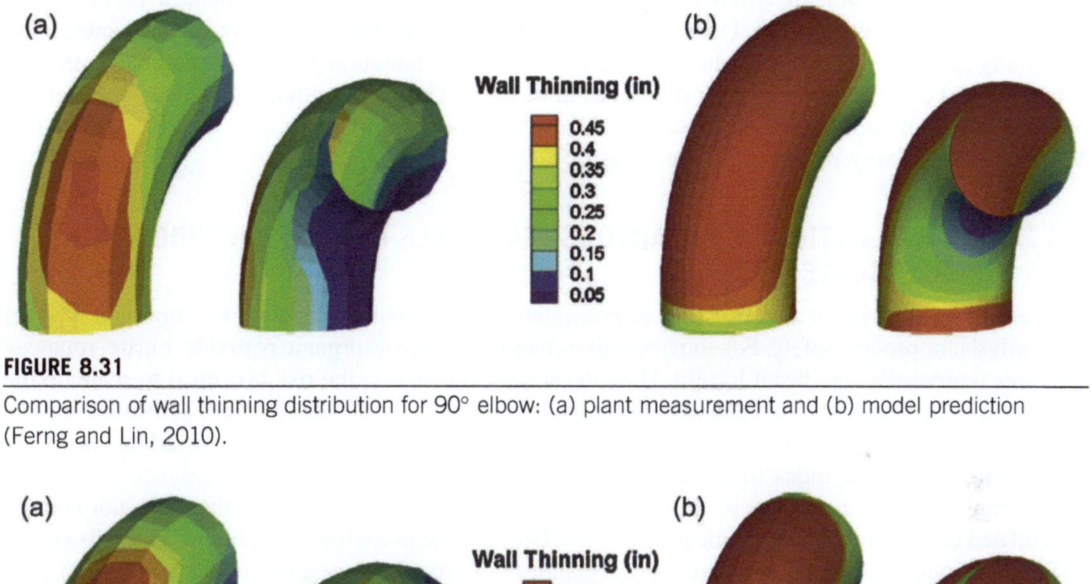


**FIGURE 8.31**

Comparison of wall thinning distribution for 90° elbow: (a) plant measurement and (b) model prediction (Ferng and Lin, 2010).

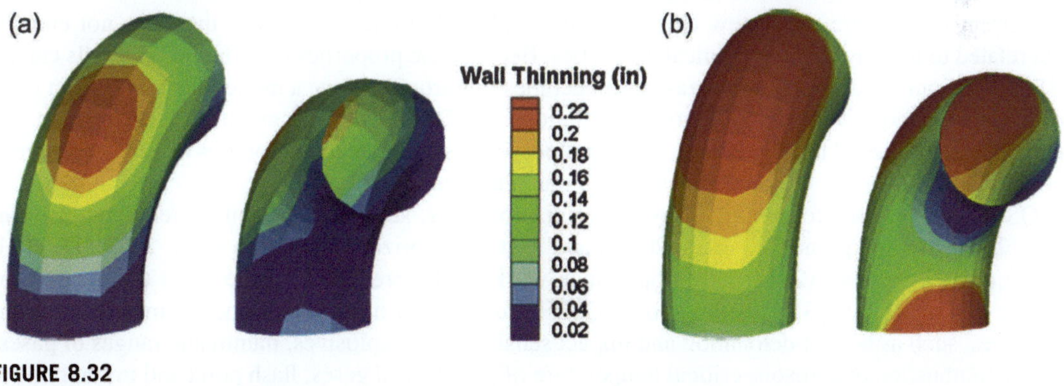


**FIGURE 8.32**

Comparison of wall thinning distribution for reducer and 90° elbow: (a) plant measurement and (b) model prediction (Ferng and Lin, 2010).

severe wall thinning. In both examples, the model correctly predicts where the most severe thinning will occur and which areas should be focused for inspection or targeted for replacement most often. Although the model seems to predict that areas higher on the elbow will be the focus of erosion/corrosion than the measurements show, these two areas are relatively close in position. The model also shows the contours of decreased wall thinning and does a good job providing a visual basis for understanding some of the dangers of erosion/corrosion.

On the negative side, the model overpredicts wall thinning throughout the elbow. While this is an issue in that it would be preferable to have a model that perfectly predicts the phenomena, it is acceptable because overprediction errs on the side of safety, which is a major goal in erosion/corrosion prediction. The other area where the model does a poor job is at the bottom of the elbow on the inside. Here, the model predicts that there will be extensive wall thinning, while the measurements show that the inside of the elbow is practically uninfluenced by erosion/corrosion. An issue such as this is more serious because it can misguide individuals in the industry such as inspectors or plant managers concerning sections of the pipe likely to experience wall thinning and can misdirect resources.

Erosion and corrosion are two phenomena that are responsible for many accidents across a variety of industries and consequences. Understanding the principles of erosion and corrosion is integral in preventing these accidents in the future. While there are many models and trends to predict the damage of erosion and corrosion, much work still needs to be done. The CFD methodology is an area that can be used for great success in the prediction of erosion/corrosion damage and the location of that damage, which is very important to the industry.

### 8.4.2 RISK EVALUATION OF HAZARDOUS MATERIALS BY QSPR METHODOLOGY IN MULTISCALE

Evaluation of the physical and/or chemical properties of hazardous materials is an important topic in risk analysis for process safety. For some reactive chemicals, such as organic peroxide, nitrite, runaway reactions represent a significant hazard. How to estimate and assess the risk is critical to scale up the process. It is also necessary to calculate the heat of the main and the side reaction, activation energy, reaction kinetics, and physical properties. For the scope of storage and transportation, the onset temperature is a simple index to judge the level of hazards of the peroxide.

Currently, there are many new chemicals invented or developed; however, there are not enough data related to the physical or chemical properties. Because the properties of different materials can be difficult, tedious, or dangerous to measure, it becomes important to have a model that can predict their properties. Quantitative structure–property relationship (QSPR) models are models that relate the properties of a material to its structure, making it possible to estimate the properties of a material under certain conditions using only a database and a mathematical model.

QSPR can be used to predict many different properties, such as autoignition temperature (Pan et al., 2008), boiling point (Dai et al., 2013), and heat of vaporization (Marino et al., 2003). Lu et al. (2010) and Ridha et al. (2012) have done much work on the prediction of the onset temperature of organic peroxides by QSPR methodology. QSPR studies also have been done on many important properties, such as heat of detonation and impact sensitivity of explosives, flammable ranges of gases, heat of combustion of aerosols, critical temperature of pressurized gases, flash point and initial boiling point of flammable liquids, and autoignition temperature of self-heating chemicals, as well as many others (Quintero et al., 2012).

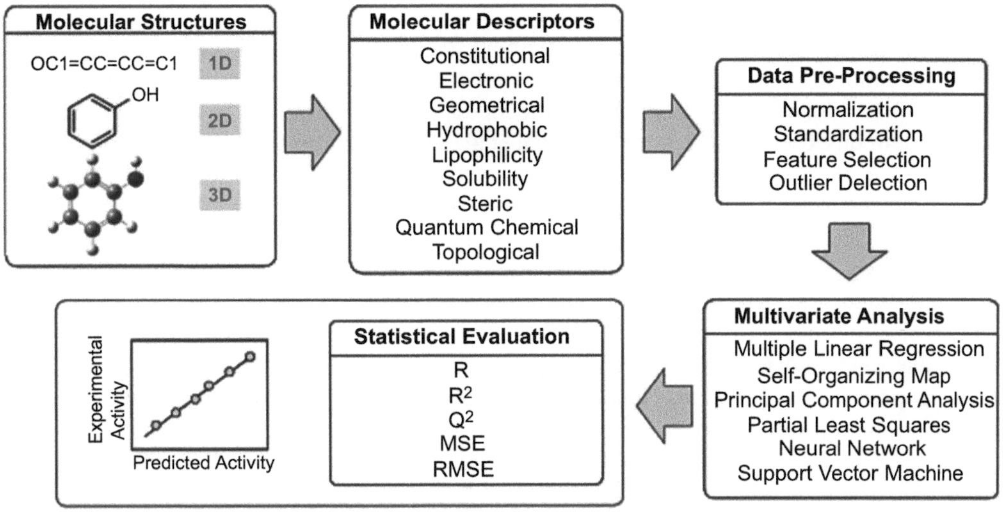


**FIGURE 8.33**

Schematic overview of the QSPR process.

Generally, there are two main types of QSPR models, classification and regression. Classification models relate a set of molecular descriptors to a categorical value of the response variable, whereas regression models relate a set of molecular descriptors directly to the response variable. Figure 8.33 shows a schematic overview of the QSPR process and the methods used for each step of a whole QSPR process. It includes five steps: building up molecular structure information, choosing proper molecular descriptors, data preprocessing, multivariate analysis, and statistical evaluation. First the molecular structure of the compound must be determined. The molecular structure is then input into a program (e.g., CODESSA) that is able to comprehend the structure of the compound and calculate molecular descriptors such as hydrogen bonding surface area, cubic root of the gravitational index, relative negative surface charge, and others. Once molecular descriptors have been calculated, they can be manipulated to form the QSPR model. Although hundreds of descriptors can be calculated, many of them are poorly correlated to the behavior being studied and discarded. The data preprocessing step is responsible for narrowing the list of descriptors to a manageable amount. Also, for model simplicity, models typically have fewer highly correlated descriptors as opposed to many loosely correlated descriptors. For example, Katritzky et al. (2001) generated a fit for flash points in their article. His best fit used only a three-parameter system. Although it may have been possible to get a better fit, it is important to keep in mind that QSPR is a guidance tool that is only meant to estimate parameters, so small errors are acceptable. Next, multivariate analysis is used to determine the correlation. There are many different multivariate analysis approaches, but their overall function is to find the dependence of the property of interest on the independent variables, the descriptors. Once the correlation has been generated, it is compared to the data and evaluated. The most well-known evaluation parameter is the $R^2$ value, but there are other important parameters as well.

#### 8.4.2.1 Molecular descriptors

Molecular descriptors are the key information about the physicochemical properties of a molecule, such as hydrophobic, electronic, geometrical, and topological descriptors (Nantasenamat et al., 2009). One of the advantages of QSPR is that the model is based solely on the chemical structure and is not dependent on experimental properties (Pan et al., 2008). On the other hand, one of the issues of QSPR is that the results are heavily reliant on the selection of molecular descriptors. Because of this issue, feature selection is a crucial preprocessing step. A wide variety of molecular descriptors can be used in QSPR models; therefore, the right features must be selected to determine the structure–property relationship. This can be done by qualitative selection by a human or by using artificial intelligence to determine important features through the use of artificial neural networks (Bertinetto et al., 2009).

Molecular descriptors are descriptors that contain important information about the physicochemical properties of a molecule (Nantasenamat et al., 2009). They are numbers that are calculated from the information of a molecule which can in turn be used to represent the molecule quantitatively. There are many types of descriptors, some of which are hydrophobic, constitutional, quantum chemical, electronic, geometrical, solvational, solubility, topological, and steric descriptors (Pan et al., 2008; Nantasenamat et al., 2009). By gathering information about the molecule, the interactions of the molecule with other molecules can be better understood allowing the properties of the molecule on the macroscale to be calculated. The molecular descriptor values are the independent variables that are used to calculate the output for QSPR models. When calculating the macroscale properties of a material, not all descriptors are equally important. Certain descriptors may be more important, meaning that they will receive a different weight in the final property calculation.

#### 8.4.2.2 Data preprocessing

Data preprocessing is an important step to prepare the data to form a QSPR model. There are many important steps in data preprocessing, such as data cleaning, data transformation, and feature selection (Nantasenamat et al., 2009). Data cleaning and transformation are methods used to remove outliers and standardize the data so that they take a form that can be easily used to create a model. This section focuses mostly on the applications of QSPR models and will emphasize the feature selection portion of data preprocessing. When trying to create a QSPR model, a data set may contain hundreds of variables (descriptors); however, many of these variables will contain redundant data. In order to simplify the dimensionality of the model, it is important to select only variables that contain unique and important information. Data mining procedures can be used to remove variables that do not contribute to the model. Another step in feature selection is feature fusion, the process by which multiple variables are combined using mathematical operations. This process is performed to reduce the number of variables in the model while still maintaining the level of information.

#### 8.4.2.3 Multivariate analysis

When the data mining operations have been completed and a set of molecular descriptors have been selected, multivariate analysis must be used to find the appropriate correlations between the descriptors and the properties. There are many approaches to multivariate analysis, such as multiple linear regression, partial least squares, neural network, and support vector machine, as well as others (Nantasenamat et al., 2009). Depending on the situation, different approaches may be more applicable.

Multiple linear regression, for instance, is a less complex method that can be used effectively for nearly linear correlations. In a multiple linear regression, the data are fit to a linear approximation of the form

$$y = a_0 + a_1x_1 + a_2x_2 + \cdots + a_nx_n$$

In this equation, $y$ is the property of the material, the coefficients ($a_1 \ldots a_n$) are the regression coefficients based on the best fit, and the independent variables ($x_1 \ldots x_n$) are the descriptors that have been selected. The $y$ intercept in this equation is $a_0$, which describes the baseline value of the function. In some applications the intercept has meaning, but in others it is simply a value. Generally, least squares analysis is used to fit the linear approximation to the data; however, there are other approaches.

Another approach for more complex systems is the use of artificial neural networks (ANNs). ANNs are set up similarly to the neuron layout of the human brain. In an ANN, the selected molecular descriptors are each given a node in the input layer. The inputs are connected to the second layer, called the hidden layer. The connections are what provide the different weights for the input variables. The hidden layer is responsible for pattern recognition processing. The algorithm then back calculates the error and makes small adjustments to the connection weights. It iterates this process by slowly adjusting weights so that it approaches the given acceptable error threshold. Once the threshold is reached, meaning a solution has been found, the algorithm stops making adjustments and passes the information onto the final, output, layer.

#### 8.4.2.4 Statistical evaluation

Once a model has been constructed, its performance must be evaluated to determine whether it is reliable. This evaluation is generally performed by leaving out some information when developing the model, then testing the model on the omitted information. The original information is split into a training set and a test set (Nantasenamat et al., 2009). The model is built by using the training set for calibration. For example, when developing a QSPR model using multiple linear regression, some of the data are used to calculate the regression coefficients to minimize error. Once the coefficients have been found, the model is then tested against the testing set of data, which was not present in the construction of the model. In dividing the data into training and testing set, commonly the $N$-fold cross-validation approach is used. In this approach, the data are divided into $N$-folds, where $N-1$ folds are used for training and onefold is used for testing. For example, if the data are presented as 20 data points, 19 data points would be used for training and the remaining data point would be used for testing. This approach provides the ability to evaluate the model on two different levels. The first is against the known data in the training set. If the multivariate analysis is done correctly and the analysis approach is good, then the model error will be within the acceptable range. The second evaluation is against the unknown data in the testing set. This information was not used in the development of the model, which makes a more realistic test of the models predictive ability.

#### 8.4.2.5 A case study of QSPR applied to the flash point

Katritzky et al. (2001, 2007) developed a method for the prediction of flash point using various multiparameter systems. In their initial work and updated work, they detail the procedure used, which can give significant insight into the development of a QSPR model. Their work on flash points is especially relevant in the field of hazardous materials, as flash points are used to determine if a liquid is flammable or combustible (NFPA30, 2012). Flash point is defined as the lowest temperature at which a material can vaporize to form an ignitable mixture with air.

In an ideal situation, experimental flash points would be known for all materials in all possible scenarios. This is not feasible; however, due to the rapid discovery of new compounds as well as the risks posed by experiments with hazardous compounds. For these reasons, it is important to have a reliable model for the prediction of flash points. There are upwards of 20 million known substances, and flash points are recorded for only a few thousands. Many approaches have been used to develop a method of determining flash point. Early methods used group contribution, which was relatively successful. More recently, methods have been developed that are based on functions of the experimental boiling point (Satyanarayana and Rao, 1992). These methods are often complicated but can give very accurate results (Keshavarz et al., 2011).

In order to begin the development of a QSPR model for flash point prediction, a data set of compounds with known flash points is required; 271 compounds were used with flash points taken from the MDL ISIS chemical database. Their structures were then analyzed using the MOPAC quantum chemical package, and molecular descriptors were calculated using CODESSA. Then, correlations were generated and analyzed, also using CODESSA.

Descriptor selection was performed using the heuristic method. This method was chosen due to the decreased time required to select descriptors. In this method, descriptors are first eliminated by looking at their variation in magnitude throughout the data set. If the value of a descriptor is fairly constant, it is eliminated. There are other statistical tests performed to eliminate descriptors as well such as an F-test and a *t*-test. The remaining descriptors are ordered by decreasing correlation coefficient. Next, they are paired with all other descriptors and selected based on the F-value in two-parameter correlations. This process is repeated for the inclusion of further descriptors. In Katritzky's work, six different correlations were generated using a variety of descriptors. Two of the more interesting correlations are presented. The first is the correlation using an experimental boiling point as a descriptor. This correlation is a three-parameter correlation with the experimental boiling point, relative negative surface charge (RCNG), and H-donors charged surface area (HDCA) as the parameters. The equation generated, as well as the plot generated from the equation compared with the actual flash point data, is shown next (Figure 8.34).

$$T_{FP} = (0.72 \pm 0.014)T_{BP} + (78.99 \pm 11.97)RCNG + (2.05 \pm 0.33)HDCA - (8.40 \pm 7.63)$$

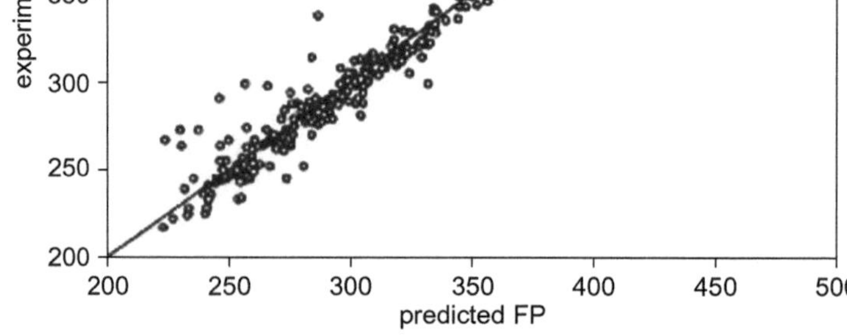


**FIGURE 8.34**

Plot of observed $T_{FP}$ versus predicted $T_{FP}$ by a three-parameter correlation using the experimental boiling point.

This equation fits the data well with an $R^2$ value of 0.9529 and is the best fit of all correlations in the article. However, by using experimental data in the prediction, it is not purely theoretical, which is the goal of QSPR analysis. Another correlation is generated that uses the predicted boiling point in the place of the experimental boiling point. This correlation is a three-parameter correlation with the predicted boiling point, hydrogen acceptor surface area (HASA and RNSG as the parameters). The equation generated, as well as the plot generated from the equation compared with actual flash point data, is shown next (Figure 8.35).

$$T_{FP} = (0.75 \pm 0.018)T_{BP} + (103.65 \pm 14.98)RCNG + (0.23 \pm 0.06)HASA - (28.11 \pm 9.77)$$

This equation also fits the data well, with an $R^2$ value of 0.9247. Although the fit is not as good as the first correlation, all of the descriptors are theoretical, meaning that flash point can be predicted for compounds that have no experimental data.

Overall, Katritzky find some very useful information and generates a correlation that fits the data fairly well. As they describe in their conclusion, many of the outliers are small molecules with low flash points that are hard to determine accurately. Additionally, their method correlates flash points within 11K, which is similar to the error range in flash point measurements. Although the QSPR model is not ideal for all situations, for compounds that do not have experimental flash point data, QSPR is a valuable tool that can give reasonable estimates for flash point values.

Katritzky et al. (2007) also write an update to their initial article on flash point prediction. This article revisits the original but expands the data set to 758 compounds. Most of the process is the same, but an important change is made to the multivariate analysis; the second article uses ANNs in addition to multilinear regression to compare the results of the two approaches. The multilinear approach found an average error of 13.9K, whereas the ANN approach found an average error of 12.6K. Both models used a four-parameter correlation where similar descriptors were selected; the main difference was the fitting model. Because ANN has a nonlinear fit, it is able to better represent the data, even when the same independent variables are used. In either case, the correlations generated by QSPR analysis represent the data closely enough to give good approximations. Additionally, even the ANN approach, which is slower than the heuristic method for descriptor selection, takes place in a realistic amount of time, making it a good alternative to multilinear approaches.

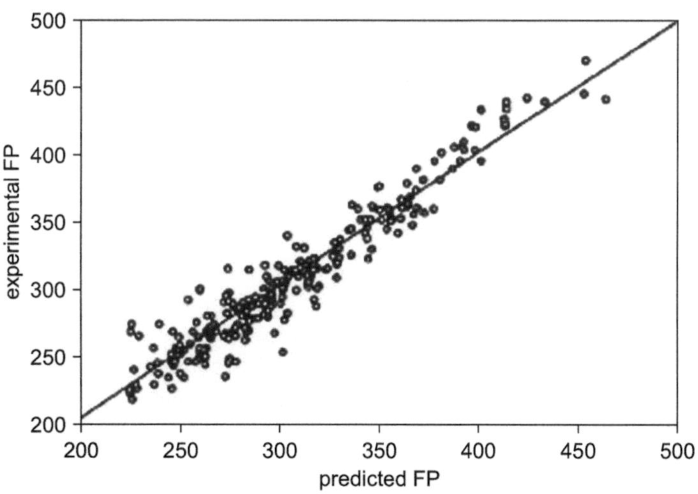


**FIGURE 8.35**

Plot of observed $T_{FP}$ versus predicted $T_{FP}$ by a three-parameter correlation using the predicted boiling point.

## 8.5 MESOSCALE REACTOR DESIGN

During last several decades, mesoscale reactors have been developed for the fabrication of a variety of materials. The size of mesoscale reactors is around a tenth of a millimeter. This kind of reactors is often called microreactors, although this might underestimate their size. In this section, mesoscale reactors are going to be treated as microreactors, given their similar behaviors regarding transport processes and fluid flow. Microreactors show different behaviors in terms of fluid flow, diffusion, and heat transfer compared to conventional reactors.

The flows in mesoscale reactors are generally laminar flow except for a few kinds of reactors. The fluid flow in small-diameter channels is mainly laminar because Reynolds number is a function of hydraulic diameter. For an isothermal wall under laminar flow, the heat transfer coefficient for forced convection flow inside a circular tube can be described as given in,

$$Nu = 3.657 + \frac{0.0018\left(RePr\frac{D}{L}\right)^{1/3}}{\left(0.04 + \left(RePr\frac{D}{L}\right)^{-2/3}\right)^2}. \tag{8.1}$$

As the diameter of the reactor/channel is very small, the Nusselt number becomes a constant of 3.657 for internal laminar flows when the wall temperature is constant.

The heat transfer to the channel walls is higher than in conventional reactors. This characteristic leads to a potential advantage – the reactors will remain isothermal (or nearly isothermal) even though the reaction is exothermic, which will result in a good selectivity.

The correlation of convective heat transfer for single-phase flow facilitates the investigation of microscale thermal devices:

$$h = Nu\frac{k}{d}, \tag{8.2}$$

where $h$ is the heat transfer coefficient, $Nu$ is the Nusselt number, $k$ is the thermal conductivity, and $d$ is the channel diameter.

This correlation illustrates that the decrease of channel diameter leads to the increase of the heat transfer coefficient. Owing to the fact that forced convection takes place in hydraulic diameter with tens to hundreds of micrometers, the heat transfer coefficient will be extremely high. The heat transfer to the channel walls is higher than in conventional reactors. This characteristic leads to a potential advantage – the reactors will remain isothermal (or nearly isothermal) even though the reaction is exothermic, which will result in a good selectivity.

### 8.5.1 ADVANTAGES OF MESOSCALE REACTORS TO IMPROVE PROCESS SAFETY

Mesoscale reactors have a number of fundamental advantages. The small size allows accurate measurements to the processes with less space, materials, and energy. In addition, reactions that occur in a microreactor can be parallel batch processes, which reduces cost and significantly improve performance by integrating a number of functional elements, such as microelectronics fabrications. Microreactors are always integrated together under continuous operations. This could avoid unstable intermediates and potential delays due to the batch workup. The absence of workup delays can prevent intermediates from decaying and provide the possibility of better selectivity.

The measurements do not depend on the size of the systems, but on the production or fabrication rate. When the size of reaction systems is reduced to mesoscale, the production rate decreases and the profit during a certain time becomes smaller as well. However, in nature, a great number of products are fabricated in mesoscale reactors. For example, the living cell, which is comparable to mesoscale chemical reactors, maintains its small size for millions of years. Therefore, to achieve high production rates, individual reactors must be combined together. This is one of the basic concepts of microreactors technology (Ehrfeld et al., 1996; Wegeng et al., 1996). The advantages of mesoscale reactors include two fundamental properties: physical size and number of units (Ehrfeld et al., 2000).

The gradients of a number of physical variables, such as pressure, concentration, temperature, or density, are dependent on dimensions of reactors. These gradients can affect the transport phenomena and fluid flow behaviors. When the dimension of a reactor decreases, the gradients of these variables will increase and therefore influence heat transfer, mass transport, and diffusional flux. The typical value of heat exchange coefficients of microreactors is at least 1 $MW/(m^3\ K)$, up to 500 $MW/(m^3\ K)$, while conventional glassware only has a few kilowatts per cubic meter per Kelvin. Therefore, heat can be removed much more effectively by microreactors than by conventional vessels and even critical reactions such as nitrations can be performed safely at high temperatures (Roberge et al., 2005). On the other hand, the volume of a reactor and corresponding materials holdup decreases if the dimension of a reactor decreases. It will result in higher exchange surface and gradients, which leads to a shorter response time. As a result, the process control of microreactors becomes easier than the conventional large reactors. Microreactors can withstand some process conditions, which cannot be realized in macroscale reactors due to the intensified transport properties and the decreased holdup and response time (Ehrfeld et al., 2000). For example, microreactors can significantly decrease hot spot temperatures as well as the duration of high temperature exposition due to exothermicity effect. Therefore, microreactors may benefit kinetic investigations. Another example is the speed of mixing. Continuous operation and mixing offers a very different profile of concentration compared to a batch process. In a batch process, reagent A is filled in when reagent B is slowly added. B initially is in a high excess of A. In a microreactor, A and B are mixed almost simultaneously, so such excess will not take place. The effect of the instant mixing needs to be determined by the reaction mechanism – it is important to be aware of such different concentration profiles. In addition, because of smaller dimension of a mesoscale reactor, the diffusion path decreases so that the speed of mixing increases. Therefore, the boundary conditions for a chemical reaction under study would be assigned almost arbitrarily within the operational range. Uniform boundary conditions also become achievable while they are not true in the macroscale reactors in the laboratory or the chemical plants. This advantage leads to a wide field of application of microreactors – development to find optimal conditions for a production plant with regard to yield and selectivity under safety operational conditions (Jäckel and Wörz, 1997; Wörz et al., 2001). Smaller size of reactors also makes heating and cooling effective and operating temperatures can be decreased as low as –100 °C. As a result of the effective heat transfer, reactions can be performed at higher temperatures compared to conventional batch reactors. For example, many low-temperature reactions as organometal chemistry can be conducted in microreactors at temperatures of –10 °C rather than –50 to –78 °C as in laboratory glassware equipment.

The holdup is small in microreactors, so it is possible to design an inherently safer microreactor. To scale up, engineers need to have a proper layout, so that inherent safety could be maintained in a large microreactor network. The short response time of control to a reaction unit and the flexibility of production capacity are also considered as advantageous to process safety. For example, the microreactor can potentially reduce the hazards involved in a runaway reaction by the high heat transfer coefficient, making the process inherently safer. In addition, stirred tank, which is one of the major sources of operational disturbances, will not be required if the plant is constituted by a large number of microreactors. The microreactors have a much better mixing effect with a negligible amount of holdup. A microreactor is usually used for the synthesis of small quantities of chemicals, but by stacking the number of microchannels, a microreactor could be simply scaled up to meet the demand of industries. Scale-up of batch processes from the research and development level to industry volume often fails due to a number of factors, such as the nonlinear effect. On the other hand, when microreactors are transferred from the laboratory to chemical industrial plants, such factors during scaling up take less time to be considered compared to batch processes.

A lot of effort to the development of microreactors technology has been made in Europe and the United States (Wegeng et al., 1996). Reactions with explosion, flammability, and toxic components and processes are mainly focused on by different research groups. Such reactions could be easily interrupted by rapid cooling to eliminate side reactions and therefore enhance the purity of products with safer operational conditions. Further, a number of reactions, such as in gas-phase catalysis, are realized in microreactors with a high degree of conversion. In addition, pressurization of materials within microreactors (and associated components) is generally easier than with traditional batch reactors. Therefore, the rate of reactions can be increased by raising the temperature higher than the boiling point of the solvent. Despite the typical Arrhenius behavior, it is conveniently facilitated in microreactors, and it should be considered as a key advantage. Pressurization may also allow dissolution of reactant gasses within the flow stream.

Microreactors have been established in the scientific, technological, and commercial fields. However, we must agree that it still needs great effort to develop. At present, there are no comprehensive theoretical basis or design regulations for microfabrication engineering. Therefore, readers should consider this section as a snapshot of work under development, which will play a vital role in the future.

## 8.5.2 POTENTIAL PROCESSES FOR THE APPLICATION OF MESOSCALE REACTOR

Mesoscale reactor is used intensively in the laboratory for different kinds of reactions, such as hydrogenation, decomposition, and oxidation. In 2009, Lan et al. (2009) reported an enhanced hydrogenation of olefins using microreactors. Figure 8.36 shows the concept of this hydrogenation reaction under organic/aqueous biphasic conditions. The olefin forms one phase, while water that contains a number of microreactors forms the other phase. Due to the diffusion effect, olefin diffuses into water and then is captured by microreactors. Hydrogenation takes place inside the microreactors. The product diffuses back into water and is extracted in the organic phase. The resultant process is of higher efficiency than other conventional hydrogenation processes of olefins.

Ganley et al. (2004) described a ruthenium-impregnated anodic aluminum catalyst for use in microreactors for the production of hydrogen from an ammonia feed. A 0.9 × 0.9 cm reactor contains anhydrous ammonia that decomposes under the catalyst of ruthenium on aluminum substrate. The possible application of this microreactor fabrication is hydrogen generation for fuel cells in mobile power production.

As discussed earlier, a lot of reactions that are considered extremely dangerous could be realized by microreactors. For example, the direct fluorination of organic compounds is too dangerous for

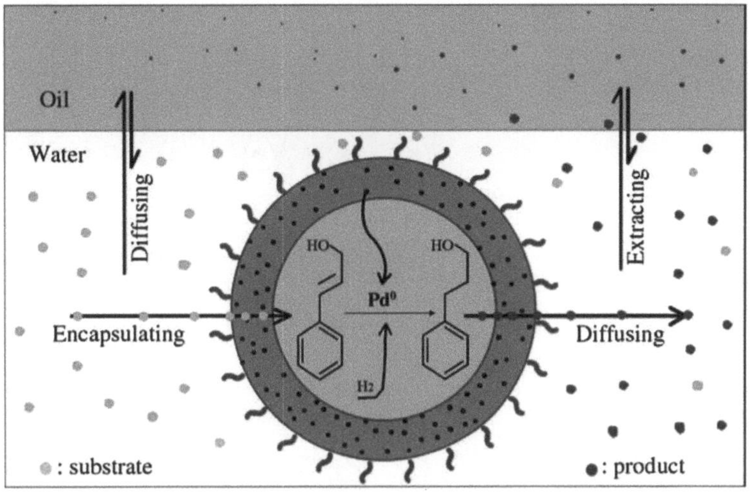


**FIGURE 8.36**

Schematic hydrogenation of olefins in the microreactor under biphasic conditions (Lan et al., 2009).

commercial applications. Many explosions have been reported on a lab-scale (Jähnisch et al., 2000). Fluorine is a chemical agent that is very toxic and corrosive, resulting in a major safety hazard. The most well-known fluorination of aromatics reaction is the Schiemman process (Balz and Schiemann, 1927). The process involves diazonium tetrafluoroborates, which is another hazard for the industrial applications. In addition, diazonium tetrafluoroborates must be kept in a dry environment. The presence of moisture leads to a runaway decomposition. The use of microreactors could allow such safety issues to be addressed effectively, as illustrated in Figure 8.37 (Chambers et al., 1999). Because of increasing contacting surface area, the heat transfer behaviors could facilitate the control of fluorination reactions with outstanding yields. Furthermore, the authors also show the potential scale-up of the single-channel device to a three-channel microreactor device.

### 8.5.3 TECHNIQUES NEEDED FOR MESOSCALE REACTOR MANUFACTURE

Although the design of microreactors has become more and more advanced, much more knowledge in this field is needed. For example, owing to the small diameter of channels in microreactors, particles often cause clogging issues. This has been considered as one of the main reasons for the intensive study to be conducted so that microreactors can be accepted as an alternative to batch reactors. Microjet reactor is the only kind of micromixer that can prevent particles from clogging in reactors. Micromixers are one of the important components in mesoscale technology. There are mainly three categories of micromixers, which are multiple microjet injector, static micromixer, and fluid layer multilamination. Multiple microjet injector was first used as an integral part of a microliquid dosing system in 1993 (Miyake et al., 1993). The basic element of the mixer contains a large number of regular holes (Figure 8.38). During operation, the mixing area is full of fluid 1, and fluid 2 is injected through the multiple channels into the mixing area. The multiple microjets are therefore produced,

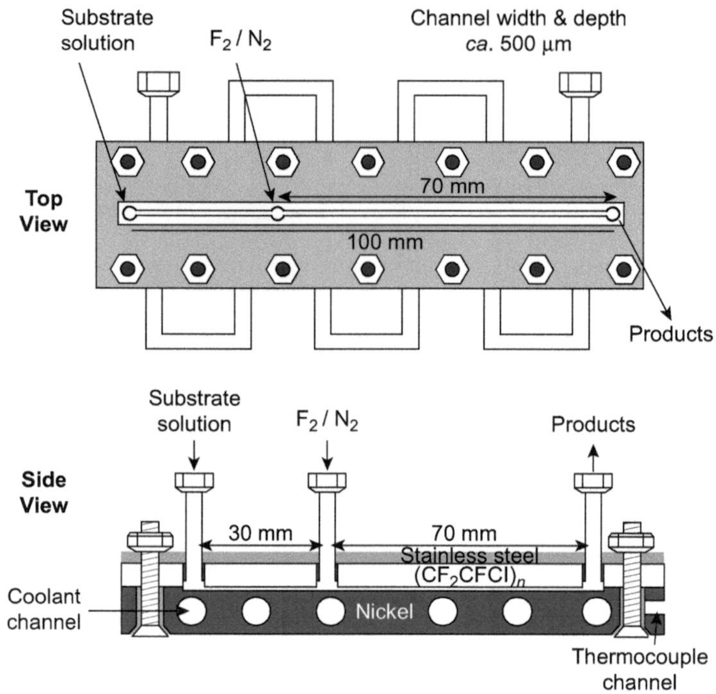


**FIGURE 8.37**

A single-channel microreactor for fluorination reactions (Chambers et al., 1999).

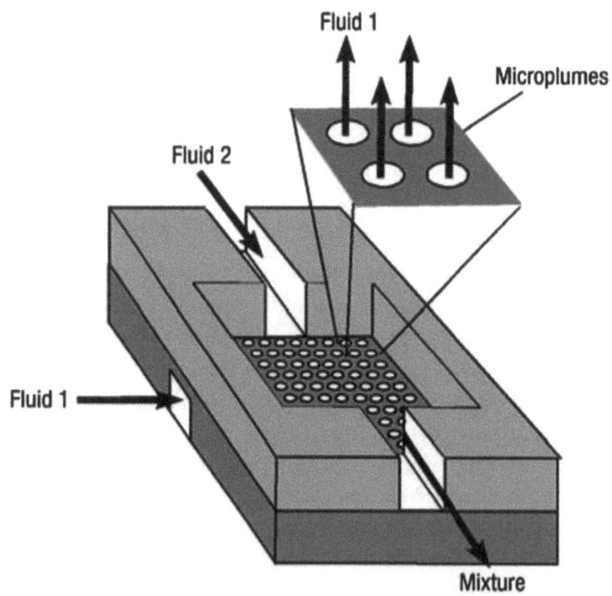


**FIGURE 8.38**

Schematic of a micromixer with multiple microjet injectors (Miyake et al., 1993).

so that the contact surface area between fluids 1 and 2 is increased. The distance between each channel ranges between 10 and 100 μm. The diffusional path is therefore shortened physically. Numerical simulation indicates the feasibility of this device. The simulation shows that multiple channels for injection accelerates the mixing process compared to a large single-channel injection. Experiments also qualitatively analyze the mixing process. Microscopic observation shows that the mixing process only takes a few seconds to establish a homogeneous mixture of fluids 1 and 2.

When the viscosity of the flow is high, the flow rate under operations would be relatively low. It results in the poor performance of mixing process. Some researchers proposed a device named static micromixer to enhance the mixing performance (Branebjerg et al., 1996). Static micromixers have mixing elements molded into the walls of the channel. Therefore, radial mixing could be enhanced. Figure 8.39 shows the design of a static micromixer, which consists of diagonally oriented grooves to promote cross-channel mixing in a mesoscale. When fluids cannot penetrate into each other's stream sufficiently deep, the connected channels become important to ensure mixing quality. Experimental observation shows that the mixing process is completed at outlet in 100–300 m. Such static micromixers have also been used in wet-chemical anisotropic etching of monocrystalline silicon (Schwesinger et al., 1996).

During fabrication, if subsequent diffusional mixing is required, micromixing devices with multilamination of flow can be used. The bottom part of Figure 8.40 shows the design of such micromixers (Ehrfeld et al., 1998). The fluid flows through the mixing element in counter flow into interdigitated channels. The typical width of the channels is 25 or 40 μm. The channel configuration leads to a periodical flow lamellae of the two fluids. Because of the thinness of the lamellae, mixing takes place by diffusion in a very short time. The top part of Figure 8.40 shows how this micromixer is assembled. Two fluids flow into the two inlets and return back from the outlet. The mixing element is covered by a top part, which contains the two inlets and the one outlet, and inserted into the bottom part. There is no general requirement for the material of mixing element. Nickel on copper or silver mixing elements are intensively used for commercial application. The top and bottom part are usually made of stainless steel or plastics.

Besides, the heat transfer phenomena are also different from that in a macroscale. Sufficiently large contact surface area could ensure the effective heat transfer from one flowing fluid across solid boundaries to another flowing fluid. Another important parameter that needs to be considered is the ratio of heat transferred to pressure loss. When fluid stream is split into many microdimensions of partial stream, this parameter becomes significant in a compact heat exchanger. Small dimensions correspond to low Reynolds numbers, so viscosity is more important than turbulent behavior. Viscosity loss is inherently inevitable for decreased dimension of flow.

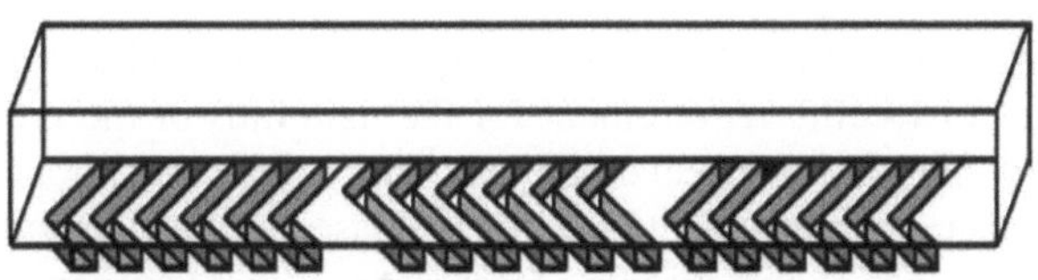

**FIGURE 8.39**

Schematic design of a static micromixer (Stroock and Whitesides, 2003).

**FIGURE 8.40**

Top: Micromixer with multilamination of flow is inserted into the bottom part and covered by the top part. Bottom: Scanning electron micrographs of a mixing element based on multilamination of flow (Ehrfeld et al., 1998).

There are two types of heat exchangers in microreactor engineering. To improve heat transfer performance, Karlsruhe Nuclear Research Center introduced a cross-flow micro-heat exchanger in the 1980s (Bier et al., 1990; Schubert et al., 1990). The typical design of a cross-flow micro-heat exchanger is shown in Figure 8.41. It consists of a stack of plat devices containing rectangular microchannels. The cross section of each microchannel is $100 \times 80$ μm. The stack is connected with inlet and outlet channels of the heat exchanger fluid. The typical volume of this kind of heat exchanger is around 1 $cm^3$. The inner surface area is about 300 $cm^2$, while the effective heat transfer area could reach 150 $cm^2$. The strong bonding and small dimensions of the channels could bear a relatively high operational pressure. Experimental results show that pressure up to 25 bar cannot affect the functions of such devices. There is no specific requirement for the materials of the cross-flow heat exchanger. Aluminum, copper, silver, and stainless steel could be used for industrial applications. The shape of the channel and the number of stacks are subjected to change based on the demand of applications (Linder et al., 1994). The main advantage shown by some researchers is that laminar flow always exists even at high flow rate in every channel (Schubert et al., 1998). The overall heat transfer coefficient is more than 20 kW/($m^2$ K), which is higher than conventional macroscale devices.

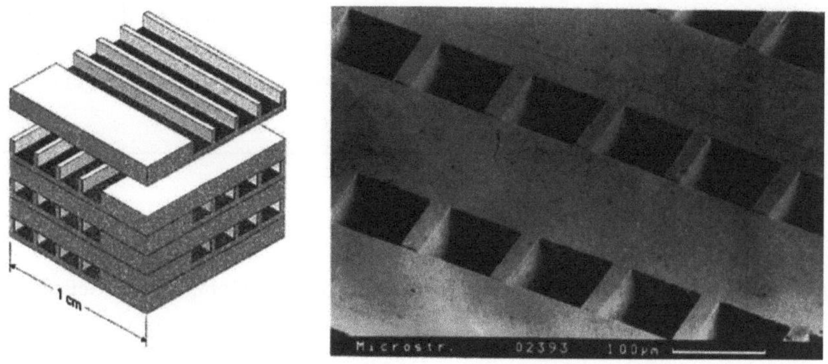


**FIGURE 8.41**

Schematic design of a cross-flow heat exchanger (Bier et al., 1990).

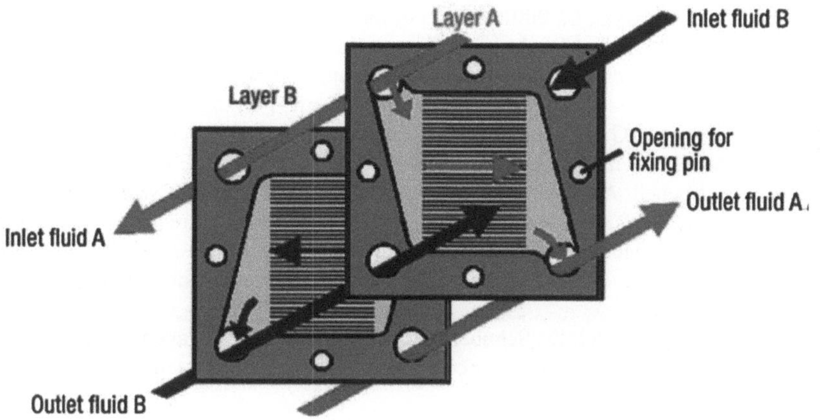


**FIGURE 8.42**

Schematic design of a counterflow heat exchanger (Ehrfeld et al. 1998).

The other popular heat exchanger is called counter flow heat exchanger. From thermodynamic view point, this kind of heat exchanger is much more efficient compared to other kinds of heat exchangers. Figure 8.42 shows the basic design of such devices. Platelets with an outer frame and an inner thin membrane stack together to form a counter flow heat exchanger. There are parallel fins on the thin membranes. These fins form a channel that fluids flow through, so that the mechanical stability is enhanced. The mechanical stability is an important criterion for operations at different high pressure. The thinness of membrane could facilitate efficient heat transfer. Fluid A flows through two diagonal openings, while fluid B flows through the other two. Therefore, the uniform countercurrent flow is established. The platelets are $20 \times 20$-mm-square copper foil with a thickness of 20–500 μm. The width of each channel is around 300 μm. The total heat transfer area is about 140 $mm^2$ per platelet. Such devices could be used in various processes, such as deep lithography and subsequent electroforming.

Experiments show that the heat transfer rate is a function of thickness of the fluid layer. When the thickness of the fluid layer decreases, the heat transfer rate will increase. The heat transfer coefficient achieved by such devices is obviously higher than the conventional heat exchangers.

## REFERENCES

AICHE/CCPS, 2003. Guidelines for Facility Siting and Layout. American Institute of Chemical Engineers, New York.

Amar, J.G., 2006. The Monte Carlo method in science and engineering. Computing in Science and Engineering 8 (2), 9–19.

Arendt, J.S., Lorenzo, D.K., 2000. Evaluating Process Safety in the Chemical Industry: A User's Guide to Quantitative Risk Analysis. Center for Chemical Process Safety, 3 Park Avenue, New York, New York 10016-5991.

Balaji, S., Du, J., White, C.M., Ydstide, B.E., 2010. Multi-scale modeling and control of fluidized beds for the production of solar grade silicon. Powder Technology 199, 23–31.

Baldea, M., Daoutidis, P., 2007. Control of integrated process networks-a multi-time scale perspective. Computers and Chemical Engineering 31, 426–444.

Baldyga, J., Henczka, M., Makowski, L., 2001. Effects of mixing on parallel chemical reactions in a continuous-flow stirred-tank reactor. Chemical Engineering Research and Design 79 (8), 895–900.

Balz, G., Schiemann, G., 1927. Über aromatische Fluorverbindungen, I.: Ein neues Verfahren zu ihrer Darstellung. Berichte der deutschen chemischen Gesellschaft (A and B Series) 60 (5), 1186–1190.

Bentner, J., Bauer, G., Obermair, G.M., Morgenstern, I., Schneider, J., 2001. Optimization of the time-dependent traveling salesman problem with Monte Carlo methods. Physical Review E 64 (3), 036701.

Bertinetto, C., Duce, C., Micheli, A., Solaro, R., Starita, A., Tine, M.R., 2009. Evaluation of hierarchical structured representations for QSPR studies of small molecules and polymers by recursive neural networks. Journal of Molecular Graphics and Modelling 27 (7), 797–802.

Bi, H.T., Li, J., 2004. Multiscale analysis and modeling of multiphase chemical reactors. Advanced Powder Technology 15 (6), 607–627.

Bier, W., Keller, W., Linder, G., Seidel, D., Schubert, K., 1990. Manufacturing and testing of compact micro heat exchangers with high volumetric heat transfer coefficients. In: Dynamic Systems and Control Conference, 19. ASME, New York.

Branebjerg, J., Gravesen, P., Krog, J.P., Nielsen, C.R., 1996. Fast mixing by lamination. In: Proceedings of IEEE, 98. MEMS, San Diego, CA, pp. 441–448.

Caldwell, A., 2013. Lecture on Markov Chains, MCMC, Metropolis-Hastings Algorithm, Optimization with Monte Carlo Methods, Simulated Annealing. Retrieved November 11, 2013, from https://www.mpp.mpg.de/~caldwell/ss13/Lecture10_final.pdf.

Cale, T.S., Bloomfield, M.O., Richards, D.F., Jansen, K.E., Gobbert, M.K., 2002. Integrated multiscale process simulation. Computational Materials Science 23 (1–4), 3–14.

Chambers, R.D., Hutchinson, J., Sandford, G., 1999. Recent studies at Durham on direct fluorination. Journal of Fluorine Chemistry 100 (1–2), 63–73.

Charlette, F., Meneveau, C., Veynante, D., 2002. A power-law flame wrinkling model for LES of premixed turbulent combustion. Part I. Non-dynamic formulation and initial tests. Combustion and Flame 131 (1–2), 159–180.

Chou, C.-C., Falk, M.L., 2006. Multiscale diffusion Monte Carlo simulation of epitaxial growth. Journal of Computational Physics 217 (2), 519–529.

ClassNK. 2008. Guidelines on Pipeline Wall Thinning- Taking into Account Flow Acclerated Corrosion of Water and Wet Steam Pipelines in Ships. Chiyoda-ku, Japan: ClassNK.

Cutler, C.R., and Ramaker, B.L., 1979. Dynamic matrix control—a computer control algorithm. AIChE National Mtg, Houston, Texas (1979); also Proc. Joint Aut. Control Conf., San Francisco, California (1980).

Dadvar, M., Sahimi, M., 2003. Pore network model of deactivation of immobilized glucose isomerase in packed-bed reactors. Part III: multiscale modelling. Chemical Engineering Science 58 (22), 4935–4951.

Dai, Y-m., Zhu, Z-p., Cao, Z., Zhang, Y-f., Zeng, J-l., Li, X., 2013. Prediction of boiling points of organic compounds by QSPR tools. Journal of Molecular Graphics and Modelling 44 (0), 113–119.

Dhib, R., Goa, J., Penlidis, A., 2000. Simulation of free radical bulk/solution homo-polymerization using mono- and bi-functional initiators. Polymer Reaction Engineering 8 (4), 299–464.

Di Sarli, V., Di Benedetto, A., Russo, G., Jarvis, S., Long, E.J., Hargrave, G.K., 2009. Large eddy simulation and PIV measurements of unsteady premixed flames accelerated by obstacles. Flow, Turbulence and Combustion 83 (2), 227–250.

Ding, J., Wang, X., Zhou, X.F., Ren, N.Q., Guo, W.Q., 2010. CFD optimization of continuous stirred-tank (CSTR) reactor for biohydrogen production. Bioresource Technology 101 (18), 7016–7024.

Dorneanu, B., Bildea, C.S., Grievink, J., 2009. On the application of model reduction to plantwide control. Computers & Chemical Engineering 33 (3), 699–711.

Efendiev, Y., Galvis, J., Gildin, E., 2012. Local–global multiscale model reduction for flows in high-contrast heterogeneous media. Journal of Computational Physics 231 (24), 8100–8113.

Ehrfeld, W., Gärtner, C., Golbig, K., Hessel, V., Konrad, R., Löwe, H., Richter, T., Schulz, C., 1998. Fabrication of components and systems for chemical and biological microreactors. In: Ehrfeld, W. (Ed.), Microreaction Technology. Springer, Berlin Heidelberg, pp. 72–90.

Ehrfeld, W., Hessel, V., Haverkamp, V., 2000. Microreactors. Ullmann's Encyclopedia of Industrial Chemistry. Wiley-VCH Verlag GmbH & Co. KGaA.

Ehrfeld, W., Hessel, V., M€obius, H., Richter, T., Russow, K., 1996. Potentials and realization of microreactors. In: Ehrfeld, W. (Ed.), Microsystem Technology for Chemical and Biological Microreactors. VCH (Verlagsgesellschaft), Weinheim, Germany, pp. 1–28.

Ferng, Y.M., Lin, B.H., 2010. Predicting the wall thinning engendered by erosion–corrosion using CFD methodology. Nuclear Engineering and Design 240 (10), 2836–2841.

Ganley, J.C., Seebauer, E.G., Masel, R.I., 2004. Porous anodic alumina microreactors for production of hydrogen from ammonia. AIChE Journal 50 (4), 829–834.

García, C.E., Prett, D.M., Morari, M., 1989. Model predictive control: theory and practice—a survey. Automatica 25 (3), 335–348.

Gerorgiadis, M.C., Macchietto, S., 1997. Layout of process plants: a novel approach. Computers and Chemical Engineering 21 (S1), S337–S342.

Graham, L., Lester, D., Wu, J., 2009. Slurry erosion in complex flows: experiment and CFD. In: Seventh International Conference on CFD in the Minerals and Process Industries. CSIRO Materials and Engineering, Melbourne, Australia.

Gupta, D., 2005. Diffusion Processes in Advanced Technological Materials. William Andrew Publishing/Noyes.

Hannemann-Tamás, R., Gábor, A., Szederkényi, G., Hangos, K.M., 2013. Model complexity reduction of chemical reaction networks using mixed-integer quadratic programming. Computers & Mathematics with Applications 65 (10), 1575–1595.

Hernandez, H.F., Tauer, K., 2008. Brownian dynamics and kinetic Monte Carlo simulation in emulsion polymerization. In: Bertrand, B., Xavier, J. (Eds.), Computer Aided Chemical Engineering, vol. 25. Elsevier, pp. 769–774.

Jäckel, K.P., Wörz, O., 1997. Winzlinge mit großer Zukunft Mikroreaktoren für die Chemie. Chemie-Technik 26 (1), 130–134.

Jähnisch, K., Baerns, M., Hessel, V., Ehrfeld, W., Haverkamp, V., Löwe, H., Wille, C., Guber, A., 2000. Direct fluorination of toluene using elemental fluorine in gas/liquid microreactors. Journal of Fluorine Chemistry 105 (1), 117–128.

Jayakumar, S., Reklaitis, G.V., 1994. Chemical plant layout via graph partitioning—I. Single level. Computers and Chemical Engineering 18 (5), 441–458.

Jayakumar, S., Reklaitis, G.V., 1996. Chemical plant layout via graph partitioning—II. Multiple levels. Computers and Chemical Engineering 20 (5), 563–578.

Jung, S., Ng, D., Lee, J.-H., Vazquez-Roman, R., Mannan, S., 2010. An approach for risk reduction (methodology) based on optimizing the facility layout and siting in toxic gas release scenarios. Journal of Loss Prevention in the Process Industries 23, 139–148.
Kahn, H., 1957. Applications of Monte Carlo. Tech Info Service, Oak Ridge, TN.
Kang, H., Weinberg, W., 1992. Dynamic Monte Carlo simulations of surface-rate processes. Accounts of Chemical Research 25 (6), 253–259.
Katritzky, A.R., Petrukhin, R., Jain, R., Karelson, M., 2001. QSPR analysis of flash points. Journal of Chemical Information and Computer Sciences 41 (6), 1521–1530.
Katritzky, A.R., Stoyanova-Slavova, I.B., Dobchev, D.A., Karelson, M., 2007. QSPR modeling of flash points: an update. Journal of Molecular Graphics and Modelling 26 (2), 529–538.
Keshavarz, M., Motamedoshariati, H., Ghanbarzadeh, M., 2011. A reliable simple method for prediction of the flash points of saturated hydrocarbons in order to improve their safety. Chemistry & History of Chemistry 6, 58–75.
Končar, B., Kljenak, I., Mavko, B., 2004. Modelling of local two-phase flow parameters in upward subcooled flow boiling at low pressure. International Journal of Heat and Mass Transfer 47 (6–7), 1499–1513.
Kunza, R.F., Bogera, D.A., Stinebringa, D.R., Chyczewskia, T.S., Lindaua, J.W., Gibelinga, H.J., Venkateswaranb, S., Govindanc, T.R., 2000. A preconditioned Navier–Stokes method for two-phase flows with application to cavitation prediction. Computers & Fluids 29 (8), 849–875.
Lan, Y., Zhang, M., Zhang, W., Yang, L., 2009. Enhanced Pd-catalyzed hydrogenation of olefins within polymeric microreactors under organic/aqueous biphasic conditions. Chemistry – A European Journal 15 (15), 3670–3673.
Lanore, J.M., Kalli, H., 1977. Usefulness of the Monte Carlo method in reliability calculations. Transactions of the American Nuclear Society 27, 374–375. Medium: X; Size.
Lewis, E.E., Zhuguo, T., 1986. Monte Carlo reliability modeling by inhomogeneous Markov processes. Reliability Engineering 16 (4), 277–298.
Lilly, D.K., 1992. A proposed modification of the Germano subgrid-scale closure method. Physics of Fluids 4 (3), 633–635.
Linder, G., Bier, W., Schaller, T., Schubert, K., Seidel, D., 1994. Mikrowärmeübertrager und Mikroreaktoren. In: Tagungsband des Symposiums "Mikrotechnik". ACHEMA, Frankfurt.
Liu, Y., Jin, F., Li, Q., Zhou, S., 2008. A fixed-grid bidirectional evolutionary structural optimization method and its applications in tunnelling engineering. International Journal for Numerical Methods in Engineering 73 (12), 1788–1810.
Lu, Y., Ng, D., Mannan, M.S., 2010. Prediction of the reactivity hazards for organic peroxides using the QSPR approach. Industrial & Engineering Chemistry Research 50 (3), 1515–1522.
Lucia, A., 2010. Multi-scale methods and complex processes: a survey and look ahead. Computers and Chemical Engineering 34, 1467–1475.
Luo, J.Y., Gosman, A.D., Issa, R.I., 1994. Prediction of impeller-induced flows in mixing vessels using multiple frames of reference. Institute of Chemical Engineering Symposium Series 136, 549–558.
Mannan, S. (Ed.), 2005. Lees' Loss Prevention in the Process Industries. Butterworth-Heinemann, Burlington.
Marino, D.J.G., Peruzzo, P.J., Krenkel, G., Castro, E.A., 2003. QSPR modeling of heat of formation and heat of vaporization of aliphatic ketones by means of electrotopological indices. Chemical Physics Letters 369 (3–4), 325–334.
Marsh, J., Teh, T., 2007. Conflicting views: $CO_2$ corrosion models, corrosion inhibitor availability philosophies, and the effect on subsea systems and pipeline design. In: Offshore Europe 2007. Society of Petroleum Engineers, Aberdeen, Scotland.
Mecklenburgh, J.C., 1973. Plant Layout: A Guide to the Layout of Process Plant and Sites. Wiley, New York.
Merchant, T.P., Gobbert, M.K., Cale, T.S., Borucki, L.J., 2000. Multiple scale integrated modeling of deposition processes. Thin Solid Films 365 (2), 368–375.

Metropolis, N., Ulam, S., 1949. The Monte Carlo method. Journal of American Statistical Society 44 (247), 335.
Miyake, R., Lammerink, T.S.J., Elwenspoek, M., Fluitman, J.H.J., 1993. Micro Mixer with Fast Diffusion. MEMS-93, Fort Lauderdale, FL, 248–253.
Molnar, D., Mukherjee, R., Choudhury, A., Mora, A., Binkele, P., Selzer, M., Nestler, B., Schmauder, S., 2012. Multiscale simulations on the coarsening of Cu-rich precipitates in $\alpha$-Fe using kinetic Monte Carlo, molecular dynamics and phase-field simulations. Acta Materialia 60 (20), 6961–6971.
Moore, J.M., 1962. Plant Layout and Design. Mcmillan, New York.
Moritz, H.U., 1989. Increase in viscosity and its influence on polymerization processes. Chemical Engineering & Technology 12 (1), 71–87.
Muth, E.J., 1970. Excess time, a measure of system repairability. Reliability, IEEE Transactions R-19 (1), 16–19.
Nantasenamat, C., Isarankura-Na-Ayudhya, C., Naenna, T., Prachayasittikul, V., 2009. A practical overview of quantitative structure-activity relationship. Excli Journal 8, 74–88.
NFPA 30, 2012. NFPA 30: Flammable and Combustible Liquids Code. NFPA, 30–39.
Nicola, S., 2012. Corrosion Detection and Prediction Studies. Master of Science. A&M University, Texas.
Nyborg, R., 2002. Overview of $CO_2$ Corrosion Models for Wells and Pipelines. Corrosion 2002. Co, NACE International, Denver.
OGP, 2010. Risk Assessment Data Directory – Consequence Modelling. The International Association of Oil and Gas Producers, London, UK.
Pan, Y., Jiang, J., Wang, R., Cao, H., 2008. Advantages of support vector machine in QSPR studies for predicting auto-ignition temperatures of organic compounds. Chemometrics and Intelligent Laboratory Systems 92 (2), 169–178.
Patel, H., Ein-Mozaffari, F., Dhib, R., 2010. CFD analysis of mixing in thermal polymerization of styrene. Computers & Chemical Engineering 34 (4), 421–429.
Patel, S.N.D.H., Jarvis, S., Ibrahim, S.S., Hargrave, G.K., 2002. An experimental and numerical investigation of premixed flame deflagration in a semiconfined explosion chamber. Proceedings of the Combustion Institute 29 (2), 1849–1854.
Pitting, F.G., 2003. Corrosion: Fundamentals, Testing, and Protection. ASM International, Materials Park, OH.
Prett, D.M., Gillette, R.D., 1979. Optimization and constrained multivariable control of a catalytic cracking unit. AIChE National Mtg, Texas, Houston.
Quintero, F.A., Patel, S.J., Muñoz, F., Sam Mannan, M., 2012. Review of existing QSAR/QSPR models developed for properties used in hazardous chemicals classification system. Industrial & Engineering Chemistry Research 51 (49), 16101–16115.
Raimondeau, S., Vlachos, D.G., 2002. Recent developments on multiscale, hierarchical modeling of chemical reactors. Chemical Engineering Journal 90, 3–23.
Reinelt, G., 1994. The Traveling Salesman: Computational Solutions for TSP Applications. Springer-Verlag.
Ridha, T., Liu, Y., Gao, X., Mannan, M.S., 2012. Prediction of onset temperature for thermal decomposition of organic peroxides using quantitative structure-properties relationship. In: 2012 AIChE Annual Meeting, Pittsburgh, PA.
Roberge, D.M., Ducry, L., Bieler, N., Cretton, P., Zimmermann, B., 2005. Microreactor technology: a revolution for the fine chemical and pharmaceutical industries? Chemical Engineering & Technology 28 (3), 318–323.
Rubinstein, R.Y., 1981. Simulation and the Monte Carlo Method. Wiley, New York.
Saeed, S., Ein-Mozaffari, F., 2008. Using dynamic tests to study the continuous mixing of xanthan gum solutions. Journal of Chemical Technology and Biotechnology 83, 559–568.
Satyanarayana, K., Rao, P.G., 1992. Improved equation to estimate flash points of organic compounds. Journal of Hazardous Materials 32 (1), 81–85.
Scenna, N.J., Cruz, A.S.M.S., 2005. Railroad risk analysis due to the transportation of chlorine in Rosario city. Reliability Engineering and System Safety 90 (1), 83–90.

Schmidt, J., Atkinson, G.T., Gant, S.E., Bockhorn, H., Hauptmanns, U., Dudlik, A., Helmsen, F., Fröhlich, R., Henk, W., Kirchner, T., Jones, A., Klein, R., Jörgensen, B., Klomfass, A., Leitl, B., Knaust, C., Mattes, T., Koch, M., Moncalvo, D., Krause, U., Morillo, A., Münch, M., Rodi, W., Niesser, K., Schälike, S., Peschel, W., Pfeil, N., Schatzmann, M., Prasser, H.-M., Schiller, F., Rabe, F., Vela, I., Wehrstedt, K.-D., Scholz, B., Wellenhofer, A., Schönbucher, A., Wozny, G., Staak, D., Wuersig, G.-M., Thoma, K., 2012. Process and Plant Safety: Applying Computational Fluid Dynamics. In: Boschstr, 12. Wiley-VCH Verlag & Co. KGaA, Weinheim, Germany, 69469.

Schubert, K., Bier, W., Brandner, J., American Institute of Chemical Engineers, 1998. Realization and Testing of Microstructure Reactors, Micro Heat Exchangers and Micromixers for Industrial Applications in Chemical Engineering. American Institute of Chemical Engineers.

Schubert, K., Bier, W., Linder, G., Seidel, D., 1990. Profiled microdiamonds for producing microstructures. Industrial Diamond Review 5.

Schwesinger, N., Frank, T., Wurmus, H., 1996. A modular microfluid system with an integrated micromixer. Journal of Micromechanics and Microengineering 6, 99–102.

Shooman, M.L., 1968. Probabilistic Reliability: An Engineering Approach. McGraw-Hill, New York.

Shreider, Y.A. (Ed.), 1964. Method of Statistical Testing: Monte Carlo Method. Elsevier, Amsterdam.

Sobol, I.M., 1974. The Monte Carlo Method. University of Chicago Press, Chicago, IL.

Soon Heung, C., Joo Young, P., Myung Ki, K., 1985. The Monte-Carlo method without sorting for uncertainty propagation analysis in PRA. Reliability Engineering 10 (4), 233–243.

Spouge, J., Kinsella, K., Morgan, C., Bridge, D., Smith, E., Haugen, S., Paterson, L., Vollen, F., Hall, S., Skudder, A., Harris, S., Morgan, B., 1999. A Guide to Quantitative Risk Assessment for Offshore Installations. University Gate Park Row, Bristol, BS1 5UB, UK (Centre for Marine and Petroleum Technology).

Stack, M.M., Abdelrahman, S.M., Jana, B.D., 2010. A new methodology for modelling erosion–corrosion regimes on real surfaces: gliding down the galvanic series for a range of metal-corrosion systems. Wear 268 (3–4), 533–542.

Stack, M.M., Stoti, F.H., 1993. An approach to modelling erosion-corrosion of alloys using erosion-corrosion maps. Corrosion Science 35 (5–8), 1027–1034.

Stephanopoulous, G., Karsligil, O., Dyer, M., 2000. Multi-scale aspects in model-predictive control. Journal of Process Control 10, 275–282.

Stroock, A.D., Whitesides, G.M., 2003. Controlling flows in microchannels with patterned surface charge and topography. Accounts of Chemical Research 36 (8), 597–604.

Turner, D.B., 1964. A diffusion model for an urban area. Journal of Applied Meteorology 3 (1), 83–91.

Ustinenko, V., Samigulin, M., Ioilev, A., Lo, S., Tentner, A., Lychagin, A., Razin, A., Girin, V., Vanyukov, Y., 2008. Validation of CFD-BWR, a new two-phase computational fluid dynamics model for boiling water reactor analysis. Nuclear Engineering and Design 238 (3), 660–670.

Vázquez-Román, R., Lee, J.-H., Jung, S., Mannan, S.M., 2010. Optimal facility layout under toxic release in process facilities: a stochastic approach. Computers and Chemical Engineering 34, 122–133.

Vazquez-Roman, R., Mannan, S.M., 2010. A new trend in designing plant layouts for the process industry. In: Cakaj, S. (Ed.), Modeling Simulation and Optimization – Tolerance and Optimal Control. InTech.

Venart, J.E.S., 2007. Flixborough: a final footnote. Journal of Loss Prevention in the Process Industries 20 (4), 621–643.

Versteeg, H.K., Malalasekera, W., 1995. An Introduction to Computational Fluid Dynamics: The Finite Volume Method. John Wiley & Sons Inc., 605 Third Avenue, New York, NY 10158, USA (Longman Scientific & Technical).

Wegeng, R.W., Call, C.J., Drost, M.K., 1996. Chemical system miniaturization. In: Proceedings of 1996 Spring National Meeting AIChE, 1. AIChE, New Orleans.

Wen, B., Zabaras, N., 2012. A multiscale approach for model reduction of random microstructures. Computational Materials Science 63 (0), 269–285.

Wörz, O., Jäckel, K.P., Richter, T., Wolf, A., 2001. Microreactors, a new efficient tool for optimum reactor design. Chemical Engineering Science 56 (3), 1029–1033.

Zhao, Y., Jiang, C., Yang, A., 2012. Towards computer-aided multiscale modelling: a generic supporting environment for model realization and execution. Computers & Chemical Engineering 40 (0), 45–57.

CHAPTER

# APPLICATION OF MODELING FOR INDUSTRIAL HYGIENE AND TOXICOLOGICAL ISSUES

# 9

Industrial hygiene deals with workplace conditions to prevent workers from injury or illness. Chemical hazards, especially toxic hazards, are one of the main health concerns in the workplace. Hazardous chemicals in the workplace should be accordingly identified, evaluated, and controlled to ensure safety and health in the workplace.

Chemicals can be hazardous due to its properties leading to toxic effects in environment and health of living beings. The Globally Harmonized System of Classification and Labeling of Chemicals (GHS) provides an internationally agreed-on system developed by the United Nations that requires the classification of chemicals according to the hazardous properties with similar categories.

## 9.1 APPLICATION OF QUANTITATIVE STRUCTURE–ACTIVITY RELATIONSHIP (QSAR) IN INDUSTRIAL HYGIENE

QSAR is a methodology that uses mathematical models to correlate the biological activity (e.g., $LD_{50}$) and descriptive parameters (descriptors) related to the structure of a molecule. QSARs are regression models that are widely applied to biological and chemical sciences. In the field of the classification of hazardous chemicals, QSAR models can be useful to predict the hazardous nature of various chemicals in cases where experimental data are not available for this purpose as stated by the GHS rule.

The basic assumption for QSAR study is that molecules that have similar structures tend to have similar properties. The objective of a QSAR model is to relate a group of predictor variables with the potency of the responsive variable. The predictors contain the information that describes theoretical molecular descriptors or physicochemical properties of chemicals; the responsive variables include various properties, such as biological activity, that are used for the classification of hazardous chemicals. The development of descriptors from molecular structure and correlating molecular descriptors with responsive activities through multivariate analysis are two steps to establish QSAR models. The process of developing QSAR models is not covered in this chapter.

The predictors contain chemical information encoded in the molecular structure from a practical perspective, and the chemical information includes constitutional, geometrical, hydrophobic, electronic, steric, and topological descriptors. More information regarding the descriptor can be found in the literature (Helguera et al., 2008a; Todeschini and Consonni, 2000). The descriptors can be obtained using the quantum chemical software. Once it is determined by the software, it can be considered as an independent variable to develop QSAR models. List of quantum chemical softwares is summarized in literature (Nantasenamat et al., 2009).

Before the implementation of multivariate analysis, data preprocessing is an essential step to ensure the integrity of the dataset contributing to QSAR studies, which usually contains data cleaning, data transformation, and feature selection. The multivariate analysis correlates predictors with responsive variables through mathematical regression. Linear regression is a common approach to establishing the correlation between the responsive variables and a number of the chemical predictors. For cases in which the correlations cannot be explained by a linear relation, nonlinear methods are used for multivariate analysis to develop QSAR models. Some of the popular nonlinear techniques include artificial neural network (Nantasenamat et al., 2009), partial least squares regression (Wold et al., 2001), and support vector machine (Wang, 2005). The developed QSAR models require validation and evaluation of the performance in predicting desirable properties through statistical methods. The validation is conducted using statistical parameters, such as Pearson's correlation coefficient ($r$) and root-mean-square (RMS) error, as well as F-test and outliers. A commonly adopted method to evaluate the ability of developed QSAR models to predict properties of interest was given in Tropsha et al.'s recommendation (Tropsha et al., 2003).

QSAR has been used to predict different health hazards. Based on the criteria specified in Appendix A of the Hazard Communication Standard (29 CFR 1910. 1200) promulgated by the Occupational Safety and Health Administration (OSHA), the health hazards of chemicals are categorized by GHS classification as acute toxicity, skin corrosion and irritation, serious eye damage and eye irritation, respiratory or skin sensitization, germ cell mutagenicity, carcinogenicity, reproductive toxicity, specific target organ toxicity due to single exposure or repeated/prolonged exposure, and aspiration hazard (OSHA, 2014).

Acute toxicity refers to "those adverse effects occurring following oral or dermal administration of a single dose of a substance, or multiple doses given within 24 h, or an inhalation exposure of 4 h" (OSHA, 2014).

Acute toxicity is characterized by $LD_{50}$ for oral and dermal intake, $LC_{50}$ for inhalation, or the acute toxicity estimates. QSAR studies have been conducted to predict $LD_{50}$, $LC_{50}$, or $EC_{50}$ for many chemicals, including pesticides, benzene derivatives, aliphatic compounds, alcohol ethoxylates, and chemical mixtures (Devillers and Flatin, 2000; Gute and Basak, 1997; Croni et al., 2000; Morrall et al., 2003; Wei et al., 2004). QSAR studies were also conducted to predict acute aquatic toxicity (Gute and Basak, 1997), which is not covered by OSHA, since it does not have the authority to regulate environmental issues.

The QSAR models predicting mutagenicity and carcinogenicity have been reviewed by Romualdo Benigni (Tatiana et al., 2007; Benigni and Bossa, 2008).

A mutation is defined as "a permanent change in the amount or structure of the genetic material in a cell. The term mutation applies both to heritable genetic changes that may be manifested at the phenotypic level and to the underlying DNA modifications when known (including, for example, specific base pair changes and chromosomal translocations). The term mutagenic and mutagen will be used for agents giving rise to an increased occurrence of mutations in populations of cells and/or organisms"(OSHA, 2014).

Carcinogen means "a substance or a mixture of substances which induce cancer or increase its incidence. Substances and mixtures which have induced benign and malignant tumors in well-performed experimental studies on animals are considered also to be presumed or suspected human carcinogens unless there is strong evidence that the mechanism of tumor formation is not relevant for humans" (OSHA, 2014).

Tatiana et al. (2007) evaluated 78 noncommercial QSAR models through two-phase analysis. Benigni and Bossa (2008) reviewed 13 local QSAR models, which concluded that 30–70% were correct for potency-generated predictions and 70–100% were correct for discriminating between active and inactive chemicals in the external prediction. Some other QSAR models predicting carcinogenicity are summarized in Table 9.1.

QSAR studies were also performed to predict skin corrosion and irritation. Skin corrosion is "the production of irreversible damage to the skin; namely, visible necrosis through the epidermis and into the dermis, following the application of a test substance for up to 4 h. Corrosive reactions are typified by ulcers, bleeding, bloody scabs, and, by the end of observation at 14 days, by discoloration due to blanching of the skin, complete areas of alopecia, and scars" (OSHA, 2014). Skin irritation is "the production of reversible damage to the skin following the application of a test substance for up to 4 h" (OSHA, 2014). The corrosivity has been correlated with log $P_{ow}$ (logarithm of octanol/water partition coefficient), melting point, molecular volume, and $pK_a$, since they characterize either skin permeability or cytotoxicity. A lower log $P_{ow}$ value, either from large molecular volumes or high melting points, indicates the tendency of low skin permeability for chemicals, that is, less corrosive, unless the $pK_a$ value shows they are either acidic or basic (Barratt, 1996). Some QSAR studies predict irritation by calculating the primary irritation index (PII) or skin irritation score, which are correlated with log $P_{ow}$, molecular volume, the dipole moment, or the molecular weight.

QSAR studies use structural fragments and chemical substructures to predict eye irritation/corrosion, and the QSAR models are reviewed by Saliner et al. (2008). Serious eye damage is "the production of tissue damage in the eye, or serious physical decay of vision, following application of a test substance to the anterior surface of the eye, which is not fully reversible within 21 days of

**Table 9.1 QSAR Studies to Predict Carcinogenicity (Quintero et al., 2012)**

| Health Hazard | Property Measured | Type and Number of Compounds | References |
|---|---|---|---|
| Carcinogenicity | Carcinogenic potential | 30 Chemical classes (fibers, polymers, metals/metalloids, organic chemicals) | Woo et al. (1995) |
| Carcinogenicity | Carcinogenic potential, CPDB mouse | 24 Organic compounds | Zhang et al. (1996) |
| Carcinogenicity | Carcinogenic activity | 46 Methylated and 32 nonmethylated polycyclic aromatic hydrocarbons (PAH) | Ivanciuc (2002) |
| Carcinogenicity | Carcinogenic activity | 148 *N*-nitroso compounds | Luan et al. (2005) |
| Carcinogenicity | Carcinogenic activity | 189 Compounds | Helguera et al. (2005) |
| Carcinogenicity | Carcinogenic potency ($TD_{50}$) | 49 Nitrocompounds (aromatic and aliphatic) | Morales et al. (2006b) |
| Carcinogenicity | Carcinogenic potency ($TD_{50}$) | 62 Nitrocompounds | Morales et al. (2006a) |
| Carcinogenicity | Carcinogenic potency ($TD_{50}$) | 49 Nitrocompounds | Helguera et al. (2008b) |
| Carcinogenicity | Carcinogenic potency | 805 Diverse organic chemicals | Fjodorova et al. (2010) |

application" (OSHA, 2014). Eye irritation is "the production of changes in the eye following the application of test substance to the anterior surface of the eye, which are fully reversible within 21 days of application" (OSHA, 2014).

Total erythema score (TES) and EC3 are simulated for the prediction of respiratory and skin sensitizer. Respiratory sensitizer means "a chemical that will lead to hypersensitivity of the airways following inhalation of the chemical. Skin sensitizer means a chemical that will lead to an allergic response following skin contact" (OSHA, 2014). TES is used to indicate the index of biological property and EC3 is considered as a quantitative measure of sensitizing potential (Barratt et al., 1994; Patlewicz et al., 2003).

QSAR model has also been developed to predict reproductive toxicity (Hewitt et al., 2007). In this QSAR model, 184 heterogeneous antivirals and tocolytics compounds were evaluated for placental clearance index or placental transfer index. Reproductive toxicity includes "adverse effects on sexual function and fertility in adult males and females, as well as adverse effects on development of the offspring. Some reproductive toxic effects cannot be clearly assigned to either impairment of sexual function and fertility or to developmental toxicity. Nonetheless, chemicals with these effects shall be classified as reproductive toxicants" (OSHA, 2014).

Currently, there are no QSAR studies to predict toxicity to specific target organs for single or repeated exposures in the literature (Quintero et al., 2012). "Specific target organ toxicity-single exposure (STOT-SE), means specific, non-lethal target organ toxicity arising from a single exposure to a chemical." "Specific target organ toxicity-repeated exposure (STOT-RE), means specific target organ toxicity arising from repeated exposure to a substance or mixture" (OSHA, 2014).

No aspiration hazard prediction using the QSAR method has been found in the literature. Aspiration toxicity includes "severe acute effects such as chemical pneumonia, varying degrees of pulmonary injury or death following aspiration" (OSHA, 2014).

## 9.2 CASE STUDIES ON QSAR

In this section, some case studies are expanded to explain the application of QSAR methodology to simulate the acute toxicity, skin irritation or corrosivity, and eye irritation.

### 9.2.1 ACUTE TOXICITY ($LC_{50}$) (GUTE AND BASAK, 1997)

The fathead minnow dataset includes 69 benzene derivatives, which have seven different substituent groups: bromo, methoxyl, nitro, methyl, chloro, hydroxyl, and amino. At least six of the molecules have all these substituents.

Topostructural, geometric, topochemical, and quantum chemical indices are the main theoretical descriptors studied in the QSAR study. The topochemical and topostructural indices are in the category of topological indices. Topostructural indices (TSIs) contain information that describes the adjacency and distances of atoms in the structures of molecules. It does not consider the chemical nature of the atoms that bond with each other and other factors (e.g., hybridization states). Topochemical indices (TCIs) represent quantitative information that describes the topology (connectivity of atoms) and chemical properties that are specific for the atoms of the molecule. The geometrical indices are given in terms of a Wiener number for hydrogen-suppressed molecular structure, hydrogen-filled molecular structure, and van der Waals volume.

The data were transformed and selected, and 23 descriptors were used to develop the model in this study. The hierarchical approach was used, and the seven sets of indices were developed. The simplest parameters (i.e., TSIs) were considered first in the hierarchy in the activity modeling process; then, the complexity in the next level was studied. The activity was modeled using the TSIs, and the indices in the best model were selected in combination of TCIs for the model development in the next level. Likewise, the model development proceeded with adding geometrical indices and quantum chemical parameters. Finally, a model was developed using TSIs, TCIs, geometrical indices, and quantum chemical parameters.

The work studied the utility of hierarchical QSAR to predict $LC_{50}$ for 69 benzene derivatives. The descriptors were algorithmically selected during the process of model development. The classes of predictors, topostructural descriptors, geometrical descriptors, topochemical indices, and semi-empirical quantum chemical indices were chosen in the final model.

The results indicate that the acute aquatic toxicity was not well correlated with any individual class of descriptors. A remarkable improvement in the model was achieved by considering the chemical nature of substituents, molecular shape and size, and quantum chemical parameters. A QSAR model was developed using the hierarchical QSAR approach, which has an explained variance ($R^2$) of 86.3% and a standard error of 0.30, and was an acceptable predictive model utilizing four classes of parameters in the study.

## 9.2.2 SKIN IRRITATION AND CORROSIVITY (BARRATT, 1996)

The relationship between skin corrosivity from chemicals and descriptors describing skin permeability and cytotoxicity was discussed in the literature (Barratt, 1995). The hypothesis was that the chemicals should penetrate into the skin first; the cell beneath the stratum corneum would die at the presence of the sufficient dose and of chemicals with sufficient cytotoxicity. $pK_a$ was the parameter used to measure cytotoxicity, which was used to model the skin corrosivity together with parameters for the permeability study, including log P, melting point, and molecular volume.

In this work, the skin irritation and corrosivity potentials were studied for neutral and electrophilic organic chemicals by applying the same methodology. Because most chemicals studied in this work are in liquid form at the ambient condition, melting point, a parameter used for the determination of aqueous solubility, was not considered for the modeling of skin permeability. The dipole moment was the parameter used to model reactivity, because some neutral molecules with low molecular weight are able to alter the electrical resistance of phospholipid membranes, in other words, ion permeability of the membranes, which has been validated by the fact that many chemicals with a significant dipole moment reduce the electrical resistance of liposomes. Molecular weight was used in this study as a compensation for the utilization of the EC Annex V method of corrosivity.

After the construction of a chemical structure with Sybyl 6.1 software, the logP value was calculated using the CHEMICALC system. The TSAR spreadsheets (Oxford Molecular Ltd., Sandford-on-Thames, UK) take the structure information as the input, and calculate the molecular volumes and dipole moments. Stepwise regression as well as discriminant and principal components analysis were used to analyze the datasets.

With respect to skin irritation, the discriminant analysis of the dataset containing 52 organic chemicals of either neutral or electrophilic nature indicated that 38 chemicals were well classified. Stepwise regression analysis was used to correlate the PII with several parameters, including logP, dipole moment, molecular volume, and molecular weight, which indicated a weak dependence on molecular weight. The predictive PII values were compared with actual values, and the results are shown in Figure 9.1, which include the actual classifications of the irritation/nonirritant chemicals, as well as misclassification of chemicals through discriminant analysis.

With respect to skin corrosivity, 39 of 42 chemicals (40 electrophiles and two proelectrophiles) were well classified using discriminant analysis of the dataset based on log P and molecular volume. The additional parameter, dipole moment, did contribute to developing a more desirable model, even though the overall predictive performance was the same, because all 12 corrosive chemicals were well classified. It was found that low volume, high dipole moment, and low log P contribute to the skin corrosivity.

A stepwise approach for the classification of untested organic chemicals, including neutral and electrophilic, is illustrated schematically in Figure 9.2. First, the chemicals should be determined in terms of compatibility with the approach developed in this work. Inorganics, organic salts, organometallics, organic acids, surfactants, bases, and phenols are excluded in this model. In the next step, knowledge of organic chemistry, or expert system, such as DEREK (Sanderson and Earnshaw, 1991), can be applied to determine whether the chemicals are electrophiles or preelectrophiles. The electrophile or preelectrophile chemicals can be examined using the QSAR model for skin corrosivity first.

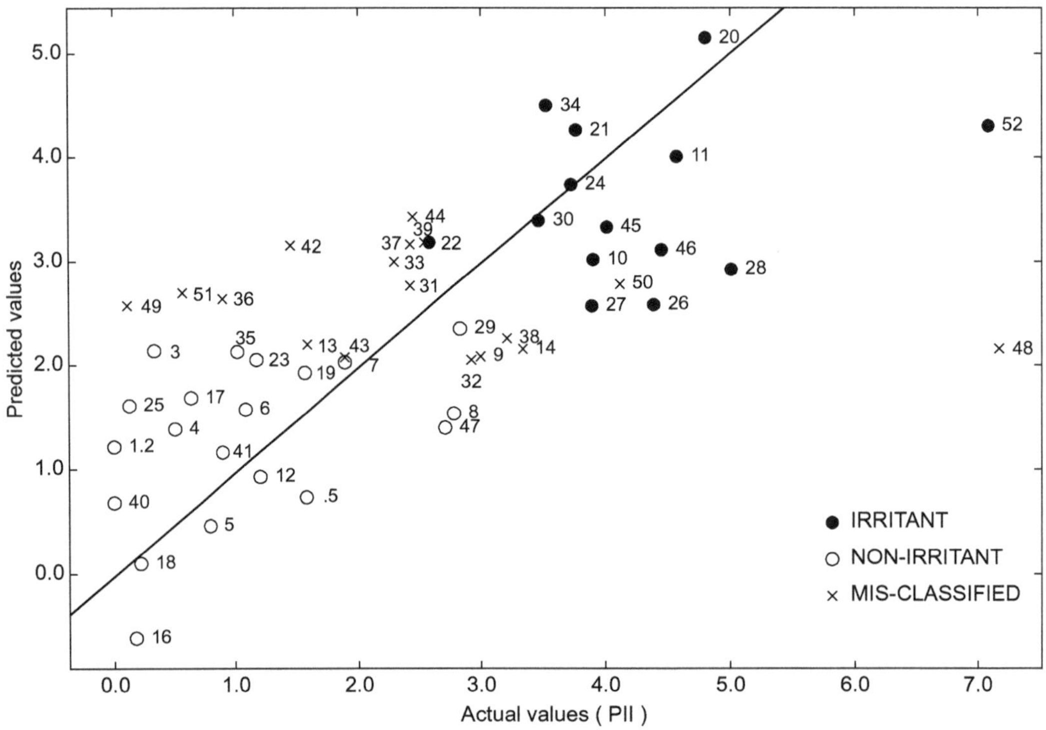


**FIGURE 9.1**

Graph of the actual and predictive values of PII (Barratt, 1996).

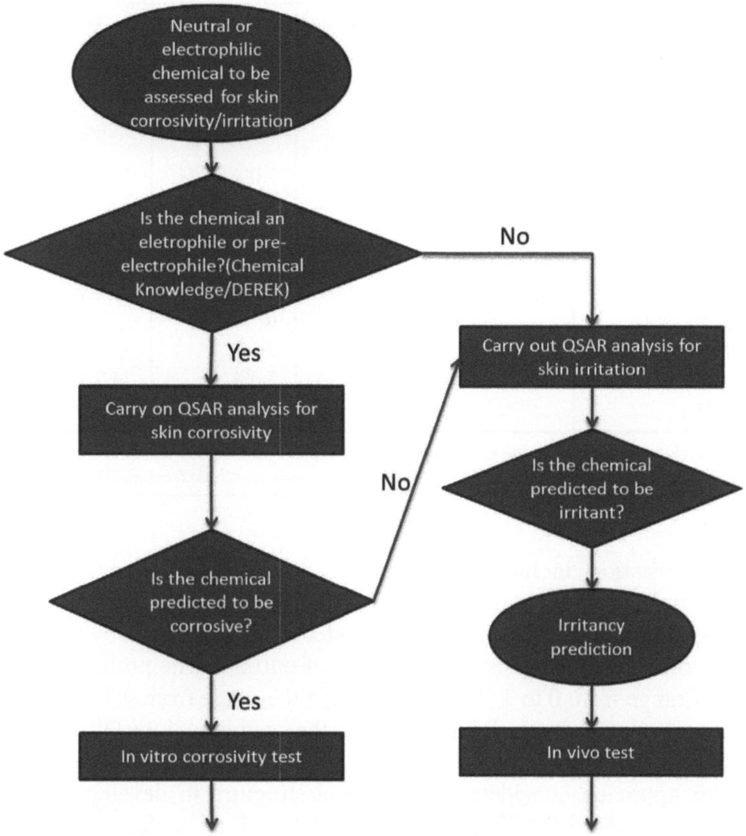


**FIGURE 9.2**

Scheme for the prediction of skin corrosivity or irritation potentials of neutral and electrophilic organic chemicals (Barratt, 1996).

The corrosive chemicals suggested by the QSAR model should be validated by the in vitro corrosivity test. The noncorrosive chemicals suggested by the QSAR model should be studied for skin irritation using the QSAR model. The results of skin irritation study should be confirmed by the results of the in vivo test.

### 9.2.3 EYE IRRITATION (SALINER ET AL., 2006)

Many QSAR models have been developed to study eye irritation. Structural fragments were used to develop the models at an early stage, which was implemented in TOPKAT. Later, the models were developed using the chemicals substructures embedded in MultiCASE (MultiCASE Inc.).

Enslein et al. (1988) developed a QSAR model to simulate eye irritation based on structural fragments. In this work, the dataset of 592 chemicals was classified into two categories: compounds without rings and compounds with rings. The descriptors that describe the structural fragments'

**Table 9.2 Classification Results of Discriminant QSAR Models for the Prediction of Eye Irritation**

| Models | Subequations | Overall Accuracy | False Positives | False Negatives | Indeterminate | F ($p < 0.001$) |
|---|---|---|---|---|---|---|
| Non-ring compounds | Negative vs Other | 89.8% | 3.2% (22.5%)[a] | 7.0% (8.2%)[a] | 15.0% | 25.3 |
| | Severe vs Other | 90.2% | 2.7% (4.3%)[a] | 9.1% (19.1%)[a] | 30.9% | 15.2 |
| Ring compounds | Negative/Mild vs Moderat/ Severe | 92.7% | 2.9% (8.1%)[a] | 4.45% (6.9%)[a] | 29.3% | 8.63 |
| | Severe vs Other | 87.8% | 6.6% (9.9%)[a] | 5.6% (16.7%)[a] | 24.3% | 10.0 |

[a]*Percentage of false predictions in relation to the experimental class.*
*© FAO, 1988. In Vitro Toxicology. pp. 1–14. http://agris.fao.org/agris-search/search.do?recordID=US19890077884 (02.02.15.).*

contribution to eye irritation included Kier and Hall molecular connectivity indices, and the MOLSTAC substructual key. The two subsets were studied through two separate models through discriminant analysis. The variables of each subset model were determined by the regression algorithm after the process of collinearity analysis and detection of outliers. The probabilities obtained through discriminant analysis range from 0 to 1. The prediction in the range from 0.3 to 0.7 should not be used, since it is considered as indeterminate. Table 9.2 shows the accuracy of the QSAR model to classify the level of eye irritation in this work.

The Multi-CASE approach considers the chemical structure in developing a predictive QSAR model for eye irritation. Multi-CASE was developed based on a database containing 186 non-congeneric organic chemicals, which were studied through a modified Draize test of eye irritation in vivo (Klopman et al., 1993). The study indicated that 37 functionalities and substructural attributes were contributing to eye irritation, which include carboxylate and sulfonate groups (in anionic surfactants), halogenated and unsaturated structures, amino groups, phenolic moieties (in nonionic surfactants), anhydrides and epoxides, as well as an ester group with a low irritation potential. Twenty-one chemicals, including mixtures and polymers, were used to examine the developed Multi-CASE program, which contains five substructures not included in the training set. The Multi-CASE program successfully predicted 18 of 21 chemicals.

In summary, QSAR is a useful methodology that can be used to predict some toxicological information in the case of lack of appropriate experimental data for industrial hygiene applications. Further study and research on the application of QSAR will facilitate progress in industrial hygiene.

## REFERENCES

Barratt, M.D., 1995. Quantitative structure activity relationships for skin corrosivity of organic acids, bases and phenols. Toxicology Letters 75, 169–176.

Barratt, M.D., 1996. Quantitative structure-activity relationships for skin irritation and corrosivity of neutral and electrophilic organic chemicals. Toxicology In Vitro 10, 247–256.

Barratt, M.D., Basketter, D.A., Roberts, D.W., 1994. Skin sensitization structure-activity relationships for phenyl benzoates. Toxicology In Vitro 8, 823–826.

Benigni, R., Bossa, C., 2008. Predictivity of QSAR. Journal of Chemical Information and Modeling 48, 971–980.

Croni, M.T., et al., 2000. Structure-toxicity relationships for aliphatic compounds encompassing a variety of mechanisms of toxic action to *Vibrio fischeri*. SAR and QSAR in Environmental Research 11, 301–312.

Devillers, J., Flatin, J., 2000. A general QSAR model for predicting the acute toxicity of pesticides to *Oncorhynchus mykiss*. SAR and QSAR in Environmental Research 11 (1), 25–43. Available at: http://dx.doi.org/10.1080/10629360008033227 (accessed 13.02.14.).

Enslein, K., et al., 1988. Estimation of rabbit eye irritation scores by structure-activity equations. In Vitro Toxicology 2 (1).

Fjodorova, N., et al., 2010. Counter propagation artificial neural network categorical models for prediction of carcinogenicity for non-congeneric chemicals. SAR and QSAR in Environmental Research 21, 57–75.

Gute, B.D., Basak, S.C., 1997. Predicting acute toxicity ($LC_{50}$) of benzene derivatives using theoretical molecular descriptors: a hierarchical QSAR approach. SAR and QSAR in Environmental Research 7, 117–131.

Helguera, A.M., et al., 2005. A topological substructural approach applied to the computational prediction of rodent carcinogenicity. Bioorganic & Medicinal Chemistry 13, 2477–2488.

Helguera, A.M., Combes, R.D., et al., 2008a. Applications of 2D descriptors in drug design: a DRAGON tale. Current Topics in Medicinal Chemistry 8, 1628–1655.

Helguera, A.M., Cordeiro, M.N.D.S., et al., 2008b. QSAR modeling of the rodent carcinogenicity of nitrocompounds. Bioorganic & Medicinal Chemistry 16, 3395–3407.

Hewitt, M., et al., 2007. Structure-based modelling in reproductive toxicology: (Q)SARs for the placental barrier. SAR and QSAR in Environmental Research 18 (1–2), 57–76. Available at: http://www.ncbi.nlm.nih.gov/pubmed/17365959 (accessed 13.02.14.).

Ivanciuc, O., 2002. Support vector machine classification of the carcinogenic activity of polycyclic aromatic hydrocarbons. Internet Electronic Journal of Molecular Design 203–218. Available at: http://pdf.easechem.com/pdf/09/ebb50ca2-efd2-4d7f-ade7-afc9e46be528.pdf (accessed 13.02.14.).

Klopman, G., et al., 1993. Multiple computer automated structure evaluation methodology as an alternative to in vivo eye irritation testing. ATLA. Alternatives to Laboratory Animals 21 (1), 14–27. Available at: http://cat.inist.fr/?aModele=afficheN&cpsidt=4658104 (accessed 13.02.14.).

Luan, F., et al., 2005. Classification of the carcinogenicity of N-nitroso compounds based on support vector machines and linear discriminant analysis. Chemical Research in Toxicology 18, 198–203.

Morales, A.H., Duchowicz, P.R., et al., 2006a. Application of the replacement method as a novel variable selection strategy in QSAR. 1. Carcinogenic potential. Chemometrics and Intelligent Laboratory Systems 81, 180–187.

Morales, A.H., Pérez, M.A.C., et al., 2006b. Quantitative structure activity relationship for the computational prediction of nitrocompounds carcinogenicity. Toxicology 220, 51–62.

Morrall, D.D., Belanger, S.E., Dunphy, J.C., 2003. Acute and chronic aquatic toxicity structure-activity relationships for alcohol ethoxylates. Ecotoxicology and Environmental Safety 56, 381–389.

Nantasenamat, C., Isarankura-na-ayudhya, C., Naenna, T., 2009. A practical overview of quantitative structure-activity relationship. EXCLI Journal 8, 74–88.

OSHA, 2014. Hazard Communication, 29 CFR Parts 1910, 1915, 1917, 1918, and 1926. No. OSHA-H022K-2006-0062. Available at: www.osha.gov/dsg/hazcom/index.html (accessed 10.02.14.).

Patlewicz, G., Roberts, D., Walker, J., 2003. QSARs for the skin sensitization potential of aldehydes and related compounds. QSAR & Combinatorial Science 22, 196–203.

Quintero, F.A., et al., 2012. Review of existing QSAR/QSPR models developed for properties used in hazardous chemicals classification system. Industrial & Engineering Chemistry Research 51 (49), 16101–16115. Available at: http://dx.doi.org/10.1021/ie301079r (accessed 13.02.14.).

Saliner, A.G., Patlewicz, G., Worth, A., 2006. Review of Literature-Based Models for Skin and Eye Irritation and Corrosion. JRC report EUR. Available at: http://scholar.google.com/scholar?hl=en&btnG=Search&q=intitle:Review+of+Literature-Based+Models+for+Skin+and+Eye+Irritation+and+Corrosion#1 (accessed 28.09.13.).

Saliner, A.G., Patlewicz, G., Worth, A.P., 2008. A review of (Q)SAR models for skin and eye irritation and corrosion. QSAR & Combinatorial Science 27 (1), 49–59. Available at: http://doi.wiley.com/10.1002/qsar.200710103 (accessed 13.02.14.).

Sanderson, D.M., Earnshaw, C.G., 1991. Computer-prediction of possible toxic action from chemical-structure–the Derek system. Human & Experimental Toxicology 10, 261–273.

Tatiana, N., et al., 2007. Collection and Evaluation of (Q)SAR Models for Mutagenicity and Carcinogenicity. Available at: http://publications.jrc.ec.europa.eu/repository/handle/111111111/5588 (accessed 13.02.14.).

Todeschini, R., Consonni, V., 2000. Handbook of Molecular Descriptors.

Tropsha, A., Gramatica, P., Gombar, V.K., 2003. The importance of being earnest: validation is the absolute essential for successful application and interpretation of QSPR models. QSAR Combinatorial Science 22, 69–77.

Wang, L., 2005. Support Vector Machines: Theory and Applications. Springer. Available at: http://www.springer.com/engineering/computational+intelligence+and+complexity/book/978-3-540-24388-5 (accessed 13.02.14.).

Wei, D.B., Zhai, L.H., Hu, H.Y., 2004. QSAR-based toxicity classification and prediction for single and mixed aromatic compounds. SAR and QSAR in Environmental Research 15, 207–216.

Wold, S., et al., 2001. Some recent developments in PLS modeling. Chemometrics and Intelligent Laboratory Systems 58, 131–150.

Woo, Y.T., et al., 1995. Development of structure-activity relationship rules for predicting carcinogenic potential of chemicals. Toxicology Letters 79, 219–228.

Zhang, Y.P., et al., 1996. Prediction of the carcinogenicity of a second group of organic chemicals undergoing carcinogenicity testing. Environmental Health Perspectives 104 (Suppl.), 1045–1050.

CHAPTER

# CONCLUSION

# 10

In many applications, incorporation of a range of time and length scales in its mathematical model formulation has become crucial. The focus on complex multi-scale phenomena and its capture in models is one of the most significant developments in modeling methodology in the past three decades (Cameron and Gani, 2011). Modeling has played and will continue to play an increasingly important role in a broad range of application areas in the industry. However, the industry has been slow to reflect a similar level of growing interest and implementation of multi-scale modeling approaches as that of academia (Cameron and Ingram, 2008). The current typical outlook in process safety applications in industry is similar and the implementation of multi-scale modeling approaches is minimal. Challenges remain in implementing effective and efficient multi-scale model development approaches, in the form of seamlessly bridging various time and length scales depending on the problem's objective along with the selection of appropriate model forms at individual scale level tailored to the problem objective. Among the two, development of realistic methods and approaches for effective coupling of length scales (CLS) is a great challenge for computer simulations (Curtin and Miller, 2003).

For a particular problem, the modeler has to decide the proper length scale and a suitable modeling methodology. For example, in assessing the thermal radiation hazard from jet or pool fire, a computational fluid dynamics–based approach is more appropriate. However, also, if one is interested in how that heat from radiation affects the structural integrity of occupied buildings, a combination of computational fluid dynamics and finite element should serve the purpose. Then, the modeler needs to focus on selecting an appropriate model within the computational fluid dynamics–based approach such as, will two-parameter k-epsilon turbulence model be suffice to model the cloud dynamics of, say, the released flammable gas cloud? Deciding the meshing size, typically variable across the domain, is another challenge a modeler faces in defining a problem at this scale. Often a facility performing a full-blown consequence analysis will require one to evaluate many potential scenarios. It is advisable to perform consequence analysis and quantitative risk assessment as initial screening studies, thereby helping in narrowing down the number of specific scenarios to be modeled further, such as CFD studies, by applying the Pareto principle (i.e., to select scenarios that cumulatively account for 80% or more of the plant's total risk).

While developing novel materials and processes, especially with chemicals and materials whose hazardous properties need to be determined, the application of molecular modeling–based approaches can help answer some questions. Molecular modeling–based approaches are powerful tools for the characterization of hazardous material properties through assessing safety-related concerns of novel chemicals under various process conditions. Molecular modeling–based approaches are promising tools for understanding the role of process chemistry in fires and explosion (Mannan et al., 2006). Several safety-related information including material reactivity, decomposition, and flammability can be estimated using molecular modeling approaches. In scenarios involving uncertainties, at the minimum, modeling approaches can reduce the requirement of much expensive experimental analysis for assessment of hazardous processes, such as understanding and estimating consequences of a thermal runaway reaction.

Several challenges, as mentioned in various chapters, remain in the context of implementing multi-scale modeling for process safety applications. However, in process safety applications, the lack of understanding of a phenomenon often introduces conservatism in a safety model. In some situations, this might even compromise safety. Additionally, owing to the nature of the problems, it is not always

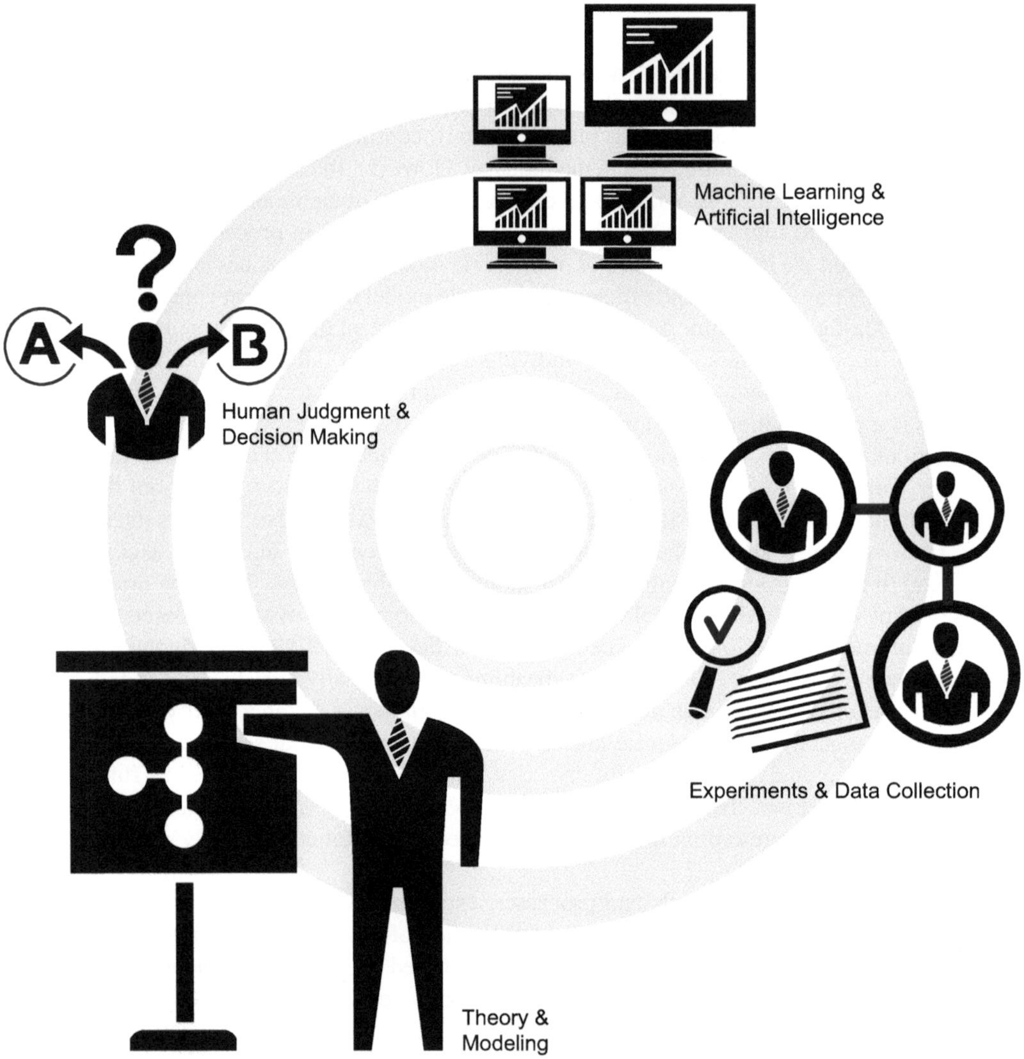


**FIGURE 10.1**

Integrated multi-scale approaches to modeling (Taylor, 2015).

feasible to determine the underlying root behind a phenomenon through experiments. It is only appropriate to extend the advancement in theory of materials at different time and length scales to address less understood or coarsely defined safety concerns through exploitation of significant increase in computational power in recent times.

There are currently many examples where definitions of safe modes of operation are identified through empirical means such as safe-operating conditions for materials used in oil and gas production system codified in ISO15156. These safe zones are not rigorously established and hence are misleading while using changed process conditions or when novel materials are introduced in the process (Taylor, 2015). Ideally, a comprehensive integrated multi-scale model will lie at the intersection of the following as illustrated in Figure 10.1:

- Underlying physics
- Experimental data
- Human judgment
- Machine learning

These activities working in synchrony have tremendous potential to improve safety, environment, and sustainability.

We hope that this book is able introduce to its readers to the effective application of multi-scale modeling approaches in process safety applications. Similar to any active area of science, the application of multi-scale modeling methodologies in process safety discussed here is incomplete with further development in the field expected. As one becomes adept at understanding of the strength and weaknesses of all these approaches, one will be in a powerful position to develop mapping between various modeling methodologies and its applicability to various process safety concerns. Hence, this book is expected to serve as an introduction toward the application of different modeling methodologies from the atomistic scale to the process level for addressing process safety concerns. Multi-scale modeling approaches are established powerful tools as proven in several areas of science, and process safety challenges are too big a concern for the industry to simply miss the boat.

## REFERENCES

Cameron, I.T., Gani, R., 2011. Product and Process Modelling. Elsevier.

Cameron, Ingram, 2008. A survey of industrial process modelling across the product and process lifecycle. Computers & Chemical Engineering 32 (3), 19. http://dx.doi.org/10.1016/j.compchemeng.2007.02.015.

Curtin, W.A., Miller, R.E., 2003. Atomistic/Continuum Coupling in Computational Materials Science. Institute of Physics Publishing.

Mannan, M.S., Aldeeb, A.A., Rogers, W.J., 2006. Understanding the role of process chemistry in fires and explosions. Process Safety Progress 21 (4), 323–328. http://dx.doi.org/10.1002/prs.680210409.

Taylor, C.D., 2015. Integrated Multiscale Modelling of Materials. https://www.dnvgl.com/technology-innovation/latest-publications.html.

CHAPTER

# EXERCISES

# 11

**SUPPORTING FILES AND UP-TO-DATE EXERCISE MATERIAL IS AVAILABLE ON THE BOOK'S COMPANION WEBSITE**

## 11.1 PROBLEMS

1. If the energy balance at the lower flammability limit (LFL) of a mixture of flammable and inert gas can be expressed as:

$$x_{LFL} \cdot H_{A,Fuel} + (1 - x_{LFL})H_{A,Inert} = x_{LFL} \cdot H_G$$

   where:

   $x_{LFL}$ is LFL value of the fuel expressed in volume fraction.
   $H_{A,Fuel}$ is the energy absorbed per unit volume of fuel.
   $H_{A,Inert}$ is the energy absorbed per unit volume of inert.
   $H_G$ is the energy generated or released per unit volume of flammable gas mixture.

   a. Write the corresponding energy balance equation at the UFL limit
   b. From oxygen calorimetry, using $H_G = C_0 H_0$, where $H_0$ is heat generated per unit volume of oxygen, derive the expression for $H_{A,Fuel}$ and $H_0$.

2. A flammable vapor is composed of 63.5% ethyl acetate, 20.8% ethyl alcohol, and rest toluene by volume.
   a. Based on the two sets of LFL values as provided in Table 11.1, estimate the LFL of the mixture using Le Chatelier's formula.
   b. The sensitivity to input data illustrates the importance of having accurate data at hand. The LFL value of the given mixture is most sensitive to the LFL value of ______________.

3. Using Antoine's equation for vapor pressure and assuming ideal vapor liquid behavior, calculate flash point of a water—ethanol liquid mixture containing 40 mol% ethanol.

4. Consider the following set of reactions as observed in the nitric acid oxidation of 2-octanol to 2-octanone (Eizenberg et al., 2006).

$$\text{A (2-octanol)} + \text{B (nitrosonium ion)} \rightarrow \text{P (2-octanone)} + \text{2B (nitrosonium ion)} \ldots \quad (11.1)$$

$$\text{P (2-octanone)} + \text{B (nitrosonium ion)} \rightarrow \text{E (carboxylic acid)} \ldots \quad (11.2)$$

**Table 11.1 LFL Values for Chemicals in Example 1**

| Chemical | LFL Values (Vol.% in Air) | |
|---|---|---|
| | Lewis and Elbe (2013) | National Fire Protection Association (1986) |
| Ethyl acetate | 2.5 | 2.0 |
| Ethyl alcohol | 4.3 | 3.3 |
| Toluene | 1.4 | 1.2 |

The overall set of reactions is exothermic in nature, and hence there is a requirement of heat removal to maintain the reactor temperature at around 0 °C. Under normal process conditions, only partial conversion (7.5%) of 2-octanone to carboxylic acid is achieved. However, if the operating temperature exceeds 5 °C, owing to increased reaction rate leading to insufficient heat removal, runaway conditions will develop. Please use the information as provided next to address the following:

a. Develop a steady state mathematical model of the reactor system involving these reactions to represent the system behavior under normal process conditions.
b. Extend the developed model to represent the dynamic system behavior with deviations from normal process conditions.
c. Develop a table for Hazard and Operability Study (HAZOP) keywords applicable for studying the given runaway reaction.
d. Implement the developed dynamic model to perform HAZOP analysis of the system.

5. Consider the following study (Švandová et al., 2005):
Hydrolysis reaction of propylene oxide: Propylene oxide hydrolyses to monopropylene glycol and on further consecutive reactions to higher glycols as illustrated in Figure 11.1.
For a three continuously stirred tank reactor (CSTR) system in series, consider $V_k$ as the volume of the $k$th reactor, $C_{i,k}$ as the molar concentration of $i$th component, $R_{j,k}$ as the reaction rate of the $j$th reaction in the $k$th CSTR, $E_j$ as the activation energy of the $j$th reaction, $K_j$ as the reaction rate constant of the $j$th reaction, and $T_k$ as the temperature of the $k$th CSTR. Also assume that all the reactions follow the Arrhenius equation and are of first order. The reaction parameters are provided in Table 11.2. Initial values are provided in Tables 11.3 and 11.4. Cooling medium as water (not shown in Figure 11.2) is flowing from the jacket of the first reactor to the second reactor and finally to the third reactor. The process moves to unsafe conditions once reactor temperature exceeds 503.15 K.
Find the following:
a. Mass and energy balance equation as a function of time, using subscripts for each reactor.
b. HAZOP study looks into potential process deviation scenarios using various guidewords. Based on that, examine the temperature in reactors if the $C_{\text{Propylene oxide},0}$ is varied from 5 to 30 mol/s, as six different scenarios.
c. In a similar vein, examine the effect of a sudden drop of cooling water flow rate from 6.8 to 5.8 mol/s.

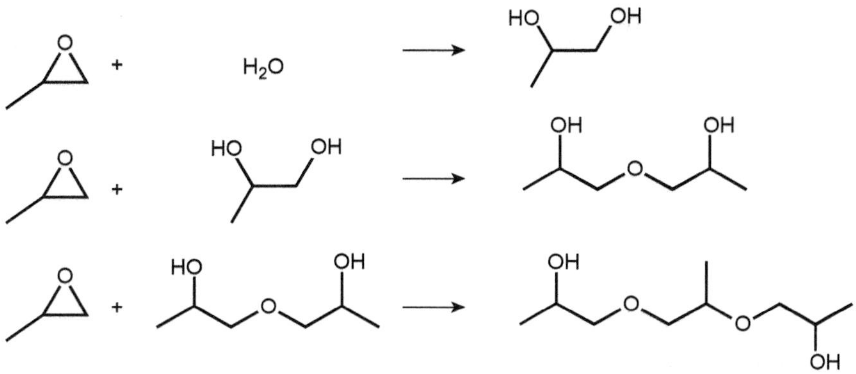


(R1)

(R2)

(R3)

**FIGURE 11.1**

Reaction set in a propylene oxide–water system.

**Table 11.2 Kinetic Data**

| Parameters | Reaction (R1) | Reaction (R2) | Reaction (R3) |
|---|---|---|---|
| K ($m^3\ mol^{-1}\ s^{-1}$) | 96,000 | 9600 | 960 |
| E ($J\ mol^{-1}$) | 75,362 | 82,899 | 91,189 |
| $\Delta_r H$ ($J\ mol^{-1}$) | −91,360 | −111,000 | −244,600 |

**Table 11.3 Initial Conditions of the Reactor**

| | |
|---|---|
| $C_{Propylene\ oxide,0}$ (mol/s) | 13 |
| $C_{Water,0}$ (mol/s) | 6.8 |
| $T_0$ (C) (feed temperature) | 26 |
| $T_{0,1}$ (K) | 470 |
| $T_{0,2}$ (K) | 425.7 |
| $T_{0,3}$ (K) | 400 |
| $V_{i=1,2,3}$ ($m^3$) | 1 |

**Table 11.4 Initial Conditions around Heat Exchanger**

| | |
|---|---|
| UA (kW/K) | 10 |
| Cooling medium | Water |
| $F_{water}$ (mol/s) | 105 |

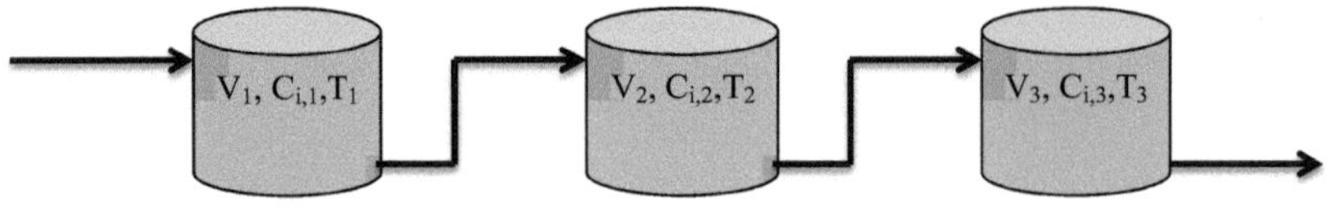


**FIGURE 11.2**

Schematic of CSTR's in series.

**Table 11.5 Quantitative Representation of Qualitative HAZOP Keywords**

| Parts/Components | Qualitative Failure Mode | Quantitative Representation |
|---|---|---|
| Valve | Open failure/closed failure | |
| Heat exchanger | Fouling, plugging | |
| Separator | Open failure/closed failure | |

6. For performing a dynamic simulation to assist an HAZOP study, explain how you will represent quantitative guidewords for the qualitative keywords in Table 11.5.
7. Correct the order of steps below for performing a multivariate statistical analysis and explain reasons behind it:
   a. Develop analysis plan.
   b. Define the multivariate technique.
   c. Estimate multivariate model.
   d. Validate the multivariate model.
   e. Evaluate the assumptions underlying the multivariate technique.
   f. Estimate multivariate model.
   g. Assess overall model fit.
   h. Interpret variate(s).
   i. Define research problem and objective.
8. Lead dioxide, a compound found in waste batteries, can be reduced to lead acetate with the addition of a reducing agent and acetic acid. A certain batch process adds both acetic acid and sulfuric acid in a stoichiometric amount to convert the initial lead dioxide slurry to lead sulfate slurry. Even though acetic acid is regenerated in the process, it is necessary to validate the same. Accordingly, between each batch, the laboratory tests for excess sulfuric acid in the solution by reacting it with a $BaCl_2$ solution. In the test, any presence of sulfuric acid is indicated by the resulting precipitation in the form of $BaSO_4$. It is also given that, if any lead acetate remains unreacted (which can only happen in absence of sulfuric acid), it will react with $BaCl_2$ and $PbCl_2$ will precipitate out of the solution. Now, if a process proceeds along the following steps in each batch and, in just one batch, less than required sulfuric acid is added, illustrate how and what problems might come up and keep growing. Also explain how would you fix these problems.

a. Measured amount of lead dioxide, acetic acid, and sulfuric acid is mixed in a CSTR.
b. After completion of the reaction, a sample is sent to the laboratory to test acid concentrations.
c. Initially, the total acid equivalent is estimated through titration.
d. Sulfuric acid concentration is estimated by the $BaCl_2$ testing.
e. Acetic acid concentration is estimated by subtracting the sulfuric acid concentration from the initial total acid equivalent.

9. Explain how you will develop a molecular dynamics–based model to understand evaporation characteristics of a droplet. Consider the initial system to be an FCC crystal of a Xe/$N_2$ system with liquid Xe (with initial diameter D) at the center of the cube surrounded by $N_2$.
10. Assume that the heat of formation for ethane, $\Delta H_{f,ethene}$, is not known and consider the following reaction, which includes ethane as a product.

$$CH_2 = CH - CH_3 + CH_4 \rightarrow CH_2 = CH_2 + CH_3 - CH_3$$

   a. Summarize how $\Delta H_{f,ethene}$ can be determined using the above reaction, indicate any needed information, and write an expression using the usual notations.
   b. Compare the reaction below with the reaction above and state why this new reaction is more useful or less useful for determining $\Delta H_{f,ethene}$.

$$CH \equiv CH + 2CH_4 \rightarrow CH_2 = CH_2 + CH_3 - CH_3$$

11. Applying the linear combination of atomic orbitals (LCAO) approach:
   a. List the 6-31G** basis functions to be used for calculations involving molecule hydrogen fluoride (HF), a toxic gas, from H($1s^2$) and F($1s^22s^22p^5$) and indicate the number of Gaussian primitive functions for each in parentheses needed for this LCAO application.
   b. State the total number of 6-31G** basis functions and the total number of Gaussian primitive functions to be used for HF.
12. Consider the following set of models: AM1, Hartree-Fock, MP2, B3LYP, and QCISD(T). Indicate for the scenarios below which model(s) you would choose and why:
   a. Equilibrium structure of a large molecule at low or moderate cost.
   b. Enthalpy of an isodesmic reaction at low or moderate cost.
   c. Accurate enthalpy of reaction for medium-sized molecule at moderate cost.
13. A room has a relatively small opening either at the ceiling or the floor level. A fire in the room develops energy at the rate of $\dot{Q}$ kW. Derive an expression for dP/dt and the mass flow rate out of the room as a function of $\dot{Q}$ and a number of other variables. Name the main assumptions. The expression you should arrive at is: $\frac{c_v \cdot V}{R}\frac{dP}{dt} + \dot{m}_e c_p T_e = \dot{Q}$.
14. Discuss how the above equation can lead to
   a. An analytical expression for the pressure build-up in a hermetically closed room.
   b. An analytical expression for the mass flow through an opening, when the opening is either at floor or ceiling level.

## 11.2 APPLICATION OF KiSThelP SOFTWARE FOR EXPLOSIVE DECOMPOSITION REACTION

**Sébastien Canneaux[a,b], Frédéric Bohr[b] and Eric Henon[c]**

Kinetic and Statistical Thermodynamical Package (KiSThelP) is a cross-platform free open-source program developed to estimate molecular and reaction properties from electronic structure data. Some key features are gas-phase molecular thermodynamic properties, thermal equilibrium constants and kinetic calculations. Theories like transition state theory (TST) and RRKM are employed. KiSThelP provides graphical front-end capabilities designed to facilitate calculations and interpreting results (data can be read directly from quantum chemistry results). These features make this program well-suited to support and enhance students learning and can serve as a very attractive courseware as well, taking the teaching content directly from results in molecular and kinetic modeling (S. Canneaux et al., 2014).

$$H + HN_3 \rightarrow NH_2 + N_2$$

### 11.2.1 CONTEXT

For many years, a large number of experimental studies in the area of gaseous detonation are carried out in order to test detonation propagation models. The experimental study of the thermal decomposition of energetic molecules is dangerous. Thus, explosive decomposition modeling meets some problems owing to the lack of kinetic data on elementary reactions. Also, kinetic data extrapolation at large temperatures is difficult from an experimental point of view.

Therefore, in the absence of experimental results, investigation by means of theoretical computations is useful in order to provide kinetic parameters and, then, significant insight into reaction mechanisms.

### 11.2.2 AIM

This example acts as a basic introduction to KiSThelP to compute a reaction heat and a reaction rate constant.

### 11.2.3 GENERAL METHODOLOGY

To determine thermodynamic and kinetic properties from quantum results, the use of statistical mechanics is required for relating the microscopic properties of individual atoms and molecules to the macroscopic bulk properties of materials that are under study in your laboratory (Figure 11.3).

[a]University of Lille1 Sciences and Technologies, 59655 Villeneuve d'Ascq Cedex, France
[b]LISM, EA 4695, University of Reims Champagne-Ardenne, Moulin de la Housse, 51687 Reims, France
[c]ICMR, UMR CNRS 7312, University of Reims Champagne-Ardenne, Moulin de la Housse, 51687 Reims, France

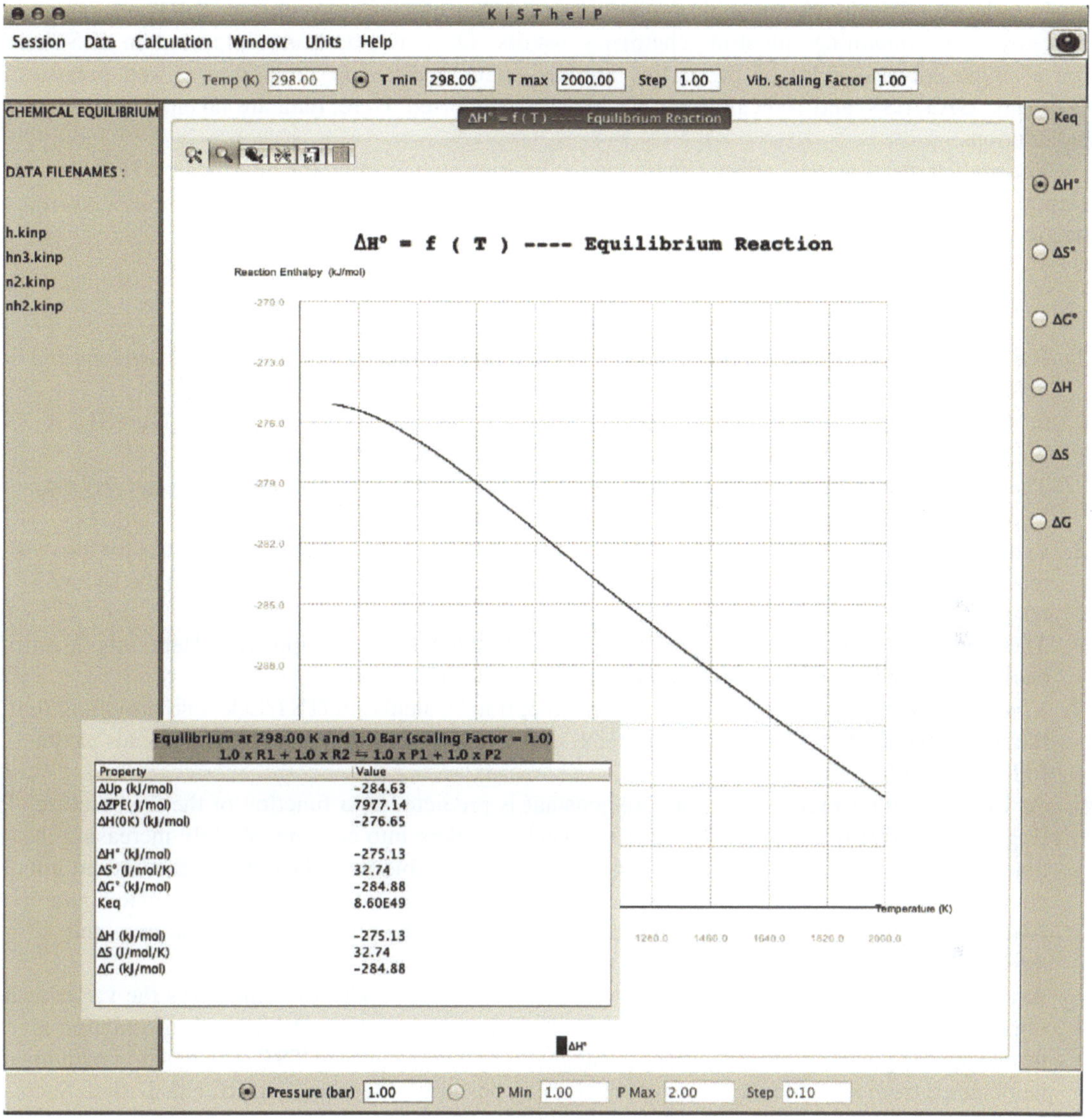


**FIGURE 11.3**

Thermodynamic properties for the reaction $H + HN_3 \rightarrow NH_2 + N_2$; KiSThelP graphical front-end designed to facilitate calculations and interpreting results.

In order to estimate macroscopic properties, a number of files are required. KiSThelP typically processes file containing quantum chemistry results. Once the data are imported in KiSThelP for a single species, a lot of properties are delivered at different temperatures and pressures. Gaussian, GAMESS, and NWChem output files are supported. Here, data are taken from Gaussian calculations.

The studied reaction mechanism (involving the limiting step forming the intermediate $H_2N_3$) is (see Henon, E., Bohr, F., 2000. Journal of Molecular Structucture (Theochem) 531, 283–299 for more details):

$$H + HN_3 \rightarrow H_2N_3 \rightarrow NH_2 + N_2$$

1. For the above-mentioned reaction, the following quantum mechanics calculations are required before using KiSThelP:
   a. Geometry optimization and vibrational frequency calculations for H, $HN_3$, $H_2N_3$, $NH_2$, $N_2$.
   b. Transition state (TS) determination for the rate determining step.

   These calculations produced the following files[1]: H.out, $HN_3$.out, $H_2N_3$.out, $NH_2$.out, $N_2$.out, TS.out. In this example, the level of theory was: DFT(M06)/6-311G**.
2. *Reaction heat*: using the menu option "Calculation/Keq", and providing the file names for the two reactants (H.out, $HN_3$.out) and for the two final products ($N_2$.out, $NH_2$.out) leads to the following result:
   Thermodynamics properties are predicted, specially the heat of reaction $\Delta H°$. Here, this quantity has been plotted in the temperature range 298 to 2000 K.
3. *Kinetic parameters prediction*: Using the menu option "Calculation/TST/Eck" and providing the file names for the two reactants (H.out, $HN_3$.out) and the transition state (TS.out) leads to the following result:
   As can be seen on Figure 11.4, the rate constant is predicted as a function of the temperature (log_10 K = f(1/T)). Here, tunneling effect has been taken into account, slightly increasing the reaction rate (red curve). Numerous possibilities are available in KiSThelP. On the bottom left panel of Figure 11.4, thermodynamics and kinetics parameters are reported for a given temperature. Also, the temperature dependency of the rate constant can be presented using the two- and three-parameter conventional Arrhenius equation.
   Available in KiSThelP, the variational transition state theory (VTST) emphasizes the variational effect on the location of the transition state at a given temperature. Additional information is needed on the potential energy surface compared to the conventional TST. The reaction path must be obtained from a quantum chemistry program by performing a so-called IRC calculation. Such a reaction path is illustrated in Figure 11.5 at 700 K.

[1]Available in the book's companion Website.

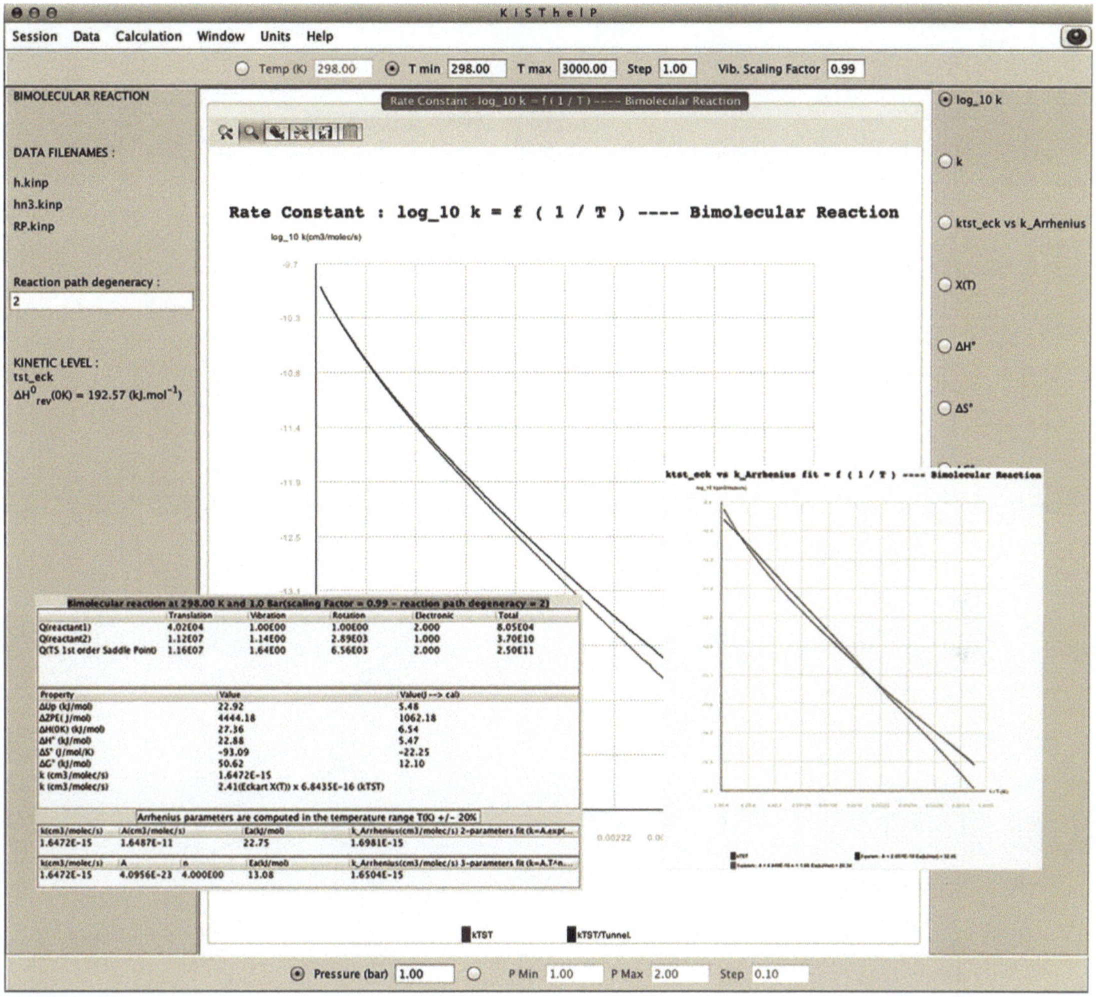

**FIGURE 11.4**

Kinetic properties for the elementary reaction $H + HN_3 \rightarrow H_2N_3$.

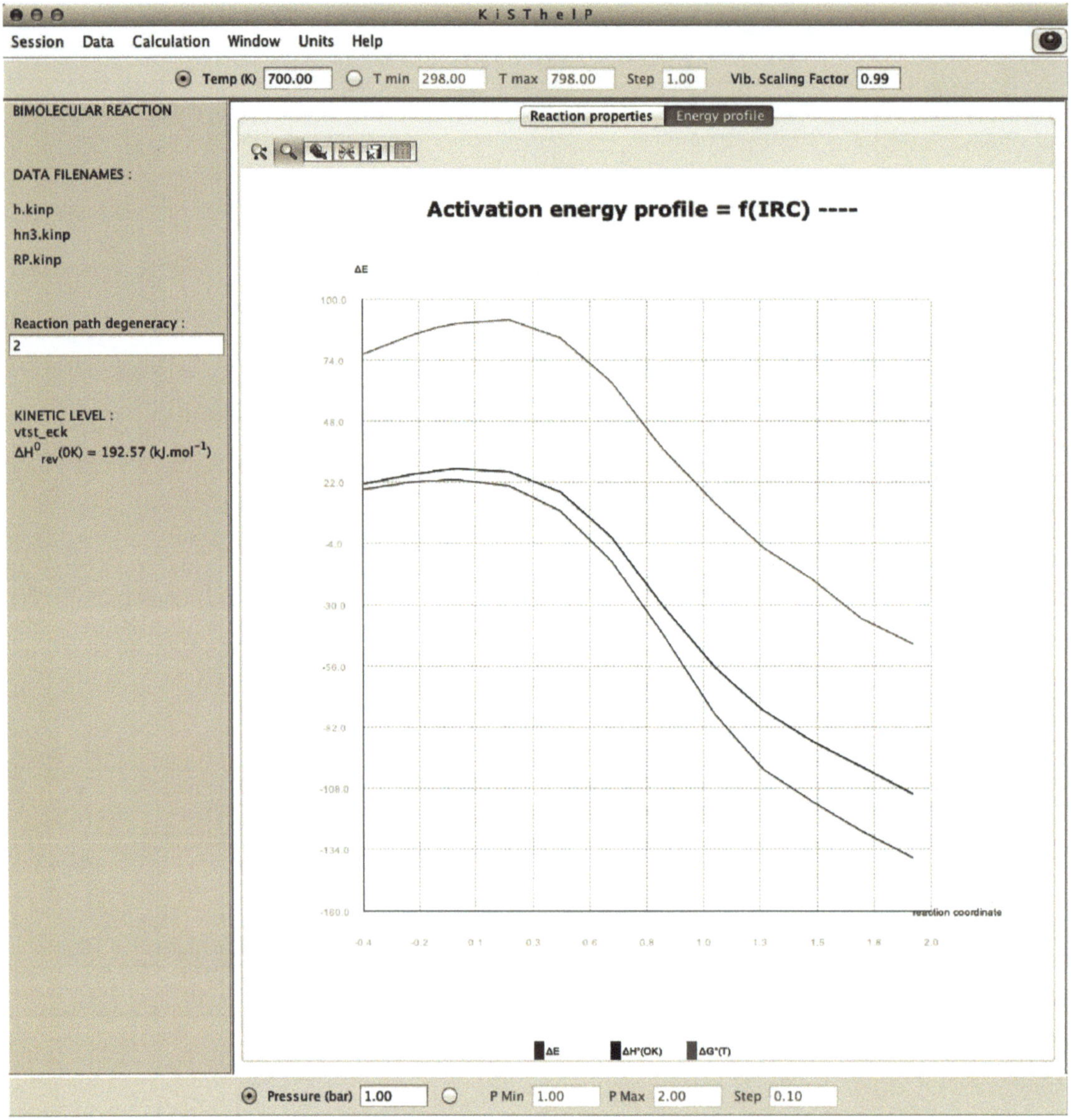


**FIGURE 11.5**

Reaction path (along the IRC) for the elementary reaction $H + HN_3 \rightarrow H_2N_3$; variational effect on the location of the transition state at a given temperature.

## 11.3 APPLICATION TO A REACTION OF ATMOSPHERIC INTEREST

$$CH_3OCH_2OCH_2O \rightarrow H + CH_3OCH_2OCHO$$

### 11.3.1 CONTEXT

Because of the increasing use as industrial solvents and fuel additives of oxygenated compounds such as ethers, their degradation in the atmosphere has received significant attention in the past few years.

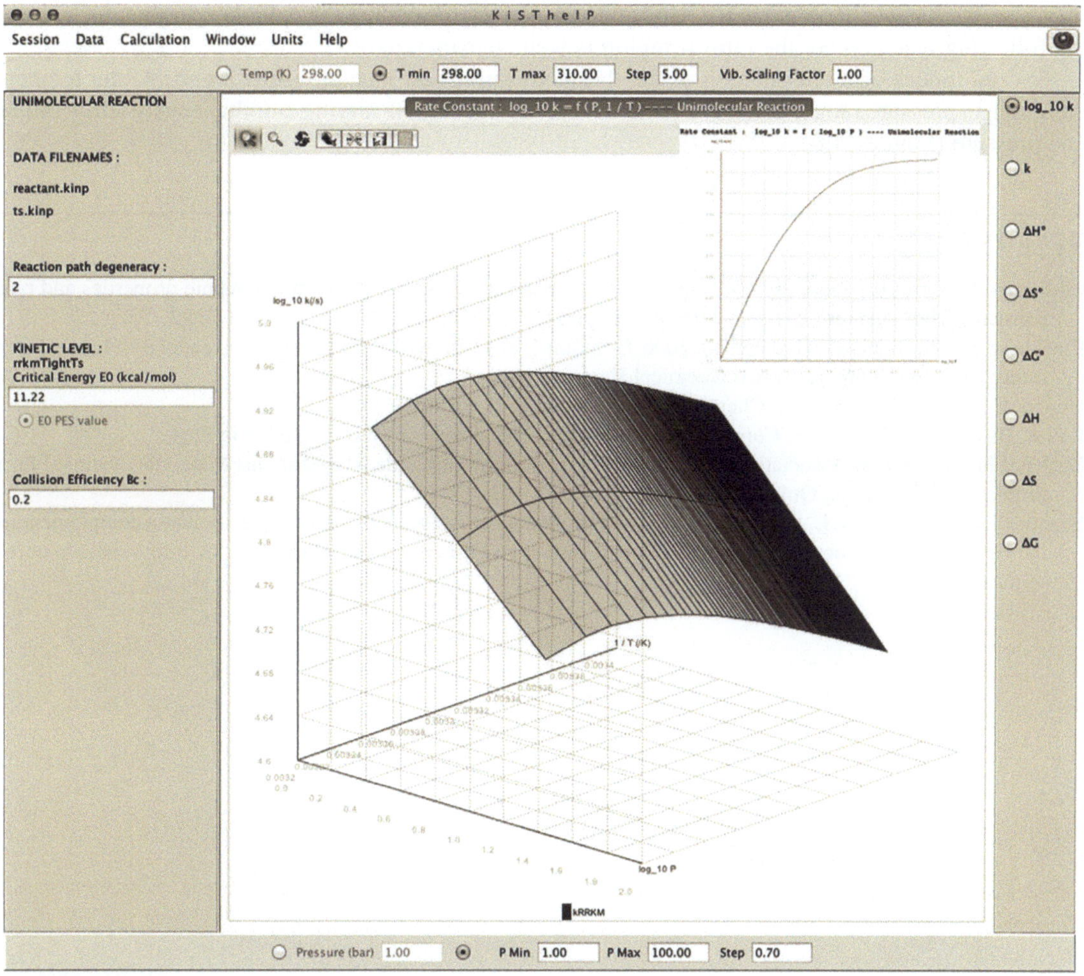


**FIGURE 11.6**

Two- and three-dimensional plots of the RRKM fall-off behavior as a function of pressure and/or temperature for the elementary reaction $CH_3OCH_2OCH_2O \rightarrow H + CH_3OCH_2OCHO$.

### 11.3.2 AIM

In polluted atmosphere the title reaction can exhibit pressure dependence. The purpose of this example is to provide an initial introduction to predict pressure-dependent rate constants.

### 11.3.3 GENERAL METHODOLOGY

The same methodology as previously illustrated in example one can be applied here. But, in addition to the required quantum chemistry data (for $CH_3OCH_2OCH_2O$, transition state), the Lennard-Jones collision parameters must be supplied. It can be noticed that a database of these collision parameters is made available in KiSThelP. The level of theory used to obtain the presented results can be found in Henon et al. (2003).

Using the KiSThelP menu "k/RRKM" and interactively providing the collision parameters leads to the following results (here the fall-off region is well characterized):

Several modes of representation of the thermodynamic and kinetic results are available: temperature range, pressure range, or a three-dimensional representation varying simultaneously temperature and pressure (Figure 11.6).

## REFERENCES

Canneaux, S., Bohr, F., Henon, E., 2014. KiSThelP: A program to predict the thermodynamic properties and rate constants from quantum chemistry results. Journal of Computational Chemistry 35 (1), 82–93.

Eizenberg, S., Shacham, M., Brauner, N., 2006. Combining HAZOP with dynamic process model development for safety analysis. Computer Aided Chemical Engineering 21.

Henon, E., et al., 2003. Physical Chemistry Chemical Physics 5, 5431–5437.

Lewis, B., Von Elbe, G., 2013. Combustion, Flames and Explosions of Gases. Academic Press.

National Fire Protection Association, August 1986. Fire Protection Guide on Hazardous Materials. National Fire Protection Association, Quincy, MA, Boston.

Švandová, Z., Jelemenský, L., Markoš, J., 2005. Steady states analysis and dynamic simulation as a complement in the HAZOP study of chemical reactors. Process Safety and Environmental Protection 83. http://kisthelp.univ-reims.fr.

# Index

*Note:* 'Page numbers followed by "f" indicate figures, "t" indicate tables'.

## N



## O



## P

Kolibri 360 GmbH
Ferdinand-Jühlke-Straße 7
99095 Erfurt, Deutschland
produktsicherheit@kolibri360.de